W9-AKW-611

INTRODUCTORY ALGEBRA
THIRD EDITION

MERVIN L. KEEDY
Purdue University

MARVIN L. BITTINGER
Indiana University
Purdue University at Indianapolis

ADDISON-WESLEY PUBLISHING COMPANY
Reading, Massachusetts • Menlo Park, California
London • Amsterdam • Don Mills, Ontario

Sponsoring Editor: Patricia Mallion
Production Editor: Martha Morong
Designer: Marshall Henrichs
Illustrator: Kater Print Company
Cover Design: Dick Hannus
Photos: Marshall Henrichs

Library of Congress Cataloging in Publication Data

Keedy, Mervin L.
  Introductory algebra.

  Includes index.
  1. Algebra.   I. Bittinger, Marvin L., joint
author. II. Title.
QA152.2.K43 1979   512.9   78-55821
ISBN 0-201-03874-9

Second printing, May 1980

Copyright © 1979 by Addison-Wesley Publishing Company, Inc.
Philippines copyright 1979 by Addison-Wesley Publishing Company,
Inc.

All rights reserved. No part of this publication may be reproduced,
stored in a retrieval system, or transmitted, in any form or by any
means, electronic, mechanical, photocopying, recording, or other-
wise, without the prior written permission of the publisher. Printed
in the United States of America. Published simultaneously in Can-
ada. Library of Congress Catalog Card No. 78-55821.

ISBN 0-201-03874-9
CDEFGHIJ-MU-89876543210

This book covers beginning algebra. It is intended for use by students who have not been exposed to the subject, or who need a review. There is strong emphasis on algebraic skills, but the "whys" of algebra also receive significant attention.

One of the important features of this book is its format. Most pages have an outer margin, used for material of several types. For each section the objectives are stated in behavioral terms. Also in the margins are sample, or developmental, exercises. These are designed to get the student involved actively in the development. These exercises are always very much like the examples in the text and the homework exercises.

TEACHING MODES

This book can be used in many ways. We list a few.

- *As an ordinary textbook.* To use the book this way, the instructor and students simply ignore the margin exercises.
- *For a modified lecture.* To bring student-centered activity into the class, the instructor stops lecturing and has the students do margin exercises at appropriate times.
- *For a no-lecture class.* The instructor makes assignments that students do on their own, including working exercise sets. The following class period the instructor answers questions, and students have an extra day or two to polish their work before handing it in. In the meantime, they are working on the next assignment. This method provides individualization while keeping a class together. It minimizes the number of instructor-hours required, and has been found to work well with large classes.
- *As a learning laboratory.* This book has a low reading level and high understandability, so it is quite suitable for use in a learning laboratory or any kind of learning situation that is essentially self-study.

SUPPLEMENTS

The following (optional) supplements are available.

- *Audio-Tape Cassettes* are available for use in audio-tutorial situations.
- *An Instructor's Manual and Test Booklet* contains a pretest, and five alternate forms of each chapter test and the final examination, with answers spaced for easy grading.
- *An Answer Booklet* contains the answers to all the exercises. This is available to the student at low cost,

PREFACE

and is especially useful when instructors want students to have all the answers.

● *A Student's Guide to Margin Exercises* contains worked-out solutions to all the margin exercises.

● *A Diagnostic Test* can be used to determine which of the texts—*Arithmetic, Introductory Algebra, Intermediate Algebra,* or *Algebra and Trigonometry,* by the same authors—should be used for a student.

WHAT'S NEW IN THE THIRD EDITION?

This edition looks different from the earlier editions because of the new design, which includes a functional use of color. The book has undergone some other changes, as suggested by users of the earlier editions, but it is still basically the same book.

Among the improvements that have been made are the following:

● *Applications.* There are more applied problems, and many of the new ones are of a more real-life nature.

● *Exercise Sets.* The tear-out exercise sets have been placed at the ends of lessons.

● *Objectives.* The objectives and the material associated with them have been made very easy to identify. Each objective is identified with a domino symbol such as this: [●●●]. The same symbol is placed in the text and also in the exercises and tests beside the corresponding material. This makes the text much easier to read, and more easily usable for self-study.

● *Exercises.* The exercise sets have been greatly improved. They are now grouped according to the applicable objectives (and marked accordingly). They are now carefully paired, so that each even-numbered exercise is very much like the one that precedes it. They are also more carefully graded, the easier ones coming first. Answers to the odd-numbered exercises are given in the back of the book.

Special, optional, calculator exercises and optional challenge exercises have also been included at the end of various exercise sets.

● *Readiness Checks.* A readiness check is a short set of exercises that can be used to determine whether a student has acquired the skills prerequisite to success with the chapter. These have been placed at the beginning of all chapters except the first one. They are entirely optional.

If a student misses an item on a readiness check, a review of the keyed material may be considered before beginning the chapter.

● *Tests.* The format of the chapter tests has been slightly changed. They are now identified as being usable as tests or as review. They have also been shortened so that they can be more easily used during a regular class period. The final exam has been retained.

● *Sets.* The material on sets is new, not having appeared in the previous editions. It has been placed in the last chapter, along with the material on inequalities. This placement of the material has the following advantages:

1. It does not change the essential character of the book in its earlier editions. The material on sets is entirely optional.

2. Sets and inequalities are quite naturally related, thus making a logical chapter.

3. The material on inequalities can easily be taught earlier, in spite of its position in the text. The same is true of the new material on sets.

● *Improved Development.* As suggested by a large number of users, to whom we are deeply grateful, the development of ideas and skills in many places has been smoothed, altered, or corrected. Hopefully this will make the material even more teachable. Notable in this respect is the de-emphasis on the use of related sentences, much more drill and explanation concerning integers and rational numbers in Chapter 2, and a completely new treatment of graphing sequenced in a more gradual and deliberate manner.

● *EXTRAS.* Optional material that will be of interest to some students has been included where space was available. Some of this material deals with real-life applications, some of it is calculator-oriented, and some is designed to appeal to curiosity. These sections can be identified by the heading "Something Extra."

We wish to express our thanks to the following people who reviewed the manuscript: Steven D. Kerr, Weber State College; Larrye Steldt, Indiana University at Kokomo; and Charles N. Walter, Brigham Young University. Our special thanks also go to Judy Beecher of Indiana University–Purdue University at Indianapolis for her precise review of the entire manuscript, Patricia Mallion, mathematics editor, for her imagination and leadership, and Martha Morong, production editor, for efforts above and beyond those normally expected.

October 1978   M. L. K.
                                            M. L. B.

CONTENTS

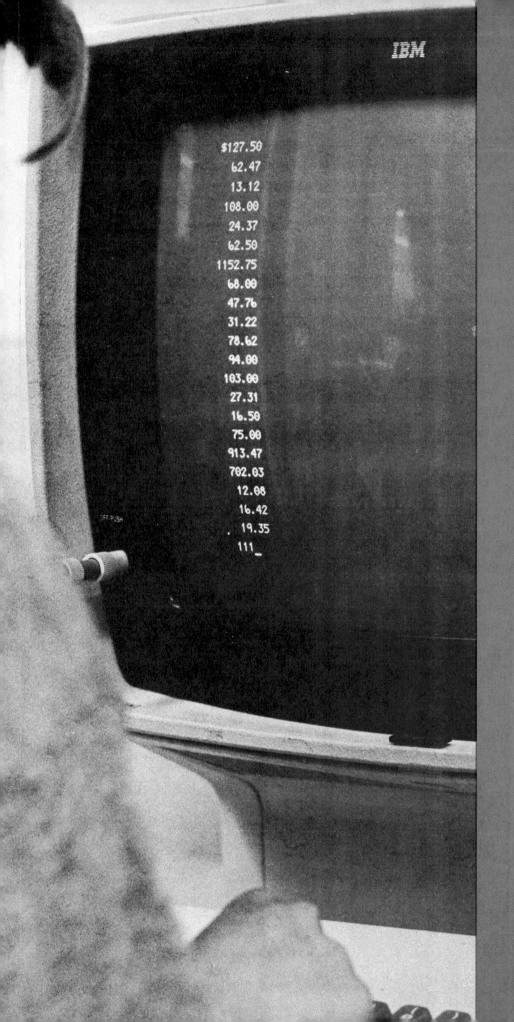

THE NUMBERS
OF ORDINARY
ARITHMETIC
AND THEIR
PROPERTIES

OBJECTIVES

After finishing Section 1.1, you should be able to:

◼• Write several numerals for any number of arithmetic by multiplying by 1.

◼•• Find simplest fractional notation for a number of arithmetic.

◼••• Add, subtract, and multiply with fractional notation.

1. Write three fractional numerals for $\frac{2}{3}$. $\frac{6}{9}$ $\frac{12}{18}$ $16|24$

2. Write three fractional numerals for $\frac{1}{2}$. $\frac{5}{10}$ $\frac{50}{100}$ $25|50$

3. Write three fractional numerals for $\frac{3}{5}$. $\frac{15}{25}$ $9|15$ $18|30$

4. Write three fractional numerals for 1. $4|4$ $3|3$ $7|7$

5. Write three fractional numerals for 4. $4|1$ $16|4$ $8|2$

1.1 NUMBERS AND ALGEBRA

The symbols we use to name numbers are called *numerals*. Thus symbols such as 4, 8, XVI, and $\frac{3}{4}$ are numerals. In algebra we also use symbols such as x, y, a, and b that represent numbers to calculate and solve problems.

◼• NUMBERS OF ORDINARY ARITHMETIC

The numbers used for counting are called *natural numbers*. They are

1,   2,   3,   4,   5,   6,   7,   8,   9,   10,   11,   and so on.

The *whole numbers* consist of the natural numbers and zero. They are

0,   1,   2,   3,   4,   5,   6,   7,   8,   9,   10,   11,   and so on.

The *numbers of ordinary arithmetic* consist of the whole numbers and the fractions such as $\frac{2}{3}$ and $\frac{9}{5}$. These numbers are sometimes called the *numbers of arithmetic*. All the numbers of arithmetic can be named by *fractional notation a/b*.

Example 1   Write three fractional numerals for the number *three fourths*.

$$\frac{3}{4}, \quad \frac{6}{8}, \quad \frac{600}{800}$$

Example 2   Write three fractional numerals for the number *one third*.

$$\frac{1}{3}, \quad \frac{2}{6}, \quad \frac{10}{30}$$

Any fractional numeral with a denominator three times the numerator names the number *one third*.

DO EXERCISES 1–3 (IN THE MARGIN AT THE LEFT).

Whole numbers can also be named with fractional notation.

Examples

3. $0 = \frac{0}{1} = \frac{0}{4} = \frac{0}{56}$,   and so on.

4. $2 = \frac{2}{1} = \frac{6}{3} = \frac{128}{64}$,   and so on.

5. $1 = \frac{2}{2} = \frac{5}{5} = \frac{642}{642}$,   and so on.

DO EXERCISES 4 AND 5 (IN THE MARGIN).

Chapter 1

To find different names, we can use the idea of multiplying by 1.

Examples

6. $\dfrac{2}{3} = \dfrac{2}{3} \times \boxed{1}$

$= \dfrac{2}{3} \times \boxed{\dfrac{5}{5}}$ Using $\dfrac{5}{5}$ for 1

$= \dfrac{10}{15}$

7. $\dfrac{2}{3} = \dfrac{2}{3} \times \boxed{1}$

$= \dfrac{2}{3} \times \boxed{\dfrac{13}{13}}$ This time we use $\dfrac{13}{13}$ for 1.

$= \dfrac{26}{39}$

DO EXERCISES 6 AND 7.

■●■ **SIMPLEST FRACTIONAL NOTATION**

The simplest fractional notation for a number has the smallest possible numerator and denominator. To simplify, we reverse the process used in Examples 6 and 7.

Examples   Simplify.

8. $\dfrac{10}{15} = \dfrac{2 \times \boxed{5}}{3 \times \boxed{5}}$ Factoring numerator and denominator
Factoring means writing as a product.

$= \dfrac{2}{3} \times \boxed{\dfrac{5}{5}}$ Factoring the fraction

$= \dfrac{2}{3}$ We "removed" a factor of 1.

9. $\dfrac{36}{24} = \dfrac{6 \times 6}{4 \times 6} = \dfrac{3 \times \boxed{2} \times \boxed{6}}{2 \times \boxed{2} \times \boxed{6}} = \dfrac{3}{2} \times \boxed{\dfrac{2 \times 6}{2 \times 6}} = \dfrac{3}{2}$

You may be in the habit of canceling. For example, you might have done Example 9 as follows. This is a shortcut for the procedure used in Examples 8 and 9.

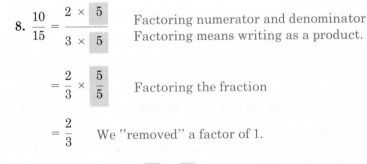

$\dfrac{\overset{3}{\cancel{\overset{\cancel{18}}{36}}}}{\underset{\cancel{\underset{2}{12}}}{\cancel{24}}}$ or $\dfrac{36}{24} = \dfrac{3 \times \cancel{12}}{2 \times \cancel{12}} = \dfrac{3}{2}$

6. Multiply by 1 to find three different names for $\dfrac{4}{5}$.

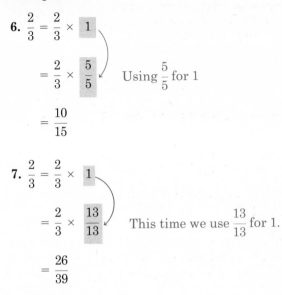

$4/5 \cdot 2/2 = 8/10$

$4/5 \cdot 3/3 = 12/15$

$4/5 \cdot 5/5 = 20/25$

7. Multiply by 1 to find three different names for $\dfrac{8}{7}$.

$8/7 \cdot 3/3 = 24/21$

$8/7 \cdot 5/5 = 40/35$

$8/7 \cdot 10/10 = 80/70$

Simplify.

8. $\dfrac{18}{27}$ = $\dfrac{6}{9}$ = $\dfrac{2}{3}$

9. $\dfrac{38}{18}$ = $\dfrac{19}{9}$ =

10. $\dfrac{56}{49}$ = $\dfrac{8}{7}$

Simplify.

11. $\dfrac{27}{54}$ = $\dfrac{9 \times}{18}$ = $\dfrac{3}{6}$ = $\dfrac{1}{2}$

12. $\dfrac{48}{12}$ = $\dfrac{4 \cdot 12}{4 \cdot 3}$ = $\dfrac{4}{4} \cdot \dfrac{3}{3} \dfrac{4}{1}$ = 4

Multiply and simplify.

13. $\dfrac{6}{5} \cdot \dfrac{25}{12}$ = $\dfrac{1}{1} \cdot \dfrac{5}{2}$ = $\dfrac{5}{2}$

14. $\dfrac{3}{8} \cdot \dfrac{5}{3} \cdot \dfrac{7}{2}$ $\dfrac{3}{8} \cdot \dfrac{5}{2} \cdot \dfrac{7}{1}$ = $\dfrac{35}{16}$

Incorrect canceling causes many errors. Remember:

If you can't factor, you can't cancel.

DO EXERCISES 8–10.

We can put a factor of 1 in the numerator or denominator.

Examples   Simplify.

10. $\dfrac{18}{72} = \dfrac{2 \times 9}{8 \times 9}$

$= \dfrac{1 \times \boxed{2} \times \boxed{9}}{4 \times \boxed{2} \times \boxed{9}}$ Putting a 1 in the numerator

$= \dfrac{1}{4} \times \boxed{\dfrac{2 \times 9}{2 \times 9}}$

$= \dfrac{1}{4}$

11. $\dfrac{72}{9} = \dfrac{8 \times 9}{1 \times 9} = \dfrac{8}{1} \times \boxed{\dfrac{9}{9}} = \dfrac{8}{1} = 8$

DO EXERCISES 11 AND 12.

●●●   MULTIPLICATION, ADDITION, AND SUBTRACTION

Example 12   Multiply and simplify.

$\dfrac{5}{6} \cdot \dfrac{9}{25} = \dfrac{5 \cdot 9}{6 \cdot 25}$ Multiplying numerators and denominators

$= \dfrac{1 \cdot 5 \cdot 3 \cdot 3}{2 \cdot 3 \cdot 5 \cdot 5}$ Factoring numerator and denominator

$= \dfrac{3 \cdot 5 \cdot 1 \cdot 3}{3 \cdot 5 \cdot 2 \cdot 5}$

$= \boxed{\dfrac{3 \cdot 5}{3 \cdot 5}} \cdot \dfrac{1 \cdot 3}{2 \cdot 5}$ Factoring the fraction

$= \dfrac{3}{10}$ "Removing" a factor of 1

In Example 12, we used a dot · instead of × for a multiplication sign. It means exactly the same thing as ×.

DO EXERCISES 13 AND 14.

We can multiply by 1 to find common denominators.

Example 13 Add and simplify.

$$\frac{3}{8} + \frac{5}{12} = \frac{3}{8} \cdot \boxed{\frac{12}{12}} + \frac{5}{12} \cdot \boxed{\frac{8}{8}} \qquad \text{Multiplying by 1}$$

$$= \frac{36}{96} + \frac{40}{96}$$

$$= \frac{76}{96} \qquad \text{Adding}$$

$$= \frac{19}{24} \cdot \boxed{\frac{4}{4}} \qquad \text{Factoring}$$

$$= \frac{19}{24} \qquad \text{``Removing'' a factor of 1}$$

Example 14 Add and simplify.

$$\frac{1}{4} + \frac{3}{5} = \frac{1}{4} \cdot \boxed{\frac{5}{5}} + \frac{3}{5} \cdot \boxed{\frac{4}{4}} = \frac{5}{20} + \frac{12}{20} = \frac{17}{20}$$

DO EXERCISES 15 AND 16.

Example 15 Subtract and simplify.

$$\frac{9}{8} - \frac{4}{5} = \frac{9}{8} \cdot \boxed{\frac{5}{5}} - \frac{4}{5} \cdot \boxed{\frac{8}{8}} = \frac{45}{40} - \frac{32}{40} = \frac{13}{40}$$

Example 16 Subtract and simplify.

$$\frac{7}{10} - \frac{1}{5} = \frac{7}{10} - \frac{1}{5} \cdot \boxed{\frac{2}{2}} = \frac{7}{10} - \frac{2}{10} = \frac{5}{10} = \frac{1}{2} \cdot \boxed{\frac{5}{5}} = \frac{1}{2}$$

DO EXERCISES 17 AND 18.

Add and simplify.

15. $\dfrac{5}{6} + \dfrac{7}{10}$

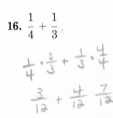

$$\frac{5}{6} \cdot \frac{10}{10} + \frac{7}{10} \cdot \frac{6}{6}$$

$$\frac{50}{60} + \frac{42}{60} = \frac{92}{60} = \frac{46}{30} \quad \frac{23}{15} \quad \frac{8}{5}$$

16. $\dfrac{1}{4} + \dfrac{1}{3}$

$$\frac{1}{4} \cdot \frac{3}{3} + \frac{1}{3} \cdot \frac{4}{4}$$

$$\frac{3}{12} + \frac{4}{12} \quad \frac{7}{12}$$

Subtract and simplify.

17. $\dfrac{4}{5} - \dfrac{4}{6}$

$$\frac{4}{5} \cdot \frac{6}{6} - \frac{4}{6} \cdot \frac{5}{5}$$

$$\frac{24}{30} - \frac{20}{30} = \cdot \frac{4}{30} = \frac{2}{15}$$

18. $\dfrac{5}{12} - \dfrac{2}{9}$

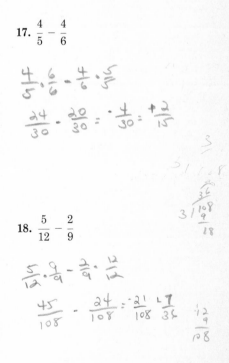

$$\frac{5}{12} \cdot \frac{9}{9} - \frac{2}{9} \cdot \frac{12}{12}$$

$$\frac{45}{108} - \frac{24}{108} = \frac{21}{108} \quad \frac{7}{36}$$

NAME _____   CLASS _____

EXERCISE SET 1.1

⚫ Write four different fractional numerals for each number.

$\frac{15}{75}$

1. $\frac{4}{3} = \frac{8}{6} = \frac{16}{12} = \frac{40}{30}$ **2.** $\frac{5}{9} = \frac{25}{45} = \frac{15}{27} = \frac{20}{36}$ **3.** $\frac{6}{11} = \frac{12}{22} = \frac{24}{44} = \frac{36}{33}$ **4.** $\frac{15}{7} = \frac{30}{14} = \frac{45}{21} = \frac{75}{35}$

5. $\frac{2}{11} = \frac{4}{22} = \frac{6}{33} = \frac{8}{44}$ **6.** $1 \cdot \frac{4}{4} = \frac{3}{3} = \frac{7}{7}$ **7.** $5 \quad \frac{25}{5} = \frac{30}{6} = \frac{35}{7}$ **8.** $0 \quad \frac{0}{3} = \frac{0}{7} = \frac{0}{21}$

⚫⚫ Simplify.

9. $\frac{8}{6} = \frac{4}{3}$ **10.** $\frac{15}{25} = \frac{3}{5}$ **11.** $\frac{17}{34} = \frac{1}{2}$ **12.** $\frac{35}{25} = \frac{7}{5}$

13. $\frac{100}{50} = \frac{4}{2} = \frac{2}{1} = 2$ **14.** $\frac{13}{39} = \frac{1}{3}$ **15.** $\frac{250}{75} = \frac{10}{3}$ **16.** $\frac{12}{18} = \frac{3}{4}$

⚫⚫⚫ Compute and simplify.

$\frac{17}{3} \atop 51$

17. $\frac{1}{4} \cdot \frac{1}{2} = \frac{1}{8}$ **18.** $\frac{11}{10} \cdot \frac{8^4}{5} = \frac{44}{55} \atop 5$ **19.** $\frac{17}{2} \cdot \frac{3}{4} = \frac{51}{8}$ **20.** $\frac{11}{12} \cdot \frac{12}{11} = 1$

$\frac{1 \cdot 3}{4 \cdot 2}$

1. _____
2. _____
3. _____
4. _____
5. _____
6. _____
7. _____
8. _____
9. _____
10. _____
11. _____
12. _____
13. _____
14. _____
15. _____
16. _____
17. _____
18. _____
19. _____
20. _____

Copyright © 1979, Philippines copyright 1979, by Addison-Wesley Publishing Company, Inc. All rights reserved.

ANSWERS

21. 1

22. $\frac{3}{4}$

23. $\frac{17}{18}$ $\frac{7}{6}$

24. 16/15

25. 5/6

26. 41/64

27. 2/4 = 1/2

28. 7/5

29. 5/18

30. 31/45

31. 31/60

32. 13/48

21. $\frac{1}{2} + \frac{1}{2} = \frac{2}{2} = 1$

22. $\frac{1}{2} + \frac{1}{4}$

$\frac{1}{2} \cdot \frac{4}{4} + \frac{1}{4} \cdot \frac{2}{2}$

$\frac{4}{8} + \frac{2}{8} = \frac{6}{8} = \frac{3}{4}$

23. $\frac{4}{9} + \frac{13}{18}$

$\frac{4}{9} \cdot \frac{18}{18} + \frac{13}{18} \cdot \frac{9}{9}$

$\frac{72}{62} + \frac{117}{62}$

$\frac{189}{62}$

$\frac{8}{18} + \frac{13}{18} = 21/18 \quad 7/6$

24. $\frac{4}{5} + \frac{8}{15}$

$\frac{12}{15} + \frac{8}{15}$

$\frac{16}{15}$

25. $\frac{3}{10} + \frac{8}{15}$

$\frac{9}{30} + \frac{16}{30} = \frac{25}{30} = \frac{5}{6}$

26. $\frac{9}{8} + \frac{7}{12}$

$\frac{9}{8} \cdot \frac{12}{12} + \frac{7}{12} \cdot \frac{8}{8}$

$\frac{108}{96} + \frac{56}{96}$

$\frac{164}{96} \quad \frac{82}{48} \quad \frac{41}{24}$

27. $\frac{5}{4} - \frac{3}{4}$

28. $\frac{12}{5} - \frac{2}{5}$

29. $\frac{13}{18} - \frac{4}{9}$

$\frac{13}{18} - \frac{8}{18} = \frac{5}{18}$

30. $\frac{13}{15} - \frac{8}{45}$

$\frac{39}{45} - \frac{8}{45}$

$31/45$

31. $\frac{11}{12} - \frac{2}{5}$

$\frac{11}{12} \cdot \frac{5}{5} - \frac{2}{5} \cdot \frac{12}{12}$

$\frac{55}{60} - \frac{24}{60} \quad 31/60$

$\frac{12}{60}$

32. $\frac{15}{16} - \frac{2}{3}$

$\frac{15}{16} \cdot \frac{3}{3} - \frac{2}{3} \cdot \frac{16}{16}$

$\frac{45}{48} - \frac{32}{48} = 13/48$

1.2   DECIMAL NOTATION

Another notation for the numbers of arithmetic is called *decimal notation*. It is based on ten. Standard notation, such as 9345 and 20,867, is decimal notation for whole numbers. We can use *expanded notation* to show meanings.

■○   EXPANDED NOTATION

Example 1   Expanded notation for 9345 is

$$9000 + 300 + 40 + 5.$$

Example 2   Write three kinds of expanded notation for 3502.

$$
\begin{aligned}
3502 &= 3000 &&+ 500 &&+ 2 \\
&= 3 \cdot 1000 &&+ 5 \cdot 100 &&+ 2 \\
&= 3 \cdot 10 \cdot 10 \cdot 10 + 5 \cdot 10 \cdot 10 + 2
\end{aligned}
$$

DO EXERCISES 1 AND 2.

■■   EXPONENTIAL NOTATION

Shorthand notation for $10 \cdot 10 \cdot 10$ is called *exponential notation*.

For $\underbrace{10 \cdot 10 \cdot 10}_{\text{3 factors}}$ we write 10^3.

This is read "ten cubed" or "ten to the third power." We call the number 3 an *exponent* and we say that 10 is the *base*. For $10 \cdot 10$ we write 10^2, read "ten squared," or "ten to the second power."

Example 3   Write exponential notation for $10 \cdot 10 \cdot 10 \cdot 10 \cdot 10$.

$$10 \cdot 10 \cdot 10 \cdot 10 \cdot 10 = 10^5$$

DO EXERCISE 3.

Example 4   Write expanded notation with exponents for 82,653.

$$82,653 = 8 \cdot 10^4 + 2 \cdot 10^3 + 6 \cdot 10^2 + 5 \cdot 10 + 3$$

DO EXERCISES 4 AND 5.

An exponent tells us how many times the base is used as a factor. The base can be a number other than 10.

Examples

5. 3^5 means $3 \cdot 3 \cdot 3 \cdot 3 \cdot 3$
6. 7^4 means $7 \cdot 7 \cdot 7 \cdot 7$
7. If we use n to stand for a number, n^4 means $n \cdot n \cdot n \cdot n$.

DO EXERCISES 6 AND 7.

OBJECTIVES

After finishing Section 1.2, you should be able to:

■ Write expanded notation without exponents.

■ Tell the meaning of exponential notation such as 5^3.

■ For expressions such as *nnn* and *xyz*, give their meanings as $n \cdot n \cdot n$, or n^3, and $x \cdot y \cdot z$.

■ Tell the meaning of expressions with exponents of 0 and 1, and write expanded notation with exponents.

■ Convert from decimal to fractional notation.

■ Convert from fractional to decimal notation.

1. Write three kinds of expanded notation for 2374.

   $2000 + 300 + 70 + 4$

   $2 \cdot 10 \cdot 10 \cdot 10 + 3 \cdot 10 \cdot 10 + 7 \cdot 10 + 4$

2. Write three kinds of expanded notation for 34,809.

   $30000 + 4000 + 800 + 9$

   $3 \cdot 10 \cdot 10 \cdot 10 \cdot 10 + 4 \cdot 10 \cdot 10 \cdot 10 + 8 \cdot 10 \cdot 10 + 9$

3. Write exponential notation for $10 \cdot 10 \cdot 10 \cdot 10. = 10^4$

4. Write expanded notation with exponents for 20,347.

   $2 \cdot 10^4 + 3 \cdot 10^2 + 4 \cdot 10 + 7$

5. Write expanded notation with exponents for 752,398.

   $7 \cdot 10^5 + 5 \cdot 10^4 + 2 \cdot 10^3 + 3 \cdot 10^2 + 9 \cdot 10 + 8$

6. What is the meaning of 5^4?

   $5 \cdot 5 \cdot 5 \cdot 5$

7. What is the meaning of x^5?

   $x \cdot x \cdot x \cdot x \cdot x$

8. Write exponential notation for *nnnnn*.

   n^5

9. What is the meaning of y^3? Do not use × or ·.

   yyy

10. What number does *Prt* represent if *P* stands for 2000, *r* stands for 0.07, and *t* stands for 1?

    $2000 \cdot 0.07 \cdot 1$

    $\begin{array}{r} 2000 \\ .07 \\ \hline 14000 \\ 0000 \\ \hline 14000 \end{array}$ 140

11. What does x^3 represent if *x* stands for 2?

    $2 \cdot 2 \cdot 2 =$
    $4 \cdot 2 = 8$

12. Write expanded notation for 3562 using exponents. Include exponents of 0 and 1.

    $3 \cdot 10^3 + 5 \cdot 10^2 + 6 \cdot 10^1 + 2 \cdot 10^0$

13. What is 4^1?

    4

14. What is 6^0?

    1

●●● VARIABLES

A letter can be used to stand for various numbers. In this case, we call it a *variable*. If we write two or more variables together, such as

   nnn   or   *lwh*   or   *Prt*,

we agree that the numbers are to be multiplied.

Examples

8. *nnn* means $n \cdot n \cdot n$. It also means n^3.
9. *lwh* means $l \cdot w \cdot h$.
10. If *P* stands for 1000, *r* stands for 0.08, and *t* stands for 3, then *Prt* means $1000 \times 0.08 \times 3$, or 240.

DO EXERCISES 8–11.

▪▪ 1 AND 0 AS EXPONENTS

The numbers 0 and 1 can also be given meaning as exponents. To see how, we can look at expanded notation.

   $8795 = 8 \cdot 10^3 + 7 \cdot 10^2 + 9 \cdot 10 + 5$

To keep the decreasing pattern of the exponents we write

   $8 \cdot 10^3 + 7 \cdot 10^2 + 9 \cdot 10^1 + 5 \cdot 10^0.$

Thus,

   10^1 should mean 10   and 10^0 should mean 1.

> **For any number n, n^1 means n.**
> **For any number n (other than zero), n^0 means 1.**

Later, we will explain why 0^0 is meaningless.

Examples

11. $5^1 = 5$ 12. $8^1 = 8$ 13. $3^0 = 1$ 14. $5^0 = 1$

DO EXERCISES 12–14.

▪▪ FROM DECIMAL TO FRACTIONAL NOTATION

Decimal notation for fractions contains *decimal points*. The place values to the right of a decimal point are tenths, hundredths, and so on. We can show this with expanded notation.

Examples

15. $34.248 = 3 \cdot 10 + 4 + 2 \cdot \dfrac{1}{10} + 4 \cdot \dfrac{1}{100} + 8 \cdot \dfrac{1}{1000}$

16. $0.9 = 9 \cdot \dfrac{1}{10}$

DO EXERCISES 15–17.

To convert from decimal to fractional notation, we can multiply by 1, as shown in the following examples.

Examples Write fractional notation.

17. $34.2 = \dfrac{34.2}{1} \cdot \dfrac{10}{10}$ Multiplying by 1

$= \dfrac{342}{10}$ 34.2 is 34 and two tenths, or 342 tenths

18. $16.453 = \dfrac{16.453}{1} \cdot \dfrac{1000}{1000}$ Multiplying by 1

$= \dfrac{16453}{1000}$ 16.453 is 16,453 thousandths

DO EXERCISES 18 AND 19.

⠿ FROM FRACTIONAL TO DECIMAL NOTATION

To convert from fractional to decimal notation we divide.

Example 19 $\frac{3}{8}$ means $3 \div 8$, so we divide

```
  0.375
8)3.000
  2 400
   600
   560
    40
    40
```

Sometimes we get a repeating decimal when we divide.

Example 20 $\frac{4}{11}$ means $4 \div 11$, so we divide

```
  0 .3636 . . .
11) 4 .00
    33
    70
    66
     4 0
     33
     70
     66
      4
```

Write expanded notation.

15. 23.678

$2 \cdot 10 + 3 + 6 \cdot \frac{1}{10} + 7 \cdot \frac{1}{100} + 8 \cdot \frac{1}{1000}.$

16. 0.27

$2 \cdot \frac{1}{10} + 7 \cdot \frac{1}{100}$

17. 4.0067

$4 + 6 \cdot \frac{1}{1000} + 7 \cdot \frac{1}{10000}$

Write fractional notation. You need not simplify.

18. 1.62

$\frac{1.62}{1} \cdot \frac{100}{100} = \frac{162}{100}$

19. 35.431

$\frac{35.431}{1} \cdot \frac{1000}{1000} = \frac{35431}{10000}$

Write decimal notation. Use a bar for repeating decimals.

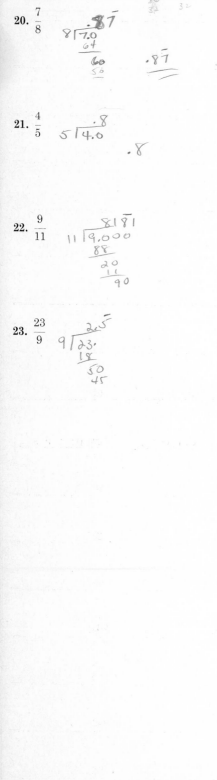

20. $\dfrac{7}{8}$

21. $\dfrac{4}{5}$

22. $\dfrac{9}{11}$

23. $\dfrac{23}{9}$

Thus $\dfrac{4}{11} = 0.363636\ldots$. Such decimals are often abbreviated by putting a bar over the repeating part, as follows.

$$\frac{4}{11} = 0.36\overline{36}$$

DO EXERCISES 20–23.

From time to time throughout this book, you will find some "extras," such as the one on this page. These are optional. You may find them of interest, but they can be skipped. Some require the use of a calculator.

SOMETHING EXTRA
CALCULATOR CORNER: NUMBER PATTERNS

There are many interesting number patterns in mathematics. Look for a pattern in the following. We can use a calculator for the computations.

$$6^2 = 36$$
$$66^2 = 4356$$
$$666^2 = 443556$$
$$6666^2 = 44435556$$

Do you see a pattern? If so, find 66666^2 without the use of your calculator.

EXERCISES

In each of the following, do the first four calculations using your calculator. Look for a pattern. Use the pattern to do the last calculation without the use of your calculator.

1. 3^2
    33^2
    333^2
    3333^2
    $\boxed{33333^2}$

2. $9 \cdot 6$
    $99 \cdot 66$
    $999 \cdot 666$
    $9999 \cdot 6666$
    $\boxed{99999 \cdot 66666}$

3. $37 \cdot 3$
    $37 \cdot 33$
    $37 \cdot 333$
    $37 \cdot 3333$
    $\boxed{37 \cdot 33333}$

4. $37 \cdot 3$
    $37 \cdot 6$
    $37 \cdot 9$
    $37 \cdot 12$
    $\boxed{37 \cdot 15}$

NAME   CLASS

EXERCISE SET 1.2

Review the objectives for this lesson.

▪ Write three kinds of expanded notation without exponents.

1. 5677   **2.** 6782   **3.** 908,563   **4.** 476,728

$5000 + 600 + 70 + 7$

$5 \times 1000 + 6 \times 100 + 7 \times 10 + 7$

$5 \cdot 10^3 + 6 \cdot 10^2 + 7 \cdot 10 + 7$

$900000 + 8000 + 500 + 60 + 3$

$9 \cdot 10^5 + 8 \cdot 10^3 + 5 \cdot 10^2 + 6 \cdot 10 + 3$

$9 \cdot 100000 + 8 \cdot 1000 + 5 \cdot 100 \cdot 6 \cdot 10 + 3$

▪▪ What is the meaning of each?

5. 5^2 **6.** 8^3 **7.** m^3 **8.** t^2

$5 \cdot 5$ $8 \cdot 8 \cdot 8$ mmm tt

9. 4^5 **10.** 9^4 **11.** x^6 **12.** n^7

$4 \cdot 4 \cdot 4 \cdot 4 \cdot 4$ $9 \cdot 9 \cdot 9 \cdot 9$ $x \cdot x \cdot x \cdot x \cdot x$ $nnnnnnn$

▪▪▪ Write exponential notation.

13. nnn **14.** $yyyy$ **15.** rr **16.** $xxxxx$

n^3 $y4$ $r2$ x^5

What is the meaning of each of the following? Do not use $\times$ or $\cdot$

17. x^3 **18.** m^2 **19.** y^4 **20.** p^5

xxx mm $yyyy$ $ppppp$

What does each of the following represent if n stands for 10?

21. n^2 **22.** n^3 **23.** n^5 **24.** n^4

$10 \cdot 10$ $10 \cdot 10 \cdot 10$ 100000 10000

$= 100$ 1000

What does each of the following represent if P stands for 1000, r stands for 0.06, and t stands for 2?

25. r^2 **26.** r^3 **27.** Prt **28.** Pr^2 [*Hint:* This is $P \cdot r^2$, not $(Pr)^2$.]

$\begin{array}{c} .06 \\ .06 \\ \hline .0036 \end{array}$ $\begin{array}{c} .2 \cdot .2 \cdot .2 \\ 4 \cdot .2 = .8 \end{array}$ Prt $1000 \cdot 0.0036$

$1000 \cdot .06 \cdot 2$ $.3600$

$\begin{array}{c} 1000 \\ .06 \\ \hline 6000 \end{array}$ $\begin{array}{c} 60 \cdot 2 \\ 120 \end{array}$

ANSWERS

1. _____

2. _____

3. _____

4. _____

5. $5 \cdot 5 = 25$

6. $8 \cdot 8 \cdot 8$

7. $m \cdot m \cdot m$

8. $t \cdot t$

9. $4 \cdot 4 \cdot 4 \cdot 4 \cdot 4$ 10. $9 \cdot 9 \cdot 9 \cdot 9$

11. $x \cdot x \cdot x \cdot x \cdot x \cdot x$ 12. $hhhhhh$

13. n^3 14. $y4$

15. $r2$ 16. x^5

17. xxx 18. mm

19. $yyyy$ 20. $ppppp$

21. 100 22. 1000

23. $100,000$ 24. $10,000$

25. $.0036$ 26. 8

27. 120 28. $.36$

Copyright © 1979, Philippines copyright 1979, by Addison-Wesley Publishing Company, Inc. All rights reserved.

ANSWERS

29. _____

30. _____

31. _____

32. _____

33. $3 \cdot 10^2 + 3 \cdot 10^1 \cdot 8 \cdot 10^0$

34. $1 \cdot 10^3 + 8 \cdot 10^2 + 7 \cdot 10^1 + 6 \cdot 10^0$

35. $8 \cdot 10^4 + 0 \cdot 10^3 + 5 \cdot 10^2 + 6 \cdot 10^1 + 4 \cdot 10^0$

36. $1 \cdot 10^4 + 2 \cdot 10^3 + 9 \cdot 10^2 \cdot 7 \cdot 10^1 \cdot 6 \cdot 10^0$

37. $\dfrac{291}{10}$

38. $\dfrac{1633}{100}$

39. $\dfrac{467}{100}$

40. $\dfrac{31415}{10,000}$

41. $3 \cdot 10^0 + 6 \cdot \dfrac{1}{10}$

42. $2 \cdot 10^1 + 9 \cdot 10^0 + 1 \cdot \dfrac{1}{10}$

43. $1 \cdot 10^1 + 8 \cdot 10^0 + 7 \cdot \dfrac{1}{10} + 8 \cdot \dfrac{1}{100} + 9 \cdot \dfrac{1}{1000}$

44. $3 \cdot 10^2 + 9 \cdot 10^0 + 6 \cdot \dfrac{1}{10} \cdot 2 \cdot \dfrac{1}{100}$

45. $.25$

46. $.50$

47. $.6$

48. 1.2

49. $.2\overline{2}$

50. $.5$

51. $.125$

52. $.625$

53. $.45\overline{45}$

54. $.63\overline{63}$

55. _____

What is the meaning of each of the following?

29. y^0 **30.** h^0 **31.** p^1 **32.** z^1

1 1 P z

Write expanded notation with exponents for each number. Use exponents of 0 and 1.

33. 378   **34.** 1876   **35.** 80,564   **36.** 12,976

$3 \cdot 10^2 +$

Write fractional notation.

37. 29.1   **38.** 16.33   **39.** 4.67   **40.** 3.1415

Write expanded notation.

41. 3.6   **42.** 29.1   **43.** 18.789   **44.** 309.62

Write decimal notation.

45. $\dfrac{1}{4}$ **46.** $\dfrac{1}{2}$ **47.** $\dfrac{3}{5}$ **48.** $\dfrac{6}{5}$

49. $\dfrac{2}{9}$ **50.** $\dfrac{4}{9}$ **51.** $\dfrac{1}{8}$ **52.** $\dfrac{5}{8}$

53. $\dfrac{5}{11}$ **54.** $\dfrac{7}{11}$

55. What power of 6 is too large for your calculator readout?

1.3 PROPERTIES OF NUMBERS OF ARITHMETIC

Some properties of the numbers of arithmetic are so simple that they may seem unimportant. However, they are important, especially in algebra.

● PARENTHESES. SYMBOLS OF GROUPING

What does $5 \times 2 + 4$ mean? If we multiply 5 by 2 and add 4, we get 14. If we add 2 and 4 and multiply by 5, we get 30. To tell which operation to do first, we use parentheses.

Examples

1. $(3 \times 5) + 6$ means $15 + 6$, or 21.
2. $3 \times (5 + 6)$ means 3×11, or 33.

DO EXERCISES 1–5.

●● GROUPING AND ORDER

What does $3 + 5 + 4$ mean? Does it mean $(3 + 5) + 4$ or $3 + (5 + 4)$? Either way the sum is 12, so it doesn't matter. In fact, if we are doing addition only, we can group numbers in any manner. This means that we really don't need parentheses if we are doing only addition. This illustrates another property.

> **For any numbers a, b, and c,**
>
> $(a + b) + c = a + (b + c)$.   **(The associative law of addition)**

Another basic property of numbers is that they can be added or multiplied in any order. For example, $3 + 2$ and $2 + 3$ are the same. Also, $5 \cdot 7$ and $7 \cdot 5$ are the same.

> **For any numbers a and b,**
>
> $a + b = b + a$;   **(The commutative law of addition)**
>
> $a \cdot b = b \cdot a$.   **(The commutative law of multiplication)**

DO EXERCISES 6–9.

The commutative and associative laws together help make addition easier.

Example 3   Add $3 + 4 + 7 + 6 + 8$. Look for combinations that make ten.

Add 3 and 7 to make 10, 4 and 6 to make 10, and then add the two 10's. Finally add the 8. The sum is 28.

DO EXERCISES 10 AND 11.

OBJECTIVES

After finishing Section 1.3, you should be able to:

■● Do calculations as shown by parentheses.

■●● Use grouping and order in addition and multiplication and tell which laws are illustrated by certain sentences.

■●●● Use the distributive law to do calculations like $(4 + 8) \cdot 5$ in two ways.

■❸❸ Use the distributive law to factor expressions like $3x + 3y$.

■❸❸ Evaluate expressions like $4x + 4y$ and $4(x + y)$ when numbers are given for the letters.

Do these calculations.

1. $(5 \times 4) + 2$

     $20 + 2 = 22$

2. $5 \times (4 + 2)$

     $5 \times 6 = 30$

3. $(4 \times 6) + 2$

     $24 + 2 = 26$

4. $5 \times (2 \times 3)$

     $5 \times 6 = 30$

5. $(6 \times 2) + (3 \times 5)$

     $12 + 15 = 27$

Do these calculations.

6. $17 + 10$

     27

7. $10 + 17$

     27

8. 26×70

     1820

9. 70×26

     1820

Add. Look for combinations that make ten.

10. $5 + 2 + 3 + 5 + 8$

     $10 \quad 13 = 23$

11. $1 + 5 + 6 + 9 + 4$

     $10 + 10 + 5 = 25$

ANSWERS ON PAGE A–2

Which laws are illustrated by these
sentences?

12. $61 \times 56 = 56 \times 61$ *Comm. mult.*

13. $(3 + 5) + 2 = 3 + (5 + 2)$ *Ass. Add*

14. $4 + (2 + 5) = (4 + 2) + 5$ *Ass. Add*

15. $7 \cdot (9 \cdot 8) = (7 \cdot 9) \cdot 8$ *Ass of mult*

Do the calculations as shown.

16. a) $(2 + 5) \cdot 4$

   $7 \cdot 4 = 28$

   b) $(2 \cdot 4) + (5 \cdot 4)$

   $8 + 20 = 28$

17. a) $(7 + 4) \cdot 7$

   $11 \cdot 7 = 77$

   b) $(7 \cdot 7) + (4 \cdot 7)$

   $49 + 28 = 77$

Compute.

18. $3 \cdot 5 + 2 \cdot 4$

   $15 + 8 = 23$

19. $4 \cdot 2 + 7 \cdot 1$

   $8 + 7 = 15$

For multiplication, does the grouping matter? For example, is
$2 \cdot (5 \cdot 3) = (2 \cdot 5) \cdot 3$? Our experience with arithmetic tells us that
multiplication is also associative.

> For any numbers a, b, and c,
>
> $a \cdot (b \cdot c) = (a \cdot b) \cdot c.$ **(The associative law of multiplication)**

Examples   Which laws are illustrated by these sentences?

   4. $3 + 5 = 5 + 3$ Commutative law of addition (order changed)
   5. $(2 + 3) + 5 = 2 + (3 + 5)$ Associative law of addition
      (grouping changed)
   6. $(3 \cdot 5) \cdot 2 = 3 \cdot (5 \cdot 2)$ Associative law of multiplication

DO EXERCISES 12–15.

●●●   THE DISTRIBUTIVE LAW

If we wish to multiply a number by a sum of several numbers, we can
either add and then multiply or multiply and then add.

Example 7   Compute in two ways: $(4 + 8) \cdot 5$.

$$
\left.
\begin{array}{c}
(4 + 8) \cdot \boxed{5} \\
12 \cdot 5 \\
60
\end{array}
\right\} \quad \text{Adding and then multiplying}
$$

$$
\left.
\begin{array}{c}
(4 \cdot \boxed{5}) + (8 \cdot \boxed{5}) \\
20 + 40 \\
60
\end{array}
\right\} \quad \text{Multiplying and then adding}
$$

DO EXERCISES 16 AND 17.

The property we are investigating is the *distributive law of multiplication over addition*. Before we state it formally, we need to make an
agreement about parentheses. We agree that in an expression like
$(4 \cdot 5) + (3 \cdot 7)$, we can omit the parentheses. Thus $4 \cdot 5 + 3 \cdot 7$ means
$(4 \cdot 5) + (3 \cdot 7)$. In other words, we do the multiplications first.

DO EXERCISES 18 AND 19.

> In an expression such as $ab + cd$, it is understood that paren-
> theses belong around ab and cd. In other words, the multiplica-
> tions are to be done first.

Using our agreement about parentheses, we now state the distributive
law.

> **For any numbers a, b, and c,**
>
> $$a(b + c) = ab + ac.$$ **(The distributive law)**

We cannot omit parentheses on the left above. If we did we would have $ab + c$, which by our agreement means $(ab) + c$.

Note that the distributive law can be extended to more than two numbers inside the parentheses:

$$a(b + c + d) = ab + ac + ad.$$

The distributive law would apply to the following situation. Someone decides to invest \$1000 in one bank at 8%, and \$2000 in another bank at 8%. At the end of one year the total interest from the two investments would be

$$(\,8\% \cdot 1000) + (\,8\% \cdot 2000).$$

The same interest would also have been made by investing the entire \$3000 in just one bank. The interest is

$$8\% \cdot (1000 + 2000), \quad \text{or} \quad 8\% \cdot 3000.$$

DO EXERCISE 20.

∷ FACTORING

Any equation can be reversed. Thus for the distributive law we could also write

$$ab + ac = a(b + c).$$

The distributive law is the basis for a process called *factoring*.

Example 8   Factor: $3x + 3y$.

By the distributive law,

$$3\,x + 3\,y = 3\,(x + y).$$

When we write $3(x + y)$, we say we have *factored* $3x + 3y$. That is, we have written it as a product.

Example 9   Factor: $5x + 5y + 5z$.

$$5\,x + 5\,y + 5\,z = 5\,(x + y + z)$$

DO EXERCISES 21–23.

∷ FACTORING AND EVALUATING

It is important to realize that when we factor an expression like $3x + 3y$ the factored expression represents the same number as the original one, no matter what x and y stand for!

Do these calculations.

20. a) $(0.08 \times 1000) + (0.08 \times 2000)$

$$80 + 160 = 240.$$

b) $0.08 \times (1000 + 2000)$

$$\begin{array}{r} 3000 \\ .08 \\ \hline 240.00 \end{array}$$

Factor.

21. $4x + 4y$

$$4(x + y)$$

22. $5a + 5b$

$$5(a + b)$$

23. $7p + 7q + 7r$

$$7(p + s + r)$$

Factor. Then evaluate both the original and factored expressions when $x = 4$ and $y = 3$.

24. $5x + 5y = 5(x+y)$

$5 \cdot 3 + 5 \cdot 4 \qquad 5(3+4)$

$\qquad\qquad\qquad 5 \cdot 7 = 35$

$15 + 20$

$= 35$

25. $7x + 7y = 7(x+y)$

$7 \cdot 4 + 7 \cdot 3 \qquad 7(4+3)$

$28 + 21 \qquad\qquad 7 \cdot 7 = 49$

$= 49$

Example 10 Factor $4x + 4y$. Then evaluate both the original and factored expressions when x stands for 2 and y stands for 3.

First, factor $4x + 4y$.

$$\boxed{4}\, x + \boxed{4}\, y = \boxed{4}\, (x + y)$$

Evaluate $4x + 4y$.

$$4x + 4y = 4 \cdot 2 + 4 \cdot 3 \qquad \text{Replacing } x \text{ by 2 and } y \text{ by 3}$$
$$= 8 + 12$$
$$= 20$$

Evaluate $4(x + y)$.

$$4(x + y) = 4(2 + 3) \qquad \text{Replacing } x \text{ by 2 and } y \text{ by 3}$$
$$= 4 \cdot 5$$
$$= 20$$

To say that x stands for 2, we may also write $x = 2$. That is, we are agreeing to use x and 2 as names of the same number.

Example 11 Factor $6x + 6y$. Then evaluate both the original and factored expressions when $x = 3$ and $y = 8$.

a) $6x + 6y = 6(x + y)$ Factoring

b) $6x + 6y = 6 \cdot 3 + 6 \cdot 8$ Evaluating $6x + 6y$
$$= 18 + 48$$
$$= 66$$

c) $6(x + y) = 6(3 + 8)$ Evaluating $6(x + y)$
$$= 6 \cdot 11$$
$$= 66$$

DO EXERCISES 24 AND 25.

NAME _____   CLASS _____   ANSWERS

EXERCISE SET 1.3

■ • Do these calculations.

1. $(10 + 4) + 8$
    $14 + 8 = 22$

2. $10 \times (9 + 4)$
    $10 \cdot 13 = 130$

3. $(10 \cdot 7) + 19$
    $70 + 19 = 89$

4. $(10 \cdot 7) + (20 \cdot 14)$
    $70 + 280 = 350$

■ ■ Do these calculations. Choose grouping and ordering to make the work easy.

5. $(8 + 4) + (5 + 2) + (6 + 15) + 1$
    $(12 + 7) + (21 + 1)$ 19
    $19 + 22 = 41$ $\frac{22}{41}$

6. $(9 + 6) + (3 + 4) + (1 + 7) + 11$
    $(15 + 7) + (8 + 11)$
    $22 + 19 = 41$

7. $(14 + 3) + (12 + 7) + (8 + 6) + 9$
    $\frac{19}{17}{36}$ $(17 + 19) + (14 + 9)$
    36 $23 = 59$

8. $(17 + 7) + (16 + 3) + (4 + 3) + 8$
    $\frac{24}{19}{43}$ $(24 + 19) + (7 + 8)$
    $43 + 15 = 58$

Which laws are illustrated by these sentences?

9. $67 + 3 = 3 + 67$ Comm. ADD

10. $15 \cdot 44 = 44 \cdot 15$ Comm. Mult

11. $6 + (9 + 5) = (6 + 9) + 5$
    Ass. ADD.

12. $8 \cdot (7 \cdot 6) = (8 \cdot 7) \cdot 6$
    Ass. Mult

■ ■ ■ Compute in two ways.

$\frac{13}{50}$ $13 \cdot 4 = 52$

13. $(6 + 7) \cdot 4$
    $6 \cdot 4 + 7 \cdot 4 = 52$
    $24 + 28 = 52$

14. $(8 + 10) \cdot 2 = 36$
    $8 \times 2 + 10 \cdot 2 =$
    $16 + 20 = 36$

⠿ Factor.

15. $9x + 9y$
    $9(x + y)$

16. $7w + 7u$
    $7(w + u)$

17. $\frac{1}{2}a + \frac{1}{2}b$
    $\frac{1}{2}(a + B)$

18. $\frac{3}{4}x + \frac{3}{4}y$
    $3/4(x + y)$

19. $1.5x + 1.5z$
    $1.5(x + z)$

20. $0.7a + 0.7b$
    $0.7(a + b)$

1. _____
2. _____
3. _____
4. _____
5. _____
6. _____
7. _____
8. _____
9. _____
10. _____
11. _____
12. _____
13. _____
14. _____
15. _____
16. _____
17. _____
18. _____
19. _____
20. _____

ANSWERS

21. _____

22. _____

23. _____

24. _____

25. _____

26. _____

27. _____

28. _____

29. _____

30. _____

31. _____

32. _____

33. _____

34. _____

35. _____

36. _____

21. $4x + 4y + 4z$

$4(x + y + z)$

22. $10a + 10b + 10c$

$10(a + b + c)$

23. $\frac{4}{7}a + \frac{4}{7}b + \frac{4}{7}c + \frac{4}{7}d$

$4/7(a + b + c + d)$

24. $\frac{3}{5}x + \frac{3}{5}y + \frac{3}{5}z + \frac{3}{5}w$

$3/5(x + y + z + w)$

Factor. Then evaluate both expressions when $x = 5$ and $y = 10$.

25. $9x + 9y$

$45 + 90 = 135$

$9(x + y)$

$9(5 + 10)$

$9 \cdot 15 = 135$

26. $8x + 8y$

$40 + 80 = 120$

$8(x + y)$

$8(5 + 10)$

$8 \cdot 15 = 120$

27. $10x + 10y$

$50 + 100 = 150$

$10(5 + 10)$

$10 \cdot 15$

$= 150$

28. $2x + 2y$

$10 + 20 = 30$

$2(x + y)$

$2(15)$

$= 30$

Factor. Then evaluate both expressions when $a = 0$ and $b = 9$.

29. $5a + 5b$

$1 + 9 = 10$

$5(a + b)$

$5(0 + 9)$

$5 \cdot 9 = 45$

30. $7a + 7b$

$7(a + b)$

$7(0 + 9)$

$7 \cdot 9 = 49$

31. $20a + 20b$

$20(a + b) =$

$20(0 + 9) =$

$20 \cdot 9 = 180$

32. $14a + 14b$

$14(9 + b)$

$14(0 + 9)$

$14 \cdot 9 = 126$

$4^2 - 3^2 = 16 - 9 = 5 \quad (x - y)^2 = (4 - 3)^2, 1 2$

33. Evaluate $x^2 - y^2$ when $x = 4$ and $y = 3$.

$(4 - 1) \cdot (4 + 1) = 3 \cdot 5 = 15$

34. Evaluate $(x - y) \cdot (x + y)$ when $x = 4$ and $y = 3$.

35. Calculate each of the following. Look for a pattern.

$$1 \cdot 9 + 1$$
$$12 \cdot 9 + 2$$
$$123 \cdot 9 + 3$$
$$1234 \cdot 9 + 4$$

Use the pattern to find the following without using your calculator.

$1234567 \cdot 9 + 7$

36. Calculate each of the following. Look for a pattern.

$$9^2 - 2^2$$
$$89^2 - 12^2$$
$$889^2 - 112^2$$
$$8889^2 - 1112^2$$

Use the pattern to find the following without using your calculator.

$88889^2 - 11112^2$

1.4 USING THE DISTRIBUTIVE LAW

The distributive law is the basis of many procedures in both arithmetic and algebra. Below are some further examples of factoring and some examples of other procedures based on this property.

▩ FACTORING

The parts of an expression such as $6x + 3y + 9z$ separated by plus signs, $6x$, $3y$, and $9z$, are called *terms* of the expression. To factor, look for a factor common to all the terms. Then "remove" it, so to speak, using the distributive law.

Examples Factor.

1. $6x + 3y + 9z = \boxed{3} \cdot 2x + \boxed{3} \cdot y + \boxed{3} \cdot 3z$ The common
                                                                          factor is 3.

$\qquad\qquad\qquad = \boxed{3}\,(2x + y + 3z)$ Using the distributive law.

2. $7y + 21z + 7 = \boxed{7} \cdot y + \boxed{7} \cdot 3z + \boxed{7} \cdot 1$ The common factor
                                                                                   is 7.

$\qquad\qquad\qquad = \boxed{7}\,(y + 3z + 1)$

Example 3 Simple interest on a principal of P dollars invested at interest rate r for t years is given by Prt. In t years, principal P will grow to the amount

$\qquad$ (Principal) + (Interest) = $P + Prt$.

Factor this expression.

$\qquad P + Prt = \boxed{P} \cdot 1 + \boxed{P}\, rt$

$\qquad\qquad\quad = \boxed{P}\,(1 + rt)$

DO EXERCISES 1–5.

▩▩ MULTIPLYING

When we reverse the factoring process, we say that we are "multiplying."

Examples Multiply.

4. $\boxed{3}\,(x + 2) = \boxed{3} \cdot x + \boxed{3} \cdot 2$ Using the distributive law
$\qquad\qquad = 3x + 6$

5. $\boxed{6}\,(s + 2t + 5w) = \boxed{6} \cdot s + \boxed{6} \cdot 2t + \boxed{6} \cdot 5w$ Using the
                                                                                        distributive law
$\qquad\qquad\qquad = 6s + 12t + 30w$

We multiplied each term inside the parentheses by the factor outside.

DO EXERCISES 6–8.

OBJECTIVES

After finishing Section 1.4, you should be able to:

▩ Factor expressions like $5x + 10$ by using the distributive law.

▩▩ Use the distributive law to multiply expressions like $5(x + 3)$.

▩▩▩ Collect like terms in expressions like $3x + 4y + 5x + 3y$.

Factor.

1. $5x + 10$

$5 \cdot x + 2 \cdot 5$

$5\,(x + 2)$

2. $12 + 3x$

$(3 \cdot 4) + (3 \cdot x)$

$3 \cdot (4 + x)$

3. $6x + 12 + 9y$

$(3 \cdot 2x) + (3 \cdot 4) + (3 \cdot 3y)$

$3\,(2x + 4 + 3y)$

4. $5x + 10y + 25$

$(5 \cdot x) + 5 \cdot 2y + 5 \cdot 5)$

$5\,(1x + 2y + 5)$

5. $Q + Qab$

$(Q \cdot 1) + (Q \cdot ab)$

$Q\,(1 + ab)$

Multiply.

6. $5(y + 3)$

$5y + 15$

7. $4(x + 2y + 5)$

$4x + 8y + 20$

8. $8(m + 3n + 4p)$

$8m + 24n + 32p)$

Collect like terms.

9. $6y + 2y$

$(6+2)y =$

$8y$

10. $4x + x$

$4 \cdot x + 1 \cdot x$

$= 5x$

11. $x + 0.03x$

$1 \cdot x + 0.03 \cdot x =$

$1.03x$

12. $10p + 8p + 4q + 5q$

$18p + 9q$

13. $7x + 3y + 4x + 5y$

$7x + 4x + 3y + 5y$

$11x + 8y$

Collect like terms.

14. $4y + 12y$

$16y$

15. $3s + 4s + 6w + 7w$

$7s + 13w$

16. $5x + 4y + 4x + 6y$

$9x + 10y$

17. $5a + b + a + 0.07b$

$6a + 1.07b$

◼◼◼ COLLECTING LIKE TERMS

If two terms have the same letters, they are called *like* terms, or *similar* terms. We can often simplify expressions by *collecting* or *combining* like terms. (We could also say *collecting* or *combining similar terms*.)

Examples   Collect like terms.

6. $3\,x\, + 4\,x\, = (3 + 4)\,x$ Using the distributive law

$\qquad = 7x$

7. $2x + 3y + 5x + 8y = 2\,x\, + 5\,x\, + 3\,y\, + 8\,y$ Regrouping and reordering using the associative and commutative laws

$\qquad\qquad = (2 + 5)\,x\, + (3 + 8)\,y$ Factoring

$\qquad\qquad = 7x + 11y$

8. $7x + x = 7 \cdot\,x\, + 1 \cdot\,x$

$\qquad = (7 + 1)\,x$

$\qquad = 8x$

9. $x + 0.05x = 1 \cdot\,x\, + 0.05\,x$

$\qquad\qquad = (1 + 0.05)\,x$

$\qquad\qquad = 1.05x$

DO EXERCISES 9–13.

With practice we can leave out some steps, collecting like terms mentally.

Examples   Collect like terms.

10. $5y + 2y + 4y = 11y$

11. $3x + 7x + 2y = 10x + 2y$

12. $3a + 5a + 8t + 2t = 8a + 10t$

13. $8p + q + p + 0.3q = 9p + 1.3q$

DO EXERCISES 14–17.

NAME

CLASS

ANSWERS

EXERCISE SET 1.4

◐ Factor.

1. $2x + 4$

$2 \cdot x + 2 \cdot 2$

$2(x + 2)$

2. $9x + 27$

$9 \cdot x + 9 \cdot 3$

$9(x + 3)$

3. $6x + 24$

$6 \cdot x + 6 \cdot 4$

$6(x + 4)$

4. $5y + 20$

$5 \cdot 4 + 5 \cdot 4$

$5(4 + 4)$

5. $9x + 3y$

$(3 \cdot 3x) + (3 \cdot 4)$

$3(3x + 4)$

6. $15x + 5y$

$(5 \cdot 3x) + (5 \cdot 4)$

$5(3x + 4)$

7. $14x + 21y$

$7 \cdot 2x + 7 \cdot 34$

$7(2x + 34)$

8. $18x + 24y$

$(6 \cdot 3x) + (6 \cdot 44)$

$6(3x + 44)$

9. $5 + 10x + 15y$

$(5 \cdot 1) + (5 \cdot 2x) + 5 \cdot 34$

$5(1 + 2x + 34)$

10. $7 + 14b + 56w$

$(7 \cdot 1) + (7 \cdot 2b) + (7 \cdot 8)$

$7(1 + 2b + 8)$

11. $8a + 16b + 64$

$(8 \cdot 1a) + (8 \cdot 2b) + (8 \cdot 8)$

$8(1a + 2b + 8)$

12. $9x + 27y + 81$

$(9 \cdot 1x) + (9 \cdot 34) + (9 \cdot 9)$

$9(1x + 34 + 9)$

13. $3x + 18y + 15z$

$(3 \cdot 1x) + (3 \cdot 64) + (3 \cdot 52)$

$3(1x + 64 + 52)$

14. $4r + 28s + 16t$

$(4 \cdot 1r) + (4 \cdot 7s) + (4 \cdot 4t)$

$4(1r + 7s + 4t)$

◐◐ Multiply.

15. $3(x + 1)$

$3x + 3$

16. $2(x + 2)$

$2x + 4$

17. $4(1 + y)$

$4 + 4y$

18. $9(s + 1)$

$9s + 9$

19. $9(4t + 3z)$

$36t + 27z$

20. $8(5x + 3y)$

$40x + 24y$

21. $7(x + 4 + 6y)$

$7x + 28 + 42y$

22. $8(9x + 5y + 8)$

$72x + 40y + 64$

23. $5(3x + 9 + 7y)$

$15x + 45 + 35y$

24. $4(5x + 8 + 3z)$

$20x + 32 + 12z$

1. ___
2. ___
3. ___
4. ___
5. ___
6. ___
7. ___
8. ___
9. ___
10. ___
11. ___
12. ___
13. ___
14. ___
15. ___
16. ___
17. ___
18. ___
19. ___
20. ___
21. ___
22. ___
23. ___
24. ___

Copyright © 1979, Philippines copyright 1979, by Addison-Wesley Publishing Company, Inc. All rights reserved.

ANSWERS

●●● Collect like terms.

25. _____

26. _____

27. _____

28. _____

29. _____

30. _____

31. _____

32. _____

33. _____

34. _____

35. _____

36. _____

37. _____

38. _____

39. _____

40. _____

41. _____

42. _____

25. $2x + 3x$

$5x$

26. $7y + 9y$

$16y$

27. $10a + a$

$11a$

28. $16x + x$

$17x$

29. $2x + 9z + 6x$

$8x + 9z$

30. $3a + 5b + 7a$

$10a + 5B$

31. $41a + 90c + 60c + 2a$

$43a + 150c$

32. $42x + 6b + 4x + 2b$

$46x + 8b$

33. $x + 0.09x + 0.2t + t$

$1.09x + 1.2t$

34. $0.01a + 0.23b + a + b$

$1.01a + 1.23b$

35. $8u + 3t + 10u + 6u + 2t$

$24u + 5t$

36. $5t + 6h + t + 8t + 9h$

$14t + 15h$

37. $23 + 5t + 7y + t + y + 27$

$50 + 6t + 8y$

38. $45 + 90d + 87 + 9d + 3 + 7d$

45
87
$135 + 106d$
135

39. $\frac{1}{2}b + \frac{1}{2}b$

$1\frac{1}{2}b$

40. $\frac{2}{3}x + \frac{1}{3}x$

$x\left(\frac{2}{3} + \frac{1}{3}\right)$

$x1$

41. $2y + \frac{1}{4}y + y$

$3\frac{1}{4}y$

42. $\frac{1}{2}a + a + 5a$

$6\frac{1}{2}a$

1.5   THE NUMBER 1 AND RECIPROCALS

The number 1 has some very special properties important in both arithmetic and algebra. When we multiply any number by 1, we get that same number.

> **For any number n,**
>
> $$n \cdot 1 = n.$$

When we divide a number by 1, we get the same number with which we started.

> **For any number n,**
>
> $$\frac{n}{1} = n.$$

When we divide a number by itself, the result is the number 1. This is true for any number except zero. We will see later why we do not divide by zero.

> **For any number n, except zero,**
>
> $$\frac{n}{n} = 1.$$

Examples   Simplify.

1. $\dfrac{3}{5} \cdot \boxed{\dfrac{7}{7}} = \dfrac{3}{5} \cdot \boxed{1} = \dfrac{3}{5}$ **2.** $\dfrac{\frac{3}{5}}{\boxed{1}} = \dfrac{3}{5}$ **3.** $\dfrac{\frac{4}{3}}{\frac{4}{3}} = \boxed{1}$

DO EXERCISES 1–4.

■·■ RECIPROCALS

Two numbers whose product is 1 are called *reciprocals* of each other. All the numbers of arithmetic, except zero, have reciprocals.

Examples

4. The reciprocal of $\dfrac{2}{3}$ is $\boxed{\dfrac{3}{2}}$ because $\dfrac{2}{3} \cdot \boxed{\dfrac{3}{2}} = \dfrac{6}{6} = 1$.

5. The reciprocal of 9 is $\boxed{\dfrac{1}{9}}$ because $9 \cdot \boxed{\dfrac{1}{9}} = \dfrac{9}{9} = 1$.

6. The reciprocal of $\dfrac{1}{4}$ is $\boxed{4}$ because $\dfrac{1}{4} \cdot \boxed{4} = 1$.

DO EXERCISES 5–8.

After finishing Section 1.5, you should be able to:

■· Find the reciprocal of any number.

■·■ Divide by multiplying by a reciprocal.

■·■·■ Graph numbers of arithmetic on a number line.

■■·■■ Use the proper symbol >, <, or = between two fractional numerals.

Simplify.

1. $\dfrac{4}{7} \cdot \dfrac{11}{11}$

2. $\dfrac{67}{67}$

3. $\dfrac{\frac{2}{3}}{1}$

4. $\dfrac{\frac{7}{5}}{\frac{7}{5}}$

Find the reciprocal of each number.

5. $\dfrac{4}{11}$

6. $\dfrac{15}{7}$

7. 5

8. $\dfrac{1}{3}$

Divide by multiplying by 1.

9. $\dfrac{\dfrac{3}{5}}{\dfrac{4}{7}}$

10. $\dfrac{\dfrac{5}{4}}{\dfrac{3}{2}}$

11. $\dfrac{\dfrac{9}{7}}{\dfrac{4}{5}}$

Divide by multiplying by the reciprocal of the divisor.

12. $\dfrac{4}{3} \div \dfrac{7}{2}$

13. $\dfrac{3}{5} \div \dfrac{7}{4}$

14. $\dfrac{\dfrac{2}{9}}{\dfrac{5}{7}}$

⬤⬤ RECIPROCALS AND DIVISION

The number 1 and reciprocals can be used to explain division of numbers of arithmetic. To divide, we can multiply by 1, choosing carefully the symbol for 1.

Example 7 Divide $\dfrac{2}{3}$ by $\dfrac{7}{5}$.

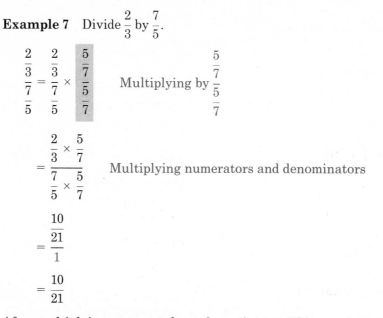

$$\dfrac{\dfrac{2}{3}}{\dfrac{7}{5}} = \dfrac{\dfrac{2}{3}}{\dfrac{7}{5}} \times \dfrac{\dfrac{5}{7}}{\dfrac{5}{7}} \qquad \text{Multiplying by } \dfrac{\dfrac{5}{7}}{\dfrac{5}{7}}$$

$$= \dfrac{\dfrac{2}{3} \times \dfrac{5}{7}}{\dfrac{7}{5} \times \dfrac{5}{7}} \qquad \text{Multiplying numerators and denominators}$$

$$= \dfrac{\dfrac{10}{21}}{1}$$

$$= \dfrac{10}{21}$$

After multiplying we got 1 for a denominator. This was because we used the reciprocal of the divisor, $\frac{7}{5}$, for both the numerator and denominator of the symbol for 1.

DO EXERCISES 9–11.

When multiplying by 1 to divide, we get a denominator of 1. What do we get in the numerator? In Example 7, we got $\frac{2}{3} \times \frac{5}{7}$. This is the product of $\frac{2}{3}$, the dividend, and $\frac{5}{7}$, the reciprocal of the divisor.

> **To divide, multiply by the reciprocal of the divisor:**
> $$\dfrac{a}{b} \div \dfrac{c}{d} = \dfrac{a}{b} \cdot \dfrac{d}{c}.$$

Example 8 Divide by multiplying by the reciprocal of the divisor.

$$\dfrac{1}{2} \div \dfrac{3}{5} = \dfrac{1}{2} \cdot \dfrac{5}{3} = \dfrac{5}{6} \qquad \dfrac{5}{3} \text{ is the reciprocal of } \dfrac{3}{5}$$

After dividing, simplification is often possible and should be done.

Example 9 Divide.

$$\dfrac{2}{3} \div \dfrac{4}{9} = \dfrac{2}{3} \cdot \dfrac{9}{4} = \dfrac{18}{12} = \dfrac{3 \cdot 6}{2 \cdot 6} = \dfrac{3}{2} \cdot \dfrac{6}{6} = \dfrac{3}{2}$$

Simplifying

DO EXERCISES 12–14.

◆◆◆ THE NUMBER LINE AND ORDER

The order of numbers of arithmetic can be shown on a number line.

DO EXERCISES 15 AND 16.

Note that $\frac{1}{2}$ is less than 1, and $\frac{1}{2}$ is to the left of 1 on the number line.

> **For any numbers a and b,**
> $$a < b \qquad \text{(read ``a is less than b'')}$$
> **means that a is to the left of b on the number line.**

The number $\frac{5}{2}$ is to the right of $\frac{1}{2}$. This means that $\frac{5}{2}$ is greater than $\frac{1}{2}$.

> **For any numbers a and b,**
> $$a > b \qquad \text{(read ``a is greater than b'')}$$
> **means that a is to the right of b on the number line.**

Sentences such as $\frac{5}{2} > \frac{1}{2}$ and $x < 2$ are called *inequalities*.

Examples   Determine whether true or false. Use the number line above.

10. $1.5 > 1$;   true   1.5 is to the right of 1

11. $\dfrac{5}{2} < \dfrac{3}{4}$;   false $\dfrac{5}{2}$ is not to the left of $\dfrac{3}{4}$

12. $\dfrac{1}{2} < 6$;   true $\dfrac{1}{2}$ is to the left of 6

DO EXERCISES 17–19.

◼◼ COMPARING NUMBERS

We want to develop a more efficient way of comparing numbers of arithmetic. Consider $\frac{4}{5}$ and $\frac{3}{5}$. The number line shows that $\frac{4}{5} > \frac{3}{5}$.

Note also that $4 > 3$. When denominators are the same, we just compare numerators.

Graph each number on a number line.

15. $\dfrac{6}{5}$

16. $\dfrac{17}{18}$

Determine whether true or false.

17. $\dfrac{1}{2} < \dfrac{3}{4}$

18. $\dfrac{3}{4} > \dfrac{1}{2}$

19. $\dfrac{13}{4} < \dfrac{5}{2}$

Use the proper symbol, $>$, $<$, or $=$, between each pair of numerals.

20. $\dfrac{3}{4} \qquad \dfrac{4}{4}$

21. $\dfrac{1}{2} \qquad \dfrac{1}{2}$

22. $\dfrac{22}{19} \qquad \dfrac{21}{19}$

ANSWERS ON PAGE A–2

Use the proper symbol $>$, $<$, or $=$.

23. $\dfrac{9}{5}$ $\dfrac{11}{7}$

24. $\dfrac{16}{12}$ $\dfrac{13}{8}$

25. $\dfrac{23}{7}$ $\dfrac{29}{9}$

26. $\dfrac{23}{13}$ $\dfrac{18}{11}$

For any numbers of arithmetic $\dfrac{a}{b}$ and $\dfrac{c}{b}$,

$$\dfrac{a}{b} > \dfrac{c}{b} \text{ when } a > c,$$

$$\dfrac{a}{b} < \dfrac{c}{b} \text{ when } a < c,$$

$$\text{and } \dfrac{a}{b} = \dfrac{c}{b} \text{ when } a = c.$$

DO EXERCISES 20–22.

It is not so easy to tell which of $\frac{5}{7}$ and $\frac{2}{3}$ is larger. Let's find a common denominator and compare numerators.

Example 13   Insert the proper symbol $>$, $<$, or $=$ between $\frac{5}{7}$ and $\frac{2}{3}$. We multiply by 1 to find a common denominator:

$$\dfrac{5}{7} = \dfrac{5}{7} \cdot \dfrac{3}{3} = \dfrac{15}{21} \quad \text{and} \quad \dfrac{2}{3} = \dfrac{2}{3} \cdot \dfrac{7}{7} = \dfrac{14}{21}.$$

Since $15 > 14$, it follows that $\frac{15}{21} > \frac{14}{21}$, so $\frac{5}{7} > \frac{2}{3}$.

Example 14   Insert the proper symbol $>$, $<$, or $=$ between $\frac{7}{10}$ and $\frac{8}{9}$.

$$\dfrac{7}{10} = \dfrac{7}{10} \cdot \dfrac{9}{9} = \dfrac{63}{90} \quad \text{and} \quad \dfrac{8}{9} = \dfrac{8}{9} \cdot \dfrac{10}{10} = \dfrac{80}{90}$$

Since $63 < 80$, it follows that $\frac{63}{90} < \frac{80}{90}$, so $\frac{7}{10} < \frac{8}{9}$.

Example 15   Insert the proper symbol $>$, $<$, or $=$ between $\frac{6}{27}$ and $\frac{2}{9}$.

$$\dfrac{6}{27} = \dfrac{6}{27} \cdot \dfrac{9}{9} = \dfrac{54}{243} \quad \text{and} \quad \dfrac{2}{9} = \dfrac{2}{9} \cdot \dfrac{27}{27} = \dfrac{54}{243}$$

Since the numerators (and denominators) are equal, $\frac{6}{27} = \frac{2}{9}$.

DO EXERCISES 23–26.

SOMETHING EXTRA
CALCULATOR CORNER

Find decimal notation for each entry in the table using your calculator. Round to six decimal places.

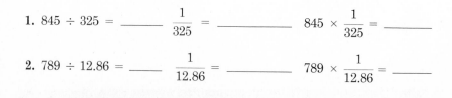

1. $845 \div 325 =$ _____ $\dfrac{1}{325} =$ _____ $845 \times \dfrac{1}{325} =$ _____

2. $789 \div 12.86 =$ _____ $\dfrac{1}{12.86} =$ _____ $789 \times \dfrac{1}{12.86} =$ _____

NAME   CLASS   ANSWERS

EXERCISE SET 1.5

● Find the reciprocal of each number.

1. $\dfrac{3}{4}$ **2.** $\dfrac{5}{8}$ **3.** $\dfrac{1}{8}$ **4.** $\dfrac{1}{10}$

5. 1   **6.** 9

●● Divide by multiplying by the reciprocal of the divisor.

7. $\dfrac{7}{6} \div \dfrac{3}{5}$ **8.** $\dfrac{7}{5} \div \dfrac{3}{4}$ **9.** $\dfrac{8}{9} \div \dfrac{4}{15}$ **10.** $\dfrac{3}{4} \div \dfrac{3}{7}$

11. $\dfrac{1}{4} \div \dfrac{1}{2}$ **12.** $\dfrac{1}{10} \div \dfrac{1}{5}$ **13.** $\dfrac{\frac{13}{12}}{\frac{39}{5}}$ **14.** $\dfrac{\frac{17}{6}}{\frac{3}{8}}$

15. $100 \div \dfrac{1}{5}$ **16.** $78 \div \dfrac{1}{6}$ **17.** $\dfrac{3}{4} \div 10$ **18.** $\dfrac{5}{6} \div 15$

●●● Graph each number on the number line.

19. $\dfrac{5}{4}$ **20.** $\dfrac{7}{6}$ **21.** $\dfrac{15}{16}$ **22.** $\dfrac{11}{12}$

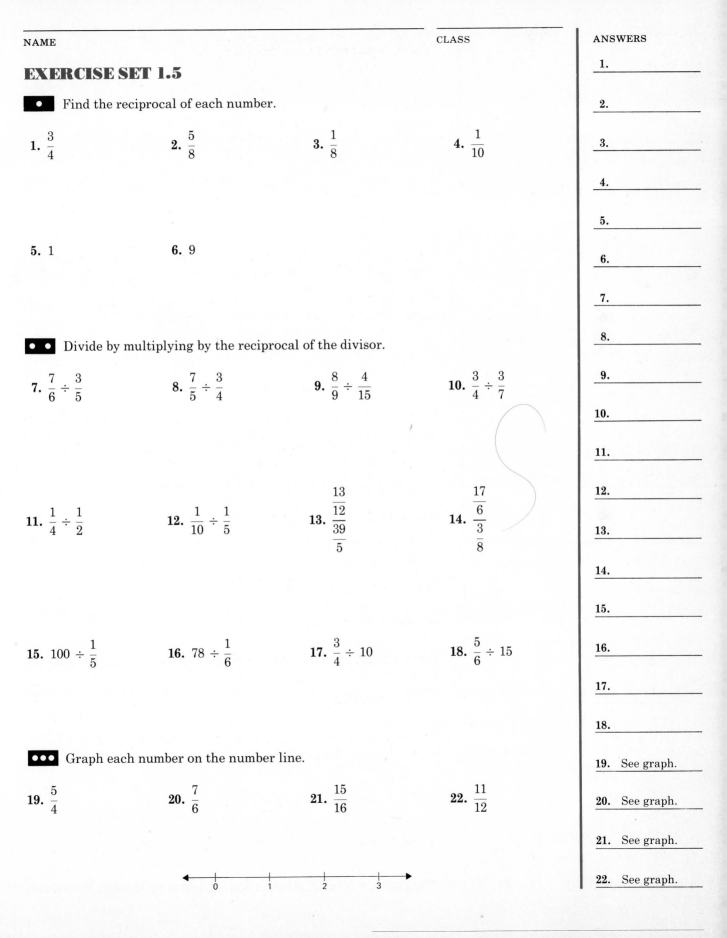

Answers

1. _____
2. _____
3. _____
4. _____
5. _____
6. _____
7. _____
8. _____
9. _____
10. _____
11. _____
12. _____
13. _____
14. _____
15. _____
16. _____
17. _____
18. _____
19. See graph.
20. See graph.
21. See graph.
22. See graph.

Copyright © 1979, Philippines copyright 1979, by Addison-Wesley Publishing Company, Inc. All rights reserved.

ANSWERS

23. _____

24. _____

25. _____

26. _____

27. _____

28. _____

29. _____

30. _____

31. _____

32. _____

33. _____

34. _____

35. _____

36. _____

:: Use the proper symbol $>$, $<$, or $=$ between each pair of numerals.

23. $\dfrac{1}{2}$ $\dfrac{2}{4}$ **24.** $\dfrac{9}{12}$ $\dfrac{3}{4}$ **25.** $\dfrac{11}{15}$ $\dfrac{13}{24}$ **26.** $\dfrac{19}{16}$ $\dfrac{5}{4}$

27. $\dfrac{13}{8}$ $\dfrac{8}{5}$ **28.** $\dfrac{21}{16}$ $\dfrac{5}{4}$ **29.** $\dfrac{4}{5}$ $\dfrac{8}{10}$ **30.** $\dfrac{8}{9}$ $\dfrac{16}{18}$

31. $\dfrac{7}{22}$ $\dfrac{1}{3}$ **32.** $\dfrac{8}{23}$ $\dfrac{1}{3}$

---

33. ▦ Find decimal notation for the reciprocal of 0.3125.

34. ▦ Find decimal notation for the reciprocal of 0.24.

35. ▦ Use the proper symbol $>$, $<$, or $=$.

$\dfrac{1439}{2007}$ $\dfrac{2359}{2876}$

36. ▦ Find decimal notation for each. Use this to order from largest to smallest.

$\dfrac{11}{13}$, $\dfrac{17}{20}$, $\dfrac{2}{3}$, $\dfrac{35}{37}$, $\dfrac{5}{6}$, $\dfrac{23}{25}$

1.6 SOLVING EQUATIONS

◼◦ EQUATIONS AND SOLUTIONS

An *equation* is a number sentence with = for its verb. For example,

$$3 + 2 = 5, \qquad 7 - 2 = 5, \qquad \text{and} \qquad x + 6 = 13.$$

Some equations are true. Some are false. Some are neither true nor false.

Examples

1. The equation $3 + 2 = 5$ is true.
2. The equation $7 - 2 = 8$ is false.
3. The equation $x + 6 = 13$ is neither true nor false because we don't know what x stands for.

An equation says that the symbols on either side of the equals sign stand for, or name, the same number. For example, $5 - 4 = 3 + 7$ says that $5 - 4$ names the same number as $3 + 7$. This equation is false.

DO EXERCISES 1–3.

The letter x in $x + 6 = 13$ is called a *variable*. Some replacements for the variable make the equation true. Some make it false.

> **The replacements that make an equation true are called** *solutions.* **When we find the solutions, we say that we have** *solved* **the equation.**

Example 4   Solve $x + 6 = 13$ by trial.

If we replace x by 2 we get a false equation: $2 + 6 = 13$.
If we replace x by 8 we get a false equation: $8 + 6 = 13$.
If we replace x by 7 we get a true equation: $7 + 6 = 13$.

No other replacement makes the equation true, so the only solution is 7.

DO EXERCISES 4–8.

◼◦◦ SOLVING EQUATIONS $x + a = b$

We want to develop enough equation-solving techniques so that you begin to get a feel for algebra. Consider the true sentence

$$2 + 3 = 5.$$

If we subtract 3 on both sides, we still get a true equation.

$2 + 3 =$	5	Two names of the same number.
-3	-3	Subtract 3 two ways using both names.
2	2	We get a true equation.

OBJECTIVES

After finishing Section 1.6, you should be able to:

◼◦ Solve simple equations by trial.

◼◦◦ Solve equations of the type $x + a = b$ by subtracting a.

◼◦◦◦ Solve equations of the type $ax = b$ by dividing by a.

1. Write three true equations.

   $4 - 1 = 2 + 1$
   $6 + 6 = 20 - 8$
   $8 - 4 = 2 + 2$

2. Write three false equations.

   $6 + 6 = 7 - 4$
   $8 - 4 = 2 + 3$
   $3 - 3 = 1 + 1$

3. Write three equations that are neither true nor false.

   $x + 5 = 20 - 8$
   $4 + 3 = 7 + 21$
   $4 + 3 = 4 + 5$

4. Find three replacements that make $x + 5 = 12$ false.

   $17 + 5 = 12$
   $3 + 5 = 12$
   $4 + 5 = 12$

5. Find the replacement that makes $x + 5 = 12$ true.

   $7 + 5 = 12$

Solve by trial.

6. $x + 4 = 10$

   $6 + 4 = 10$

7. $3x = 12$

   $4 \cdot 3 = 12$

8. $3y + 1 = 16$

   $(3 \cdot 5) + 1 = 16$

ANSWERS ON PAGE A–2

Solve. Be sure to check.

9. $x + 9 = 17$

$x = 17 \cdot 9$

$x = 8$

$8 + 9 = 17$

$17 = 17$

10. $x + 234 = 507$

$x = 507 - 234$

$x = 273$

$273 + 234 = 507$

11. $x + 2.78 = 7.65$

$x = 7.65 - 2.78$

$x = 4.87$

Suppose we want to solve $x + 3 = 5$. We subtract 3 on both sides.

$$\begin{array}{c|c} x + 3 = & 5 \\ \underline{ - 3} & \underline{-3} \\ x & 2 \end{array}$$

We find that $x = 2$. This equation is true when x is replaced by 2. We check to see if $x + 3 = 5$ is true when x is replaced by 2:

$$\begin{array}{c|c} x + 3 & = 5 \\ \hline 2 + 3 & 5 \\ 5 & \end{array}$$

Thus, 2 is a solution.

> **To solve $x + a = b$, subtract a on both sides.**

Example 5 Solve: $x + 6 = 13$.

$$x = 13 - 6 \qquad \text{Subtracting 6}$$
$$x = 7$$

The solution to $x = 7$ is obviously the number 7. It is also the solution of the original equation. We check to find out. To do this we replace x by 7. When we simplify, we get 13 on both sides.

$Check:$
$$\begin{array}{c|c} x + 6 & = 13 \\ \hline 7 + 6 & 13 \\ 13 & \end{array}$$

The solution is 7.

Example 6 Solve: $x + 1.64 = 9.08$.

$$x = 9.08 - 1.64 \qquad \text{Subtracting 1.64}$$
$$x = 7.44$$

$Check:$
$$\begin{array}{c|c} x + 1.64 & = 9.08 \\ \hline 7.44 + 1.64 & 9.08 \\ 9.08 & \end{array}$$

The solution is 7.44.

DO EXERCISES 9–11.

Example 7 Solve: $x + \dfrac{3}{4} = \dfrac{5}{6}$.

$$x = \dfrac{5}{6} - \dfrac{3}{4} \qquad \text{Subtracting } \dfrac{3}{4}$$

$$x = \dfrac{5}{6} \cdot \dfrac{4}{4} - \dfrac{3}{4} \cdot \dfrac{6}{6} \qquad \begin{array}{l}\text{Multiplying by 1}\\ \text{to get a common}\\ \text{denominator}\end{array}$$

$$x = \dfrac{20}{24} - \dfrac{18}{24} = \dfrac{2}{24}$$

$$x = \dfrac{1}{12} \cdot \dfrac{2}{2} = \dfrac{1}{12} \qquad \text{Simplifying}$$

Check: $x + \dfrac{3}{4} = \dfrac{5}{6}$

$$\begin{array}{c|c} \dfrac{1}{12} + \dfrac{3}{4} & \dfrac{5}{6} \\[2ex] \dfrac{1}{12} + \dfrac{9}{12} & \\[2ex] \dfrac{10}{12} & \\[2ex] \dfrac{5}{6} & \end{array}$$

The solution is $\frac{1}{12}$.

DO EXERCISE 12.

⬤⬤⬤ SOLVING EQUATIONS $ax = b$

Consider the true equation

$\qquad 5 \cdot 9 = 45.$

If we divide by 5 on both sides, we still get a true equation:

$\qquad \dfrac{5 \cdot 9}{5} = \dfrac{45}{5}, \quad \text{or} \quad 9 = 9.$

Suppose we want to solve $5x = 45$. We divide by 5 on both sides:

$\qquad \dfrac{5x}{5} = \dfrac{45}{5}$

or $\quad x = 9.$

The solution of $x = 9$ is 9. We check to see if 9 is a solution of $5x = 45$:

$$\begin{array}{c|c} 5x = 45 & \\ \hline 5 \cdot 9 & 45 \\ 45 & \end{array}$$

Thus, 9 is a solution.

Solve.

12. $x + \dfrac{11}{8} = \dfrac{5}{2}$

$x = \dfrac{5}{2} - \dfrac{11}{8}$

$x = \dfrac{5}{2} \cdot \dfrac{4}{4} - \dfrac{11}{8} \cdot \dfrac{2}{2}$

$x = \dfrac{40}{16} - \dfrac{22}{16}$

$x = \dfrac{18}{16} \ \ \overset{9}{}$

$x = \dfrac{9}{8}$

ANSWER ON PAGE A–2

Solve. Be sure to check.

13. $5x = 35$

$$x = \frac{35}{5}$$

$$x = 6$$

14. $8x = 36$

$$x = \frac{36}{8}$$

$$x = 4.5$$

$8\overline{)36.}$
$\quad \underline{32}$
$\quad\ \ 40$
(4.5)

15. $1.2x = 6.72$

$$x = \frac{6.72}{1.2}$$

$1.2\overline{)6.72}$
(5.6)
$\quad \underline{62}$
$\quad\ 52$
$\quad\ 50$

Which of these divisions are possible?

16. $\dfrac{21}{6}$ yes

17. $\dfrac{13}{0}$ no

18. $\dfrac{0}{8}$ no

19. $\dfrac{11}{10 - 10}$ no

20. $\dfrac{9}{12 - (4 \cdot 3)}$ no

21. $\dfrac{P}{x - x}$ no

To solve $ax = b$ when a is nonzero, divide on both sides by a.

Example 8 Solve: $5x = 16$.

$$x = \frac{16}{5}, \quad \text{or } 3.2 \qquad \text{Dividing by 5}$$

Check:
$$\begin{array}{c|c} 5x = 16 \\ \hline 5(3.2) & 16 \\ 16 & \end{array}$$

The solution is 3.2. (It could also be left as $\frac{16}{5}$.)

Example 9 Solve: $4.7x = 40.42$.

$$x = \frac{40.42}{4.7} \qquad \text{Dividing by 4.7}$$

$$x = 8.6$$

Check:
$$\begin{array}{c|c} 4.7x = 40.42 \\ \hline 4.7(8.6) & 40.42 \\ 40.42 & \end{array}$$

The solution is 8.6.

DO EXERCISES 13–15.

DIVISION BY ZERO

We cannot use the preceding method to solve an equation $ax = b$ when $a = 0$, because this would result in division by zero. But, why? $b/0$ would be some number r such that $0 \cdot r = b$. But $0 \cdot r = 0$, so the only possible number b that could be divided by 0 is 0. Look for a pattern.

a) $\dfrac{0}{0} = 5$ because $0 = 0 \cdot 5$ **b)** $\dfrac{0}{0} = 789$ because $0 = 0 \cdot 789$

c) $\dfrac{0}{0} = 17$ because $0 = 0 \cdot 17$ **d)** $\dfrac{0}{0} = \dfrac{1}{2}$ because $0 = 0 \cdot \dfrac{1}{2}$

It looks as if $\frac{0}{0}$ could be any number at all. This would be very confusing, getting any answer we want when we divide 0 by 0. Thus we agree to exclude division by zero.

We never divide by zero.

DO EXERCISES 16–21.

NAME   CLASS   ANSWERS

EXERCISE SET 1.6

⚫ Solve by trial.

1. $x + 8 = 10$

$x = 10 - 8$

$x = 2$

2. $x + 5 = 13$

$x = 13 - 5$

$x = 8$

3. $x - 4 = 5$

$x = 5 + 4$

$x = 9$

4. $x - 6 = 12$

$x = 12 + 6$

$x = 18$

5. $5x = 25$

$x = \frac{25}{5}$

$x = 5$

6. $7x = 42$

$x = \frac{42}{7}$

$x = 6$

7. $5y + 7 = 107$

$5y = 107 - 7$

$5y = 100$

$y = \frac{100}{5} = 20$

8. $9x + 5 = 86$

$9x = 86 - 5$

$9x = 81$

$x = \frac{81}{9} = 9$

9. $7x - 1 = 48$

$7x = 48 + 1$

$7x = 49$

$x = \frac{49}{7} = 7$

10. $4y - 2 = 10$

$4y = 10 + 2$

$4y = 12$

$y = \frac{12}{4} = 3$

⚫⚫ Solve. Be sure to check.

11. $x + 17 = 22$

$x = 22 - 17$

$x = 5$

12. $x + 18 = 32$

$x = 32 - 18$

$x = 14$

13. $x + 56 = 75$

$x = 75 - 56$

$x = 19$

14. $x + 47 = 83$

$x = 83 - 47$

$x = 36$

15. $x + 2.78 = 8.44$

$x = 8.44 - 2.78$

$x = 5.66$

16. $x + 3.04 = 4.69$

$x = 4.69 - 3.04$

$x = 1.65$

17. $x + 5064 = 7882$

$x = 7882 - 5064$

$x = 2818$

18. $x + 4112 = 8007$

$x = 8007 - 4112$

$x = 3895$

19. $x + \frac{1}{4} = \frac{2}{3}$

$x = \frac{2}{3} - \frac{1}{4}$

$x = \frac{2}{3} \cdot \frac{4}{4} - \frac{1}{4} \cdot \frac{3}{3}$

$x = \frac{8}{12} - \frac{3}{12} \quad x = \frac{5}{12}$

20. $x + \frac{1}{3} = \frac{4}{5}$

$x = \frac{4}{5} - \frac{1}{3}$

$x = \frac{4}{5} \cdot \frac{3}{3} - \frac{1}{3} \cdot \frac{5}{5}$

$x = \frac{12}{15} - \frac{5}{15} \quad x = \frac{7}{15}$

1. _____

2. _____

3. _____

4. _____

5. _____

6. _____

7. _____

8. _____

9. _____

10. _____

11. _____

12. _____

13. _____

14. _____

15. _____

16. _____

17. _____

18. _____

19. _____

20. _____

Copyright © 1979, Philippines copyright 1979, by Addison-Wesley Publishing Company, Inc. All rights reserved.

ANSWERS

21. $\frac{1}{6}$

22. $\frac{1}{8}$

23. $x = 4$

24. $x = 8$

25. $x = 1.2$

26. $x = 4.5$

27. $.024$

28. $x = 4$

29. 3.13

30. 6.25

31. 8.5

32. 2.41

33. $\frac{140}{3}$ $46\frac{2}{3}$

34. $135/4$

35.

36.

21. $x + \dfrac{2}{3} = \dfrac{5}{6}$

$x = \frac{5}{6} - \frac{2}{3}$

$x = \frac{5}{6} \cdot \frac{3}{3} - \frac{2}{3} \cdot \frac{6}{6}$

$x = \frac{15}{18} - \frac{12}{18}$

$x = \frac{3}{18} \quad x = \frac{1}{6}$

22. $x + \dfrac{3}{4} = \dfrac{7}{8}$

$x = \frac{7}{8} - \frac{3}{4}$

$x = \frac{7}{8} \cdot \frac{4}{4} - \frac{3}{4} \cdot \frac{8}{8}$

$x = \frac{28}{32} - \frac{24}{32}$

$x = \frac{4}{32}$

$x = \frac{1}{8}$

●●● Solve. Be sure to check.

23. $6x = 24$

$x = \frac{24}{6}$

$x = 4$

24. $4x = 32$

$x = \frac{32}{4}$

$x = 8$

25. $4x = 5$

$x = \frac{5}{4}$

$x = 1.2$

26. $6x = 27$

$x = \frac{27}{6}$

$x = 4.5$

27. $10y = 2.4$

$y = \frac{2.4}{10}$

$y = .24$

28. $9x = 3.6$

$x = \frac{3.6}{9}$

$x = 4$

29. $2.9y = 8.99$

$y = \frac{8.99}{2.9}$

$y = 3.13$

30. $5.5y = 34.1$

$y = \frac{34.1}{5.5}$

$y = 6.25$

31. $6.2x = 52.7$

$x = \frac{52.7}{6.2}$

$x = 8.5$

32. $9.4x = 23.5$

$x = \frac{23.5}{9.4}$

$x = 2.41$

33. $\dfrac{3}{4}x = 35$

$x = \frac{35}{\frac{3}{4}}$

34. $\dfrac{4}{5}x = 27$

$x = \frac{27}{1} \cdot \frac{5}{4}$

Solve.

35. ▦ $x + 506{,}233 = 976{,}421$

36. ▦ $0.1265x = 1065.636$

1.7 SOLVING PROBLEMS

▪ Why learn to solve equations? Because applied problems can be solved using equations. To do this, we first translate the problem situation to an equation and then solve the equation. The problem situation may be explained in words (as in a textbook) or may come from an actual situation in the real world.

Example 1   Solve this problem.

$$\underbrace{\text{Three-fourths}}_{\dfrac{3}{4}} \;\underbrace{\text{of}}_{\cdot}\; \underbrace{\text{what number}}_{x} \;\underbrace{\text{is}}_{=}\; \underbrace{\text{thirty-five}}_{35}?$$

The translation gives us the equation

$$\frac{3}{4}x = 35.$$

We solve it:

$$x = \frac{35}{\dfrac{3}{4}} \qquad \text{Dividing by } \frac{3}{4}$$

$$x = 35 \cdot \frac{4}{3}$$

$$x = \frac{140}{3}.$$

To check, we find out if $\frac{3}{4}$ of this number is 35:

$$\frac{3}{4} \cdot \frac{140}{3} = \frac{3 \cdot 140}{4 \cdot 3} = \frac{3 \cdot 35 \cdot 4}{4 \cdot 3} = 35.$$

> **Note that in translating, *is* translates to = . The word *of* translates to ⋅ , and the unknown number translates to a variable.**

DO EXERCISES 1 AND 2.

Sometimes it helps to reword a problem before translating.

Example 2   Hugh Diddy's salary is $12,000. This is 1.5 times Q. Cumber's salary. What is Cumber's salary?

    Rewording: $12,000 is 1.5 times Cumber's salary.
    Translating: $12,000 = 1.5y$

DO EXERCISES 3 AND 4.

OBJECTIVE

After finishing Section 1.7, you should be able to:

▪ Solve applied problems by translating to equations and solving.

Translate to equations. Then solve and check.

1. What number plus thirty-seven is seventy-three?

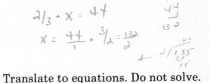

2. Two-thirds of what number is forty-four?

Translate to equations. Do not solve.

3. Bob R. Ooky's batting average is one and a half times that of George Hurler. Ooky's average is 0.320. What is Hurler's average?

4. There were 224 washing machines in a warehouse. When some of them were removed, three-fourths of them were left. How many were removed?

NAME

CLASS

ANSWERS

EXERCISE SET 1.7

■●■ Translate to equations. Then solve and check.

1. Two-thirds of what number is forty-eight?

1. _____

2. One-eighth of what number is fifty-six?

2. _____

3. What number plus five is twenty-two?

3. _____

4. What number plus eight is sixty-three?

4. _____

5. What number is 4 more than 5?

5. _____

6. What number is 7 less than 10?

6. _____

Copyright © 1979, Philippines copyright 1979, by Addison-Wesley Publishing Company, Inc. All rights reserved.

Translate to equations. Do not solve.

7. The area of Lake Superior is four times the area of Lake Ontario. The area of Lake Superior is 78,114 km². What is the area of Lake Ontario?

7. _____

8. The area of Alaska is about 483 times the area of Rhode Island. The area of Alaska is 1,519,202 km². What is the area of Rhode Island?

8. _____

9. Izzi Zlow's typing speed is 35 words per minute. This is two-fifths of Ty Preitter's speed. What is Preitter's speed?

9. _____

10. Walter Logged's body contains 57 kg of water. This is two-thirds of his weight. What is Logged's weight?

10. _____

11. The boiling point of ethyl alcohol is 78.3°C. This is 13.5°C more than the boiling point of methyl alcohol. What is the boiling point of methyl alcohol?

11. _____

12. The height of the Eiffel Tower is 295 m. This is about 203 m more than the height of the Statue of Liberty. What is the height of the Statue of Liberty?

12. _____

13. A color television set with tubes uses about 640 kilowatt hours of electricity in a year. This is 1.6 times that used by a solid-state color set. How many kilowatt hours does the solid-state color model use each year?

13. _____

14. The distance from the earth to the sun is about 150,000,000 km. This is about 391 times the distance from the earth to the moon. What is the distance from the earth to the moon?

14. _____

15. Recently, the average cost of having a baby in one Ohio hospital was $1175. This was about 1.8 times the average cost for a certain California hospital. What was the cost of having a baby in the California hospital?

15. _____

16. It takes a 60-watt bulb about 16.6 hours to use one kilowatt hour of electricity. This is about 2.5 times as long as it takes a 150-watt bulb to use one kilowatt hour. How long does it take a 150-watt bulb to use one kilowatt hour?

16. _____

17. Solve the equation in Exercise 7.

18. Solve the equation in Exercise 8.

19. Solve the equation in Exercise 9.

20. Solve the equation in Exercise 10.

21. Solve the equation in Exercise 11.

22. Solve the equation in Exercise 12.

17. _____

18. _____

19. _____

20. _____

21. _____

22. _____

Copyright © 1979, Philippines copyright 1979, by Addison-Wesley Publishing Company, Inc. All rights reserved.

23. Solve the equation in Exercise 13.   **24.** Solve the equation in Exercise 14.

23. _____

24. _____

25. Solve the equation in Exercise 15.   **26.** Solve the equation in Exercise 16.

25. _____

26. _____

27. _____

These problems are impossible to solve because some piece of information is missing. Tell what you would need to know to solve the problem.

27. A person makes three times the salary of ten years ago. What was the salary ten years ago?

28. Records were on sale for 75¢ off the marked price. After buying four records, a person has $8.72 left. How much was there to begin with?

28. _____

1.8 PERCENT NOTATION

● CONVERTING TO DECIMAL NOTATION

There are other ways to name numbers of arithmetic besides using fractional and decimal notation. One other kind of notation uses the percent symbol %, which means "per hundred." We can regard the percent symbol as part of a numeral. For example,

$$37\% \quad \text{is defined to mean} \quad 37 \times 0.01 \quad \text{or} \quad 37 \times \frac{1}{100}$$

In general,

$$n\% \text{ means } \quad n \times 0.01 \quad \textbf{or} \quad n \times \frac{1}{100}.$$

Example 1 Find decimal notation for 78.5%.

$$78.5\% = 78.5 \times 0.01 \qquad \text{Replacing \% by } \times 0.01$$
$$= 0.785$$

DO EXERCISES 1 AND 2.

●● CONVERTING TO FRACTIONAL NOTATION

Example 2 Find fractional notation for 88%.

$$88\% = 88 \times \frac{1}{100} \qquad \text{Replacing \% by } \times \frac{1}{100}$$

$$= \frac{88}{100} \qquad \text{You need not simplify.}$$

Example 3 Find fractional notation for 34.7%.

$$34.7\% = 34.7 \times \frac{1}{100} \qquad \text{Replacing \% by } \times \frac{1}{100}$$

$$= \frac{34.7}{100}$$

$$= \frac{34.7}{100} \cdot \frac{10}{10} \qquad \text{Multiplying by 1 to get a whole number in the numerator}$$

$$= \frac{347}{1000}$$

DO EXERCISES 3–5.

OBJECTIVES

After finishing Section 1.8, you should be able to:

● Convert from percent notation to decimal notation.

●● Convert from percent notation to fractional notation.

●●● Convert from decimal to percent notation.

●● Convert from fractional to percent notation.

●●● Solve applied problems involving percents.

Find decimal notation.

1. 46.2%

2. 100%

Find fractional notation.

3. 67%

4. 45.6%

5. $\frac{1}{4}\%$

Find percent notation.

6. 6.77

7. 0.9944

Find percent notation.

8. $\dfrac{1}{4}$

9. $\dfrac{3}{8}$

10. $\dfrac{2}{3}$

●●● CONVERTING FROM DECIMAL TO PERCENT NOTATION

By applying the definition of % in reverse, we can convert from decimal notation to percent notation. We multiply by 1, naming it 100×0.01 .

Example 4   Find percent notation for 0.93.

$$
\begin{aligned}
0.93 &= 0.93 \times 1 \\
&= 0.93 \times (\; 100 \times 0.01 \;) \qquad \text{Replacing 1 by } 100 \times 0.01 \\
&= (0.93 \times 100) \times 0.01 \qquad \text{Using associativity} \\
&= 93 \times 0.01 \\
&= 93\% \qquad \text{Replacing } \times 0.01 \text{ by } \%
\end{aligned}
$$

Example 5   Find percent notation for 0.002.

$$
\begin{aligned}
0.002 &= 0.002 \times (\; 100 \times 0.01 \;) \\
&= (0.002 \times 100) \times 0.01 \\
&= 0.2 \times 0.01 \\
&= 0.2\% \qquad \text{Replacing } \times 0.01 \text{ by } \%
\end{aligned}
$$

DO EXERCISES 6 AND 7.

●●● CONVERTING FROM FRACTIONAL TO PERCENT NOTATION

We can also convert from fractional to percent notation. Again, we multiply by 1, but this time we use $100 \times \frac{1}{100}$.

Example 6   Find percent notation for $\frac{5}{8}$.

$$
\begin{aligned}
\frac{5}{8} &= \frac{5}{8} \times \left(100 \times \frac{1}{100} \right) \\
&= \left(\frac{5}{8} \times 100 \right) \times \frac{1}{100} \\
&= \frac{500}{8} \times \frac{1}{100} \\
&= \frac{500}{8}\%, \quad \text{or} \quad 62.5\%
\end{aligned}
$$

DO EXERCISES 8–10.

●●● SOLVING PROBLEMS INVOLVING PERCENTS

Let's solve some applied problems involving percents. Again, it is helpful to translate the problem situation to an equation and then solve the equation.

Example 7　What is 12% of 59?

$$x = 12\% \cdot 59 \qquad \text{Translating}$$

Solve:　　　　　$x = 12 \times 0.01 \times 59$

$$x = 0.12 \times 59$$

$$x = 7.08$$

The solution is 7.08, so 7.08 is 12% of 59.

DO EXERCISES 11 AND 12.

Example 8　What percent of 45 is 15?

$$x \quad \% \quad \cdot \ 45 = 15 \qquad \text{Translating}$$

Solve:　　　$x \times 0.01 \times 45 = 15$

$$x(0.45) = 15$$

$$x = \frac{15}{0.45} \qquad \text{Dividing by 0.45}$$

$$x = \frac{15}{0.45} \times \boxed{\frac{100}{100}} = \frac{1500}{45} = 33\frac{1}{3}$$

The solution is $33\frac{1}{3}$, so $33\frac{1}{3}\%$ of 45 is 15.

Example 9　7.5 is what percent of 1.5?

$$7.5 = \quad x \quad \% \quad \cdot \ 1.5 \qquad \text{Translating}$$

Solve:　　　$7.5 = x \times 0.01 \times 1.5$

$$7.5 = x(0.015)$$

$$\frac{7.5}{0.015} = x \qquad \text{Dividing by 0.015}$$

$$x = \frac{7.5}{0.015} \times \boxed{\frac{1000}{1000}} = \frac{7500}{15} = 500$$

The solution is 500, so 7.5 is 500% of 1.5.

Equations are reversible. Thus, in Example 9, $x = 500$ and $500 = x$ mean the same thing.

DO EXERCISES 13 AND 14.

Translate and solve.

11. What is 23% of 48?

12. 25% of 40 is what?

Translate and solve.

13. What percent of 50 is 16?

14. 15 is what percent of 60?

ANSWERS ON PAGE A–3

Translate and solve.

15. 45 is 20 percent of what?

16. 120 percent of what is 60?

Translate to equations. Then solve.

17. The area of Arizona is 19% of the area of Alaska. The area of Alaska is 586,400 sq mi. What is the area of Arizona?

18. An investment is made at 7% simple interest for one year. It grows to $8988. How much was originally invested (the principal)?

Example 10   3 is 16 percent of what?

$$3 = 16 \quad \% \quad \cdot \quad y \qquad \text{Translating}$$

Solve:
$$3 = 16 \times 0.01 \times y$$
$$3 = 0.16y$$
$$0.16y = 3$$
$$y = \frac{3}{0.16} \qquad \text{Dividing by 0.16}$$
$$y = \frac{3}{0.16} \cdot \frac{100}{100} = \frac{300}{16} = 18.75$$

16% of 18.75 is 3, so the solution is 18.75.

DO EXERCISES 15 AND 16.

Sometimes it is helpful to reword the problem before translating.

Example 11   Blood is 90% water. The average adult has 5 quarts of blood. How much water is in the average adult's blood?

Rewording: 90% of 5 is what?

Translating: $90\% \cdot 5 = x$

Solve: $90 \times 0.01 \times 5 = x$
$$0.90 \times 5 = x$$
$$4.5 = x$$

The number 4.5 checks in the problem. Thus, there are 4.5 quarts of water in the average adult's blood.

Example 12   An investment is made at 8% simple interest for 1 year. It grows to $783. How much was originally invested (the principal)?

Rewording: (Principal) + (Interest) = Amount

Translating: $x \quad + \quad 8\%x \quad = \quad 783$ Interest is 8% of the principal

Solve:
$$x + 8\%x = 783$$
$$x + 0.08x = 783$$
$$1.08x = 783 \qquad \text{Collecting like terms}$$
$$x = \frac{783}{1.08} \qquad \text{Dividing by 1.08}$$
$$x = 725$$

The original investment (principal) was $725.

DO EXERCISES 17 AND 18.

NAME	CLASS	ANSWERS

EXERCISE SET 1.8

Note: If you need extra practice on percent, do Exercise Set 1.8A on pp. 49–52.

◖●◗ Find decimal notation.

1. 76%   **2.** 54%   **3.** 54.7%   **4.** 96.2%

◖●●◗ Find fractional notation.

5. 20%   **6.** 80%   **7.** 78.6%   **8.** 12.5%

◖●●●◗ Find percent notation.

9. 4.54   **10.** 1   **11.** 0.998   **12.** 0.751

◖⋮⋮◗ Find percent notation.

13. $\dfrac{1}{8}$ **14.** $\dfrac{1}{3}$ **15.** $\dfrac{17}{25}$ **16.** $\dfrac{11}{20}$

◖⋰⋱◗ Translate and solve.

17. What is 65% of 840?

18. 34% of 560 is what?

19. 24 percent of what is 20.4?

20. 45 is 30 percent of what?

21. What percent of 80 is 100?

22. 30 is what percent of 125?

1. _____

2. _____

3. _____

4. _____

5. _____

6. _____

7. _____

8. _____

9. _____

10. _____

11. _____

12. _____

13. _____

14. _____

15. _____

16. _____

17. _____

18. _____

19. _____

20. _____

21. _____

22. _____

Copyright © 1979, Philippines copyright 1979, by Addison-Wesley Publishing Company, Inc. All rights reserved.

Translate to equations. Then solve.

23. On a test of 88 items, a student got 76 correct. What percent were correct?

23. _____

24. A baseball player got 13 hits in 25 times at bat. What percent were hits?

24. _____

25. A family spent $208 one month for food. This was 26% of its income. What was their monthly income?

25. _____

26. The weight of the human brain is 2.7% of the body weight. A human's brain weighs 2.1 kilograms. What is its body weight?

26. _____

27. The sales tax rate in New York City is 5%. How much would be charged on a purchase of $428.86? How much would the total cost of the purchase be?

27. _____

28. Water volume increases 9% when it freezes. If 400 cubic centimeters of water is frozen, how much would its volume increase? What would be the volume of the ice?

28. _____

29. An investment is made at 9% simple interest for 1 year. It grows to $8502. How much was originally invested?

29. _____

30. An investment is made at 8% simple interest for 1 year. It grows to $7776. How much was originally invested?

30. _____

31. A person earned $9600 one year. An 8% increase in salary was received, but the cost of living rose 7.4%. How much additional earning power was actually received?

31. _____

32. Due to inflation the price of an item rose 8%, which was 12¢. What was the old price? the new price?

32. _____

NAME   CLASS

EXERCISE SET 1.8A

This exercise set supplements Exercise Set 1.8 for those who need it.

■● Find decimal notation.

1. 38%   **2.** 35%   **3.** 72.1%   **4.** 35.6%

5. 65.4%   **6.** 82.6%   **7.** 3.25%   **8.** 5.32%

9. 8.24%   **10.** 9.45%   **11.** 0.61%   **12.** 0.83%

13. 0.43%   **14.** 0.73%   **15.** 0.012%   **16.** 0.023%

17. 0.045%   **18.** 0.053%   **19.** 0.0035%   **20.** 0.0041%

21. 125%   **22.** 135%   **23.** 240%   **24.** 320%

■●● Find fractional notation.

25. 30%   **26.** 40%   **27.** 70%   **28.** 80%

ANSWERS

1. _____
2. _____
3. _____
4. _____
5. _____
6. _____
7. _____
8. _____
9. _____
10. _____
11. _____
12. _____
13. _____
14. _____
15. _____
16. _____
17. _____
18. _____
19. _____
20. _____
21. _____
22. _____
23. _____
24. _____
25. _____
26. _____
27. _____
28. _____

Copyright © 1979, Philippines copyright 1979, by Addison-Wesley Publishing Company, Inc. All rights reserved.

ANSWERS

29. _____

30. _____

31. _____

32. _____

33. _____

34. _____

35. _____

36. _____

37. _____

38. _____

39. _____

40. _____

41. _____

42. _____

43. _____

44. _____

45. _____

46. _____

47. _____

48. _____

49. _____

50. _____

51. _____

52. _____

53. _____

54. _____

55. _____

56. _____

29. 13.5%   **30.** 17.8%   **31.** 73.4%   **32.** 82.5%

33. 3.2%   **34.** 4.8%   **35.** 8.4%   **36.** 7.6%

37. 120%   **38.** 140%   **39.** 250%   **40.** 370%

41. 0.35%   **42.** 0.53%   **43.** 0.48%   **44.** 0.59%

45. 0.042%   **46.** 0.035%   **47.** 0.083%   **48.** 0.74%

●●● Find percent notation.

49. 0.62   **50.** 0.73   **51.** 0.85   **52.** 0.91

53. 0.623   **54.** 0.741   **55.** 0.812   **56.** 0.732

57. 7.2

58. 8.3

59. 3.5

60. 2.6

61. 2

62. 3

63. 4

64. 5

65. 0.072

66. 0.085

67. 0.013

68. 0.045

69. 0.0013

70. 0.0057

71. 0.0073

72. 0.0068

⠒⠒ Find percent notation.

73. $\dfrac{17}{100}$

74. $\dfrac{35}{100}$

75. $\dfrac{119}{100}$

76. $\dfrac{173}{100}$

77. $\dfrac{7}{10}$

78. $\dfrac{3}{10}$

79. $\dfrac{8}{10}$

80. $\dfrac{9}{10}$

81. $\dfrac{7}{20}$

82. $\dfrac{11}{20}$

83. $\dfrac{7}{25}$

84. $\dfrac{12}{25}$

ANSWERS

57. _____

58. _____

59. _____

60. _____

61. _____

62. _____

63. _____

64. _____

65. _____

66. _____

67. _____

68. _____

69. _____

70. _____

71. _____

72. _____

73. _____

74. _____

75. _____

76. _____

77. _____

78. _____

79. _____

80. _____

81. _____

82. _____

83. _____

84. _____

Copyright © 1979. Philippines copyright 1979 by Addison-Wesley Publishing Company, Inc. All rights

ANSWERS

85. _____

86. _____

87. _____

88. _____

89. _____

90. _____

91. _____

92. _____

93. _____

94. _____

95. _____

96. _____

97. _____

98. _____

99. _____

100. _____

101. _____

102. _____

103. _____

104. _____

85. $\dfrac{1}{2}$ **86.** $\dfrac{3}{2}$ **87.** $\dfrac{1}{4}$ **88.** $\dfrac{3}{4}$

89. $\dfrac{3}{5}$ **90.** $\dfrac{4}{5}$ **91.** $\dfrac{17}{50}$ **92.** $\dfrac{31}{50}$

93. $\dfrac{1}{3}$ **94.** $\dfrac{2}{3}$ **95.** $\dfrac{3}{8}$ **96.** $\dfrac{5}{8}$

⣏⣹   Translate and solve.

97. What is 38% of 250?

98. What is 47% of 320?

99. 37.2% of 85 is what?

100. 17.6% of 70 is what?

101. What percent of 80 is 20?

102. What percent of 60 is 20?

103. 35 is 20 percent of what?

104. 16 is 25 percent of what?

1.9 GEOMETRIC FORMULAS

● RECTANGLES

In this section we review some formulas from geometry. We first consider rectangles. The area of a rectangle is the number of *unit* squares it takes to fill it up. Unit squares look like these.

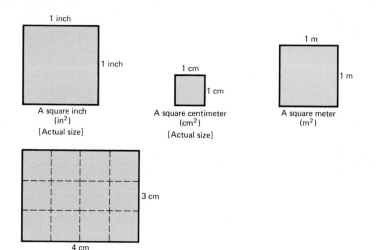

A square inch
(in²)
[Actual size]

A square centimeter
(cm²)
[Actual size]

A square meter
(m²)

The rectangle has length 4 cm and width 3 cm. It takes 12 unit squares to fill it up. Therefore, the area is 12 cm² (or 12 sq cm). In any rectangle we can find the area by multiplying the length by the width.

> **If a rectangle has length l and width w, the area is given by**
>
> $$A = lw. \quad \textbf{(Area is length times width)}$$

The *perimeter* of a rectangle is the distance around it. We can find the perimeter by adding the lengths of the four sides. Or, we can double the length and the width and then add.

> **The perimeter of a rectangle of length l and width w is given by**
>
> $$P = 2l + 2w, \quad \textbf{or} \quad 2(l + w).$$

Example 1   Find the area and perimeter of this rectangle.

$$A = lw = (4.7 \text{ cm}) \cdot (2.4 \text{ cm})$$
$$= 4.7 \times 2.4 \text{ cm} \cdot \text{cm}$$
$$= 11.28 \text{ cm}^2$$
$$\text{(or 11.28 square centimeters)}$$

2.4 cm

4.7 cm

$$P = 2(l + w)$$
$$= 2(4.7 \text{ cm} + 2.4 \text{ cm})$$
$$= 2 \times 7.1 \text{ cm} = 14.2 \text{ cm}$$

OBJECTIVES

After finishing Section 1.9, you should be able to:

● Find the area and perimeter of a rectangle, given the length and width.

●● Find the area of a parallelogram, trapezoid, or triangle.

●●● Given two angle measures of a triangle, find the measure of the third angle.

◉◉ Find the volume of a rectangular solid.

◉◉◉ Given the radius of a circle, find the diameter and the circumference.

▦ Given the radius of a circle, find the area.

1. Find the area and perimeter of this rectangle.

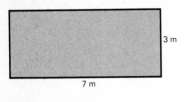

3 m

7 m

2. Find the area and perimeter of this square.

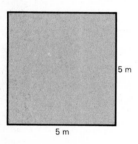

5 m

5 m

3. Find the area of this parallelogram.

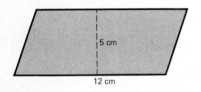

5 cm

12 cm

When we make computations with dimension symbols, such as inches (in.) or centimeters (cm), we can treat them as if they were numerals or variables.

Compare:

a) $3x \cdot 4x = 3 \cdot 4 \cdot x \cdot x = 12x^2$ with 3 in. $\cdot$ 4 in. $= 3 \cdot 4$ in. $\cdot$ in. $= 12$ in^2

b) $2x + 5x = (2 + 5)x = 7x$ with 2 cm $+$ 5 cm $= (2 + 5)$ cm $= 7$ cm

DO EXERCISE 1.

If the length and width of a rectangle are the same, then the rectangle is a square. In a square, all four sides are the same length. Suppose a square has sides of length s. Then its area is $s \cdot s$ or s^2 and its perimeter is $s + s + s + s$, or $4s$.

> If a square has sides of length s, then the area is given by $A = s^2$ and the perimeter is given by $P = 4s$.

DO EXERCISE 2.

●● PARALLELOGRAMS, TRAPEZOIDS, AND TRIANGLES

A *parallelogram* is a four-sided figure with two pairs of parallel sides. To find the area of a parallelogram, we can think of cutting off a part of it, as shown below. We can then place the part as shown to form a rectangle. The area of the rectangle is $b \cdot h$ (length of the base times height). This is also the area of the parallelogram.

> The area of a parallelogram is given by $A = b \cdot h$, where b is the length of the base and h is the height.

DO EXERCISE 3.

A *trapezoid* is a four-sided figure with at least one pair of parallel sides. To find the area of a trapezoid, we can think of cutting out another one just like the given one and placing the two of them together, as shown below. This forms a parallelogram, whose area is $h \cdot (a + b)$.

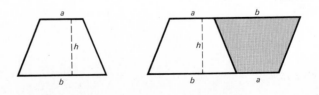

The trapezoid is half of this parallelogram, so its area is $\frac{1}{2}h \cdot (a + b)$. We call the parallel sides of the trapezoid the *bases*, so the area is half the product of the height and the sum of the bases.

If a trapezoid has bases of lengths a and b and has height h, its area is given by $A = \frac{1}{2} \cdot h \cdot (a + b)$.

Example 2 Find the area of this trapezoid.

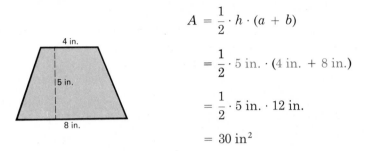

$$A = \frac{1}{2} \cdot h \cdot (a + b)$$

$$= \frac{1}{2} \cdot 5 \text{ in.} \cdot (4 \text{ in.} + 8 \text{ in.})$$

$$= \frac{1}{2} \cdot 5 \text{ in.} \cdot 12 \text{ in.}$$

$$= 30 \text{ in}^2$$

DO EXERCISE 4.

To find the area of a triangle, we can think of cutting out another one like the one given and placing the two of them together, as shown below.

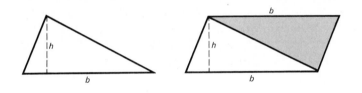

This forms a parallelogram, whose area is $b \cdot h$. The triangle is half of the parallelogram, so the area is half of $b \cdot h$.

If a triangle has a base of length b and has height h, then the area is given by $A = \frac{1}{2} \cdot b \cdot h$.

Example 3 Find the area of this triangle.

$$A = \frac{1}{2} bh$$

$$= \frac{1}{2} \cdot 6\frac{1}{4} \text{ cm} \cdot 5\frac{1}{2}$$

$$= \frac{1}{2} \cdot \frac{25}{4} \cdot \frac{11}{2} \text{ cm}^2$$

$$= \frac{275}{16} \text{ cm}^2, \quad \text{or } 17\frac{3}{16} \text{ cm}^2$$

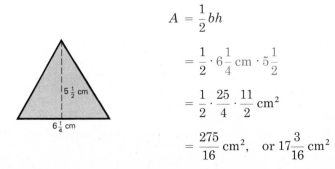

DO EXERCISE 5.

4. Find the area of this trapezoid.

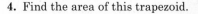
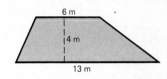

5. Find the area of this triangle.

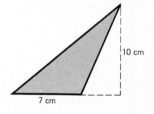

6. Find the missing angle measure.

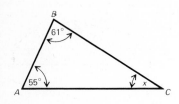

••• ANGLES OF A TRIANGLE

The measures of the angles of a triangle add up to 180°. Thus, if we know the measures of two angles of a triangle we can calculate the third.

> **In any triangle, the sum of the measures of the angles is 180°:**
> $$m(\angle A) + m(\angle B) + m(\angle C) = 180°.$$

Example 4   Find the missing angle measure.

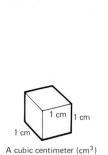

$$m(\angle A) + m(\angle B) + m(\angle C) = 180°$$
$$x + 65° + 24° = 180°$$
$$x + 89° = 180° \qquad \text{Collecting like terms}$$
$$x = 180° - 89°$$
$$x = 91°$$

DO EXERCISE 6.

⋅⋅ VOLUMES OF RECTANGULAR SOLIDS

The volume of a rectangular solid is the number of *unit* cubes it takes to fill it up. Unit cubes look like these.

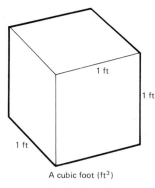

A cubic centimeter (cm³)   A cubic foot (ft³)

This rectangular solid has length 4 cm, width 3 cm, and height 2 cm. There are thus two layers of cubes, each having 3×4, or 12 cubes. The total volume is thus 24 cm³ (cubic centimeters).

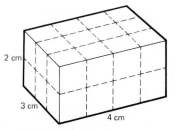

In any rectangular solid we can find the volume by multiplying the length by the width by the height.

> **If a rectangular solid has length *l*, width *w*, and height *h*, then the volume is given by $V = l \cdot w \cdot h$.**

Example 5   Find the volume of this solid.

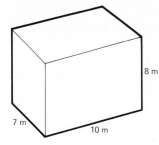

$$V = lwh$$
$$= 10 \text{ m} \cdot 7 \text{ m} \cdot 8 \text{ m}$$
$$= 10 \cdot 56 \text{ m}^3 \text{ (cubic meters)}$$
$$= 560 \text{ m}^3$$

DO EXERCISE 7.

⠿ CIRCLES

Here is a circle, with center O. Also shown are a diameter d and a radius r. A diameter is twice as long as a radius.

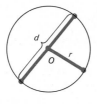

In any circle, if d is the diameter and r is the radius, then $d = 2 \cdot r$.

The *circumference* of a circle is the distance around it. If we divide the circumference of a circle by the diameter, we always get the same number. This number is not a number of arithmetic, so we use the Greek letter π (pi) to name it. This number is approximately 3.14 or $\frac{22}{7}$. We use the letter C for circumference. Then $C/d = \pi$ in any circle, or $C = \pi d$.

If C is the circumference of a circle and r is the radius, then $C = \pi d$, or since $d = 2r$, $C = 2\pi r$.

Example 6   Find the length of a diameter of a circle whose radius is 14 ft.

$$d = 2r$$
$$= 2 \cdot 14 \text{ ft} = 28 \text{ ft}$$

Example 7   Find the circumference of a circle whose radius is 14 ft. Use $\frac{22}{7}$ for π.

$$C = 2 \cdot \pi \cdot r$$

$$\approx 2 \cdot \frac{22}{7} \cdot 14 \text{ ft} \qquad \approx \text{ means "approximately equal to"}$$

$$\approx 88 \text{ ft}$$

DO EXERCISES 8 AND 9.

---

7. Find the volume of this solid.

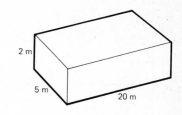

8. Find the length of a diameter of a circle whose radius is 9 meters.

9. Find the circumference of a circle whose radius is 9 cm. Use 3.14 for π.

10. Find the area of a circle with a 9-ft radius. Use 3.14 for π.

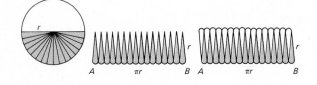 **AREAS OF CIRCLES**

Suppose we take half a circular region, cut it into small wedges, and arrange the wedges as shown below. If the wedges are very small, the distance from A to B will be very nearly πr (half the circumference of the circle). Now if we cut the other half of the circle the same way and place the wedges together, we get what is almost a parallelogram, with area $\pi r \cdot r$.

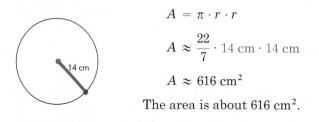

The area of a circle of radius r is given by $A = \pi r^2$.

Example 8   Find the area of this circle. Use $\frac{22}{7}$ for π.

$$A = \pi \cdot r \cdot r$$

$$A \approx \frac{22}{7} \cdot 14 \text{ cm} \cdot 14 \text{ cm}$$

$$A \approx 616 \text{ cm}^2$$

14 cm

The area is about 616 cm².

DO EXERCISE 10.

SOMETHING EXTRA   ▦

CALCULATOR CORNER: ESTIMATING π

Each of the following is an approximation for π. Find decimal notation for each to seven decimal places, using your calculator. Compare with the following approximation, correct to nine decimal places:

$$\pi \approx 3.141592653.$$

[Answers may vary due to capacity of calculator and possible rounding.]

1. $\pi \approx \dfrac{355}{113}$

2. $\pi \approx \dfrac{62{,}832}{20{,}000}$

3. $\pi \approx 2 \cdot \dfrac{2 \cdot 2 \cdot 4 \cdot 4 \cdot 6 \cdot 6 \cdot 8}{1 \cdot 3 \cdot 3 \cdot 5 \cdot 5 \cdot 7 \cdot 7}$

4. $\pi \approx 4\left(\dfrac{1}{2} + \dfrac{1}{3} - \dfrac{1}{15} + \dfrac{1}{35} - \dfrac{1}{63} + \dfrac{1}{99}\right)$

NAME   CLASS   ANSWERS

EXERCISE SET 1.9

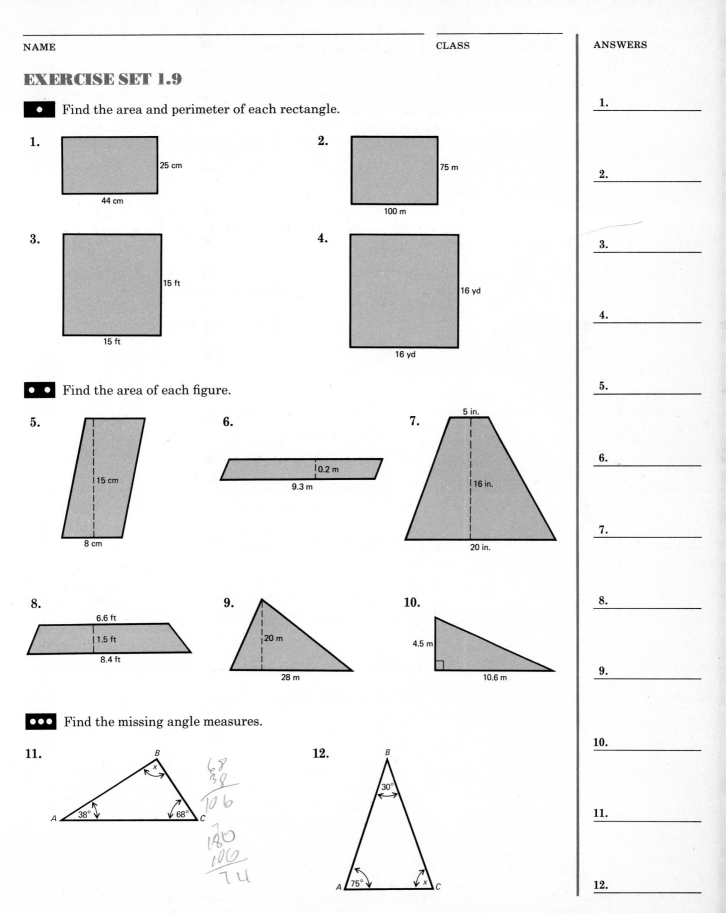

■● Find the area and perimeter of each rectangle.

1.

25 cm
44 cm

2.

75 m
100 m

3.

15 ft
15 ft

4.

16 yd
16 yd

●● Find the area of each figure.

5.

15 cm
8 cm

6.

0.2 m
9.3 m

7.

5 in.
16 in.
20 in.

8.

6.6 ft
1.5 ft
8.4 ft

9.

20 m
28 m

10.

4.5 m
10.6 m

●●● Find the missing angle measures.

11.

B
x
A 38°
68° C

68
38
106

180
106
74

12.

B
30°
A 75°
x C

ANSWERS

1. _____

2. _____

3. _____

4. _____

5. _____

6. _____

7. _____

8. _____

9. _____

10. _____

11. _____

12. _____

Copyright © 1979, Philippines copyright 1979, by Addison-Wesley Publishing Company, Inc. All rights reserved.

ANSWERS

13. _____

14. _____

15. _____

16. _____

17. _____

18. _____

19. _____

20. _____

21. _____

22. _____

23. _____

24. _____

25. _____

26. _____

27. _____

28. _____

13.

14.

⠶ Find the volume of each solid.

15.

16.

⠶ Find the diameter and circumference of each circle.

17. Use 3.14 for π.

18. Use 3.14 for π.

19. Use $\dfrac{22}{7}$ for π.

20. Use $\dfrac{22}{7}$ for π.

⠿ Find the area of the circle with given radius. Use 3.14 for π.

21. $r = 2.4$ m

22. $r = 10.2$ cm

23. $r = 100$ ft

24. $r = 200$ yd

25. The length of a diameter of a circle is 24.6 cm. What is the length of a radius?

26. The area of a rectangle is 49.2 ft^2 and the length is 10 ft. What is the width?

27. The circumference of a circle is 22.608 yd. What is the length of a diameter? Use 3.14 for π.

28. The area of a parallelogram is 254.8 square meters and the base is 52 meters long. What is the height?

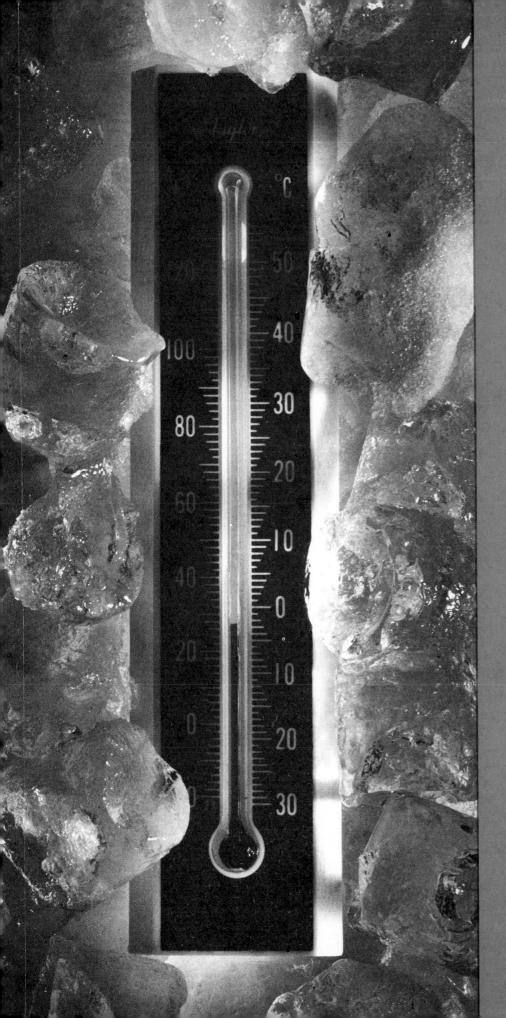

INTEGERS
AND RATIONAL
NUMBERS

READINESS CHECK: SKILLS FOR CHAPTER 2

1. Multiply and simplify.

$$\frac{21}{5} \cdot \frac{1}{7}$$

2. What is the meaning of w^4?

3. What law is illustrated by this sentence?

$$6 \cdot (4 + 8) = 6 \cdot 4 + 6 \cdot 8$$

4. Factor: $3x + 9 + 12y$.

5. Multiply: $7(3z + y + 2)$.

6. Divide and simplify: $\frac{7}{2} \div \frac{3}{8}$.

OBJECTIVES

After finishing Section 2.1, you should be able to:

■ Tell which of two numbers is greater, using < or >.

■■ Find the absolute value of any integer.

■■■ Add integers without using the number line.

2.1 INTEGERS AND THE NUMBER LINE

The set of *integers* is shown below.

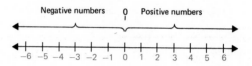

The integers include the whole numbers, but also include *negative* numbers. The negative numbers are shown to the left of 0. We read -2 as "negative two." 0 is neither negative nor positive.

Negative numbers can be associated with many situations.

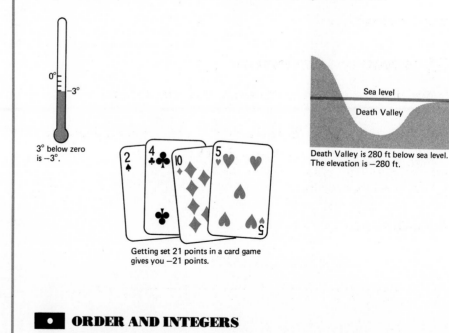

3° below zero is -3°.

Getting set 21 points in a card game gives you -21 points.

Death Valley is 280 ft below sea level. The elevation is -280 ft.

■ ORDER AND INTEGERS

One integer is greater than another if it is to the right of the first on the number line.

Examples

1. $-2 < 1$ **2.** $-8 < -3$ **3.** $-5 < 0$ **4.** $5 > -5$

All negative integers are less than zero. All positive integers are greater than zero.

DO EXERCISES 1–6.

●● ABSOLUTE VALUE

Let us consider the number line.

How far is 5 from 0? How far is -5 from 0? Since distance is always considered to be a nonnegative number (positive or zero), it follows that 5 is 5 units from 0 and -5 is 5 units from 0.

The *absolute value* of an integer can be thought of as its distance from 0 on the number line. The absolute value of an integer n can be named $|n|$.

Examples

5. $|-3| = 3$ **6.** $|25| = 25$ **7.** $|0| = 0$

The absolute value of an integer is never negative.

DO EXERCISES 7–11.

●●● ADDITION OF INTEGERS

To explain addition of integers we can use the number line.

To do the addition $a + b$, we start at a, and move according to b.
a) If b is positive, we move to the right.
b) If b is negative, we move to the left.
c) If b is 0, we stay at a.

Example 8   Add $3 + (-5)$.

$3 + (-5) = -2$

Make true sentences using $<$ or $>$.

1. $-5 \quad -7$ $-5 > -7$

2. $0 \quad -3$ $0 > -3$

3. $-5 \quad -2$ $-5 < -2$

4. $10 \quad 0$ $10 > 0$

5. $-3 \quad 4$ $-3 < 4$

6. $-5 \quad 5$ $-5 < 5$

Find the absolute value of each integer. Think of the number line and distance.

7. -8 8

8. 10 10

Simplify.

9. $|-29|$ 29

10. $|6|$ 6

11. $|0|$ 0

ANSWERS ON PAGE A–4

Add, using a number line.

12. $1 + (-4)$

-3

13. $-3 + (-5)$

-8

14. $-3 + 7$

4

15. $-5 + 5$

0

Write an addition sentence for each.

16.

$4 + -5$

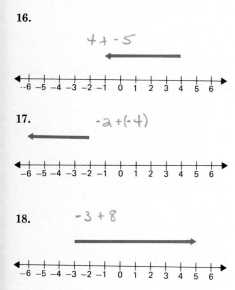

17.

$-2 + (-4)$

18.

$-3 + 8$

Add. Do not use a number line except as a check.

19. $-5 + (-6)$

-11

20. $-9 + (-3)$

-12

21. $-20 + (-14)$

-34

22. $-11 + (-11)$

-22

Example 9 Add $-4 + (-3)$.

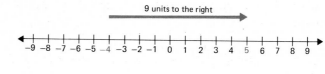

$-4 + (-3) = -7$

Example 10 Add $-4 + 9$.

$-4 + 9 = 5$

Example 11 Add $-4 + 0$.

$-4 + 0 = -4$

DO EXERCISES 12–18.

You may have noticed some patterns in the previous examples and exercises. When we add positive integers and 0, the additions are, of course, additions of whole numbers. When we add two negative integers, the result is negative. We generalize as follows.

> **To add two negative integers, we add their absolute values and make the answer negative.**

Examples Add.

12. $-5 + (-7) = -12$

13. $-8 + (-2) = -10$

DO EXERCISES 19–22.

What happens when we add a positive and a negative integer with different absolute values? The answer is sometimes negative and sometimes positive, depending on which number has the greater absolute value (greater distance from 0). We can generalize as follows.

> **To add a positive and a negative integer, find the difference of their absolute values.**
> a) **If the negative integer has the greater absolute value, the answer is negative.**
> b) **If the positive integer has the greater absolute value, the answer is positive.**

Examples Add.

14. $3 + (-5) = -2$

15. $-7 + 4 = -3$

16. $5 + (-10) = -5$

17. $7 + (-3) = 4$

18. $-6 + 10 = 4$

19. $5 + (-2) = 3$

DO EXERCISES 23–26.

What happens when we add a positive and a negative integer with the same absolute value?

> **The sum of a positive and a negative integer with the same absolute value is always 0.**

Examples Add.

20. $3 + (-3) = 0$

21. $-14 + 14 = 0$

DO EXERCISES 27–30.

To add several integers, some positive and some negative, we add the positive ones. We add the negative ones. Then we add the results.

Example 22

25	First, add the positives.	Next, add the negatives.	Now add the results.
-14			
-127	25	-14	-259
45	45	-127	189
32	32	-118	-70
-118	87	-259	
87	189		

DO EXERCISE 31.

Add. Do not use a number line except as a check.

23. $-4 + 6$

2

24. $-7 + 3$

-4

25. $5 + (-7)$

-2

26. $10 + (-7)$

3

Add. Do not use a number line except as a check.

27. $5 + (-5)$

0

28. $-6 + 6$

0

29. $-10 + 10$

0

30. $89 + (-89)$

0

Add.

31. -35
   17
   14
   -27
   31
   -12

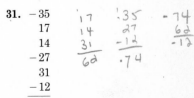

NAME   CLASS

ANSWERS
1.
2.
3.
4.
5.
6.
7.
8.
9.
10.
11.
12.
13.
14.
15.
16.
17.
18.
19.
20.
21.
22.
23.
24.
25.
26.
27.
28.
29.
30.
31.
32.
33.
34.
35.
36.

EXERCISE SET 2.1

● Make true sentences using $<$ or $>$.

1. $5 > 0$

2. $-9 < 0$

3. $-9 < 5$

4. $8 > -3$

5. $-6 < 6$

6. $-7 < 7$

7. $-8 < -5$

8. $-3 < -1$

9. $-5 > -11$

10. $-3 > -4$

11. $-6 < -5$

12. $-10 > -14$

●● Simplify.

13. $|-3|$
3

14. $|-7|$
7

15. $|10|$
10

16. $|11|$
11

17. $|0|$
0

18. $|-4|$
4

19. $|-24|$
24

20. $|-36|$
36

21. $|53|$
53

22. $|54|$
54

23. $|-8|$
8

24. $|0|$
0

25. $|-6|$
6

26. $|-9|$
9

27. $|6|$
6

28. $|9|$
9

●●● Add. Do not use a number line except as a check.

29. $8 + (-5)$
3

30. $-7 + 8$
1

31. $-4 + (-5)$
-9

32. $0 + (-3)$
-3

33. $0 + (-5)$
-5

34. $10 + (-12)$
-2

35. $13 + (-6)$
7

36. $-3 + 14$
11

Copyright © 1979, Philippines copyright 1979, by Addison-Wesley Publishing Company, Inc. All rights reserved.

ANSWERS

37. _____

38. _____

39. _____

40. _____

41. _____

42. _____

43. _____

44. _____

45. _____

46. _____

47. _____

48. _____

49. _____

50. _____

51. _____

52. _____

53. _____

54. _____

55. _____

56. _____

57. _____

58. _____

59. _____

60. _____

61. _____

62. _____

37. $-10 + 7$

-3

38. $0 + (-9)$

-9

39. $-6 + 6$

0

40. $-8 + 1$

-7

41. $11 + (-16)$

-5

42. $-7 + 15$

8

43. $-15 + (-6)$

-21

44. $-8 + 8$

0

45. $11 + (-9)$

2

46. $-14 + (-19)$

-33

47. $-20 + (-6)$

-26

48. $17 + (-17)$

0

49. $-15 + (-7)$

-22

50. $23 + (-5)$

18

51. $40 + (-8)$

32

52. $-23 + (-9)$

-32

53. $-25 + 25$

0

54. $40 + (-40)$

0

55. $63 + (-18)$

45

56. $85 + (-65)$

20

57. $\begin{array}{r} -35 \\ -63 \\ -27 \\ -14 \\ -59 \\ \hline -198 \end{array}$

58. $\begin{array}{r} -24 \\ -37 \\ -19 \\ -45 \\ -35 \\ \hline -160 \end{array}$

59. $\begin{array}{r} 27 \\ -54 \\ -32 \\ 65 \\ 46 \\ \hline \end{array}$

60. $\begin{array}{r} -62 \\ 53 \\ -87 \\ 14 \\ -28 \\ \hline \end{array}$

Add.

61. 🖩 $\begin{array}{r} -64374 \\ -27159 \\ 53690 \\ 39087 \\ 41646 \\ -11953 \\ \hline \end{array}$

62. 🖩 $\begin{array}{r} -49752 \\ 28351 \\ -92265 \\ -40870 \\ 35649 \\ \hline \end{array}$

2.2 SUBTRACTION OF INTEGERS

▪ ADDITIVE INVERSES

Numbers like 6 and -6 are called *additive inverses** of each other. Their sum is 0.

> **If the sum of two numbers is 0, they are additive inverses of each other.**

Examples   Find the additive inverse of each integer.

1. 3: -3 is the additive inverse of 3, because $-3 + 3 = 0$
2. -5:   5 is the additive inverse of -5, because $5 + (-5) = 0$
3. 0:   0 is the additive inverse of 0, because $0 + 0 = 0$

Every integer has an additive inverse.

DO EXERCISES 1–8.

We use the notation $-a$ to name the additive inverse of a. We also read $-a$ simply as "the inverse of a."

Examples   Simplify.

4. $-(-3) = 3$ The additive inverse of -3 is 3.
5. $-(-10) = 10$ The additive inverse of -10 is 10.
6. $-0 = 0$ The additive inverse of 0 is 0.
7. $-(14) = -14$ This tells us that -14 can be read two ways as "the additive inverse of 14" or "negative 14."

DO EXERCISES 9–14.

▪▪ SUBTRACTION USING A NUMBER LINE

We can subtract integers using a number line and the definition of subtraction. $a - b$ is the number which when added to b gives a.

Example 8   Subtract: $10 - 4$.

We know that $10 - 4$ is the number which when added to 4 gives 10. So we start at 4 and go to 10. We go 6 units to the right (or $+6$) so the answer is 6.

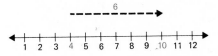

$10 - 4 = 6$

DO EXERCISE 15.

_____
* Additive inverses are sometimes called *opposites*.

OBJECTIVES

After finishing Section 2.2, you should be able to:

▪ Find the additive inverse of a number, and simplify expressions like $-(-3)$.

▪▪ Subtract using a number line.

▪▪▪ Subtract integers by adding the inverse of the subtrahend.

Add.

1. $-6 + 6$ 0

2. $-8 + 8$ 0

3. $7 + (-7)$ 0

4. $0 + 0$ 0

Find the additive inverse.

5. $-6 + 6$

6. $8 + -8$

7. $-7 + 7$

8. $0 + 0$

Simplify.

9. $-(-4) = -4$

10. $-(-20) = -20$

11. $-(9) = -9$

12. $-(12) = -12$

13. $-0 = 0$

14. $-(-38) = 38$

Subtract. Use a number line.

15. $9 - 5 = 4$

ANSWERS ON PAGE A–4

Subtract. Use a number line.

16. $4 - 10$ $= -6$

Subtract. Use a number line.

17. $-2 - 6$ $= -8$

Subtract. Use number lines.

18. $-2 - (-3)$

1

19. $-7 - (-3)$

-4

20. $-4 - 4$

-8

21. $-6 - 2$

-8

Example 9   Subtract: $6 - 8$.

$6 - 8$ is the number which when added to 8 gives 6. So we start at 8 and go to 6. We go 2 units to the left (or -2), so the answer is -2.

$6 - 8 = -2$

DO EXERCISE 16.

Example 10   Subtract: $-1 - 5$.

$-1 - 5$ is the number which when added to 5 gives -1. So we start at 5 and go to -1. We go 6 units to the left (or -6), so the answer is -6.

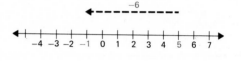

$-1 - 5 = -6$

DO EXERCISE 17.

Example 11   Subtract: $-3 - (-7)$.

$-3 - (-7)$ is the number which when added to -7 gives -3. So we start at -7 and go to -3. We go 4 units to the right (or $+4$), so the answer is 4.

$-3 - (-7) = 4$

DO EXERCISES 18–21.

●●●   INVERSES AND SUBTRACTION

We want to find a faster way to subtract. Look for a pattern in the following examples.

Example 12

a) $3 - 8 = -5$ **b)** $3 + (-8) = -5$

Example 13

a) $-5 - 4 = -9$ **b)** $-5 + (-4) = -9$

Example 14

a) $-7 - (-10) = 3$ **b)** $-7 + 10 = 3$

Example 15

a) $-7 - (-2) = -5$ **b)** $-7 + 2 = -5$

DO EXERCISES 22–25.

Perhaps you have noticed in the preceding examples and exercises that we can subtract by adding an additive inverse. This can always be done.

> **For any integers a and b,**
>
> $$a - b = a + (-b).$$ **(To subtract, we can add the inverse of the subtrahend.)**

We will verify this, but let's do some examples first.

Examples   Subtract.

16. $2 - 6 = 2 + (-6) = -4$ The additive inverse of 6 is -6.

17. $-3 - 8 = -3 + (-8) = -11$

18. $10 - 7 = 10 + (-7) = 3$

19. $18 - 20 = 18 + (-20) = -2$

20. $-4 - (-9) = -4 + 9 = 5$ The additive inverse of -9 is 9.

21. $-4 - (-3) = -4 + 3 = -1$

DO EXERCISES 26–30.

PROOF OF $a - b = a + (-b)$ (Optional)

We can prove that we can subtract by adding an inverse. We want to show that $a - b$ is the same as $a + (-b)$. By definition of subtraction, $a - b$ is that number which when added to b gives a. Let's add $a + (-b)$ to b and see what we get:

$$a + (-b) + b = a + 0 = a.$$

When $a + (-b)$ is added to b, the answer is a. Thus, $a + (-b)$ is $a - b$.

Add or subtract. Look for a pattern.

22. a) $4 - 6$ -2

   b) $4 + (-6)$ -2

23. a) $-3 - 8$ -15

   b) $-3 + (-8)$ -11

24. a) $-5 - (-9)$ 4

   b) $-5 + 9$ 4

25. a) $-5 - (-3)$ -2

   b) $-5 + 3$ -2

Subtract by adding the inverse of the subtrahend.

26. $2 - 8$

   $2 + (-8) = -6$

27. $10 - 3$

   $10 + (-3) = 7$

28. $-4 - 7$

   $-4 + (-7) = -11$

29. $-5 - (-6)$

   $-5 + 6 = 1$

30. $-5 - (-1)$

   $-5 + 1 = -4$

SOMETHING EXTRA

CALCULATOR CORNER: NUMBER PATTERNS

In each of the following, do the first four calculations using your calculator. Look for a pattern. Then do the last calculation without using your calculator.

1. 1^2
    11^2
    111^2
    1111^2
    $11{,}111^2$

2. 6×4
    66×44
    666×444
    6666×4444
    $66{,}666 \times 44{,}444$

3. 8×6
    68×6
    668×6
    6668×6
    $66{,}668 \times 6$

4. 9^2
    99^2
    999^2
    9999^2
    $99{,}999^2$

NAME   CLASS

EXERCISE SET 2.2

◼ • Find the additive inverse.

1. -7
   7

2. -8
   8

3. 1
   -1

4. 2
   -2

5. -12
   12

6. -19
   19

7. 70
   -70

8. 80
   -80

9. 0
   0

10. -6
   6

Simplify.

11. $-(-1)$
   1

12. $-(-7)$
   7

13. $-(7)$
   -7

14. $-(10)$
   -10

15. $-(-14)$
   14

16. $-(-22)$
   22

17. -0
   0

18. $-(1)$
   -1

◼◼ Subtract using a number line.

19. $3-7$
   -4

20. $4-9$
   $4+(-9)=-5$

21. $0-7$
   0

22. $0-10$
   0

23. $-8-(-2)$
   $-8+2=-6$

24. $-6-(-8)$
   $-6+8=2$

25. $-10-(-10)$
   $-10+10=0$

26. $-8-(-8)$
   0

◼◼◼ Subtract by adding the inverse of the subtrahend.

27. $2-9$
   $2+-9=-7$

28. $2-8$
   $2+(-8)=-6$

29. $-6-(-5)$
   $-6+5=-1$

30. $-4-(-3)$
   $-4+3=-1$

31. $8-(-10)$
   $8+10=18$

32. $5-(-6)$
   $5+6=11$

33. $0-5$
   0

34. $0-6$
   0

ANSWERS

1. _____
2. _____
3. _____
4. _____
5. _____
6. _____
7. _____
8. _____
9. _____
10. _____
11. _____
12. _____
13. _____
14. _____
15. _____
16. _____
17. _____
18. _____
19. _____
20. _____
21. _____
22. _____
23. _____
24. _____
25. _____
26. _____
27. _____
28. _____
29. _____
30. _____
31. _____
32. _____
33. _____
34. _____

Copyright © 1979, Philippines copyright 1979, by Addison-Wesley Publishing Company, Inc. All rights reserved.

ANSWERS

35. _____

36. _____

37. _____

38. _____

39. _____

40. _____

41. _____

42. _____

43. _____

44. _____

45. _____

46. _____

47. _____

48. _____

49. _____

50. _____

51. _____

52. _____

53. _____

54. _____

55. _____

56. _____

35. $-5 - (-2)$

$-5 + 2 = -3$

36. $-3 - (-1)$

$-3 + 1 = -2$

37. $-7 - 14$

$-7 + +14)$
$= -21$

38. $-9 - 16$

$-9 + (-16) = -25$

39. $0 - (-5)$

0

40. $0 - (-1)$

0

41. $-8 - 0$

0

42. $-9 - 0$

0

43. $7 - (-5)$

$7 + 5 = 12$

44. $8 - (-3)$

$8 + 3 = 11$

45. $2 - 25$

$2 + (-25)$
$= 23$

46. $18 - 63$

$18 + (-63)$
-48

$\begin{array}{r} 6\,8 \\ ,8 \\ \hline 45 \end{array}$

47. $-42 - 26$

$-42 + (-26)$
$= -68$

48. $-18 - 63$

$-18 + (-63)$
$= -81$

49. $-71 - 2$

$-71 + -2$
-73

50. $-49 - 3$

$-49 + -3$
$= -52$

51. $24 - (-92)$

$24 + 92 =$
$= 116$

52. $48 - (-73)$

$48 + 73$
$= 121$

53. $-50 - (-50)$

$-50 + 50$
0

54. $-70 - (-70)$

0

Subtract.

55. ▦ $123,907 - 433,789$

56. ▦ $23,011 - (-60,432)$

2.3 MULTIPLICATION AND DIVISION OF INTEGERS

◻ MULTIPLICATION

Multiplication of integers is very much like multiplication of whole numbers. The only difference is that we must determine whether the answer is positive or negative. To see how this is done, consider the pattern in the following.

This number decreases by 1 each time.   This number decreases by 5 each time.

$$4 \cdot 5 = 20$$
$$3 \cdot 5 = 15$$
$$2 \cdot 5 = 10$$
$$1 \cdot 5 = 5$$
$$0 \cdot 5 = 0$$
$$-1 \cdot 5 = -5$$
$$-2 \cdot 5 = -10$$
$$-3 \cdot 5 = -15$$

DO EXERCISE 1.

According to this pattern, it looks as though the product of a negative and a positive integer is negative. This is the case.

> **To multiply a positive and a negative integer, multiply their absolute values. The answer is negative.**

Examples Multiply.

1. $-4 \cdot 7 = -28$ **2.** $8 \cdot (-5) = -40$

3. $10 \cdot (-2) = -20$ **4.** $-7 \cdot 5 = -35$

DO EXERCISES 2–4.

How do we multiply two negative numbers? To see this we can again look for a pattern.

This number decreases by 1 each time.   This number increases by 5 each time.

$$4 \cdot (-5) = -20$$
$$3 \cdot (-5) = -15$$
$$2 \cdot (-5) = -10$$
$$1 \cdot (-5) = -5$$
$$0 \cdot (-5) = 0$$
$$-1 \cdot (-5) = 5$$
$$-2 \cdot (-5) = 10$$
$$-3 \cdot (-5) = 15$$

DO EXERCISE 5.

OBJECTIVES

After finishing Section 2.3, you should be able to:
◻ Multiply two integers.
◻◻ Divide two integers, where possible.

1. Complete, as in the example.

$$4 \cdot 10 = 40$$
$$3 \cdot 10 = 30$$
$$2 \cdot 10 = 20$$
$$1 \cdot 10 = 10$$
$$0 \cdot 10 = 0$$
$$-1 \cdot 10 = -10$$
$$-2 \cdot 10 = -20$$
$$-3 \cdot 10 = -30$$

Multiply.

2. $-3 \cdot 6 = -18$

3. $20 \cdot (-5) = -100$

4. $4 \cdot (-20) = -80$

5. Complete, as in the example.

$$3 \cdot (-10) = -30$$
$$2 \cdot (-10) = -20$$
$$1 \cdot (-10) = -10$$
$$0 \cdot (-10) = 0$$
$$-1 \cdot (-10) = 10$$
$$-2 \cdot (-10) = 20$$
$$-3 \cdot (-10) = 30$$

ANSWERS ON PAGE A–4

Multiply.

6. $-3 \cdot (-4)$

12

7. $-16 \cdot (-2)$

32

8. $-7 \cdot (-5)$

35

Divide.

9. $\dfrac{6}{-3}$ $= -2$

10. $\dfrac{-15}{-3}$ $= 5$

Divide.

11. $\dfrac{-24}{8}$ -3

12. $\dfrac{-32}{-4}$ $= 8$

13. $\dfrac{30}{-5}$ $= 6$

> **To multiply two negative integers, multiply their absolute values. The answer is positive.**

DO EXERCISES 6–8.

 DIVIDE

To divide integers we use the definition of division. a/b is the number which when multiplied by b gives a. Consider the division

$$\frac{14}{-7}.$$

We look for a number n such that $-7 \cdot n = 14$. We know from multiplication that -2 is the solution. That is,

$$-7 \cdot (-2) = 14 \quad \text{so} \quad \frac{14}{-7} = -2.$$

Examples   Divide.

5. $\dfrac{-8}{2} = -4$ because $2 \cdot (-4) = -8$

6. $\dfrac{-10}{-2} = 5$ because $-2 \cdot 5 = -10$

DO EXERCISES 9 AND 10.

From these examples we see how to handle signs in division.

> **When we divide a positive integer by a negative, or a negative integer by a positive, the answer is negative. When we divide two negative integers, the answer is positive.**

Examples   Divide.

7. $\dfrac{-24}{6} = -4$

8. $\dfrac{-30}{-6} = 5$

9. $\dfrac{18}{-9} = -2$

There are some divisions of integers, such as $18/-5$, for which we do not get integers for answers. To get an answer, we need an expanded number system, which we will consider in the next section.

DO EXERCISES 11–13.

NAME

CLASS

EXERCISE SET 2.3

● Multiply.

1. $-8 \cdot 2$

    -16

2. $-2 \cdot 5$

    -10

3. $-7 \cdot 6$

    -42

4. $-9 \cdot 2$

    -18

5. $8 \cdot (-3)$

    -24

6. $9 \cdot (-5)$

    -45

7. $-9 \cdot 8$

8. $-10 \cdot 3$

    -30

9. $-8 \cdot (-2)$

    16

10. $-2 \cdot (-5)$

    10

11. $-7 \cdot (-6)$

    42

12. $-9 \cdot (-2)$

    18

13. $-8 \cdot (-3)$

    24

14. $-9 \cdot (-5)$

    45

15. $-9 \cdot (-8)$

16. $-10 \cdot (-3)$

    30

17. $15 \cdot (-8)$

    -120

18. $12 \cdot (-10)$

    -120

19. $-25 \cdot (-40)$

20. $-22 \cdot (-50)$

1. _____

2. _____

3. _____

4. _____

5. _____

6. _____

7. _____

8. _____

9. _____

10. _____

11. _____

12. _____

13. _____

14. _____

15. _____

16. _____

17. _____

18. _____

19. _____

20. _____

Copyright © 1979, Philippines copyright 1979, by Addison-Wesley Publishing Company, Inc. All rights reserved.

ANSWERS

21. _____

22. _____

23. _____

24. _____

25. _____

26. _____

27. _____

28. _____

29. _____

30. _____

31. _____

32. _____

33. _____

34. _____

35. _____

36. _____

37. _____

38. _____

39. _____

40. _____

41. _____

21. $-6 \cdot (-15)$ **22.** $-8 \cdot (-22)$ **23.** $-25 \cdot (-8)$ **24.** $-35 \cdot (-4)$

●● Divide.

25. $\dfrac{36}{-6}$ **26.** $\dfrac{28}{-7}$ **27.** $\dfrac{26}{-2}$ **28.** $\dfrac{26}{-13}$

29. $\dfrac{-16}{8}$ **30.** $\dfrac{-22}{-2}$ **31.** $\dfrac{-48}{-12}$ **32.** $\dfrac{-63}{-9}$

33. $\dfrac{-72}{9}$ **34.** $\dfrac{-50}{25}$ **35.** $\dfrac{-100}{-50}$ **36.** $\dfrac{-200}{8}$

37. $\dfrac{-108}{9}$ **38.** $\dfrac{-64}{-8}$ **39.** $\dfrac{200}{-25}$ **40.** $\dfrac{-300}{-75}$

41. ▦ Calculate each of the following. Look for a pattern.

$(-1 \cdot 8) - 1$
$(-12 \cdot 8) - 2$
$(-123 \cdot 8) - 3$
$(-1234 \cdot 8) - 4$

Use the pattern to find the following without using your calculator.

$(-12345 \cdot 8) \ - 5$

2.4 RATIONAL NUMBERS

The set of *rational numbers* consists of all the numbers of arithmetic and their additive inverses. The rational numbers include all of the integers. There are positive and negative rational numbers, the same as for integers, but between any two integers there are many rational numbers.

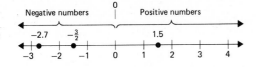

INVERSES AND ABSOLUTE VALUE

Everything we have said about integers so far holds true for rational numbers. For example, all rational numbers have additive inverses, and absolute value means the same as for integers.

Examples   Find the additive inverse.

1. 4.2   The additive inverse is -4.2.

2. $-\dfrac{3}{2}$ The additive inverse is $\dfrac{3}{2}$.

3. -9.35 The additive inverse is 9.35.

DO EXERCISES 1–3.

Examples   Simplify.

4. $-\left(-\dfrac{4}{5}\right) = \dfrac{4}{5}$

5. $-(-9.9) = 9.9$

6. $-\left(\dfrac{3}{8}\right) = -\dfrac{3}{8}$

DO EXERCISES 4–6.

Example 7   Find $-a$ when a stands for -6.

$-a$ is the inverse of a. When a stands for -6,

$$-(-6) = 6.$$

Example 8   Find $-a$ when a stands for $-\frac{7}{8}$.

$$-a = -\left(-\dfrac{7}{8}\right) = \dfrac{7}{8}$$

DO EXERCISES 7–9.

OBJECTIVES

After finishing Section 2.4, you should be able to:

■ Find the additive inverse and absolute value of any rational number and simplify additive inverse symbols.

■■ Add rational numbers.

■■■ Subtract rational numbers.

Find the additive inverse.

1. 8.7 $= -8.7$

2. $-\dfrac{8}{9}$ $= \dfrac{8}{9}$

3. -7.74 $= 7.74$

Simplify.

4. $-\left(-\dfrac{10}{3}\right) = \dfrac{10}{3}$

5. $-(-8.32)$ $= 8.32$

6. $-\left(\dfrac{5}{4}\right)$ $-\dfrac{5}{4}$

Find $-x$ when x is

7. -12 $-(-12) = 12$

8. $-\dfrac{5}{6}$ $-\left(-\dfrac{5}{6}\right) = \dfrac{5}{6}$

9. 17.2 $-(17.2) = -17.2$

ANSWERS ON PAGE A–5

Simplify.

10. $|4.1|$ 4.1

11. $\left|-\dfrac{8}{3}\right|$ $\dfrac{8}{3}$

12. $|-3.5|$ 3.5

Add.

13. $-8.6 + 2.4$ $\begin{array}{r} 8.6 \\ -2.4 \\ \hline 6.2 \end{array}$

14. $\dfrac{5}{9} + \left(-\dfrac{7}{9}\right)$ $= -\dfrac{2}{9}$

15. $-\dfrac{1}{5} + \left(-\dfrac{3}{4}\right)$ $-\dfrac{4}{20} \left(-\dfrac{15}{20}\right) = -\dfrac{19}{20}$

16. A lab technician had 500 milliliters of acid in a beaker. After pouring out 16.5 milliliters and then adding 27.3, how much was in the beaker?

$500 - 16.5 + 27.3$ $\begin{array}{r} 527.3 \\ 16.5 \\ \hline 510.8 \end{array}$

Subtract.

17. $-3 - 6$

$-3 + (-6) = -9$

18. $\dfrac{7}{10} - \dfrac{9}{10}$ $\dfrac{7}{10} + \dfrac{-9}{10} = \dfrac{-2}{10} = -\dfrac{1}{5}$

19. $-8.8 - (-1.3)$ $-9.8 + 1.3 = -7.5$

20. $-\dfrac{2}{3} - \left(-\dfrac{5}{8}\right)$ $-\dfrac{2}{3} + \dfrac{5}{8}$ $\dfrac{-16 + 15}{24} = -\dfrac{1}{24}$

ANSWERS ON PAGE A–5

Examples   Simplify.

9. $|8.7| = 8.7$ **10.** $\left|-\dfrac{3}{4}\right| = \dfrac{3}{4}$ **11.** $|-17.3| = 17.3$

DO EXERCISES 10–12.

●● ADDITION

Addition is done as with integers.

Examples   Add.

12. $-9.2 + 3.1 = -6.1$ **13.** $-\dfrac{3}{2} + \dfrac{9}{2} = \dfrac{6}{2} = 3$

14. $-\dfrac{2}{3} + \dfrac{5}{8} = -\dfrac{16}{24} + \dfrac{15}{24} = -\dfrac{1}{24}$

15. On a recent day the value of a share of IBM stock was $253\frac{1}{4}$. That day it rose in value $\$\frac{5}{8}$. The next day it lost $\$\frac{3}{8}$. What was the value of the stock at the end of the two days?

$$253\dfrac{1}{4} + \dfrac{5}{8} + \left(-\dfrac{3}{8}\right) = 253\dfrac{2}{8} + \dfrac{5}{8} + \left(-\dfrac{3}{8}\right)$$

$$= 253\dfrac{4}{8}, \quad \text{or} \quad \$253\dfrac{1}{2}$$

DO EXERCISES 13–16.

●●● SUBTRACTION

Subtraction is done as with integers.

> **For any rational numbers a and b,**
>
> $$a - b = a + (-b).$$ **(To subtract we can add the inverse of the subtrahend.)**

Examples   Subtract.

16. $3 - 5 = 3 + (-5) = -2$

17. $\dfrac{1}{8} - \dfrac{7}{8} = \dfrac{1}{8} + \left(-\dfrac{7}{8}\right) = -\dfrac{6}{8}, \quad \text{or} \quad -\dfrac{3}{4}$

18. $-4.6 - (-9.8) = -4.6 + 9.8 = 5.2$

19. $-\dfrac{3}{4} - \dfrac{7}{5} = -\dfrac{15}{20} + \left(-\dfrac{28}{20}\right) = -\dfrac{43}{20}$

DO EXERCISES 17–20.

Addition of rational numbers is associative and commutative. Each rational number has an additive inverse.

NAME   CLASS

EXERCISE SET 2.4

▪ Find the additive inverse.

1. -4.7

   4.7

2. -5.2

   5.2

3. $\dfrac{7}{2}$

   $-\dfrac{7}{2}$

4. $\dfrac{8}{3}$

   $-\dfrac{8}{3}$

5. -7

   7

6. -9

   9

7. -26.9

   26.9

8. -17.83

   17.83

Simplify.

9. $-\left(-\dfrac{1}{3}\right)$

   $\dfrac{1}{3}$

10. $-\left(-\dfrac{1}{4}\right)$

   $\dfrac{1}{4}$

11. $-\left(\dfrac{7}{6}\right)$

   $-\dfrac{7}{6}$

12. $-\left(\dfrac{9}{8}\right)$

   $-\dfrac{9}{8}$

13. $-(-9.3)$

   9.3

14. $-(-8.6)$

   8.6

15. $-(90.3)$

   -90.3

16. $-(78.8)$

   -78.8

Find $-x$ when x is

17. 12.4

   $-(12.4)$

18. 23.8

   $-(23.8)$

19. $-\dfrac{9}{10}$

   $-\left(-\dfrac{9}{10}\right)$

20. $-\dfrac{11}{12}$

   $-\left(-\dfrac{11}{12}\right)$

21. -34.8

   $-(-34.8)$

22. -44.1

   $-(-44.1)$

23. 567

   $-(567)$

24. 890

   $-(890)$

Simplify.

25. $|19.2|$

   19.2

26. $|16.8|$

   16.8

27. $\left|-\dfrac{2}{3}\right|$

   $\dfrac{2}{3}$

28. $\left|-\dfrac{5}{6}\right|$

   $\dfrac{5}{6}$

29. $|-89.3|$

   89.3

30. $|-17.4|$

   17.4

31. $\left|\dfrac{14}{3}\right|$

   $\dfrac{14}{3}$

32. $\left|\dfrac{16}{5}\right|$

   $\dfrac{16}{5}$

▪▪ Add.

33. $-6.5 + 4.7$

   $\begin{array}{r} -6.5 \\ 4.7 \\ \hline -1.8 \end{array}$

34. $-3.6 + 1.9$

   $\begin{array}{r} -3.6 \\ 1.9 \\ \hline -1.7 \end{array}$

35. $-2.8 + (-5.3)$

   $\begin{array}{r} -2.8 \\ 5.3 \\ \hline -8.1 \end{array}$

36. $-7.9 + (-6.5)$

   $\begin{array}{r} -7.9 \\ -6.5 \\ \hline -14.4 \end{array}$

ANSWERS
1.
2.
3.
4.
5.
6.
7.
8.
9.
10.
11.
12.
13.
14.
15.
16.
17.
18.
19.
20.
21.
22.
23.
24.
25.
26.
27.
28.
29.
30.
31.
32.
33.
34.
35.
36.

Copyright © 1979. Philippines copyright 1979, by Addison-Wesley Publishing Company, Inc. All rights reserved.

ANSWERS

37. _____

38. _____

39. _____

40. _____

41. _____

42. _____

43. _____

44. _____

45. _____

46. _____

47. ___-6.3___

48. ___- 14.7___

49. ___- 4.3___

50. ___13___

51. ___- 4___

52. ___-2___

53. ___-2/8___

54. ___-6/9___

55. ___1/12___

56. ___-1/8___

57. ___-17/12___

58. ___-11/8___

59. ___1/8___

60. ___-1/12___

61. ___19.9___

62. ___5___

63. ___-9___

64. ___-8.6___

65. ___-1.99___

66. ___-1.87___

67. _____

37. $-\dfrac{3}{5} + \dfrac{2}{5}$

$-\dfrac{1}{5}$

38. $-\dfrac{4}{3} + \dfrac{2}{3}$

$-\dfrac{2}{3}$

39. $-\dfrac{3}{7} + \left(-\dfrac{5}{7}\right)$

$\dfrac{2}{7}$

40. $-\dfrac{4}{9} + \left(-\dfrac{6}{9}\right)$

$\dfrac{2}{9}$

41. $-\dfrac{5}{8} + \dfrac{1}{4}$

$\dfrac{-(5+2)}{8} = -\dfrac{3}{8}$

42. $-\dfrac{5}{6} + \dfrac{2}{3}$

$\dfrac{-5+4}{6} = -\dfrac{1}{6}$

43. $-\dfrac{3}{7} + \left(-\dfrac{2}{5}\right)$

$\dfrac{-15+14}{35} = -\dfrac{29}{35}$

44. $-\dfrac{5}{8} + \left(-\dfrac{1}{3}\right)$

$\dfrac{-5+3}{24} = \dfrac{8}{24} - \dfrac{1}{3}$

45. $-\dfrac{3}{5} + \left(-\dfrac{2}{15}\right)$

$\dfrac{-9+-2}{15} = \dfrac{11}{15}$

46. $-\dfrac{5}{9} + \left(-\dfrac{1}{18}\right)$

$\dfrac{-10+-1}{18} = -\dfrac{11}{18}$

47. $-5.7 + (-7.2) + 6.6$

$\begin{array}{r} 5.7 \\ 7.2 \\ \hline -12.9 \end{array}$ $\begin{array}{r} 12.9 \\ 6.6 \\ \hline -6.3 \end{array} = -6.3$

48. $-10.3 + (-7.5) + 3.1$

$\begin{array}{r} -7.5 \\ \hline 17.8 \end{array}$ $\begin{array}{r} 17.8 \\ 3.1 \\ \hline -14.7 \end{array}$

49. $-8.5 + 7.9 + (-3.7)$

$\begin{array}{r} -37 \\ \hline -12.2 \end{array}$ $\begin{array}{r} 12.2 \\ 7.9 \\ \hline -4.3 \end{array}$

50. $-9.6 + 8.4 + (-11.8)$

$\begin{array}{r} 11.8 \\ \hline 21.4 \\ 8.4 \\ \hline 13.0 \end{array}$

●●● Subtract.

51. $-9 - (-5)$

$-9 + 5 = -4$

52. $-8 - (-6)$

$-8 + 6 = -2$

53. $\dfrac{3}{8} - \dfrac{5}{8}$

$\dfrac{3}{8} - \dfrac{5}{8} = -\dfrac{2}{8}$

54. $\dfrac{3}{9} - \dfrac{9}{9}$

$-\dfrac{6}{9}$

55. $\dfrac{3}{4} - \dfrac{2}{3}$

$\dfrac{9-8}{12} = \dfrac{1}{2}$

56. $\dfrac{5}{8} - \dfrac{3}{4}$

$\dfrac{5-6}{8} = -\dfrac{1}{8}$

57. $-\dfrac{3}{4} - \dfrac{2}{3}$

$-3/4 +$
$\dfrac{-9 + (-8)}{12} = -\dfrac{17}{12}$

58. $-\dfrac{5}{8} - \dfrac{3}{4}$

$\dfrac{-5 + (-6)}{8} = -\dfrac{11}{8}$

59. $-\dfrac{5}{8} - \left(-\dfrac{3}{4}\right)$

$\dfrac{-5+6}{8} =$

60. $-\dfrac{3}{4} - \left(-\dfrac{2}{3}\right)$

$\dfrac{-9+8}{12} =$

61. $6.1 - (-13.8)$

$6.1 + 13.8 =$
19.9

62. $1.5 - (-3.5)$

63. $-3.2 - 5.8$

$-3.2 + (-5.8)$
-9.0

64. $-2.7 - 5.9$

$-2.7 + (-5.9)$
-8.6

65. $0.99 - 1$

$.99 + (-1) =$

66. $0.87 - 1$

$.87 + (-1) =$

Calculate.

67. 📱 $9.976 + (-8.732) + (-16.999) + 83.605$

2.5 MULTIPLICATION AND DIVISION OF RATIONAL NUMBERS

▪ MULTIPLICATION

Multiplication of rational numbers is like multiplication of integers.

> **To multiply a negative and a positive number, multiply their absolute values. The answer is negative.**

Examples Multiply.

1. $-9 \cdot 4 = -36$

2. $-\dfrac{1}{3} \cdot \dfrac{5}{7} = -\dfrac{5}{21}$

3. $\dfrac{5}{6} \cdot \left(-\dfrac{1}{5}\right) = -\dfrac{5}{30} = -\dfrac{1}{6}$

4. $-8.66 \times 4.3 = -37.238$

DO EXERCISES 1–4.

> **To multiply two negative numbers, multiply their absolute values. The answer is positive.**

Examples Multiply.

5. $-6(-8) = 48$

6. $-9.5(-7.4) = 70.3$

7. $-\dfrac{1}{2}\left(-\dfrac{1}{3}\right) = \dfrac{1}{6}$

8. $-\dfrac{3}{4}\left(-\dfrac{8}{9}\right) = \dfrac{24}{36} = \dfrac{2}{3}$

DO EXERCISES 5–8.

▪▪ MULTIPLYING MORE THAN TWO NUMBERS

To multiply several numbers, we can multiply two at a time. We can do this because multiplication of rational numbers is also commutative and associative. This means we can group and order as we please.

Examples Multiply.

9. $3 \cdot (-2) \cdot 4 = (3 \cdot 4) \cdot (-2)$ Changing grouping and ordering
$= 12 \cdot (-2) = -24$

10. $-5 \times (-3.8) \times 2.4 = 19 \times 2.4$ Multiplying the negatives
$= 45.6$

11. $-\dfrac{1}{3} \cdot \left(-\dfrac{3}{5}\right) \cdot \left(-\dfrac{15}{7}\right) = \dfrac{1}{5} \cdot \left(-\dfrac{15}{7}\right)$ Multiplying the first two numbers

$= -\dfrac{3}{7}$

12. $-5 \cdot (-2) \cdot (-3) \cdot (-6) = 10 \cdot 18$ Multiplying the first two numbers and the last two numbers
$= 180$

The product of three negative numbers is negative. The product of four negative numbers is positive.

OBJECTIVES

After finishing Section 2.5, you should be able to:

▪ Multiply two rational numbers.

▪▪ Multiply several positive and negative numbers.

▪▪▪ Find the reciprocal of a rational number.

▪▪ Divide by multiplying by a reciprocal.

Multiply.

1. $6(-5)$ -30

2. $\dfrac{2}{3}\left(-\dfrac{5}{9}\right)$ $-\dfrac{10}{27}$

3. $-\dfrac{4}{5} \cdot \dfrac{7}{8}$ $-\dfrac{28}{40}$

4. -4.23×7.1
$$\begin{array}{r} 4.23 \\ 7.1 \\ \hline .423 \\ 2961 \\ \hline 30.033 \end{array}$$

Multiply.

5. $-16(-4)$
$$\begin{array}{r} 16 \\ 4 \\ \hline 64 \end{array}$$

6. $-\dfrac{4}{7}\left(-\dfrac{5}{9}\right)$ $\dfrac{20}{63}$

7. $-\dfrac{3}{2}\left(-\dfrac{4}{9}\right)$ $\dfrac{12}{18}$

8. $-3.25(-4.14)$
$$\begin{array}{r} 4.14 \\ 3.25 \\ \hline 2070 \\ 828 \\ 1242 \\ \hline 13.4550 \end{array}$$

Multiply.

9. $5 \cdot (-3) \cdot 2$

$$10(-3) = -30$$

10. $-3 \times (-4.1) \times (-2.5)$

$$\frac{4.1}{\times 3} \quad \frac{12.3}{\times 2.5} \quad \frac{605}{246}$$
$$12.3 \qquad \qquad \quad 30.65$$

11. $-\dfrac{1}{2} \cdot \left(-\dfrac{4}{3}\right) \cdot \left(-\dfrac{5}{2}\right)$

$$\frac{4}{6} \cdot \frac{5}{2} = \frac{20}{12} = \frac{5}{3}$$

12. $-2 \cdot (-5) \cdot (-4) \cdot (-3)$

$$10 \cdot -4 = 40 \quad -120$$

13. Write three fractional numerals for 1.

Find the reciprocal.

14. $-\dfrac{5}{4}$

15. -3

16. $-\dfrac{1}{5}$

Divide.

17. $\dfrac{4}{7} \div \left(-\dfrac{3}{5}\right)$

18. $-\dfrac{8}{5} \div \dfrac{2}{3}$

19. $-\dfrac{12}{7} \div \left(-\dfrac{3}{4}\right)$

DO EXERCISES 9–12.

●●● RECIPROCALS

The number 1 can be named in many ways.

Example 13

$$1 = \frac{-3}{-3} = \frac{\frac{4}{7}}{\frac{4}{7}} = \frac{-\frac{3}{5}}{-\frac{3}{5}}$$

Properties of 1:
For any rational number a, $a \cdot 1 = a$.
For any nonzero rational number a, $a/a = 1$.

DO EXERCISE 13.

Two numbers are reciprocals of each other if their product is 1. (See p. 25.) What is the reciprocal of a negative number?

Examples

14. The reciprocal of $-\dfrac{2}{3}$ is $-\dfrac{3}{2}$ because $-\dfrac{2}{3}\left(-\dfrac{3}{2}\right) = \dfrac{6}{6} = 1$.

15. The reciprocal of a nonzero number a is $\dfrac{1}{a}$ because $a \cdot \dfrac{1}{a} = \dfrac{a}{a} = 1$.

Any nonzero number a **has a reciprocal** $1/a$. **The reciprocal of a negative number is negative. The reciprocal of a positive number is positive.**

DO EXERCISES 14–16.

▪▪ DIVISION AND RECIPROCALS

We can divide by multiplying by the reciprocal of the divisor.

Examples Divide.

16. $\dfrac{2}{3} \div \left(-\dfrac{5}{4}\right) = \dfrac{2}{3} \cdot \left(-\dfrac{4}{5}\right) = -\dfrac{8}{15}$

17. $-\dfrac{5}{6} \div \left(-\dfrac{3}{4}\right) = -\dfrac{5}{6} \cdot \left(-\dfrac{4}{3}\right) = \dfrac{20}{18} = \dfrac{10 \cdot 2}{9 \cdot 2} = \dfrac{10}{9} \cdot \dfrac{2}{2} = \dfrac{10}{9}$

DO EXERCISES 17–19.

NAME   CLASS

EXERCISE SET 2.5

⚫ Multiply.

1. $9 \cdot (-8)$ **2.** $7 \cdot (-9)$ **3.** $4 \cdot (-3.1)$ **4.** $3 \cdot (-2.2)$

-72   -56

$$\begin{array}{r} 3.1 \\ \underline{4} \\ -12.4 \end{array}$$

-6.6

5. $-6 \cdot (-4)$ **6.** $-5 \cdot (-6)$ **7.** $-7 \cdot (-3.1)$ **8.** $-4 \cdot (-3.2)$

24   36   21.7   12.8

9. $\frac{2}{3} \cdot \left(-\frac{3}{5}\right)$ **10.** $\frac{5}{7} \cdot \left(-\frac{2}{3}\right)$ **11.** $-\frac{3}{4} \cdot \left(-\frac{2}{9}\right) = \frac{1}{12}$ **12.** $-\frac{5}{8} \cdot \left(-\frac{2}{5}\right)$

$-\frac{6}{15} = -\frac{2}{5}$ $-\frac{10}{21}$ $\frac{6}{72}$ $\frac{10}{8}$ $\frac{2}{8}$

13. -6.3×2.7 **14.** -4.1×9.5 **15.** $-\frac{5}{9} \cdot \frac{3}{4} = \frac{-5}{12}$ **16.** $-\frac{8}{3} \cdot \frac{9}{4} = \frac{-6}{1}$

$$\begin{array}{r} 6.3 \\ 2.7 \\ \hline 441 \\ 126 \\ \hline -17.01 \end{array}$$

$$\begin{array}{r} 9.5 \\ 4.1 \\ \hline 95 \\ 380 \\ \hline -38.95 \end{array}$$

-6

⚫⚫ Multiply.

17. $6 \cdot (-5) \cdot 3$ **18.** $8 \cdot (-7) \cdot 6$

$18 \cdot (-5) = -90$ $48 \cdot (-7)$

$$\begin{array}{r} 18 \\ 5 \\ \hline 90 \end{array}$$

$$336 \quad \boxed{-336}$$

19. $7 \cdot (-4) \cdot (-3) \cdot 5$ **20.** $9 \cdot (-2) \cdot (-6) \cdot 7$

-28 $18 \quad 108$

$84 \quad +20$ $-108 \quad 756 \quad \boxed{756}$

21. $-\frac{2}{3} \cdot \frac{1}{2} \cdot \left(-\frac{6}{7}\right)$ **22.** $-\frac{1}{8} \cdot \left(-\frac{1}{4}\right) \cdot \left(-\frac{3}{5}\right)$

$-\frac{1}{12} = \frac{6}{7} = \frac{2}{7}$ $\frac{1}{32} \cdot -\left(\frac{3}{5}\right) = \frac{3}{160}$

23. $-3 \cdot (-4) \cdot (-5)$ **24.** $-2 \cdot (-5) \cdot (-7)$

$12 \cdot (-5)$ $10 \cdot (-7)$

$= -60$ $= -70$

1. _____

2. _____

3. _____

4. _____

5. _____

6. _____

7. _____

8. _____

9. _____

10. _____

11. _____

12. _____

13. _____

14. _____

15. _____

16. _____

17. _____

18. _____

19. _____

20. _____

21. _____

22. _____

23. _____

24. _____

Copyright © 1979, Philippines copyright 1979, by Addison-Wesley Publishing Company, Inc. All rights reserved.

ANSWERS

25. _____

26. ___ -30 ___

27. _____

28. _____

29. _____

30. _____

31. _____

32. _____

33. _____

34. _____

35. _____

36. _____

37. _____

38. _____

39. _____

40. _____

41. _____

42. _____

43. _____

44. _____

45. _____

46. _____

47. _____

48. _____

25. $-2 \cdot (-5) \cdot (-3) \cdot (-5)$

10 (3)
    30 · (-5) = -150

26. $-3 \cdot (-5) \cdot (-2) \cdot (-1)$

15 · (-2)
    30 · (-1)
        = -30

●●● Find the reciprocal.

27. $-\dfrac{5}{1} \cdot \dfrac{1}{5}$

28. -7

$-\dfrac{7}{1} = -\dfrac{1}{7}$

29. $\dfrac{1}{4}$

$\dfrac{1}{4} = \dfrac{4}{1}$

30. $\dfrac{1}{8}$

$\dfrac{1}{8} = \dfrac{8}{1}$

31. $-\dfrac{7}{5}$ $-\dfrac{7}{5} = \dfrac{5}{7}$

32. $-\dfrac{5}{3}$ $\dfrac{3}{5}$

33. $-\dfrac{4}{11}$ $-\dfrac{11}{4}$

34. $-\dfrac{7}{12}$ $\dfrac{12}{7}$

●● Divide by multiplying by the reciprocal of the divisor.

35. $\dfrac{3}{4} \div \left(-\dfrac{2}{3}\right)$

$\dfrac{3}{4} \cdot -\dfrac{3}{2} = -\dfrac{9}{8}$

36. $\dfrac{7}{8} \div \left(-\dfrac{1}{2}\right)$

$\dfrac{7}{48} \cdot \left(\dfrac{3}{1}\right) = \dfrac{7}{4}$

37. $-\dfrac{5}{4} \div \left(-\dfrac{3}{4}\right)$

$-\dfrac{5}{4} \cdot \left(\dfrac{4}{3}\right) = \dfrac{15}{3}$

38. $-\dfrac{5}{9} \div \left(-\dfrac{5}{6}\right)$

$-\dfrac{5}{9} \cdot \left(-\dfrac{6}{5}\right) = \dfrac{6}{9} = \dfrac{2}{3}$

39. $-\dfrac{2}{7} \div \left(-\dfrac{4}{9}\right)$

$-\dfrac{2}{7} \cdot \left(-\dfrac{9}{4}\right) = \dfrac{9}{14}$

40. $-\dfrac{3}{5} \div \left(-\dfrac{5}{8}\right)$

$-\dfrac{3}{5} \cdot \left(-\dfrac{8}{5}\right) = \dfrac{24}{25}$

41. $-\dfrac{3}{8} \div \left(-\dfrac{8}{3}\right)$

$-\dfrac{3}{8} \cdot \left(\dfrac{3}{8}\right) = \dfrac{9}{64}$

3 · 24

42. $-\dfrac{5}{6} \div \left(-\dfrac{6}{5}\right)$

$-\dfrac{5}{6} \cdot \left(-\dfrac{5}{6}\right) = \dfrac{25}{36}$

43. $\dfrac{5}{7} \div \left(-\dfrac{5}{7}\right)$

$\dfrac{5}{7} \cdot \left(-\dfrac{7}{5}\right) = -1$

44. $\dfrac{7}{8} \div \left(-\dfrac{7}{8}\right)$

$\dfrac{7}{8} \cdot \left(\dfrac{8}{7}\right) = -1$

Evaluate.

45. $(-1)^4$

$-1 \cdot 1 \cdot 1 \cdot 1$

46. $(-1)^3$

$1 \cdot 1 \cdot 1$

47. $(-2)^3$

48. $(-2)^2$

2.6 DISTRIBUTIVE LAWS AND THEIR USE

⬤ THE DISTRIBUTIVE LAWS

For rational numbers the distributive law of multiplication over addition holds. There is also another distributive law. It is the distributive law of multiplication over subtraction. When we multiply a number by a difference, we can either subtract and then multiply or multiply and then subtract.

> **For any rational numbers a, b, and c,**
>
> $a(b - c) = ab - ac.$ **(The distributive law of multiplication over subtraction)**

Examples   Compute.

1. a) $3(4 - 2) = 3 \cdot 2$
$\quad\quad\quad\quad\quad = 6$

b) $3 \cdot 4 - 3 \cdot 2 = 12 - 6$
$\quad\quad\quad\quad\quad\quad\quad = 6$

2. a) $6(3 - 8) = 6 \cdot (-5)$
$\quad\quad\quad\quad\quad = -30$

b) $6 \cdot 3 - 6 \cdot 8 = 18 - 48$
$\quad\quad\quad\quad\quad\quad\quad = -30$

DO EXERCISES 1–4.

In an expression like $(ab) - (ac)$, we can omit the parentheses. In other words, in $ab - ac$ we *agree* that the multiplications are to be done first. The distributive laws are used in algebra in several ways. The most basic ones are factoring, multiplying, and collecting like terms.

⬤⬤ TERMS

What do we mean by the *terms* of an expression? When they are all separated by plus signs, it is easy to tell. If there are subtraction signs, we can rewrite using addition signs.

Example 3   What are the terms of $3x - 4y + 2z$?

$3x - 4y + 2z = 3x + (-4y) + 2z$

Thus, the terms are $3x$, $-4y$, and $2z$.

DO EXERCISES 5 AND 6.

⬤⬤⬤ FACTORING

Factors are numbers to be multiplied. We also say that the factors of 6 are 1, -1, 2, -2, 3, -3, 6, and -6. Each is a factor because we can *multiply* it by some number to get 6. When all the terms of an expression have a common factor, we can factor it out, using the distributive laws. We usually remove positive factors.

ANSWERS ON PAGE A–5

OBJECTIVES

After finishing Section 2.6, you should be able to:

⬤ Apply the distributive law of multiplication over subtraction in computing.

⬤⬤ Identify the terms of an expression.

⬤⬤⬤ Factor when terms have a common factor.

⬤⬤ Multiply, where one expression has several terms.

⬤⬤⬤ Collect like terms.

Compute.

1. a) $4(5 - 3) = 8$ *(handwritten: 2 above, = 8)*

b) $4 \cdot 5 - 4 \cdot 3$ *(handwritten: 20 - 12 = 8)*

2. a) $-2 \cdot (5 - 3)$ *(handwritten: -2 · 2 = -4)*

b) $-2 \cdot 5 - (-2) \cdot 3$ *(handwritten: -10 - (-6) = -4)*

3. a) $5(2 - 7)$ *(handwritten: 5 · (-5) = -25)*

b) $5 \cdot 2 - 5 \cdot 7$ *(handwritten: 10 - 35 = -25)*

4. a) $-4[3 - (-2)]$ *(handwritten: 3 + 2 = -20)*

b) $-4 \cdot 3 - (-4)(-2)$ *(handwritten: -12 (- 8) = -20)*

What are the terms of each?

5. $5x - 4y + 3$ *(handwritten: 5x + -(4y) + 3)*

6. $-4y - 2x + 3z$ *(handwritten: -4y + (-2x) + 3z)*

Factor.

7. $4x - 8$

$4 \cdot 4x - 4 \cdot 8 = 4(x-2)$

8. $3x - 6y + 9$

$3 \cdot x - 3 \cdot 2y + 3 \cdot 3 = 3(x - 2y + 3)$

9. $bx + by - bz$

$b(x + y - z)$

Multiply.

10. $3(x - 5)$

$3x - 15$

11. $5(x - y + 4)$

$5x - 5y + 20$

12. $-2(x - 3)$

$-2x - (6)$

13. $b(x - 2y + 4z)$

$bx - b(2y) + b(4z)$

Collect like terms.

14. $6x - 3x$

$\overset{3}{\cancel{6}}(x-x)$

15. $7x - 1x$

$\cancel{6(x=x)}$ $6x$

16. $x - 0.41x$

$.59x$

17. $5x + 4y - 2x - y$

$5x - 2x + 4y - y$

$3x + 3y$

18. $3x - 7x + 2y - 3z$

$3x - 7x$

$-4x + 2y - 3z$

Examples Factor.

4. $5x - 10 = \boxed{5} \cdot x - \boxed{5} \cdot 2 = \boxed{5}(x - 2)$

5. $\boxed{a}\,x - \boxed{a}\,y + \boxed{a}\,z = \boxed{a}(x - y + z)$

6. $9x + 27y - 9 = \boxed{9} \cdot x + \boxed{9} \cdot 3y - \boxed{9} \cdot 1 = \boxed{9}(x + 3y - 1)$

DO EXERCISES 7–9.

⠛ MULTIPLYING

Multiplying is the reverse of factoring.

Examples Multiply.

7. $\boxed{4}(x - 2) = \boxed{4} \cdot x - \boxed{4} \cdot 2 = 4x - 8$

8. $\boxed{b}(s - 3t + 2w) = \boxed{b} \cdot s - \boxed{b} \cdot 3t + \boxed{b} \cdot 2w = bs - 3bt + 2bw$

9. $\boxed{-4}(x - 2y + 3z) = \boxed{-4} \cdot x - (\boxed{-4}) \cdot 2y + (\boxed{-4}) \cdot 3z$

$$= -4x + 8y - 12z$$

DO EXERCISES 10–13.

⠿ COLLECTING LIKE TERMS

The process of collecting like terms is also based on the distributive laws.

Examples Collect like terms.

10. $4\boxed{x} - 2\boxed{x} = (4 - 2)\boxed{x} = 2x$

11. $2\boxed{x} + 3y - 5\boxed{x} - 2y = 2\boxed{x} - 5\boxed{x} + 3y - 2y$

$$= (2 - 5)\boxed{x} + (3 - 2)y = -3\boxed{x} + y$$

12. $3\boxed{x} - \boxed{x} = (3 - 1)\boxed{x} = 2x$

13. $x - 0.24x = 1 \cdot x - 0.24x = (1 - 0.24)x = 0.76x$

14. $x - 6x = 1 \cdot x - 6 \cdot x = (1 - 6)x = -5x$

DO EXERCISES 14–18.

NAME   CLASS   ANSWERS

EXERCISE SET 2.6

▪ ▫ Multiply.

1. $-3(3 - 7)$ **2.** $-5(9 - 14)$ **3.** $8(9 - 10)$ **4.** $6(13 - 14)$

$-3(-4)$ $-5(-5)$ $8(-1)$ $6(-1)$

$+12$ $+25$ -8 $= -6$

▪▪ What are the terms of each?

5. $8x - 1.4y$ **6.** $12x - 1.2y$

$8x + (-1.4y)$ $12x + (-1.2y)$

7. $-5x + 3y - 14z$ **8.** $-8a - 10b + 18z$

$-5x + 3y + (-14z)$ $-8a + (-10b) + 18z$

▪▪▪ Factor.

9. $8x - 24$ **10.** $10x - 50$ **11.** $32 - 4x$ **12.** $24 - 6x$

$8 \cdot x - 8 \cdot 3$ $10 \cdot x - 10 \cdot 5$ $4 \times 8 - 4 \cdot x$ $6 \cdot 4 - 6 \cdot x$

$8(x-3)$ $10(x-5)$ $4(x-8)$ $= 6(4-x)$

13. $8x + 10y - 22$ **14.** $9x + 6y - 15$ **15.** $ax - 7a$ **16.** $bx - 9b$

$2 \cdot 4x + 2 \cdot 5y - 2 \cdot 11$ $3 \cdot 3x + 3 \cdot 2y - 3 \cdot 5$ $a(x-7)$ $b(x-9)$

$2(4x + 5y - 11)$ $3(3x + 3y - 5)$

17. $ax + ay - az$ **18.** $cx - cy - cz$

$a(x+y-z)$ $c(x-y-z)$

▪▪ ▪▪ Multiply.

19. $7(x - 2)$ **20.** $5(x - 8)$ **21.** $-7(y - 2)$ **22.** $-9(y - 7)$

$7x - 14$ $5x - 40$ $-7y - 14$ $-9y - 63$

1. _____

2. _____

3. _____

4. _____

5. _____

6. _____

7. _____

8. _____

9. _____

10. _____

11. _____

12. _____

13. _____

14. _____

15. _____

16. _____

17. _____

18. _____

19. _____

20. _____

21. _____

22. _____

Copyright © 1979, Philippines copyright 1979, by Addison-Wesley Publishing Company, Inc. All rights reserved.

ANSWERS

23. _____

24. _____

25. _____

26. _____

27. _____

28. _____

29. _____

30. _____

31. _____

32. _____

33. _____

34. _____

35. _____

36. _____

37. _____

38. _____

39. _____

40. _____

41. _____

42. _____

43. _____

44. _____

45. _____

46. _____

23. $-3(7 - t)$

$-21 - t$

24. $-5(14 - n)$

$-70 - n$

25. $-4(x + 3y)$

$-4x + (-12y)$

26. $-3(a + 2b)$

$-3a + (-6b)$

27. $7(-2x - 4y + 3)$

$-14x + (-28y) + 21$

28. $9(-5x - 6y + 8)$

$-45x + -54y + 72$

⠿ Collect like terms.

29. $11x - 3x$

$x(11-3)$

$8x$

30. $17t - 9t$

$t(17-9)$

$8t$

31. $17y - y$

$16y$

32. $6n - n$

$5n$

33. $x - 12x$

$-11x$

34. $y - 14y$

$-13y$

35. $x - 0.83x$

$.17x$

36. $t - 0.02t$

$.98t$

37. $9x + 2y - 5x$

$9x - 5x + 2y$

$4x + 2y$

$2.2x + 2.4 = 2(2x \cdot 4)$

38. $8y + 3z - 4y$

$8y - 4y + 3z$

$4y + 3z$

39. $11x + 2y - 4x - y$

$11x + 4x + 2y - y$

$8x + y$

40. $13a + 9b - 2a - 4b$

$13a - 2a + 9b - 4b$

$11a + 5b$

41. $2.7x + 2.3y - 1.9x - 1.8y$

$2.7x - 1.9x + 2.3y - 1.8y$

$.8x + .5y$

42. $6.7a + 4.3b - 4.1a - 2.9b$

$6.7a - 4.1a + 4.3b - 2.9b$

$2.6a + 1.4b$

43. $\frac{1}{5}x + \frac{4}{5}y + \frac{2}{5}x - \frac{1}{5}y$

$\frac{1}{5}x + \frac{2}{5}x + \frac{4}{5}y - \frac{1}{5}y$

$\frac{3}{5}x + \frac{3}{5}y$

44. $\frac{7}{8}x + \frac{5}{8}y + \frac{1}{8}x - \frac{3}{8}y$

$\frac{7}{8}x + \frac{1}{8}x + \frac{5}{8}y - \frac{3}{8}y$

$\frac{8}{8}x + \frac{2}{8}y \quad x + \frac{1}{4}y$

45. A principal of P dollars was invested in a savings account at 8% simple interest. How much was in the account after 1 year?

46. The population of a town is P. After a 6% increase, what was the new population?

2.7 MULTIPLYING BY −1 AND SIMPLIFYING

■ MULTIPLYING BY −1

What happens when we multiply a rational number by −1?

Examples

1. $-1 \cdot 7 = -7$ **2.** $-1 \cdot (-5) = 5$ **3.** $-1 \cdot 0 = 0$

When we multiply a number by −1, the result is the additive inverse of that number.

> **For any rational number a,**
>
> $-1 \cdot a = -a.$
>
> **That is, negative one times a is the additive inverse of a.**

This fact enables us to rename an additive inverse when an expression has one or more terms.

Examples   Rename each additive inverse without parentheses.

4. $-(3 + x) = -1 \cdot (3 + x)$ Taking the additive inverse is the same as multiplying by −1. We replace − by −1.

$\qquad = -1 \cdot 3 + (-1) \cdot x$

$\qquad = -3 - x$ Replacing −1 by −

5. $-(3x - 2y + 4) = -1 \cdot (3x - 2y + 4)$

$\qquad\qquad\qquad = -1 \cdot 3x - (-1) \cdot 2y + (-1) \cdot 4$

$\qquad\qquad\qquad = -3x + 2y - 4$

The above examples show that we can rename an additive inverse by multiplying each term inside by −1. We sometimes call this "changing the sign" of every term inside the parentheses.

DO EXERCISES 1 AND 2.

■■ REMOVING CERTAIN PARENTHESES

When parentheses follow a minus sign, they can be removed by renaming.

Examples   Remove parentheses and simplify.

6. $3x - (4x + 2) = 3x + [-(4x + 2)]$ Subtracting is adding an inverse.

$\qquad\qquad\qquad = 3x + (-1) \cdot (4x + 2)$ Taking the inverse is the same as multiplying by −1.

$\qquad\qquad\qquad = 3x + (-4x) + (-2)$ Multiplying

$\qquad\qquad\qquad = 3x - 4x - 2$

$\qquad\qquad\qquad = -x - 2$ Collecting like terms

OBJECTIVES

After finishing Section 2.7, you should be able to:

■ Rename an additive inverse without parentheses, where an expression has several terms.

■■ Simplify expressions by removing parentheses and collecting like terms.

■■■ Simplify expressions with parentheses inside parentheses.

Rename each additive inverse without parentheses.

1. $-(x + 2)$

2. $-(5x - 2y - 8)$

ANSWERS ON PAGE A−5

Remove parentheses and simplify.

3. $5x - (3x + 9)$

4. $5y - 2 - (2y - 4)$

Simplify.

5. $[24 \div (-2)] \div (-2)$

Simplify.

6. $3(4 + 2) - \{7 - [4 - (6 + 5)]\}$

$3 \cdot 6 - \{7 - [4 - (11)]\}$

$\{7 - 4\}$

Simplify.

7. $[3(x + 2) + 2x]$
    $- [4(y + 2) - 3(y - 2)]$

$[3(x+2)+2x] - [4(y+2) - 3(y-2)]$

$[3x+6+2x) - [4y+8-3y+6]$

$(5x+6) - (y+14)$

$5x-y+20$

7. $3y - 2 - (2y - 4) = 3y - 2 - 2y + 4$
            $= y + 2$

When we remove parentheses that follow a subtraction or additive inverse sign, we change the sign of each term inside.

DO EXERCISES 3 AND 4.

●●● PARENTHESES WITHIN PARENTHESES

Sometimes parentheses occur within parentheses. When this happens we may use parentheses of different shapes, such as [ ], called "brackets," or { }, called "braces."

When parentheses occur within parentheses, the computations in the inner ones are to be done first.

Example 8 Simplify.

$$\left[(-4) \div \left(-\frac{1}{4}\right)\right] \div \frac{1}{4} = [(-4) \cdot (-4)] \div \frac{1}{4} \quad \text{Working from the inside out}$$

$$= 16 \div \frac{1}{4}$$
$$= 16 \cdot 4$$
$$= 64$$

DO EXERCISE 5.

Example 9 Simplify.

$$4(2 + 3) - \{6 - [3 - (7 + 3)]\}$$
$$= 4 \cdot 5 - \{6 - [3 - 10]\} \quad \text{Working from the inside out}$$
$$= 20 - \{6 - [-7]\}$$
$$= 20 - \{13\}$$
$$= 7$$

DO EXERCISE 6.

Example 10 Simplify.

$$[5(x + 2) - 3x] - [3(y + 2) - 4(y + 2)]$$
$$= [5x + 10 - 3x] - [3y + 6 - 4y - 8] \quad \text{Working from the inside out}$$
$$= [2x + 10] - [-y - 2] \quad \text{Collecting like terms inside}$$
$$= 2x + 10 + y + 2 \quad \text{Removing parentheses by renaming}$$
$$= 2x + y + 12 \quad \text{Collecting like terms}$$

DO EXERCISE 7.

NAME   CLASS

EXERCISE SET 2.7

▆▉ Rename each additive inverse without parentheses.

1. $-(2x + 7)$

2. $-(3x + 5)$

3. $-(5x - 8)$

4. $-(6x - 7)$

5. $-(4a - 3b + 7c)$

6. $-(5x - 2y + 3z)$

7. $-(6x + 8y + 5)$

8. $-(8x + 3y + 9)$

9. $-(3x + 5y - 6)$

10. $-(6a + 4b - 7)$

11. $-(-8x - 6y - 43)$

12. $-(-2a - 9b - 5c)$

▆▉▆▉ Remove parentheses and simplify.

13. $9x - (4x + 3)$

14. $7y - (2y + 9)$

15. $2a + (5a - 9)$

16. $11n + (3n - 7)$

17. $2x + 7x - (4x + 6)$

$9x + ^-4x \cdot 6$

$5x - 6$

18. $3a + 2a - (4a + 7)$

$5a + ^-4a - 7$

$1a - 7$

19. $2x - 4y - (7x + 2y)$

$2x - 4y + ^-7x + 2y$

$^-5x + 2y = -5x - 2y$

20. $3a - 7b - (4a - 3b)$

$3a - 7b + ^-4a + 3b$

$^-1a - 4b$

21. $3x - y - (3x - 2y + 5z)$

$3x - 1y - 3x + 2y - 5z)$

$1x + y - 5z$

22. $4a - b - (5a - 7b + 8c)$

$4a - 1b + ^-5a + 7b - 8c$

$1a - 6b - 8c$

ANSWERS
1.
2.
3.
4.
5.
6.
7.
8.
9.
10.
11.
12.
13.
14.
15.
16.
17.
18.
19.
20.
21.
22.

Copyright © 1979, Philippines copyright 1979, by Addison-Wesley Publishing Company, Inc. All rights reserved.

ANSWERS

●●● Remove parentheses and simplify.

23. _____

23. $[(-24) \div (-3)] \div \left(-\dfrac{1}{2}\right)$

24. _____

24. $[32 \div (-2)] \div (-2)$

25. _____

25. $8 \cdot [9 - 2(5 - 4)]$

26. _____

26. $10 \cdot [6 - 5(8 - 4)]$

27. _____

27. $[4(9 - 6) + 11] - [14 - (6 + 4)]$

28. $[7(8 - 4) + 16] - [15 - (7 + 3)]$

28. _____

29. _____

29. $[3(8 - 4) + 12] - [10 - (3 + 5)]$

30. $[9(7 - 3) + 13] - [11 - (6 + 9)]$

30. _____

31. _____

31. $[10(x + 3) - 4] + [2(x - 17) + 6]$

32. $[9(x + 5) - 7] + [4(x - 12) + 9]$

32. _____

33. $[4(2x - 5) + 7] + [3(x + 3) + 5x]$

34. $[8(3x - 2) + 9] + [7(x + 4) + 6x]$

33. _____

34. _____

35. $[7(x + 5) - 19] - [4(x - 6) + 10]$

36. $[6(x + 4) - 12] - [5(x - 8) + 11]$

35. _____

36. _____

37. $3\{[7(x - 2) + 4] - [2(2x - 5) + 6]\}$

38. $4\{[8(x - 3) + 9] - [4(3x - 7) + 2]\}$

37. _____

38. _____

39. $4\{[5(x - 3) + 2] - 3[2(x + 5) - 9]\}$

40. $3\{[6(x - 4) + 5] - 2[5(x + 8) - 10]\}$

39. _____

40. _____

2.8 INTEGERS AS EXPONENTS

◦ NEGATIVE EXPONENTS

Negative integers can be used as exponents. To see how, we consider expanded notation.

$$4256.387 = 4 \cdot 10^3 + 2 \cdot 10^2 + 5 \cdot 10^1 + 6 \cdot 10^0 + 3 \cdot 10^{-1}$$
$$+ 8 \cdot 10^{-2} + 7 \cdot 10^{-3}$$

To keep the decreasing pattern of exponents past the decimal point, we make the exponents negative. Thus, we agree that 10^{-1} means $\frac{1}{10}$ and 10^{-2} means $\frac{1}{100}$, and so on. We make a similar agreement for other bases.

> **If n is any positive integer,**
>
> $$b^{-n} \text{ is given the meaning } \frac{1}{b^n}.$$
>
> **In other words, b^n and b^{-n} are reciprocals.**

Example 1 Explain the meaning of 3^{-4} without using negative exponents.

$$3^{-4} \quad \text{means} \quad \frac{1}{3^4}, \quad \text{or} \quad \frac{1}{3 \cdot 3 \cdot 3 \cdot 3}, \quad \text{or} \quad \frac{1}{81}.$$

Example 2 Rename $\frac{1}{5^2}$ using a negative exponent.

$$\frac{1}{5^2} = 5^{-2}$$

Example 3 Rename 4^{-3} using a positive exponent.

$$4^{-3} = \frac{1}{4^3}$$

DO EXERCISES 1–9.

◦◦ MULTIPLYING USING EXPONENTS

Consider an expression with exponents, such as $a^3 \cdot a^2$. To simplify it, recall the definition of exponents:

$$a^3 \cdot a^2 \quad \text{means} \quad (a \cdot a \cdot a)(a \cdot a)$$

and $(a \cdot a \cdot a)(a \cdot a) = a$. The exponent in a^5 is the sum of those in $a^3 \cdot a^2$. Suppose one exponent is positive and one is negative.

$$a^5 \cdot a^{-2} \quad \text{means} \quad (a \cdot a \cdot a \cdot a \cdot a) \cdot \left(\frac{1}{a \cdot a}\right),$$

which simplifies as follows:

$$a \cdot a \cdot a \cdot \left(\frac{a \cdot a}{1}\right)\left(\frac{1}{a \cdot a}\right) = a \cdot a \cdot a \cdot \boxed{\frac{a \cdot a}{a \cdot a}} = a \cdot a \cdot a = a^3.$$

OBJECTIVES

After finishing Section 2.8, you should be able to:

◦ Rename a number with or without negative exponents.

◦◦ Use exponents in multiplying (adding the exponents).

◦◦◦ Use exponents in dividing (subtracting the exponents).

▪▪ Use exponents in raising a power to a power (multiplying the exponents).

▪▪▪ Solve problems involving interest compounded annually.

Explain the meaning of each of the following without using negative exponents.

1. 4^{-3} $\quad \frac{1}{4^3} = \frac{1}{4 \cdot 4 \cdot 4} = \frac{1}{64}$

2. 5^{-2} $\quad \frac{1}{5 \cdot 5} = \frac{1}{25}$

3. 2^{-4} $\quad \frac{1}{2^4} \quad \frac{1}{2 \cdot 2 \cdot 2 \cdot 2} \quad \frac{1}{16}$

Rename, using negative exponents.

4. $\frac{1}{3^2}$ $\quad 3^{-2}$

5. $\frac{1}{5^4}$ $\quad 5^{-4}$

6. $\frac{1}{7^3}$ $\quad 7^{-3}$

Rename, using positive exponents.

7. 5^{-3} $\quad \frac{1}{5 \cdot 5 \cdot 5} = \frac{1}{125}$

8. 7^{-5} $\quad \frac{1}{7^5}$

9. 10^{-4} $\quad \frac{1}{10^4}$

ANSWERS ON PAGE A–6

Multiply and simplify.

10. $3^5 \cdot 3^3$

3^8 $9 \cdot 27$

6561

11. $5^{-2} \cdot 5^4$

5^2

$5 \cdot 5 = 25$

12. $6^{-3} \cdot 6^{-4}$

6^{-7}

$1/6^7$

13. $5^0 \cdot 5^{-5}$

5^{-5}

$1/5^5$

14. $y^2 \cdot y^{-4}$

y^{-2}

15. $x^{-2} \cdot x^{-6}$

x^{-8}

$1/x^8$

If we add the exponents, we again get the correct result. Next suppose that both exponents are negative:

$$a^{-3} \cdot a^{-2} \quad \text{means} \quad \frac{1}{a \cdot a \cdot a} \cdot \frac{1}{a \cdot a}.$$

This is equal to

$$\frac{1}{a \cdot a \cdot a \cdot a \cdot a}, \quad \text{or} \quad \frac{1}{a^5}, \quad \text{or} \quad a^{-5}.$$

Again, adding the exponents gives the correct result. The same is true if one or both exponents are zero.

> **In multiplication with exponential notation, we can add exponents if the bases are the same:**
> $$a^m \cdot a^n = a^{m+n}.$$

Examples Multiply and simplify.

4. $8^4 \cdot 8^3 = 8^{4+3}$ Adding exponents
     $= 8^7$

5. $7^{-3} \cdot 7^6 = 7^{-3+6}$
     $= 7^3$

6. $4^0 \cdot 4^{-9} = 4^{0+(-9)}$
     $= 4^{-9}$

7. $x^4 \cdot x^{-3} = x^{4+(-3)}$
     $= x^1$
     $= x$

DO EXERCISES 10–15.

⚌ DIVIDING USING EXPONENTS

Consider dividing with exponential notation.

$$\frac{5^4}{5^2} \text{ means } \frac{5 \cdot 5 \cdot 5 \cdot 5}{5 \cdot 5}, \text{ which is } 5 \cdot 5 \cdot \frac{5 \cdot 5}{5 \cdot 5} \text{ and we have } 5 \cdot 5, \text{ or } 5^2.$$

> **In division, we subtract exponents if the bases are the same:**
> $$\frac{a^m}{a^n} = a^{m-n}.$$

This is true whether the exponents are positive, negative, or zero.

Examples Divide and simplify.

8. $\dfrac{5^4}{5^{-2}} = 5^{4-(-2)}$ Subtracting exponents

     $= 5^6$

9. $\dfrac{a^{-2}}{a^3} = a^{-2-3}$

     $= a^{-5}$

10. $\dfrac{b^{-4}}{b^{-5}} = b^{-4-(-5)}$

     $= b^1$
     $= b$

If you have trouble, follow this step-by-step procedure.

$\dfrac{x^{-3}}{x^{-5}} = x^{-3}$ ① Write the top exponent.

$= x^{-3-}$ ② Write a minus sign.

$= x^{-3-(\ \ )}$ ③ Write parentheses.

$= x^{-3-(-5)}$ ④ Put the bottom exponent in the parentheses.

$= x^{2}$ ⑤ Carry out the subtraction.

DO EXERCISES 16–21.

⊞ RAISING A POWER TO A POWER

Consider raising a power to a power.

Example 11

$(\ 3^2\ )^4$ means $3^2 \cdot 3^2 \cdot 3^2 \cdot 3^2$.

This is $3 \cdot 3 \cdot 3 \cdot 3 \cdot 3 \cdot 3 \cdot 3 \cdot 3$,   or 3^8.

We could have multiplied the exponents in $(3^2)^4$. Suppose the exponents are not positive.

Example 12

$(\ 5^{-2}\ )^3$ means $\dfrac{1}{5^2} \cdot \dfrac{1}{5^2} \cdot \dfrac{1}{5^2}$.

This is $\dfrac{1}{5 \cdot 5} \cdot \dfrac{1}{5 \cdot 5} \cdot \dfrac{1}{5 \cdot 5}$,   or $\dfrac{1}{5^6}$,   or 5^{-6}.

Again, we could have multiplied the exponents. This works for any integer exponents.

> **To raise a power to a power we can multiply the exponents. For any exponents m and n,**
> $$(a^m)^n = a^{mn}.$$

Examples Simplify.

13. $(3^5)^4 = 3^{5 \cdot 4}$
   $= 3^{20}$

14. $(y^{-5})^7 = y^{-5 \cdot 7}$
   $= y^{-35}$

15. $(x^4)^{-2} = x^{4(-2)}$
   $= x^{-8}$

16. $(a^{-4})^{-6} = a^{(-4)(-6)}$
   $= a^{24}$

DO EXERCISES 22–25.

Divide and simplify.

16. $\dfrac{4^5}{4^2}$

17. $\dfrac{7^{-2}}{7^3} = 7^{-2-3} = 7^{-5}$
   $\dfrac{1}{7^5}$

18. $\dfrac{a^2}{a^{-5}}$ $a^{2-5} = a^{-3}$ $\dfrac{1}{a^3}$

19. $\dfrac{b^{-2}}{b^{-3}}$

20. $\dfrac{x^0}{x^{-3}}$

21. $\dfrac{y^{-7}}{y^0}$

Simplify.

22. $(3^4)^5$

23. $(x^{-3})^4$

24. $(y^{-5})^{-3}$

25. $(x^{-4})^8$ x^{-32} $\dfrac{1}{x^{32}}$

ANSWERS ON PAGE A–6

Simplify.

26. $(2x^5y^{-3})^4$

$2x^{20}y^{-12}$

$(2^4)(x^5)^4(y^{-3})^4$

$16 \cdot x^{20} \cdot y^{-12}$

27. $(5x^5y^{-6}z^{-3})^2$

28. $(3y^{-2}x^{-5}z^8)^3$

29. Suppose $2000 is invested at 6%, compounded annually. How much is in the account at the end of 3 years?

There may be several factors in parentheses.

Examples Simplify.

17. $(5x^2y^{-2})^3 = 5^3(x^2)^3(y^{-2})^3 = 125x^6y^{-6}$

18. $(3x^3y^{-5}z^2)^4 = 3^4(x^3)^4(y^{-5})^4(z^2)^4 = 81x^{12}y^{-20}z^8$

DO EXERCISES 26–28.

⠸⠼ APPLICATION: INTEREST COMPOUNDED ANNUALLY

Suppose we invest P dollars at an interest rate of 8%, compounded annually. The amount to which this grows at the end of one year is given by

$$P + 8\%P = P + 0.08P \qquad \text{By definition of percent}$$
$$= (1 + 0.08)P \qquad \text{Factoring}$$
$$= 1.08P. \qquad \text{Simplifying}$$

Going into the second year, the new principal is $(1.08)P$ dollars since interest has been added to the account. By the end of the second year, the following amount will be in the account:

$$(1.08)[\ (1.08)P\ ], \quad \text{or} \quad (1.08)^2P. \qquad \text{New principal}$$

Going into the third year, the principal will be $(1.08)^2P$ dollars. At the end of the third year, the following amount will be in the account:

$$(1.08)[\ (1.08)^2P\ ], \quad \text{or} \quad (1.08)^3P. \qquad \text{New principal}$$

Note the pattern: At the end of years 1, 2, and 3 the amount is

$$(1.08)P, \quad (1.08)^2P, \quad (1.08)^3P, \quad \text{and so on.}$$

In general,

> If principal P is invested at interest rate r, compounded annually, in t years it will grow to the amount A given by
> $$A = P(1 + r)^t.$$

Compare this formula $A = P(1 + r)^t$ with the formula for simple interest:

$$A = P(1 + rt).$$

Example 19 Suppose $1000 is invested at 8%, compounded annually. How much is in the account at the end of 3 years?

Substituting 1000 for P, 0.08 for r, and 3 for t, we get

$$A = P(1 + r)^t$$
$$= 1000(1 + 0.08)^3 = 1000(1.08)^3 = 1000(1.259712)$$
$$\approx \$1259.71.$$

DO EXERCISE 29.

NAME _____   CLASS _____

EXERCISE SET 2.8

■   Explain the meaning of each without using negative exponents.

1. 3^{-2} **2.** 2^{-3} **3.** 10^{-4} **4.** 5^{-6}

Rename, using negative exponents.

5. $\dfrac{1}{4^3}$ **6.** $\dfrac{1}{5^2}$ **7.** $\dfrac{1}{x^3}$ **8.** $\dfrac{1}{y^2}$

9. $\dfrac{1}{a^4}$ **10.** $\dfrac{1}{t^5}$ **11.** $\dfrac{1}{p^n}$ **12.** $\dfrac{1}{m^n}$

Rename, using positive exponents.

13. 7^{-3} **14.** 5^{-2} **15.** a^{-3} **16.** x^{-2}

17. y^{-4} **18.** t^{-7} **19.** z^{-n} **20.** h^{-m}

■ ■   Multiply and simplify.

21. $2^4 \cdot 2^3$ **22.** $3^5 \cdot 3^2$ **23.** $3^{-5} \cdot 3^8$ **24.** $5^{-8} \cdot 5^9$

25. $4^{-2} \cdot 4^0$ **26.** $8^{-3} \cdot 8^0$ **27.** $x^4 \cdot x^3$ **28.** $x^9 \cdot x^4$

29. $x^{-7} \cdot x^{-6}$ **30.** $y^{-5} \cdot y^{-8}$ **31.** $t^8 \cdot t^{-8}$ **32.** $m^{10} \cdot m^{-10}$

ANSWERS

1. _____
2. _____
3. _____
4. _____
5. _____
6. _____
7. _____
8. _____
9. _____
10. _____
11. _____
12. _____
13. _____
14. _____
15. _____
16. _____
17. _____
18. _____
19. _____
20. _____
21. _____
22. _____
23. _____
24. _____
25. _____
26. _____
27. _____
28. _____
29. _____
30. _____
31. _____
32. _____

Copyright © 1979, Philippines copyright 1979, by Addison-Wesley Publishing Company, Inc. All rights reserved.

ANSWERS

33. _____

34. _____

35. _____

36. _____

37. _____

38. _____

39. _____

40. _____

41. _____

42. _____

43. _____

44. _____

45. _____

46. _____

47. _____

48. _____

49. _____

50. _____

51. _____

52. _____

53. _____

54. _____

55. _____

56. _____

57. _____

58. _____

59. _____

60. _____

●●● Divide and simplify.

33. $\dfrac{7^5}{7^2}$

34. $\dfrac{4^7}{4^3}$

35. $\dfrac{8^2}{8^6}$

36. $\dfrac{9^3}{9^7}$

37. $\dfrac{x^7}{x^{-2}}$

38. $\dfrac{t^8}{t^{-3}}$

39. $\dfrac{z^{-6}}{z^{-2}}$

40. $\dfrac{y^{-7}}{y^{-3}}$

41. $\dfrac{x^{-5}}{x^{-8}}$

42. $\dfrac{y^{-4}}{y^{-9}}$

43. $\dfrac{m^{-9}}{m^{-9}}$

44. $\dfrac{x^{-8}}{x^{-8}}$

▪▪ Simplify.

45. $(2^3)^2$

46. $(3^4)^3$

47. $(5^2)^{-3}$

48. $(9^3)^{-4}$

49. $(x^{-3})^{-4}$

50. $(a^{-5})^{-6}$

51. $(x^4 y^5)^{-3}$

52. $(t^5 x^3)^{-4}$

53. $(x^{-6} y^{-2})^{-4}$

54. $(x^{-2} y^{-7})^{-5}$

55. $(3x^3 y^{-8} z^{-3})^2$

56. $(2a^2 y^{-4} z^{-5})^3$

▪▪▪ Solve.

57. Suppose $2000 is invested at 7%, compounded annually. How much is in the account at the end of 2 years?

58. Suppose $2000 is invested at 8%, compounded annually. How much is in the account at the end of 3 years?

_____

59. 🖩 Suppose $10,400 is invested at 8%, compounded annually. How much is in the account at the end of 5 years?

60. 🖩 Suppose $20,800 is invested at 7.5%, compounded annually. How much is in the account at the end of 6 years?

CHAPTER

3

SOLVING
EQUATIONS

OBJECTIVE

After finishing Section 3.1, you should be able to:

■ Solve equations using the addition principle.

Solve, using the addition principle.

1. $x + 7 = 2$

Solve, using the addition principle.

2. $n - 4.5 = 8.7$

3. $x + \dfrac{1}{2} = -\dfrac{3}{2}$

4. $7.6 = -3.2 + y$

READINESS CHECK: SKILLS FOR CHAPTER 3

1. Add: $-3 + (-8)$.

2. Subtract: $-7 - (-23)$.

3. Multiply: $-\dfrac{2}{3} \cdot \dfrac{5}{8}$.

4. Divide: $-\dfrac{3}{7} \div \left(-\dfrac{9}{7}\right)$.

5. Multiply: $3(x - 5)$.

6. Collect like terms: $x - 8x$.

7. Remove parentheses and simplify: $7w - 3 - (4w - 8)$.

3.1 THE ADDITION PRINCIPLE

■ In this chapter we will develop some more efficient principles for solving equations.

An equation $a = b$ says that a and b stand for the same number. Suppose this is true and then add a number c to the number a. We get the same answer if we add c to b, because a and b are the same number.

> *The Addition Principle:* **If an equation $a = b$ is true, then**
> $$a + c = b + c \text{ is true for any number } c.$$

Example 1 Solve: $x + 5 = -7$.

$$x + 5 + (\boxed{-5}) = -7 + (\boxed{-5})$$
Using the addition principle; adding -5 on both sides

$$x + 0 = -7 + (-5)$$ Simplifying

$$x = -12$$

Check:

$$\begin{array}{c|c} x + 5 = -7 \\ \hline -12 + 5 & -7 \\ -7 & \end{array}$$

The solution is -12.

In Example 1, to get x alone, we added the inverse of 5. This "got rid of" the 5 on the left.

DO EXERCISE 1.

Example 2 Solve: $y - 8.4 = -6.5$.

$$y - 8.4 + \boxed{8.4} = -6.5 + \boxed{8.4}$$
Adding 8.4 to get rid of -8.4 on the left

$$y = 1.9$$

Check:

$$\begin{array}{c|c} y - 8.4 = -6.5 \\ \hline 1.9 - 8.4 & -6.5 \\ -6.5 & \end{array}$$

The solution is 1.9.

DO EXERCISES 2–4.

NAME	CLASS

EXERCISE SET 3.1

● Solve, using the addition principle. Don't forget to check.

1. $x + 2 = 6$

2. $x + 5 = 8$

3. $x + 15 = -5$

Check: _____   Check: _____   Check: _____

4. $y + 25 = -6$
$-25 \quad -25$
$y = -31$

5. $r + \dfrac{2}{3} = 1$

6. $t + \dfrac{3}{4} = 1$

Check: _____   Check: _____   Check: _____

7. $x - \dfrac{5}{6} = \dfrac{7}{8}$

8. $x - \dfrac{2}{3} = \dfrac{7}{3}$

9. $8 + y = 12$

Check: _____   Check: _____   Check: _____

10. $5 + t = 7$

11. $\dfrac{1}{3} + a = \dfrac{5}{6}$

12. $-\dfrac{1}{5} + z = -\dfrac{1}{4}$

Check: _____   Check: _____   Check: _____

13. $x - 2.3 = -7.4$

14. $x - 3.7 = 8.4$

15. $-2.6 + x = 8.3$

Check: _____   Check: _____   Check: _____

ANSWERS
1. _____
2. _____
3. _____
4. _____
5. _____
6. _____
7. _____
8. _____
9. _____
10. _____
11. _____
12. _____
13. _____
14. _____
15. _____

Copyright © 1979, Philippines copyright 1979, by Addison-Wesley Publishing Company, Inc. All rights reserved.

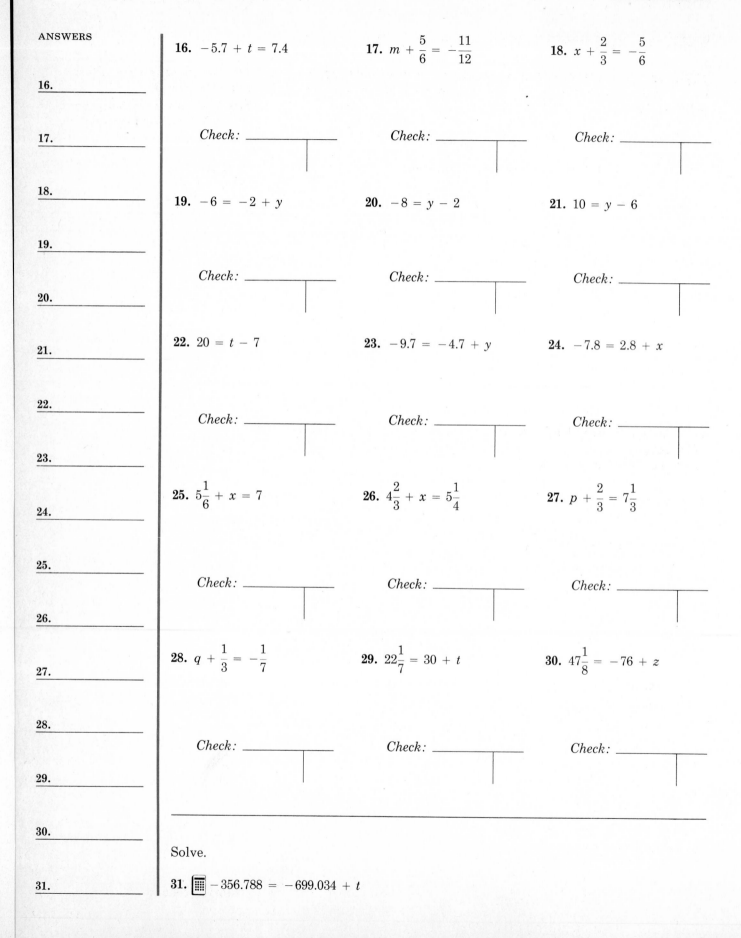

ANSWERS

16. _____

17. _____

18. _____

19. _____

20. _____

21. _____

22. _____

23. _____

24. _____

25. _____

26. _____

27. _____

28. _____

29. _____

30. _____

31. _____

16. $-5.7 + t = 7.4$

Check: _____

19. $-6 = -2 + y$

Check: _____

22. $20 = t - 7$

Check: _____

25. $5\frac{1}{6} + x = 7$

Check: _____

28. $q + \frac{1}{3} = -\frac{1}{7}$

Check: _____

17. $m + \frac{5}{6} = -\frac{11}{12}$

Check: _____

20. $-8 = y - 2$

Check: _____

23. $-9.7 = -4.7 + y$

Check: _____

26. $4\frac{2}{3} + x = 5\frac{1}{4}$

Check: _____

29. $22\frac{1}{7} = 30 + t$

Check: _____

18. $x + \frac{2}{3} = -\frac{5}{6}$

Check: _____

21. $10 = y - 6$

Check: _____

24. $-7.8 = 2.8 + x$

Check: _____

27. $p + \frac{2}{3} = 7\frac{1}{3}$

Check: _____

30. $47\frac{1}{8} = -76 + z$

Check: _____

Solve.

31. ▦ $-356.788 = -699.034 + t$

3.2   THE MULTIPLICATION PRINCIPLE

⬛ Suppose $a = b$ is true and we multiply a by some number c. We get the same answer if we multiply b by c because a and b are the same number.

> *The Multiplication Principle:*   **If an equation $a = b$ is true, then**
> $$a \cdot c = b \cdot c \text{ is true for any number } c.$$

Example 1   Solve: $3x = 9$.

$$\frac{1}{3} \cdot 3x = \frac{1}{3} \cdot 9 \qquad \text{Using the multiplication principle, multiplying by } \tfrac{1}{3}$$

$$1 \cdot x = 3 \qquad \text{Simplifying}$$

$$x = 3$$

Check: $\dfrac{3x = 9}{\begin{array}{c|c} 3 \cdot 3 & 9 \\ 9 & \end{array}}$

The solution is 3.

In Example 1, to get x alone, we multiplied by the reciprocal of 3. When we multiplied we got $1 \cdot x$, which simplified to x. This enabled us to "get rid of" the 3 on the left.

DO EXERCISE 1.

Example 2   Solve: $\dfrac{3}{8} = -\dfrac{5}{4}x$.

$$-\frac{4}{5} \cdot \frac{3}{8} = -\frac{4}{5} \cdot \left(-\frac{5}{4}x \right) \qquad \text{Multiplying by } -\tfrac{4}{5} \text{ to get rid of } -\tfrac{5}{4} \text{ on the right}$$

$$-\frac{3}{10} = x \qquad \text{Simplifying}$$

Check: $\dfrac{3}{8} = -\dfrac{5}{4}x$

$$\begin{array}{c|c} \dfrac{3}{8} & -\dfrac{5}{4}\left(-\dfrac{3}{10} \right) \\[2ex] & \dfrac{3}{8} \end{array}$$

The solution is $-\frac{3}{10}$.

DO EXERCISES 2–4.

ANSWERS ON PAGE A-6

After finishing Section 3.2, you should be able to:

⬛ Solve equations using the multiplication principle.

Solve.

1. $5x = 25$

Solve.

2. $4 = -\dfrac{1}{3}y$

$$(-3)4 = (-3)\left(-\tfrac{1}{3}\right)y$$

$$-12 = y$$

3. $4x = -7$

4. $-2.1y = 6.3$

SOMETHING EXTRA

CALCULATOR CORNER

Find the shortest path from *A* to *G*.

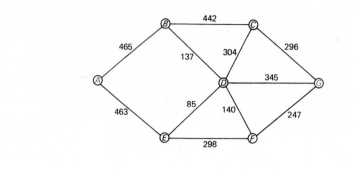

NAME

CLASS

EXERCISE SET 3.2

● Solve using the multiplication principle. Don't forget to check!

1. $6x = 36$

2. $3x = 39$

3. $9x = -36$

Check: _____

Check: _____

Check: _____

4. $7x = -49$

5. $-12x = 72$

6. $-15x = -105$

Check: _____

Check: _____

Check: _____

7. $\frac{1}{7}t = 9$

8. $\frac{1}{8}y = 11$

9. $\frac{1}{5} = \frac{1}{3}t$

Check: _____

Check: _____

Check: _____

10. $\frac{1}{9} = \frac{1}{7}z$

11. $-2.7y = 54$

12. $-3.1y = 21.7$

Check: _____

Check: _____

Check: _____

13. $\frac{3}{4}x = 27$

14. $\frac{4}{5}x = 16$

15. $-\frac{3}{5}r = -\frac{9}{10}$

Check: _____

Check: _____

Check: _____

1. _____

2. _____

3. _____

4. _____

5. _____

6. _____

7. _____

8. _____

9. _____

10. _____

11. _____

12. _____

13. _____

14. _____

15. _____

Copyright © 1979, Philippines copyright 1979, by Addison-Wesley Publishing Company, Inc. All rights reserved.

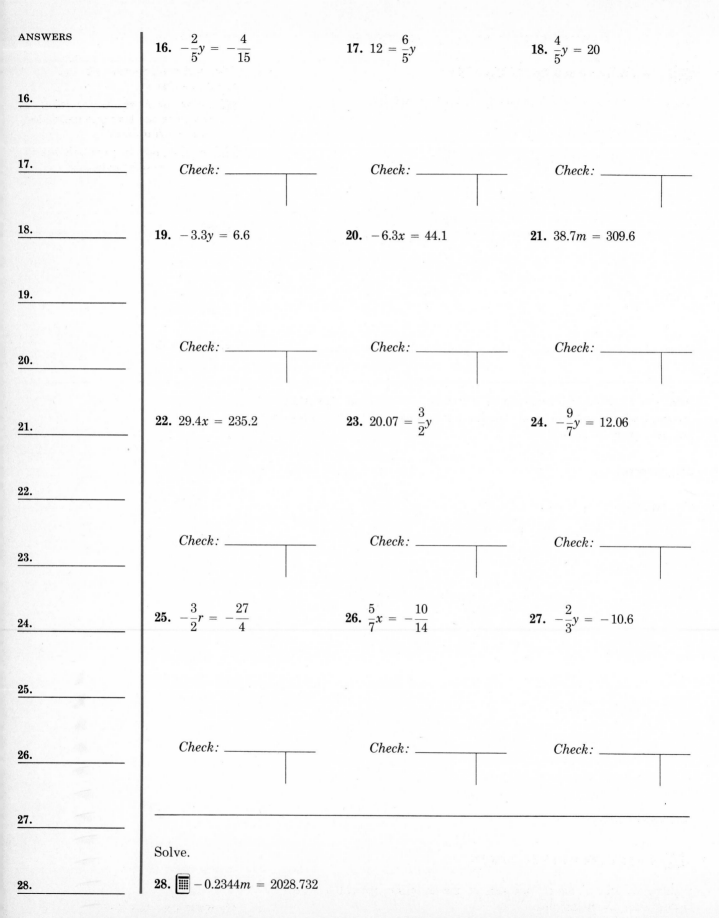

ANSWERS

16. _____

17. _____

18. _____

19. _____

20. _____

21. _____

22. _____

23. _____

24. _____

25. _____

26. _____

27. _____

28. _____

16. $-\dfrac{2}{5}y = -\dfrac{4}{15}$

17. $12 = \dfrac{6}{5}y$

18. $\dfrac{4}{5}y = 20$

Check: _____

Check: _____

Check: _____

19. $-3.3y = 6.6$

20. $-6.3x = 44.1$

21. $38.7m = 309.6$

Check: _____

Check: _____

Check: _____

22. $29.4x = 235.2$

23. $20.07 = \dfrac{3}{2}y$

24. $-\dfrac{9}{7}y = 12.06$

Check: _____

Check: _____

Check: _____

25. $-\dfrac{3}{2}r = -\dfrac{27}{4}$

26. $\dfrac{5}{7}x = -\dfrac{10}{14}$

27. $-\dfrac{2}{3}y = -10.6$

Check: _____

Check: _____

Check: _____

Solve.

28. 🖩 $-0.2344m = 2028.732$

3.3   USING THE PRINCIPLES TOGETHER

■ APPLYING BOTH PRINCIPLES

Let's solve an equation where we apply both principles.

Example 1   Solve: $3x + 4 = 13$.

$$3x + 4 + (\boxed{-4}) = 13 + (\boxed{-4}) \qquad \text{Using the addition}$$
$$\text{principle, adding } -4$$

$$3x = 9 \qquad \text{Simplifying}$$

$$\boxed{\frac{1}{3}} \cdot 3x = \boxed{\frac{1}{3}} \cdot 9 \qquad \text{Using the multiplication}$$
$$\text{principle, multiplying by } \tfrac{1}{3}$$

$$x = 3 \qquad \text{Simplifying}$$

Check:
$$\frac{3x + 4 = 13}{3 \cdot 3 + 4 \mid 13}$$
$$9 + 4$$
$$13$$

The solution is 3.

In Example 1, we used the addition principle first. This is usually better, although there might be equations for which we would use the multiplication principle first.

DO EXERCISE 1.

Example 2   Solve: $-5x - 6 = 16$.

$$-5x - 6 + \boxed{6} = 16 + \boxed{6} \qquad \text{Adding 6}$$

$$-5x = 22$$

$$\boxed{-\frac{1}{5}} \cdot (-5x) = \boxed{-\frac{1}{5}} \cdot 22 \qquad \text{Multiplying by } -\frac{1}{5}$$

$$x = -\frac{22}{5}, \quad \text{or} \quad -4\frac{2}{5}$$

Check:
$$\frac{-5x - 6 = 16}{-5\left(-\dfrac{22}{5}\right) - 6 \mid 16}$$
$$22 - 6$$
$$16$$

The solution is $-\frac{22}{5}$.

DO EXERCISES 2–4.

■■ COLLECTING LIKE TERMS

If there are like terms on one side of the equation, we collect them before using the principles.

ANSWERS ON PAGE A–7

OBJECTIVES

After finishing Section 3.3, you should be able to:

■ Solve simple equations using both the addition and multiplication principles.

■■ Solve equations in which like terms are to be collected.

Solve.

1. $9x + 6 = 51$

Solve.

2. $8x - 4 = 28$

3. $-\dfrac{1}{2}x + 3 = 1$

4. $-4 - 8x = 8$

Solve.

5. $4x + 3x = -21$

$\frac{1}{7} \cdot \frac{7x}{1} = \frac{-21}{1} \cdot \frac{1}{7} =$

$x = \frac{21}{7}$

$x = 3$

6. $x - 0.09x = 728$

$.91x = 728$

$\frac{100}{91} \cdot \frac{91}{100} x = \frac{728}{1} \cdot \frac{100}{91} =$

Solve.

7. $7y + 5 = 2y + 10$

Solve.

8. $5 - 2y = 3y - 5$

9. $7x - 17 + 2x = 2 - 8x + 15$

10. $3x - 15 = 5x + 2 - 4x$

Example 3   Solve: $3x + 4x = -14$.

$$7x = -14 \qquad \text{Collecting like terms}$$

$$\boxed{\frac{1}{7}} \cdot 7x = \boxed{\frac{1}{7}} \cdot (-14)$$

$$x = -2$$

The number -2 checks, so the solution is -2.

DO EXERCISES 5 AND 6.

If there are like terms on opposite sides of the equation, we get them on the same side using the addition principle. Then we collect them.

Example 4   Solve: $2x - 2 = -3x + 3$.

$$2x - 2 + \boxed{2} = -3x + 3 + \boxed{2} \qquad \text{Adding 2}$$

$$2x = -3x + 5 \qquad \text{Simplifying}$$

$$2x + \boxed{3x} = -3x + \boxed{3x} + 5 \qquad \text{Adding } 3x$$

$$5x = 5 \qquad \text{Collecting like terms and simplifying}$$

$$\boxed{\frac{1}{5}} \cdot 5x = \boxed{\frac{1}{5}} \cdot 5 \qquad \text{Multiplying by } \frac{1}{5}$$

$$x = 1 \qquad \text{Simplifying}$$

Check:

$2x - 2 = -3x + 3$	
$2 \cdot 1 - 2$	$-3 \cdot 1 + 3$
$2 - 2$	$-3 + 3$
0	0

The solution is 1.

In Example 4, the addition principle was used to get all terms with one variable on one side and all other terms on the other side. Then like terms were collected and we proceeded as before.

DO EXERCISE 7.

Example 5   Solve: $6x + 5 - 7x = 10 - 4x + 3$.

$$6x - 7x = 10 - 4x + 3 - 5 \qquad \text{Adding } -5$$

$$4x + 6x - 7x = 10 + 3 - 5 \qquad \text{Adding } 4x$$

$$3x = 8 \qquad \text{Collecting like terms and simplifying}$$

$$\frac{1}{3} \cdot 3x = \frac{1}{3} \cdot 8 \qquad \text{Multiplying by } \frac{1}{3}$$

$$x = \frac{8}{3} \qquad \text{Simplifying}$$

The number $\frac{8}{3}$ checks, so it is the solution.

DO EXERCISES 8–10.

NAME   CLASS   ANSWERS

EXERCISE SET 3.3

● | Solve and check.

1.

1. $5x + 6 = 31$ **2.** $3x + 6 = 30$ **3.** $4x - 6 = 34$

2.

3.

Check: _____   *Check:* _____   *Check:* _____

4.

4. $6x - 3 = 15$ **5.** $7x + 2 = -54$ **6.** $5x + 4 = -41$

5.

6.

●● | Solve and check.

7. $5x + 7x = 72$ **8.** $4x + 5x = 45$ **9.** $4y - 2y = 10$

7.

8.

9.

10. $8y - 5y = 15$ **11.** $10.2y - 7.3y = -58$ **12.** $6.8y - 2.4y = -88$

10.

11.

12.

Copyright © 1979. Philippines copyright 1979. by Addison-Wesley Publishing Company. Inc. All rights reserved.

ANSWERS

13.

14.

15.

16.

17.

18.

19.

20.

21.

22.

23.

24.

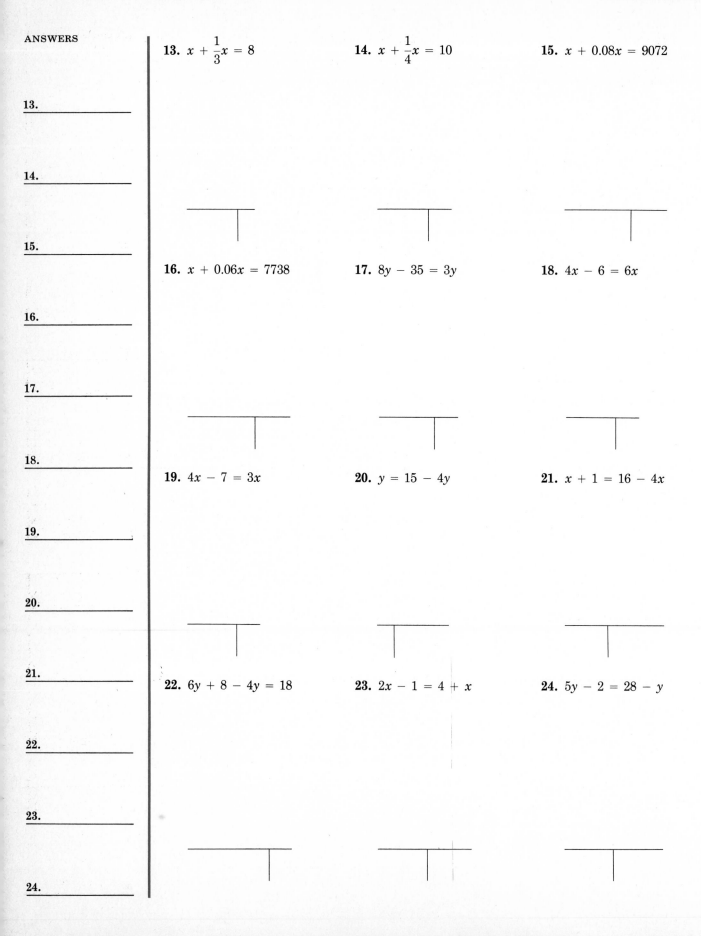

13. $x + \dfrac{1}{3}x = 8$

14. $x + \dfrac{1}{4}x = 10$

15. $x + 0.08x = 9072$

16. $x + 0.06x = 7738$

17. $8y - 35 = 3y$

18. $4x - 6 = 6x$

19. $4x - 7 = 3x$

20. $y = 15 - 4y$

21. $x + 1 = 16 - 4x$

22. $6y + 8 - 4y = 18$

23. $2x - 1 = 4 + x$

24. $5y - 2 = 28 - y$

25. $5x + 2 = 3x + 6$

26. $6x + 3 = 2x + 11$

25. _____

27. $5 - 2x = 3x - 7x + 25$

28. $10 - 3x = 2x - 8x + 40$

26. _____

27. _____

29. $4 + 3x - 6 = 3x + 2 - x$

30. $5 + 4x - 7 = 4x + 3 - x$

28. _____

29. _____

31. $4y - 4 + y = 6y + 20 - 4y$

32. $5y - 7 + y = 7y + 21 - 5y$

30. _____

31. _____

32. _____

Copyright © 1979, Philippines copyright 1979, by Addison-Wesley Publishing Company, Inc. All rights reserved.

ANSWERS

33. $\dfrac{5}{2}x + \dfrac{1}{2}x = 3x + \dfrac{3}{2} + \dfrac{5}{2}x$

34. $\dfrac{7}{8}x - \dfrac{1}{4} + \dfrac{3}{4}x = \dfrac{1}{16} + x$

33. _____

34. _____

35. $2.1x + 45.2 = 3.2 - 8.4x$

36. $7\dfrac{1}{2}y - \dfrac{1}{2}y = \dfrac{15}{4}y + 39$

35. _____

36. _____

37. $\dfrac{1}{5}t - 0.4 + \dfrac{2}{5}t = 0.6 - \dfrac{1}{10}t$

38. $1.7t + 8 - 1.62t = 0.4t - 0.32 + 8$

37. _____

38. _____

Solve.

39. ⊞ $0.008 + 9.62x - 42.8 = 0.944x + 0.0083 - x$

39. _____

3.4   EQUATIONS CONTAINING PARENTHESES

▪️   Some equations containing parentheses can be solved by first multiplying to remove parentheses and then proceeding as before.

Example 1   Solve: $4x = 2(12 - 2x)$.

$$4x = 24 - 4x \qquad \text{Multiplying to remove parentheses}$$

$$4x + 4x = 24 \qquad \text{Adding } 4x \text{ to get all } x\text{-terms on one side}$$

$$8x = 24 \qquad \text{Collecting like terms}$$

$$x = 3 \qquad \text{Multiplying by } \tfrac{1}{8}$$

Check:

$4x = 2(12 - 2x)$	
$4 \cdot 3$	$2(12 - 2 \cdot 3)$
12	$2(12 - 6)$
	$2 \cdot 6$
	12

The solution is 3.

DO EXERCISES 1 AND 2.

Example 2   Solve: $3(x - 2) - 1 = 2 - 5(x + 5)$.

$$3x - 6 - 1 = 2 - 5x - 25 \qquad \text{Multiplying to remove parentheses}$$

$$3x - 7 = -5x - 23 \qquad \text{Simplifying}$$

$$3x + 5x = -23 + 7 \qquad \text{Adding } 5x \text{ and also 7, to get all } x\text{-terms on one side and all other terms on the other side}$$

$$8x = -16 \qquad \text{Simplifying}$$

$$x = -2 \qquad \text{Multiplying by } \tfrac{1}{8}$$

Check:

$3(x - 2) - 1 = 2 - 5(x + 5)$	
$3(-2 - 2) - 1$	$2 - 5(-2 + 5)$
$3 \cdot (-4) - 1$	$2 - 5(3)$
$-12 - 1$	$2 - 15$
-13	-13

The solution is -2.

DO EXERCISES 3 AND 4.

OBJECTIVE

After finishing Section 3.4, you should be able to:

▪️ Solve simple equations containing parentheses.

Solve.

1. $2(2y + 3) = 14$

2. $5(3x - 2) = 35$

Solve.

3. $3(7 + 2x) = 30 + 7(x - 1)$

4. $4(3 + 5x) - 4 = 3 + 2(x - 2)$

SOMETHING EXTRA

HANDLING DIMENSION SYMBOLS (PART I)

Speed is often measured by measuring a distance and a time, and then dividing the distance by the time (this is *average* speed). If a distance is measured in kilometers (km) and the time required to travel that distance is measured in hours, the speed will be computed in *kilometers per hour* (km/hr). For example, if a car travels 100 km in 2 hr, the average speed is

$$\frac{100 \text{ km}}{2 \text{ hr}}, \quad \text{or} \quad 50\frac{\text{km}}{\text{hr}}.$$

The standard notation for km/hr is km/h.

The symbol

$$\frac{100 \text{ km}}{2 \text{ hr}}$$

makes it look as though we are dividing 100 km by 2 hr. It may be argued that we cannot divide 100 km by 2 hr (we can only divide 100 by 2). Nevertheless, it is convenient to treat dimension symbols such as *kilometers*, *meters*, *hours*, *feet*, *seconds*, and *pounds* much like numerals or variables, for the reason that correct results can thus be obtained mechanically. Compare, for example,

$$\frac{100x}{2y} = \frac{100}{2} \cdot \frac{x}{y} = 50\frac{x}{y} \quad \text{with} \quad \frac{100 \text{ km}}{2 \text{ hr}} = \frac{100}{2} \cdot \frac{\text{km}}{\text{hr}} = 50\frac{\text{km}}{\text{hr}}.$$

The analogy holds in other situations.

Example 1   Compare

$$3 \boxed{\text{ft}} + 2 \boxed{\text{ft}} = (3 + 2) \boxed{\text{ft}} = 5 \text{ ft}$$

with

$$3x + 2x = (3 + 2)x = 5x.$$

Example 2   Compare   3 ft + 2 yd   with $3x + 2y$.

Note in this case that we cannot simplify further since the units are different. We could change units, but we will not do that here. (See pp. 334 and 338 for more on dimension symbols.)

EXERCISES

Distances and times are given. Use them to compute speed.

1. 45 mi, 9 hr   **2.** 680 km, 20 hr

3. 6.6 meters(m), 3 sec   **4.** 76 ft, 4 min

Add these measures.

5. 45 ft, 7 ft   **6.** 85 sec, 17 sec   **7.** 17 m, 14 m

8. 3 hr, 29 hr   **9.** $\frac{3}{4}$ lb, $\frac{2}{5}$ lb   **10.** 5 km, 7 km

11. 18 g, 4 g   **12.** 70 m/sec, 35 m/sec

NAME   CLASS   ANSWERS

EXERCISE SET 3.4

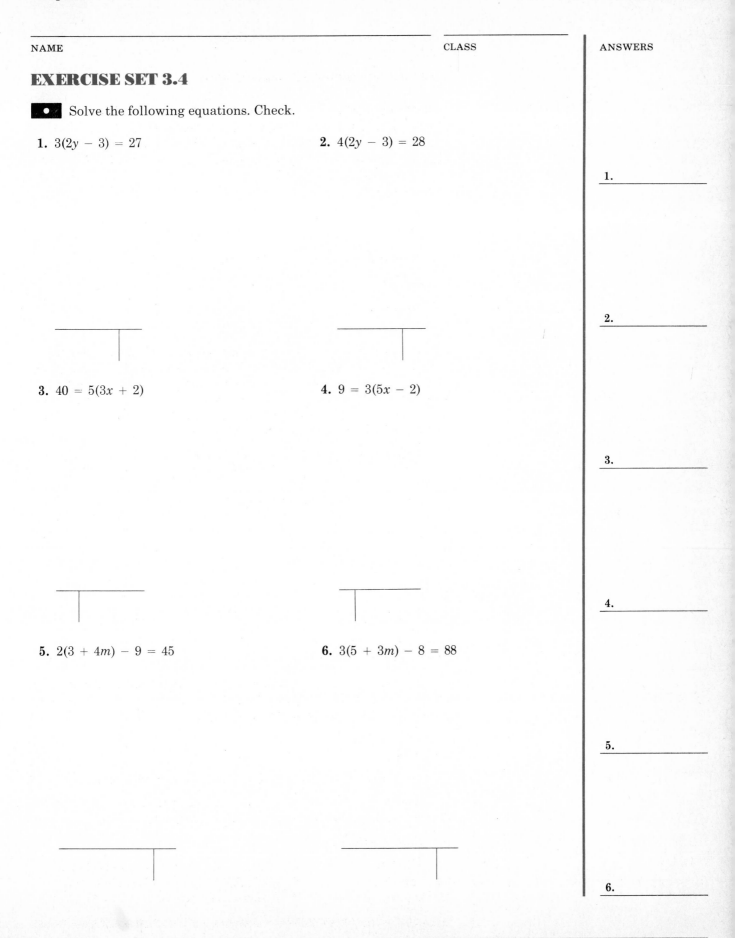 Solve the following equations. Check.

1. $3(2y - 3) = 27$ **2.** $4(2y - 3) = 28$

1. _____

2. _____

3. $40 = 5(3x + 2)$ **4.** $9 = 3(5x - 2)$

3. _____

4. _____

5. $2(3 + 4m) - 9 = 45$ **6.** $3(5 + 3m) - 8 = 88$

5. _____

6. _____

Copyright © 1979, Philippines copyright 1979, by Addison-Wesley Publishing Company, Inc. All rights reserved.

7. $5r - (2r + 8) = 16$

8. $6b - (3b + 8) = 16$

7. _____

8. _____

9. $3g - 3 = 3(7 - g)$

10. $3d - 10 = 5(d - 4)$

9. _____

10. _____

11. $6 - 2(3x - 1) = 2$

12. $10 - 3(2x - 1) = 1$

11. _____

12. _____

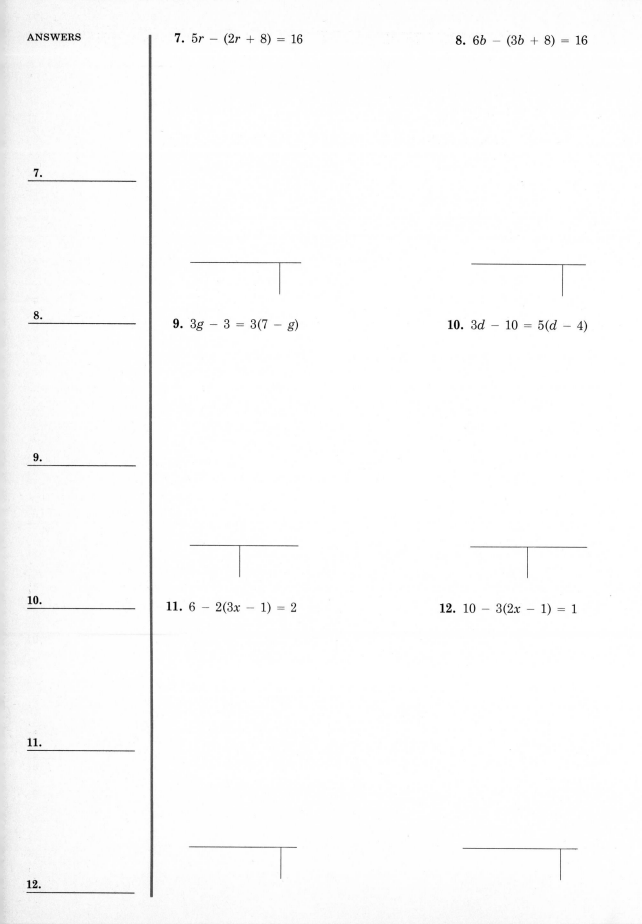

13. $5(d + 4) = 7(d - 2)$

14. $3(t - 2) = 9(t + 2)$

15. $3(x - 2) = 5(x + 2)$

16. $5(y + 4) = 3(y - 2)$

17. $8(2t + 1) = 4(7t + 7)$

$16t + 8 = 24t + 7)$

$16t - 16t + 8 = 24t - 16t + 7$

$8 = 8t + 7$

$-7 - 8 = 8t$

$1 = 8t$

18. $7(5x - 2) = 6(6x - 1)$

13. _____

14. _____

15. _____

16. _____

17. _____

18. _____

Copyright © 1979, Philippines copyright 1979, by Addison-Wesley Publishing Company, Inc. All rights reserved.

ANSWERS

19. $3(r - 6) + 2 = 4(r + 2) - 21$

20. $5(t + 3) + 9 = 3(t - 2) + 6$

19. _____

20. _____

21. $19 - (2x + 3) = 2(x + 3) + x$

22. $13 - (2c + 2) = 2(c + 2) + 3c$

21. _____

22. _____

23. $\frac{1}{4}(8y + 4) - 17 = -\frac{1}{2}(4y - 8)$

24. $\frac{1}{3}(6x + 24) - 20 = -\frac{1}{4}(12x - 72)$

23. _____

24. _____

Solve.

25. 🔢 $475(54x + 7856) + 9762 = 402(83x + 975)$

25. _____

3.5   THE PRINCIPLE OF ZERO PRODUCTS

● The product of two numbers is 0 if one of the numbers is 0. Furthermore, *if any product is 0, then a factor must be 0.*

After finishing Section 3.5, you should be able to:

● Solve equations (already factored), using the principle of zero products.

Example 1   Solve $(x + 3)(x - 2) = 0$.

We have a product of 0. This equation will be true when either factor is 0. Hence it is true when

$$x + 3 = 0 \quad \text{or} \quad x - 2 = 0.$$

Here we have two simple equations which we know how to solve,

$$x = -3 \quad \text{or} \quad x = 2.$$

There are two solutions, -3 and 2.

Solve, using the principle of zero products.

1. $(x - 3)(x + 4) = 0$

> *The Principle of Zero Products:* **An equation with 0 on one side and with factors on the other can be solved by finding those numbers that make the factors 0.**

Example 2   Solve: $(5x + 1)(x - 7) = 0$.

$5x + 1 = 0 \quad \text{or} \quad x - 7 = 0$ Using the principle of zero products

$5x = -1 \quad \text{or} \qquad x = 7$

$x = -\dfrac{1}{5} \quad \text{or} \qquad x = 7$ Solving the two equations separately

2. $(x - 7)(x - 3) = 0$

Check: For $-\frac{1}{5}$:

$(5x + 1)(x - 7) = 0$	
$(5(-\frac{1}{5}) + 1)(-\frac{1}{5} - 7)$	0
$(-1 + 1)(-7\frac{1}{5})$	
$0(-7\frac{1}{5})$	
	0

For 7:

$(5x + 1)(x - 7) = 0$	
$(5 \cdot 7 + 1)(7 - 7)$	0
$(35 + 1) \cdot 0$	
	0

3. $(4t + 1)(3t - 2) = 0$

The solutions are $-\frac{1}{5}$ and 7.

The "possible solutions" we get by using the principle of zero products are actually always solutions, unless we have made an error in solving. Thus, when we use this principle, a check is not necessary, except to detect errors.

Caution!   Do not make the mistake of using this principle when there is not a zero on one side.

4. $y(3y - 17) = 0$

Example 3   Solve: $x(2x - 9) = 0$

$x = 0 \quad \text{or} \quad 2x - 9 = 0$ Using the principle of zero products

$x = 0 \quad \text{or} \qquad 2x = 9$

$x = 0 \quad \text{or} \qquad x = \frac{9}{2}$

The solutions are 0 and $\frac{9}{2}$.

DO EXERCISES 1–4.

ANSWERS ON PAGE A–7

SOMETHING EXTRA

CALCULATOR CORNER: NUMBER PATTERNS

1. Calculate each of the following using your calculator. Look for a pattern.

$$1$$
$$1 + 3$$
$$1 + 3 + 5$$
$$1 + 3 + 5 + 7$$
$$1 + 3 + 5 + 7 + 9$$
$$1 + 3 + 5 + 7 + 9 + 11$$

Use the pattern to find each of the following without using your calculator.

$$1 + 3 + 5 + 7 + 9 + 11 + 13$$
$$1 + 3 + 5 + 7 + 9 + 11 + 13 + 15$$

2. Verify each of the following using your calculator.

$$1 = \frac{1 \cdot 2}{2}$$

$$1 + 2 = \frac{2 \cdot 3}{2}$$

$$1 + 2 + 3 = \frac{3 \cdot 4}{2}$$

$$1 + 2 + 3 + 4 = \frac{4 \cdot 5}{2}$$

$$1 + 2 + 3 + 4 + 5 = \frac{5 \cdot 6}{2}$$

$$1 + 2 + 3 + 4 + 5 + 6 = \frac{6 \cdot 7}{2}$$

$$1 + 2 + 3 + 4 + 5 + 6 + 7 = \frac{7 \cdot 8}{2}$$

Use the pattern to find each of the following without using your calculator.

$$1 + 2 + 3 + 4 + 5 + 6 + 7 + 8$$
$$1 + 2 + 3 + 4 + 5 + 6 + 7 + 8 + 9$$

Can you give an expression for

$$1 + 2 + 3 + 4 + \cdots + n?$$

EXERCISE SET 3.5

● Solve.

1. $(x + 8)(x + 6) = 0$

2. $(x + 3)(x + 2) = 0$

3. $(x - 3)(x + 5) = 0$

4. $(x + 9)(x - 3) = 0$

5. $(x - 12)(x - 11) = 0$

6. $(x - 13)(x - 53) = 0$

7. $y(y - 13) = 0$

8. $x(x - 4) = 0$

9. $0 = x(x + 21)$

10. $0 = y(y + 10)$

11. $(2x + 5)(x + 4) = 0$

12. $(2x + 9)(x + 8) = 0$

13. $(3x - 1)(x + 2) = 0$

14. $(5x + 1)(x - 3) = 0$

15. $2x(3x - 2) = 0$

16. $5x(8x - 9) = 0$

ANSWERS

1. _____

2. _____

3. _____

4. _____

5. _____

6. _____

7. _____

8. _____

9. _____

10. _____

11. _____

12. _____

13. _____

14. _____

15. _____

16. _____

Copyright © 1979, Philippines copyright 1979, by Addison-Wesley Publishing Company, Inc. All rights reserved.

ANSWERS

17. _____

18. _____

19. _____

20. _____

21. _____

22. _____

23. _____

24. _____

25. _____

26. _____

27. _____

28. _____

29. _____

30. _____

17. $\frac{1}{2}x\left(\frac{2}{3}x - 12\right) = 0$

18. $\frac{5}{7}x\left(\frac{3}{4}x - 6\right) = 0$

19. $0 = \left(\frac{1}{3} - 3x\right)\left(\frac{1}{5} - 2x\right)$

20. $0 = \left(\frac{1}{5} - 2x\right)\left(\frac{1}{9} - 3x\right)$

21. $\left(\frac{1}{3}y - \frac{2}{3}\right)\left(\frac{1}{4}y - \frac{3}{2}\right) = 0$

22. $\left(\frac{7}{4}x - \frac{1}{12}\right)\left(\frac{2}{3}x - \frac{12}{11}\right) = 0$

23. $2.5x(2.5x - 5) = 0$

24. $8.3x(4.5x - 9) = 0$

25. $(0.03x - 0.01)(0.05x - 1) = 0$

26. $(0.01x - 0.03)(0.04x - 2) = 0$

27. $\left(0.01x + \frac{1}{10}\right)\left(0.03x - \frac{1}{10}\right) = 0$

28. $\left(0.02x + \frac{1}{10}\right)\left(0.04x - \frac{1}{10}\right) = 0$

---

29. Check the numbers found below. What's wrong with the methods used? Try to find the solutions.

a) $(x - 3)(x + 4) = 8$
$\quad x - 3 = 8 \quad$ or $\quad x + 4 = 8$
$\quad\quad x = 11 \quad$ or $\quad\quad x = 4$

b) $(x - 3)(x + 4) = 8$
$\quad x - 3 = 2 \quad$ or $\quad x + 4 = 4$
$\quad\quad x = 5 \quad$ or $\quad\quad x = 0$

30. ▦ Solve $(0.00005x + 0.1)(0.0097x + 0.5) = 0$

3.6 SOLVING PROBLEMS

■•■   The first step in solving a problem is to translate it to mathematical language. Very often this means translating to an equation. Drawing a picture usually helps a great deal. Then we solve the equation and check to see if we have a solution to the problem.

Example 1   A 6-ft board is cut into two pieces, one twice as long as the other. How long are the pieces?

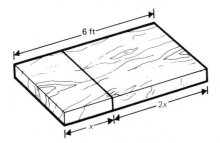

Drawing a picture

The picture can help in translating. Here is one way to do it.

Length of one piece plus length of other is 6
$$x \qquad + \qquad 2x \qquad = 6$$

(We use x for the length of one piece and $2x$ for the length of the other.) Now we solve:

$$x + 2x = 6$$
$$3x = 6 \qquad \text{Collecting like terms}$$
$$x = 2 \qquad \text{Multiplying by } \tfrac{1}{3}$$

Do we have an answer to the *problem*? If one piece is 2 ft long, the other, to be twice as long, must be 4 ft long, and the lengths of the pieces add up to 6 ft. This checks.

Note that we did not have to check the solution in the equation. Rather, we checked in the problem itself.

> *Steps to use in solving a problem:*
> 1. **Translate to an equation. (Always draw a picture if it makes sense to do so.)**
> 2. **Solve the equation.**
> 3. **Check the answer in the original problem.**

DO EXERCISE 1.

Example 2   Five plus three more than a number is nineteen. What is the number?

This time it does not make sense to draw a picture.

① Five plus three more than a number is nineteen
$$5 \quad + \qquad\qquad (x + 3) \qquad\qquad = \qquad 19$$

We have used x to represent the unknown number.

ANSWER ON PAGE A–7

After finishing Section 3.6, you should be able to:

■•■ **Solve problems by translating to equations.**

1. An 8-ft board is cut into two pieces. One piece is 2 ft longer than the other. How long are the pieces?

2. If five is subtracted from three times a certain number, the result is 10. What is the number?

② Now solve: $5 + (x + 3) = 19$

$$x + 8 = 19 \quad \text{Collecting like terms}$$

$$x = 11 \quad \text{Adding } -8$$

③ *Check:* Three more than 11 is 14. Adding 5 to 14, we get 19. This checks and the answer is 11.

DO EXERCISE 2.

Example 3   Money is invested in a savings account at 8% simple interest. After 1 year there is $8208 in the account. How much was originally invested?

$$\underbrace{\text{Original investment}}_{x} \underbrace{\text{plus}}_{+} \underbrace{\text{interest}}_{8\%x} \underbrace{\text{is}}_{=} \underbrace{8208}_{8208}$$

We have used x to represent the original investment.

$$x + 8\%x = 8208$$

$$1x + 0.08x = 8208$$

$$1.08x = 8208$$

$$x = \frac{8208}{1.08}$$

$$x = 7600$$

3. Money is borrowed at 9% simple interest. After 1 year $6213 pays off the loan. How much was originally borrowed?

Check: 8% of 7600 is 608. Adding this to 7600, we get 8208. This checks so the original investment was $7600.

DO EXERCISE 3.

Example 4   The sum of two consecutive integers is 29. What are the integers?

(*Consecutive* integers are next to each other, such as 3 and 4. The larger is 1 plus the smaller.)

$$\underbrace{\text{First integer}}_{x} + \underbrace{\text{second integer}}_{(x + 1)} = \underbrace{29}_{29} \quad \text{Rewording}$$

$$\qquad\qquad\qquad\qquad\qquad\qquad\quad \text{Translating}$$

We have let x represent the first integer. Then $x + 1$ represents the second.

4. The sum of two consecutive even integers is 38. (Consecutive even integers are next to each other, such as 4 and 6. The larger is 2 plus the smaller.) What are the integers?

We solve: $x + (x + 1) = 29$

$$2x + 1 = 29$$

$$2x = 28$$

$$x = 14.$$

Check: Our answers are 14 and 15. These are consecutive integers. Their sum is 29, so the answers check in the *problem*.

DO EXERCISE 4.

Example 5   Acme Rent-a-Car rents an intermediate-size car (such as a Chevrolet, Ford, or Plymouth) at a daily rate of $14.95 plus 19¢ per mile. A businessperson is not to exceed a daily car rental budget of $50. What mileage will allow the businessperson to stay within budget?

We translate to an equation, using $0.19 for 19¢.

14.95 plus 19¢ times number of miles driven is budget

$$14.95 + 0.19 \cdot m = 50$$

We solve:

$$14.95 + 0.19m = 50$$
$$0.19m = 35.05 \qquad \text{Adding } -14.95$$
$$m = \frac{35.05}{0.19}$$
$$m \approx 184.5. \qquad \text{Rounded to the nearest tenth}$$

This checks in the original problem. In this case it is an approximation. At least the businessperson now knows to stay under this mileage.

DO EXERCISE 5.

Example 6   The perimeter of a rectangle is 150 cm. The length is 15 cm greater than the width. Find the dimensions.

We first draw a picture.

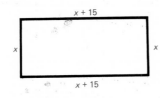

Width + Width + Length + Length = 150   Rewording

$$x + x + (x + 15) + (x + 15) = 150 \qquad \text{Translating}$$

We have let x represent the width. Then $x + 15$ represents the length

$$4x + 30 = 150$$
$$4x = 120$$
$$x = 30$$

The length is given by $x + 15 = 45$.

Check: The perimeter is $30 + 30 + 45 + 45$, which is 150. This checks, so the width is 30 cm and the length is 45 cm.

DO EXERCISE 6.

5. Acme also rents compact cars at $14.95 plus 17¢ per mile. What mileage will allow the businessperson to stay within a budget of $50?

6. The length of a rectangle is twice the width. The perimeter is 60 m. Find the dimensions.

7. The second angle of a triangle is 3 times as large as the first. The third angle measures 30° more than the first angle. Find the measures of the angles.

Example 7   The second angle of a triangle is twice as large as the first. The measure of the third angle is 20° greater than that of the first angle. How large are the angles?

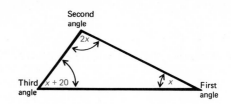

We draw a picture. We use x for the measure of the first angle. The second is twice as large, so its measure will be $2x$. The third angle is 20° greater than the first angle so its measure will be $x + 20$.

Now, to translate we need to recall a geometric fact. The measures of the angles of any triangle add up to 180°.

$$\underbrace{\text{Measure of first angle}}_{x} + \underbrace{\text{Measure of second angle}}_{2x} + \underbrace{\text{Measure of third angle}}_{(x + 20)} = \underbrace{180°}_{180}$$

Now we solve:

$$x + 2x + (x + 20) = 180$$
$$4x + 20 = 180$$
$$4x = 160$$
$$x = 40.$$

The angles will have measures as follows:

First angle: $x = 40°$
Second angle: $2x = 80°$
Third angle: $x + 20 = 60°$.

These add up to 180° so they give the answer to the *problem*.

DO EXERCISE 7.

NAME _____   CLASS _____

EXERCISE SET 3.6

▮● Solve.

1. When 18 is subtracted from six times a certain number, the result is 96. What is the number?

2. When 28 is subtracted from five times a certain number, the result is 232. What is the number?

3. If you double a number and then add 16, you get $\frac{2}{5}$ of the original number. What is the original number?

4. If you double a number and then add 85, you get $\frac{3}{4}$ of the original number. What is the original number?

5. If you add two-fifths of a number to the number itself, you get 56. What is the number?

6. If you add one-third of a number to the number itself, you get 48. What is the number?

Copyright © 1979, Philippines copyright 1979, by Addison-Wesley Publishing Company, Inc. All rights reserved.

7. A 180-m rope is cut into three pieces. The second piece is twice as long as the first. The third piece is three times as long as the second. How long is each piece of rope?

7. _____

8. A 480-m wire is cut into three pieces. The second piece is three times as long as the first. The third piece is four times as long as the second. How long is each piece?

8. _____

9. Consecutive odd integers are next to each other, such as 5 and 7. The larger is 2 plus the smaller. The sum of two consecutive odd integers is 76. What are the integers?

9. _____

10. The sum of two consecutive odd integers is 84. What are the integers?

10. _____

11. Consecutive even integers are next to each other, like 6 and 8. The larger is 2 plus the smaller. The sum of two consecutive even integers is 114. What are the integers?

11. _____

12. The sum of two consecutive even integers is 106. What are the integers?

12. _____

13. The sum of three consecutive integers is 108. What are the integers?

14. The sum of three consecutive integers is 126. What are the integers?

15. The sum of three consecutive odd integers is 189. What are the integers?

16. The sum of three consecutive odd integers is 255. What are the integers?

17. The perimeter of a rectangle is 310 m. The length is 25 m greater than the width. Find the width and length of the rectangle.

18. The perimeter of a rectangle is 304 cm. The length is 40 cm greater than the width. Find the width and length of the rectangle.

ANSWERS

13. _____

14. _____

15. _____

16. _____

17. _____

18. _____

Copyright © 1979, Philippines copyright 1979, by Addison-Wesley Publishing Company, Inc. All rights reserved.

ANSWERS

19. _____

20. _____

21. _____

22. _____

23. _____

24. _____

19. The perimeter of a rectangle is 152 m. The width is 22 m less than the length. Find the width and the length.

20. The perimeter of a rectangle is 280 m. The width is 26 m less than the length. Find the width and the length.

21. The second angle of a triangle is four times as large as the first. The third angle is 45° less than the sum of the other two angles. Find the measure of the first angle.

22. The second angle of a triangle is three times as large as the first. The third angle is 25° less than the sum of the other two angles. Find the measure of the first angle.

23. Money is invested in a savings account at 7% simple interest. After 1 year there is $4708 in the account. How much was originally invested?

24. Money is borrowed at 10% simple interest. After 1 year $7194 pays off the loan. How much was originally borrowed?

25. After a 40% reduction, a shirt is on sale at $9.60. What was the marked price (the price before reduction)?

26. After a 34% reduction, a blouse is on sale at $9.24. What was the marked price?

25. _____

27. Badger Rent-a-Car rents an intermediate-size car at a daily rate of $14.95 plus 10¢ per mile. A businessperson is not to exceed a daily car rental budget of $60. What mileage will allow the businessperson to stay within budget?

26. _____

28. Badger also rents compact cars at $13.95 plus 10¢ per mile. What mileage will allow the businessperson to stay within the budget of $60?

27. _____

29. The second angle of a triangle is three times as large as the first. The measure of the third angle is 40° greater than that of the first angle. How large are the angles?

28. _____

30. The second angle of a triangle is five times as large as the first. The measure of the third angle is 10° greater than that of the first angle. How large are the angles?

29. _____

30. _____

Copyright © 1979, Philippines copyright 1979, by Addison-Wesley Publishing Company, Inc. All rights reserved.

ANSWERS

31. The population of the world in 1976 was 4 billion. This was a 2% increase over the population 1 year earlier. What was the former population?

31. _____

32. The population of the United States in 1977 was 216 million. This was a 1% increase over the population 1 year earlier. What was the former population?

32. _____

33. The formula

$$R = -0.028t + 20.8$$

can be used to predict the world record in the 200-meter dash t years after 1920. In what year will the record be 19.0 seconds?

33. _____

34. The formula

$$F = \frac{1}{4}N + 40$$

can be used to determine Fahrenheit temperatures F from the number of chirps per minute N of a cricket. Determine the chirps per minute necessary for the temperature to be 80°.

34. _____

35. _____

35. Consumer experts advise us never to pay the sticker price for a car. A rule of thumb is to pay the sticker price minus 20% of the sticker price, plus $200. A car is purchased for $5320 using the rule. What was the sticker price?

36. In Indiana a sales tax of 4% is added on to the price of gasoline as registered on the pump. A driver pulls into a station and asks for $10 worth of regular. The attendant figures the pump should read $9.60, fills the tank with that amount, and charges the driver $10. Something is wrong! Use algebra to correct the error.

36. _____

3.7 FORMULAS

■ A formula is a kind of recipe for doing a certain kind of calculation. Formulas are often given as equations. A familiar formula from electricity is

$$E = IR.$$

This formula tells us that to calculate the voltage E in an electric circuit we multiply the current I by the resistance R. Suppose we know the voltage E and the current I. We wish to calculate the resistance R. We can get R alone on one side, or "solve" the formula for R.

Example 1 Solve for R.

$E = IR$ We want this letter alone.

$\frac{1}{I} \cdot E = \frac{1}{I} \cdot IR$ Multiplying on both sides by $\frac{1}{I}$

$\frac{E}{I} = R$

Multiplying by $1/I$ is the same as dividing by I. The formula now says that we can find R by dividing E by I.

DO EXERCISE 1.

Example 2 Solve for r: $C = 2\pi r$.

This is a formula for the circumference C of a circle of radius r.

$C = 2\pi r$ We want this letter alone.

$\frac{1}{2\pi} \cdot C = \frac{1}{2\pi} \cdot 2\pi r$ Multiplying by $\frac{1}{2\pi}$

$\frac{C}{2\pi} = \frac{2\pi}{2\pi} \cdot r$

$\frac{C}{2\pi} = r$

DO EXERCISE 2.

OBJECTIVE

After finishing Section 3.7, you should be able to:

■ Solve formulas for a specified letter.

1. Solve for I: $E = IR$.

2. Solve for D: $C = \pi D$. (This is a formula for the circumference of a circle of diameter D.)

3. Solve for c: $A = \dfrac{a + b + c + d}{4}$.

Example 3 Solve for a: $A = \dfrac{a + b + c}{3}$.

This is a formula for the average A of three numbers a, b, and c.

$$A = \frac{a + b + c}{3} \qquad \text{We want this letter alone.}$$

$$3A = a + b + c \qquad \text{Multiplying by 3}$$

$$3A - b - c = a$$

DO EXERCISE 3.

Example 4 Solve for C: $Q = \dfrac{100M}{C}$.

This is a formula used in psychology for intelligence quotient Q. M is mental age, and C is chronological, or actual, age.

$$Q = \frac{100M}{C} \qquad \text{We want this letter alone.}$$

$$CQ = 100M \qquad \text{Multiplying by } C$$

$$C = \frac{100M}{Q} \qquad \text{Multiplying by } \frac{1}{Q}$$

DO EXERCISE 4.

4. Solve for I: $A = \dfrac{9R}{I}$.

(This is a formula for computing the earned run average A of a pitcher who has given up R earned runs in I innings of pitching.)

NAME _____   CLASS _____

EXERCISE SET 3.7

● Solve:

1. $A = bh$, for b. (*an area formula*)   2. $A = bh$, for h.

3. $d = rt$, for r. (*a distance formula*)   4. $d = rt$, for t.

5. $I = Prt$, for P. (*an interest formula*)   6. $I = Prt$, for t.

7. $F = ma$, for a. (*a physics formula*)   8. $F = ma$, for m.

9. $P = 2l + 2w$, for w. (*a perimeter formula*)   10. $P = 2l + 2w$, for l.

11. $A = \pi r^2$, for r^2. (*an area formula*)   12. $A = \pi r^2$, for π.

13. $A = \frac{1}{2}bh$, for b. (*an area formula*)   14. $A = \frac{1}{2}bh$, for h.

Copyright © 1979, Philippines copyright 1979, by Addison-Wesley Publishing Company, Inc. All rights reserved.

ANSWERS

15. $E = mc^2$, for m. (*a relativity formula*)   16. $E = mc^2$, for c^2.

15. _____

16. _____

17. $A = \dfrac{a + b + c}{3}$, for b.   18. $A = \dfrac{a + b + c}{3}$, for c.

17. _____

18. _____

19. $v = \dfrac{3k}{t}$, for t.   20. $P = \dfrac{ab}{c}$, for c.

19. _____

20. _____

21. $A = \frac{1}{2}ah + \frac{1}{2}bh$, for b.   22. $A = \frac{1}{2}ah + \frac{1}{2}bh$, for a.

21. _____

22. _____

23. The formula

$$H = \frac{D^2 N}{2.5}$$

is used to find the horsepower H of an N-cylinder engine. Solve for D^2.

24. Solve for N:

$$H = \frac{D^2 N}{2.5}.$$

23. _____

24. _____

25. The area of a sector of a circle is given by

$$A = \frac{\pi r^2 S}{360},$$

where r is the radius and S is the angle measure of the sector. Solve for S.

26. Solve for r^2:

$$A = \frac{\pi r^2 S}{360}.$$

25. _____

26. _____

27. The formula

$$R = -0.0075t + 3.85$$

can be used to estimate the world record in the 1500-meter run t years after 1930. Solve for t.

28. The formula

$$F = \frac{9}{5}C + 32$$

can be used to convert from Celsius, or Centigrade, temperature C to Fahrenheit temperature F. Solve for C.

27. _____

28. _____

POLYNOMIALS

READINESS CHECK: SKILLS FOR CHAPTER 4

1. Multiply: $3(s + t + 8)$.

2. Multiply: $-7(x + 4)$.

3. Collect like terms:
$9x + 2y - 4x - 2y$.

4. Add: $-2 + (-8)$.

5. Simplify: $-(-6)$.

6. Subtract: $7 - (-5)$.

7. Multiply: $(-6)(-5)$.

8. What are the terms of $6x - 5y + 9z$?

9. Rename the additive inverse without parentheses:
$-(4x - 7y + 2)$.

10. Remove parentheses and simplify:
$5y - 8 - (9y - 6)$.

11. Multiply and simplify: $x^7 \cdot x^3$.

OBJECTIVES

After finishing Section 4.1, you should be able to:

◖● Evaluate a polynomial for a given value of a variable.

◖●● Identify the terms of a polynomial.

◖●●● Identify the like terms of a polynomial.

◖◗◖◗ Collect the like terms of a polynomial.

1. Write three polynomials.

Evaluate each polynomial for $x = 3$.

2. $-4x - 7 =$ $-4(3)-7=$
   $-12-9=\boxed{-21}$

3. $-5x^3 + 7x + 10 =$ $-5(3)^3+7(3)+10=$
$-5(27)+21+10=-135+31=\boxed{104}$

Evaluate each polynomial for $x = -4$.

4. $5x + 7 = 5(-4)+7= -20+7=\boxed{-13}$

5. $2x^2 + 5x - 4 = 2(-4)^2+5(-4)-4=$
  $2(16)+(-20)-4=$
  $32+(-24)=$
  $\boxed{8}$

4.1 POLYNOMIALS

Expressions like these are called *polynomials*:

$$3x + 2, \quad 2x^2 + 7x - 2, \quad a^4 - 3a^2 + 0.5a + 7.$$

Each part to be added or subtracted is a number or a number times a variable to some power.

A polynomial can also be just one of the parts. These are also polynomials:

$$4x^2, \quad -7a, \quad \frac{3}{4}y^5, \quad x, \quad 5, \quad -2, \quad 0.$$

DO EXERCISE 1.

◖● EVALUATING POLYNOMIALS

When we replace the variable in a polynomial by a number, the polynomial then represents a number. Finding that number is called *evaluating the polynomial*.

Examples   Evaluate each polynomial for $x = 2$.

1. $3x + 5$: $3 \cdot 2 + 5 = 6 + 5 = 11$

2. $2x^2 + 7x + 3$: $2 \cdot 2^2 + 7 \cdot 2 + 3 = 2 \cdot 4 + 14 + 3 = 8 + 14 + 3 = 25$

DO EXERCISES 2–5.

Example 3   The cost, in cents per mile, of operating an automobile at speed s, in mph, is approximated by the polynomial

$$0.005s^2 - 0.35s + 15.$$

Evaluate the polynomial for $s = 50$ to find the cost of operating an automobile at 50 mph.

$$0.005(50)^2 - 0.35(50) + 15 = 0.005(2500) - 17.5 + 15$$
$$= 12.5 - 17.5 + 15 = 10¢$$

The cost is 10¢ per mile.

DO EXERCISE 6.

●● IDENTIFYING TERMS

Subtractions can be rewritten as additions. We showed this on p. 73. Any polynomial can be rewritten using only additions.

Examples Write each polynomial using only additions.

4. $-5x^2 - x = -5x^2 + (-x)$
5. $4x^5 - 2x^6 - 4x = 4x^5 + (-2x^6) + (-4x)$

DO EXERCISES 7 AND 8.

When a polynomial has only additions, the parts being added are called *terms*.

Example 6 Identify the terms of the polynomial

$$4x^3 + 3x + 12 + 8x^3 + 5x.$$

Terms: $4x^3$, $3x$, 12, $8x^3$, $5x$

If there are subtractions you can think of the subtractions as additions.

Example 7 Identify the terms of the polynomial

$$3t^4 - 5t^6 - 4t + 2.$$

Terms: $3t^4$, $-5t^6$, $-4t$, 2

DO EXERCISES 9 AND 10.

●●● LIKE TERMS

Terms that have the same variable and the same exponent are called *like terms*, or *similar terms*.

Examples Identify the like terms in each polynomial.

8. $4x^3 + 5x - 4x^2 + 2x^3 + x^2$

   Like terms: $4x^3$ and $2x^3$ Same exponent and variable
   Like terms: $-4x^2$ and x^2 Same exponent and variable

9. $6 - 3a^2 + 8 - a - 5a$

   Like terms: 6 and 8
   Like terms: $-a$ and $-5a$

DO EXERCISES 11 AND 12.

6. Evaluate the polynomial in Example 3 for $s = 70$ to find the cost of operating an automobile at 70 mph.

$0.005(70)^2 - 0.35(70) + 15$
$0.005(4900) - 0.35(70) + 15$
$0. \quad 24.50 - 24.50 + 15$
$= \#15 \text{ per mile}$

Write each polynomial using only additions.

7. $-9x^3 - 4x^5$

$-9x^3 + (-4x^5)$

8. $-2x^3 + 3x^7 - 7x$

$-2x^3 + 3x^7 + (-7x)$

Identify the terms of each polynomial.

9. $3x^2 + 6x + \dfrac{1}{2}$

$3x^2, 6x, \frac{1}{2}$

10. $-4y^5 + 7y^2 - 3y - 2$

$-4y^5, 7y^2, -3y, -2$

Identify the like terms in each polynomial.

11. $4x^3 - x^3 + 2$

$4x^3, -x^3$

12. $4t^4 - 9t^3 - 7t^4 + 10t^3$

$4t^4, 7t^3$
$-9t^3, -10t^3$

Collect like terms.

13. $3x^2 + 5x^2$

$(3+5)x^2$

$8x^2$

14. $4x^3 - 2x^3 + 2$

$(4-2)x^3 + 2$

$2x^3 + 2$

15. $\frac{1}{2}x^5 - \frac{3}{4}x^5 + 4x^2 - 2x^2$

$(\frac{2}{4} - \frac{3}{4})x^5 + (4-2)x^2$

$-\frac{1}{4}x^5 + 2x^2$

Collect like terms.

16. $24 - 4x^3 - 24$

17. $5x^3 - 8x^5 + 8x^5$

$5x^3 - (8+8)x^5$

$5x^3 - 0x^5 = 5x^3$

18. $-2x^4 + 16 + 2x^4 + 9 - 3x^5$

$(-2+2)x^4 + 16 + 9 - 3x^5$

$0x^4 + 25 - 3x^5 = 25 - 3x^5$

Collect like terms.

19. $7x - x$

$(7-1)x = 6x$

20. $5x^3 - x^3 + 4$

$(5-1)x^3 + 4$

$4x^3 + 4$

21. $\frac{3}{4}x^3 + 4x^2 - x^3 + 7$

$(\frac{3}{4} - 1)x^3 + 4x^2 + 7$

$-\frac{1}{4}x^3 + 4x^2 + 7$

22. $8x^2 - x^2 + x^3 - 1 - 4x^2$

$(8-1-4)x^2 + x^3 - 1$

$3x^2 + x^3 - 1$

COLLECTING LIKE TERMS

We can often simplify polynomials by *collecting like terms*, or *combining similar terms*. To do this we use the distributive laws.

Examples Collect like terms.

10. $2\boxed{x^3} - 6\boxed{x^3} = (2 - 6)\boxed{x^3}$ Using a distributive law

$= -4x^3$

11. $5x^2 + 4x^4 + 2x^2 - 2x^4 = (5 + 2)x^2 + (4 - 2)x^4$

$= 7x^2 + 2x^4$

DO EXERCISES 13–15.

In collecting like terms we may get zero.

Examples Collect like terms.

12. $5\boxed{x^3} - 5\boxed{x^3} = (5 - 5)\boxed{x^3}$

$= 0x^3$

$= 0$

13. $3x^4 - 3x^4 + 2x^2 = (3 - 3)x^4 + 2x^2$

$= 0x^4 + 2x^2$

$= 2x^2$

DO EXERCISES 16–18.

Sometimes we multiply a term by 1. This does not change the polynomial.

Examples Collect like terms.

14. $5x^2 + x^2 = 5x^2 + \boxed{1}x^2$ Multiplying x^2 by 1

$= (5 + 1)x^2$ Using a distributive law

$= 6x^2$

15. $5x^4 - 6x^3 - x^4 = 5x^4 - 6x^3 - 1x^4$

$= (5 - 1)x^4 - 6x^3$

$= 4x^4 - 6x^3$

DO EXERCISES 19–22.

NAME   CLASS

EXERCISE SET 4.1

■● Evaluate each polynomial for $x = 4$.

1. $-5x + 2 =$
$-5(4)+2=$
$-20+2=$
-18

2. $-3x + 1$
$-3(4)+1=$
$-12+1=$
-11

3. $2x^2 - 5x + 7 =$
$2(4)^2-5(4)+7=$
$32-20+7=19$

4. $3x^2 + x + 7$
$3(4)^2+4+7=$
$3(16)+11 = 48+11= 59$

5. $x^3 - 5x^2 + x$
$(4)^3-5(4)^2+4=$
$64-5(16)+4=$
$64-80+4=12$

6. $7 - x + 3x^2$
$7-4+3(4)^2$
$7-4+3(16)$
$3+48=51$

The daily number of accidents involving a driver of age a is approximated by the polynomial

$$0.4a^2 - 40a + 1039.$$

7. Evaluate the polynomial for $a = 18$ to find the number of daily accidents involving an 18-year-old driver.

8. Evaluate the polynomial for $a = 20$ to find the number of daily accidents involving a 20-year-old driver.

Evaluate each polynomial for $x = -1$.

9. $3x + 5$
$3(-1)+5$
$-3+5=2$

10. $6 - 2x$
$6-2(-1)+$
$6+2=8$

11. $x^2 - 2x + 1$
$-1^2-2\cdot(-1)+1$
$1+2+1=4$

12. $5x - 6 + x^2$
$5\cdot+1)-6+ -1^2$
$-5-6+1$

13. $-3x^3 + 7x^2 - 3x - 2$
$-3(-1)^2+$

14. $-2x^3 - 5x^2 + 4x + 3$

■■ Identify the terms of each polynomial.

15. $2 - 3x + x^2$
$2, -3x, x^2$

16. $2x^2 + 3x - 4$
$2x^2, 3x, -4$

■■■ Identify the like terms in each polynomial.

17. $5x^3 + 6x^2 - 3x^2$
$6x^2, -3x^2$

18. $3x^2 + 4x^3 - 2x^2$
$3x^2, -2x^2$

19. $2x^4 + 5x - 7x - 3x^4$
$(2x^4, -3x^4) (5x, -7x)$

20. $-3t + t^3 - 2t - 5t^3$
$(-3t, -2t)(+t^3, -5t^3)$

■■ Collect like terms.

21. $2x - 5x$
$(2-5)x = -3x$

22. $2x^2 + 8x^2$
$(2+8)x^2 = 10x^2$

23. $x - 9x$
$(1-9)x = -8x$

24. $x - 5x$
$(1-5)x$
$= -4x$

	ANSWERS
1.	-18
2.	-11
3.	19
4.	59
5.	-12
6.	51
7.	$448,6$
8.	399
9.	2
10.	8
11.	4
12.	-10
13.	11
14.	-4
15.	
16.	
17.	
18.	
19.	
20.	
21.	
22.	
23.	
24.	

Copyright © 1979 Philippines copyright 1979 by Addison-Wesley Publishing Company, Inc. All rights reserved

ANSWERS

25. _____

26. _____

27. _____

28. _____

29. _____

30. _____

31. _____

32. _____

33. _____

34. _____

35. _____

36. _____$\frac{3}{10}$_____

25. $5x^3 + 6x^3 + 4$

$(5+6)x3 + 4$

$11X3 + 4$

26. $6x^4 - 2x^4 + 5$

$(6-2)x4 + 5$

$4x^4 + 5$

27. $5x^3 + 6x - 4x^3 - 7x$

$(5 + 4)x3 + (6-7)x$

$x3 + -x$

28. $3a^4 - 2a + 2a + a^4$

$(3+1)a4 - (2+2)a$

$4a4 + (2-2)a$

$4a4 - 4a$

29. $6b^5 + 3b^2 - 2b^5 - 3b^2$

$(6-2)b2 + (3-3)b2$

$4b2$

30. $2x^2 - 6x + 3x + 4x^2$

$(2+4)x2 + (6+3)x$

$6x2 - 9x$

31. $\frac{1}{4}x^5 - 5 + \frac{1}{2}x^5 - 2x$

$(\frac{1}{4} + \frac{1}{2})x5 - 5 - 2x$

$\frac{1}{4} + \frac{2}{4} =$

$\frac{3}{4}x5 - 5 - 2x$

32. $\frac{1}{3}x^3 + 2x - \frac{1}{6}x^3 + 4$

$(\frac{1}{3} - \frac{1}{6})x3 + 2x + 4$

$\frac{2-1}{6}$

$\frac{1}{6}x3 + 2x + 4$

33. $6x^2 + 2x^4 - 2x^2 - x^4 - 4x^2$

$(6-2-4)x2 + (2-1)x4$

$x4$

34. $8x^2 + 2x^3 - 3x^3 - 4x^2 - 4x^2$

$(8-4-4)x2 + (2-3)x3$

$x3$

35. $\frac{1}{4}x^3 - x^2 - \frac{1}{6}x^2 + \frac{3}{8}x^3 + \frac{5}{16}x^3$

$(\frac{1}{4} + \frac{3}{8} + \frac{5}{16})x3 + (-x - \frac{1}{6})x2$

$(\frac{4+6+5}{16})x3$

$\frac{15}{16}x3 + (\frac{6-1}{6}) - \frac{7}{6}x2 \mid \frac{15}{16}x3 + (-\frac{7}{6}x2)$

36. $\frac{1}{5}x^4 + \frac{1}{5} - 2x^2 + \frac{1}{10} - \frac{3}{15}x^4 + 2x^2$

$(\frac{1}{5} - \frac{3}{15})x4 + (\frac{1}{5} + \frac{1}{10}) + (-2+2)x2$

$\frac{3-3}{15}$

$\frac{2+1}{10}$

$+ \frac{3}{10}$

4.2 MORE ON POLYNOMIALS

◨ DESCENDING ORDER

This polynomial is arranged in *descending order:*

$$8x^4 - 2x^3 + 5x^2 - x + 3.$$ *constant*

The term with the largest exponent is first. The term with the next largest exponent is second, and so on. The associative and commutative laws allow us to arrange the terms of a polynomial in descending order.

Examples   Arrange each polynomial in descending order.

1. $4x^5 + 4x^7 + x^2 + 2x^3 = 4x^7 + 4x^5 + 2x^3 + x^2$

2. $3 + 4x^5 - 4x^2 + 5x + 3x^3 = 4x^5 + 3x^3 - 4x^2 + 5x + 3$

We usually arrange polynomials in descending order. The opposite order is called *ascending*.

DO EXERCISES 1–3.

◨◨ COLLECTING LIKE TERMS AND DESCENDING ORDER

Example 3   Collect like terms and then arrange in descending order.

$$2x^2 - 4x^3 + 3 - x^2 - 2x^3 = x^2 - 6x^3 + 3 \qquad \text{Collecting like terms}$$
$$= -6x^3 + x^2 + 3 \qquad \text{Descending order}$$

DO EXERCISES 4 AND 5.

◨◨◨ DEGREES

The *degree* of a term is its exponent.

Example 4   Identify the degree of each term of $8x^4 + 3x + 7$.

The degree of $8x^4$ is 4.
The degree of $3x$ is 1.   Recall that $x = x^1$.
The degree of 7 is 0.   Think of 7 as $7x^0$.

The *degree of a polynomial* is its largest exponent.

Example 5   Identify the degree of $3x^4 - 6x^3 + 7$.

$3x^4 - 6x^3 + 7$ The largest exponent is 4.

The degree of the polynomial is   4 .

DO EXERCISE 6.

OBJECTIVES

After finishing Section 4.2, you should be able to:

◨ Arrange a polynomial in descending order.

◨◨ Collect the like terms of a polynomial and arrange in descending order.

◨◨◨ Identify the degrees of terms of polynomials and degrees of polynomials.

▊▊ Identify the coefficients of the terms of a polynomial.

▊▊▊ Identify the missing terms of a polynomial.

▦ Tell whether a polynomial is a monomial, binomial, trinomial, or none of these.

Arrange each polynomial in descending order.

1. $x + 3x^5 + 4x^3 + 5x^2 + 6x^7 - 2x^4$

$6x^7 + 3x^5 - 2x^4 + 4x^3 + 5x^2 + x$

2. $4x^2 - 3 + 7x^5 + 2x^3 - 5x^4$

$7x^5 - 5x^4 + 2x^3 + 4x^2 - 3$

3. $-14 + 7t^2 - 10t^5 + 14t^7$

$14t^7 - 10t^5 + 7t^2 - 14$

Collect like terms and then arrange in descending order.

4. $3x^2 - 2x + 3 - 5x^2 - 1 - x$

$(3-5)x^2 + (-2-1)x + 3 - 1$

$-2x^2 + (-3x) + 2$

5. $-x + \dfrac{1}{2} + 14x^4 - 7x - 1 - 4x^4$

$(14-4)x^4 + (1-7)x + \dfrac{1}{2} - 1$

$10x^4 + (-6x) + \dfrac{1}{2}$

Identify the degree of each term and the degree of the polynomial.

6. $-6x^4 + 8x^2 - 2x + 9$

$-6x^4 = 4$

$8x^2 = 2$ Degree $= 4$

$2x = 1$

$9 = 0$

Identify the coefficient of each term.

7. $5x^9 + 6x^3 + x^2 - x + 4$

5
6   1
4

Identify the missing terms in each polynomial.

8. $2x^3 + 4x^2 - 2$

$2x^3 + 4x^2 + 0x^1 - 2$

9. $-3x^4$

$-3x^4 + 0x^3 + 0x^2 + 0x^1$

10. $x^3 + 1$

$1x^3 + 0x^2 + 0x^1 + 1$

11. $x^4 - x^2 + 3x + 0.25$

$x^4 + 0x^3 - x^2 + 3x^1 + 0.25$

Tell whether each polynomial is a monomial, binomial, trinomial, or none of these.

12. $5x^4$

TRI

13. $4x^3 - 3x^2 + 4x + 2$

14. $3x^2 + x$

now

15. $3x^2 + 2x - 4$

⠒ COEFFICIENTS

In this polynomial the colored numbers are the *coefficients:*

$$3x^5 - 2x^3 + 5x + 4.$$

Example 6   Identify the coefficient of each term in the polynomial

$$3x^4 - 4x^3 + 7x^2 + x - 8.$$

The coefficient of the first term is $\boxed{3}$.

The coefficient of the second term is $\boxed{-4}$.

The coefficient of the third term is $\boxed{7}$.

The coefficient of the fourth term is $\boxed{1}$.

The coefficient of the fifth term is $\boxed{-8}$.

DO EXERCISE 7.

⠿ MISSING TERMS

If a coefficient is 0, we usually do not write the term. We say that we have a *missing term.*

Example 7   In

$$8x^5 - 2x^3 + 5x^2 + 7x + 8,$$

there is no term with x^4. We say that the x^4 term (or the *fourth-degree term*) is missing.

We could write missing terms with zero coefficients. For example,

$$3x^2 + 9 = 3x^2 + 0x + 9,$$

but it is shorter not to write missing terms.

DO EXERCISES 8–11.

⠿ MONOMIALS, BINOMIALS, AND TRINOMIALS

Polynomials with just one term are called *monomials.* Polynomials with just two terms are called *binomials.* Those with just three terms are called *trinomials.*

Example 8

Monomials	*Binomials*	*Trinomials*
$4x^2$	$2x + 4$	$3x^3 + 4x + 7$
9	$3x^5 + 6x$	$6x^7 - 7x^2 + 4$
$-23x^{19}$	$-9x^7 - 6$	$4x^2 - 6x - \frac{1}{2}$

DO EXERCISES 12–15.

NAME _____   CLASS   ANSWERS

EXERCISE SET 4.2

| • | Arrange each polynomial in descending order.

1. $x^5 + x + 6x^3 + 1 + 2x^2$

$x^5 + 6x^3 + 2x^2 + x + 1$

2. $3 + 2x^2 - 5x^6 - 2x^3 + 3x$

$-5x^6 - 2x^3 + 2x^2 + 3x + 3$

3. $5x^3 + 15x^9 + x - x^2 + 7x^8$

$15x^9 + 7x^8 + 5x^3 - x^2 + x$

4. $9x - 5 + 6x^3 - 5x^4 + x^5$

$x^5 - 5x^4 + 6x^3 + 9x - 5$

5. $8y^3 - 7y^2 + 9y^6 - 5y^8 + y^7$

$-5y^8 + y^7 + 9y^6 + 8y^3 - 7y^2$

6. $p^8 - 4 + p + p^2 - 7p^4$

$p^8 - 7p^4 + p^2 + p - 4$

| • • | Collect like terms and then arrange in descending order.

7. $3x^4 - 5x^6 - 2x^4 + 6x^6$

$(3-2)x^4 + (-5+6)x^6$

$x^4 + x^6 = x^6 + x^4$

8. $-1 + 5x^3 - 3 - 7x^3 + x^4 + 5$

$(5-7)x^3 + (-1-3+5) - x^4$

$x^4 - 2x^3 + 1$

9. $-2x + 4x^3 - 7x + 9x^3 + 8$

$(4+9)x^3 + (-2-7)x + 8$

$13x^3 + (-9x) + 8$

10. $-6x^2 + x - 5x + 7x^2 + 1$

$(-6+7)x^2 + (1-5)x + 1$

$x^2 - 4x + 1$

11. $3x + 3x + 3x - x^2 - 4x^2$

$(3+3+3)x + (-1-4)x^2$

$-5x^2 + 9x$

12. $-2x - 2x - 2x + x^3 - 5x^3$

$(-2-2-2)x + (1-5)x^3$

$-4x^3 + (-6x)$

13. $-x + \dfrac{3}{4} + 15x^4 - x - \dfrac{1}{2} - 3x^4$

$(15-3)x^4 + (-1-1)x + (\frac{3}{4} - \frac{2}{4})$

$12x^4 - 2x + \frac{1}{4}$

14. $2x - \dfrac{5}{6} + 4x^3 + x + \dfrac{1}{3} - 2x$

$4x^3 + (2-2)x + (-\frac{5}{6} + \frac{2}{6})$

$4x^3 + x - \frac{3}{6}$

| • • • | Identify the degree of each term of each polynomial and the degree of the polynomial.

15. $-7x^3 + 6x^2 + 3x + 7$

$-7x^3 = ③$ $7 = 0$
$6x^2 = ②$
$3x = ①$ 3 Degree

16. $5x^4 + x^2 - x + 2$

$4 \quad 2 \quad 1 \quad 0$

4 Degree

17. $x^2 - 3x + x^6 - 9x^4$

$2 \quad 1 \quad 6^{th} \quad 4^{a}$

6 Degree

18. $8x - 3x^2 + 9 - 8x^3$

$1 \quad 2 \quad 0 \quad 3$

3 Degree

1. _____
2. _____
3. _____
4. _____
5. _____
6. _____
7. _____
8. _____
9. _____
10. _____
11. _____
12. _____
13. _____
14. _____
15. _____
16. _____
17. _____
18. _____

Copyright © 1979, Philippines copyright 1979, by Addison-Wesley Publishing Company, Inc. All rights reserved.

ANSWERS

19. _____

20. _____

21. _____

22. _____

23. _____

24. _____

25. _____

26. _____

27. _____

28. _____

29. _____

30. _____

31. _____

32. _____

33. _____

34. _____

35. _____

36. _____

37. _____

38. _____

Identify the coefficient of each term of each polynomial.

19. $-3x + 6$
-3

20. $3x^2 - 5x + 2$
$3 \quad 5$

21. $6x^3 + 7x^2 - 8x - 2$
$6 \quad 7 \quad \cdot 8$

22. $2 + 8x - 3x^2 + 6x^3 - 5x^4$
$8 \quad \cdot 3 \quad 6 \quad -5$

Identify the missing terms in each polynomial.

23. $x^3 - 27$
$0x^2 \; 0x^1$

24. $x^5 + x$
$0x^4 + 0x^3 + 0x^2 + 0x^1$

25. $x^4 - x$
$0x^3 + 0x^2 + 0x^1$

26. $5x^4 - 7x + 2$
$0x^3 + 0x^2$

27. $2x^3 - 5x^2 + x - 3$
0

28. $-6x^3$
$0x^2 + 0x^1$

Tell whether each polynomial is a monomial, binomial, trinomial, or none of these.

29. $x^2 - 10x + 25$
none

30. $-6x^4$
Tri

31. $x^3 - 7x^2 + 2x - 4$
none

32. $x^2 - 9$
mon

33. $4x^2 - 25$
mon

34. $2x^4 - 7x^3 + x^2 + x - 6$
none

35. $40x$
none

36. $4x^2 + 12x + 9$
mon

37. A 4-ft by 4-ft sandbox is placed on a square lawn x ft on a side. Express the area left over as a polynomial.

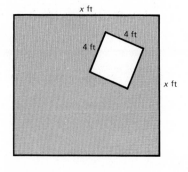

38. Express the colored area in the figure below as a polynomial.

4.3 ADDITION OF POLYNOMIALS

● To *add* polynomials we can write a plus sign between them and collect like terms.

Example 1   Add $-3x^3 + 2x - 4$ and $4x^3 + 3x^2 + 2$.

$$(-3x^3 + 2x - 4) + (4x^3 + 3x^2 + 2)$$

$$= (-3 + 4)x^3 + 3x^2 + 2x + (-4 + 2) \quad \text{Collecting like terms}$$

$$= x^3 + 3x^2 + 2x - 2$$

We still arrange the terms in descending order.

Example 2   Add $\frac{2}{3}x^4 + 3x^2 - 2x + \frac{1}{2}$ and $-\frac{1}{3}x^4 + 5x^3 - 3x^2 + 3x - \frac{1}{2}$.

$$\left(\frac{2}{3}x^4 + 3x^2 - 2x + \frac{1}{2}\right) + \left(-\frac{1}{3}x^4 + 5x^3 - 3x^2 + 3x - \frac{1}{2}\right)$$

$$= \left(\frac{2}{3} - \frac{1}{3}\right)x^4 + 5x^3 + (3 - 3)x^2$$

$$\quad + (-2 + 3)x + \left(\frac{1}{2} - \frac{1}{2}\right) \quad \text{Collecting like terms}$$

$$= \frac{1}{3}x^4 + 5x^3 + x$$

DO EXERCISES 1–4.

After some practice you will be able to add mentally.

Example 3   Add $3x^2 - 2x + 2$ and $5x^3 - 2x^2 + 3x - 4$.

$$(3x^2 - 2x + 2) + (5x^3 - 2x^2 + 3x - 4)$$

$$= 5x^3 + (3 - 2)x^2 + (-2 + 3)x + (2 - 4) \quad \text{You might do this step mentally.}$$

$$= 5x^3 + x^2 + x - 2 \quad \text{Then you would write only this.}$$

DO EXERCISES 5 AND 6.

●● **APPLIED PROBLEMS**

Example 4   Express the sum of the areas of these rectangles as a polynomial.

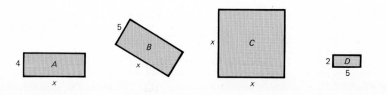

OBJECTIVES

After finishing Section 4.3, you should be able to:

● Add polynomials.

●● Solve applied problems using addition of polynomials.

Add.

1. $3x^2 + 2x - 2$ and $-2x^2 + 5x + 5$

$3x^2 + 2x - 2 \ + (-2x^2 + 5x + 5)$

$(3-2)x^2 + (2+5)x + (-2+5)$

$\qquad x^2 + 7x + 3$

2. $-4x^5 + 3x^3 + 4$ and $7x^4 + 2x^2$

$(-4x^5 + 3x^3 + 4) + 7x^4 + 2x^2$

$-4x^5 + 7x^4 + 3x^3 + 2x^2 + 4$

3. $31x^4 + x^2 + 2x - 1$ and $-7x^4 + 5x^3 - 2x + 2$

$(31x^4 + x^2 + 2x - 1) + (-7x^4 + 5x^3 - 2x + 2)$

$(31-7)x^4 + 5x^3 + x^2 + (2-2)x + 2 - 1$

$24x^4 + 5x^3 + x^2 + 2 - 1$

4. $17x^3 - x^2 + 3x + 4$ and

$\qquad -15x^3 + x^2 - 3x - \frac{2}{3}$

Add mentally. Try to just write the answer.

5. $(4x^2 - 5x + 3) + (-2x^2 + 2x - 4)$

6. $(3x^3 - 4x^2 - 5x + 3) +$

$\qquad \left(5x^3 + 2x^2 - 3x - \frac{1}{2}\right)$

ANSWERS ON PAGE A–9

7. Express the sum of the areas of these rectangles as a polynomial.

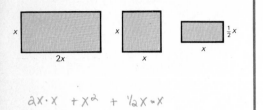

$2x \cdot x + x^2 + \frac{1}{2}x \cdot x$

area of A plus area of B plus area of C plus area of D

$$4x \quad + \quad 5x \quad + \quad x \cdot x \quad + \quad 2 \cdot 5$$

$$= 4x + 5x + x^2 + 10$$

$$= x^2 + 9x + 10$$

DO EXERCISE 7.

SOMETHING EXTRA

APPLICATIONS

The following are some further applications of polynomials.

The volume V of a sphere of radius r:

$$V = \frac{4}{3}\pi r^3.$$

The number N of possible handshakes within a group of n people:

$$N = \frac{1}{2}n^2 - \frac{1}{2}n.$$

The amount A that \$1000 will grow to at interest rate r, compounded annually for 3 years:

$$A = 1000r^3 + 3000r^2 + 3000r + 1000.$$

EXERCISES

Evaluate the polynomial

$$1000r^3 + 3000r^2 + 3000r + 1000$$

to find the amount to which \$1000 will grow at interest rates of

1. 7%   **2.** 8%   **3.** 7.5%   **4.** 7.75%

Use a calculator if you wish.

NAME

CLASS

ANSWERS

EXERCISE SET 4.3

Do odd

● Add.

1. $3x + 2$ and $-4x + 3$

$(3x + 2) + (-4x + 3)$

$(3-4)x + (2+3) = -x + 5$

2. $5x^2 + 6x + 1$ and $-7x + 2$

$5x^2 + (6-7)x + (1+2)$

$5x^2 - x + 3$

3. $-6x + 2$ and $x^2 + x - 3$

$x^2 + (-6+1)x + (2-3)$

$x^2 - 5x - 1)$

4. $6x^4 + 3x^3 - 1$ and $4x^2 - 3x + 3$

$6x^4 + 3x^3 + 4x^2 - 3x' + 2$

5. $3x^5 + 6x^2 - 1$ and $7x^2 + 6x - 2$

$3x^5 + (6+7)x^2 + 6x + (-1-2)$

$5x^5 + 13x^2 + 6x - 3$

6. $7x^3 + 3x^2 + 6x$ and $-3x^2 - 6$

$7x^3 + (3-3)x^2 + 6x - 6$

$7x^3 + 6x - 6$

7. $-4x^4 + 6x^2 - 3x - 5$ and $6x^3 + 5x + 9$

8. $5x^3 + 6x^2 - 3x + 1$ and $5x^4 - 6x^3 + 2x - 5$

9. $(7x^3 + 6x^2 + 4x + 1) + (-7x^3 + 6x^2 - 4x + 5)$

10. $(3x^4 - 5x^2 - 6x + 5) + (-4x^3 + 6x^2 + 7x - 1)$

1. _____

2. _____

3. _____

4. _____

5. _____

6. _____

7. _____

8. _____

9. _____

10. _____

Copyright © 1979, Philippines copyright 1979, by Addison-Wesley Publishing Company, Inc. All rights reserved.

ANSWERS

11. _____

12. _____

13. _____

14. _____

15. _____

16. _____

17. _____

18. _____

19. _____

20. _____

21. _____

11. $5x^4 - 6x^3 - 7x^2 + x - 1$ and $4x^3 - 6x + 1$

12. $8x^5 - 6x^3 + 6x + 5$ and $-4x^4 + 3x^3 - 7x$

13. $9x^8 - 7x^4 + 2x^2 + 5$ and $8x^7 + 4x^4 - 2x$

14. $4x^5 - 6x^3 - 9x + 1$ and $6x^3 + 9x^2 + 9x$

15. $\frac{1}{4}x^4 + \frac{2}{3}x^3 + \frac{5}{8}x^2 + 7$ and $-\frac{3}{4}x^4 + \frac{3}{8}x^2 - 7$

16. $\left(\frac{1}{3}x^9 + \frac{1}{5}x^5 - \frac{1}{2}x^2 + 7\right) + \left(-\frac{1}{5}x^9 + \frac{1}{4}x^4 - \frac{3}{5}x^5 + \frac{3}{4}x^2 + \frac{1}{2}\right)$

17. $0.02x^5 - 0.2x^3 + x + 0.08$ and $-0.01x^5 + x^4 - 0.8x - 0.02$

18. $(0.03x^6 + 0.05x^3 + 0.22x + 0.05) + \left(\frac{7}{100}x^6 - \frac{3}{100}x^3 + 0.5\right)$

■ ■ Solve.

19. Express the sum of the areas of these rectangles as a polynomial.

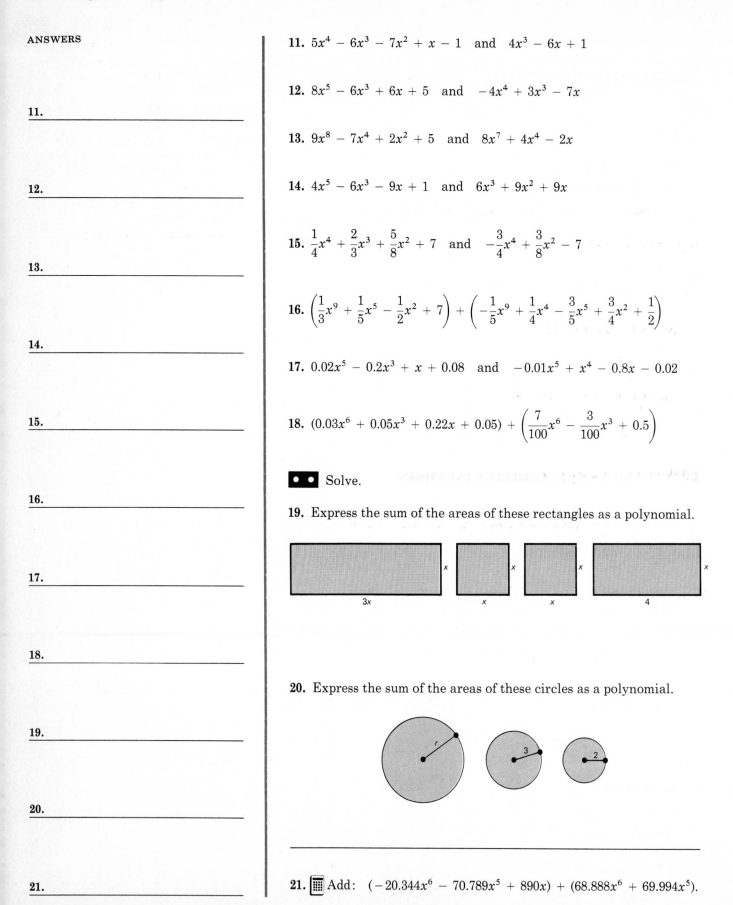

20. Express the sum of the areas of these circles as a polynomial.

21. ▦ Add: $(-20.344x^6 - 70.789x^5 + 890x) + (68.888x^6 + 69.994x^5)$.

4.4 SUBTRACTION OF POLYNOMIALS

● ADDITIVE INVERSES

Look for a pattern:

a) $2x + (-2x) = 0$;

b) $-6x^2 + 6x^2 = 0$;

c) $(5t^3 + 2) + (-5t^3 - 2) = 0$;

d) $(7x^3 - 6x^2 - x + 4) + (-7x^3 + 6x^2 + x - 4) = 0$.

If the sum of two polynomials is 0, they are additive inverses of each other.

The additive inverse of a polynomial is found by replacing each coefficient by its additive inverse.

Example 1 Find the additive inverse of

$$4x^5 - 7x^3 - 8x + \frac{5}{6}.$$

The inverse is $-4x^5 + 7x^3 + 8x - \frac{5}{6}$.

DO EXERCISES 1–4.

● ● SYMBOLS FOR ADDITIVE INVERSES

The additive inverse of the polynomial $8x^2 - 4x + 3$ is the polynomial obtained by replacing each coefficient with its additive inverse. Thus, the inverse is

$$-8x^2 + 4x - 3.$$

We can also represent the additive inverse of $8x^2 - 4x + 3$ as follows:

$$-(8x^2 - 4x + 3).$$

Thus,

$$-(8x^2 - 4x + 3) = -8x^2 + 4x - 3.$$

Example 2 Simplify: $-\left(-7x^4 - \frac{5}{9}x^3 + 8x^2 - x + 67\right)$.

$$-\left(-7x^4 - \frac{5}{9}x^3 + 8x^2 - x + 67\right) = 7x^4 + \frac{5}{9}x^3 - 8x^2 + x - 67$$

DO EXERCISES 5–7.

OBJECTIVES

After finishing Section 4.4, you should be able to:

● Find the additive inverse of a polynomial.

● ● Rename an additive inverse such as $-(3x^2 + x - 2)$, by replacing each coefficient by its inverse.

● ● ● Subtract polynomials.

Find the additive inverse of each polynomial.

1. $12x^4 - 3x^2 + 4x$

2. $-4x^4 + 3x^2 - 4x$

3. $-13x^6 + 2x^4 - 3x^2 + x - \dfrac{5}{13}$

4. $-7y^3 + 2y^2 - y + 3$

Simplify.

5. $-(4x^3 - 6x + 3)$

6. $-(5x^4 + 3x^2 + 7x - 5)$

7. $-\left(14x^{10} - \dfrac{1}{2}x^5 + 5x^3 - x^2 + 3x\right)$

Subtract.

8. $(5x^2 + 4) - (2x^2 + 1)$

9. $(7x^3 + 2x + 4) - (5x^3 - 4)$

10. $(-3x^2 + 5x - 4) - (-4x^2 + 11x - 2)$

Subtract mentally. Try to write just the answer.

11. $(-6x^4 + 3x^2 + 6) - (2x^4 + 5x^3 - 5x^2 + 7)$

12. $(-2x + 5) - (2x^2 + 3)$

13. $\left(\dfrac{3}{2}x^3 - \dfrac{1}{2}x^2 + 0.3\right) - \left(\dfrac{1}{2}x^3 + \dfrac{1}{2}x^2 + \dfrac{4}{3}x + 1.2\right)$

◻◻◻ SUBTRACTION OF POLYNOMIALS

We can subtract a rational number by adding its inverse. This also works for polynomials.

Example 3   Subtract.

$(9x^5 + x^3 - 2x^2 + 4) - (2x^5 + x^4 - 4x^3 - 3x^2)$

$= (9x^5 + x^3 - 2x^2 + 4)$

$\quad + [-(2x^5 + x^4 - 4x^3 - 3x^2)]$ Adding an inverse

$= (9x^5 + x^3 - 2x^2 + 4) + (-2x^5 - x^4 + 4x^3 + 3x^2)$

$= 7x^5 - x^4 + 5x^3 + x^2 + 4$ Collecting like terms

DO EXERCISES 8–10.

After some practice you will be able to subtract mentally.

Example 4   Subtract: $(9x^5 + x^3 - 2x) - (-2x^5 + 5x^3 + 6)$.

$(9x^5 + x^3 - 2x) - (-2x^5 + 5x^3 + 6)$

$= (9x^5 + 2x^5) + (x^3 - 5x^3) - 2x - 6$ Subtract the like terms mentally.

$= 11x^5 - 4x^3 - 2x - 6$ Write only this.

DO EXERCISES 11–13.

NAME   CLASS

EXERCISE SET 4.4   *Do*

▪ Find the additive inverse of each polynomial.

1. $-5x$

2. $x^2 - 3x$

3. $-x^2 + 10x - 2$

4. $-4x^3 - x^2 - x$

5. $12x^4 - 3x^3 + 3$

6. $4x^3 - 6x^2 - 8x + 1$

▪▪ Simplify.

7. $-(3x - 7)$

8. $-(-2x + 4)$

9. $-(4x^2 - 3x + 2)$

10. $-(-6a^3 + 2a^2 - 9a + 1)$

11. $-\left(-4x^4 - 6x^2 + \dfrac{3}{4}x - 8\right)$

12. $-(-5x^4 + 4x^3 - x^2 + 0.9)$

▪▪▪ Subtract.

13. $(5x^2 + 6) - (3x^2 - 8)$

14. $(7x^3 - 2x^2 + 6) - (7x^2 + 2x - 4)$

15. $(6x^5 - 3x^4 + x + 1) - (8x^5 + 3x^4 - 1)$

16. $\left(\dfrac{1}{2}x^2 - \dfrac{3}{2}x + 2\right) - \left(\dfrac{3}{2}x^2 + \dfrac{1}{2}x - 2\right)$

1. _____

2. _____

3. _____

4. _____

5. _____

6. _____

7. _____

8. _____

9. _____

10. _____

11. _____

12. _____

13. _____

14. _____

15. _____

16. _____

Copyright © 1979, Philippines copyright 1979, by Addison-Wesley Publishing Company, Inc. All rights reserved.

17. $(6x^2 + 2x) - (-3x^2 - 7x + 8)$

17. _____

18. $7x^3 - (-3x^2 - 2x + 1)$

18. _____

19. $\left(\dfrac{5}{8}x^3 - \dfrac{1}{4}x - \dfrac{1}{3}\right) - \left(-\dfrac{1}{8}x^3 + \dfrac{1}{4}x - \dfrac{1}{3}\right)$

19. _____

20. $\left(\dfrac{1}{5}x^3 + 2x^2 - 0.1\right) - \left(-\dfrac{2}{5}x^3 + 2x^2 + 0.01\right)$

20. _____

21. $(0.08x^3 - 0.02x^2 + 0.01x) - (0.02x^3 + 0.03x^2 - 1)$

21. _____

22. $(0.8x^4 + 0.2x - 1) - \left(\dfrac{7}{10}x^4 + \dfrac{1}{5}x - 0.1\right)$

22. _____

4.5   ADDING AND SUBTRACTING IN COLUMNS (Optional)

● ADDING IN COLUMNS

We can also add polynomials by writing like terms in columns.

Example 1   Add: $9x^5 - 2x^3 + 6x^2 + 3$ and $5x^4 - 7x^2 + 6$ and $3x^6 - 5x^5 + x^2 + 5$.

Arrange the polynomials with like terms in columns.

$$
\begin{array}{l}
9x^5 \qquad\quad -\ 2x^3 + 6x^2 + 3 \\
\qquad\quad 5x^4 \qquad\qquad -\ 7x^2 + 6 \\
\underline{3x^6 - 5x^5 \qquad\qquad\qquad +\ x^2 + 5} \\
3x^6 + 4x^5 + 5x^4 - 2x^3 \qquad\quad +\ 14
\end{array}
$$

We leave spaces for missing terms.

DO EXERCISES 1 AND 2.

●● SUBTRACTING IN COLUMNS

We can use columns to subtract. We replace coefficients by their inverses, as shown in Example 2, Section 4.4. You can also do it mentally.

Example 2   Subtract: $(5x^2 - 3x + 6) - (9x^2 - 5x - 3)$.

a) $\begin{array}{l} 5x^2 - 3x + 6 \\ \underline{9x^2 - 5x - 3} \end{array}$ Writing similar terms in columns

b) $\begin{array}{l} 5x^2 - 3x + 6 \\ \underline{-9x^2 \pm 5x \pm 3} \end{array}$ Changing signs

c) $\begin{array}{l} 5x^2 - 3x + 6 \\ \underline{-9x^2 + 5x + 3} \\ -4x^2 + 2x + 9 \end{array}$ Adding

If you can, you should skip step (b). Just write the answer.

DO EXERCISES 3 AND 4.

OBJECTIVES

After finishing Section 4.5, you should be able to:

● Add using columns.

●● Subtract using columns.

1. Add

$$
\begin{array}{l}
-2x^3 + 5x^2 - 2x + 4 \\
x^4 + 6x^2 + 7x - 10 \\
\underline{-9x^4 + 6x^3 + x^2 - 2}
\end{array}
$$

2. Add $-3x^3 + 5x + 2$ and $x^3 + x^2 + 5$ and $x^3 - 2x - 4$.

Subtract the second polynomial from the first. Use columns.

3. $4x^3 + 2x^2 - 2x - 3,\quad 2x^3 - 3x^2 + 2$

4. $2x^3 + x^2 - 6x + 2,$
    $x^5 + 4x^3 - 2x^2 - 4x$

ANSWERS ON PAGE A–9

NAME   CLASS   ANSWERS

EXERCISE SET 4.5

●   Add.

1. $\begin{aligned} -3x^4 + 6x^2 + 2x - 1 \\ -3x^2 + 2x + 1 \end{aligned}$

2. $\begin{aligned} -4x^3 + 8x^2 + 3x - 2 \\ -4x^2 + 3x + 2 \end{aligned}$

3. $\begin{aligned} 3x^5 \quad\; - 6x^3 \quad\;\; + 3x \\ -3x^4 + 3x^3 + x^2 \end{aligned}$

4. $\begin{aligned} 4x^5 \quad\; - 5x^3 \quad\;\; + 2x \\ -4x^4 + 2x^3 + 2x^2 \end{aligned}$

5. $\begin{aligned} -3x^2 + x \\ 5x^3 - 6x^2 \quad\;\; + 1 \\ 3x - 8 \end{aligned}$

6. $\begin{aligned} -4x^2 + 2x \\ 3x^3 - 5x^2 \quad\;\; + 3 \\ 5x - 5 \end{aligned}$

7. $\begin{aligned} -\tfrac{1}{2}x^4 - \tfrac{3}{4}x^3 \quad\quad + 6x \\ \tfrac{1}{2}x^3 + x^2 + \tfrac{1}{4}x \\ \tfrac{3}{4}x^4 \quad\quad + \tfrac{1}{2}x^2 + \tfrac{1}{2}x + \tfrac{1}{4} \end{aligned}$

8. $\begin{aligned} -\tfrac{1}{4}x^4 - \tfrac{1}{2}x^3 \quad\quad + 2x \\ \tfrac{3}{4}x^3 - x^2 + \tfrac{1}{2}x \\ \tfrac{1}{2}x^4 \quad\quad + \tfrac{1}{2}x^2 + \tfrac{1}{2}x + \tfrac{1}{2} \end{aligned}$

9. $\begin{aligned} -4x^2 \\ 4x^4 - 3x^3 + 6x^2 + 5x \\ 6x^3 - 8x^2 \quad\quad + 1 \\ -5x^4 \\ 6x^2 - 3x \end{aligned}$

10. $\begin{aligned} 3x^2 \\ 5x^4 - 2x^3 + 4x^2 + 5x \\ 5x^3 - 5x^2 \quad\quad + 2 \\ -7x^4 \\ 3x^2 - 2x \end{aligned}$

11. $\begin{aligned} 3x^4 - 6x^2 + 7x \\ 3x^2 - 3x + 1 \\ -2x^4 + 7x^2 + 3x \\ 5x - 2 \end{aligned}$

12. $\begin{aligned} 5x^4 - 8x^2 + 4x \\ 5x^2 - 2x + 3 \\ -3x^4 + 3x^2 + 5x \\ 3x - 5 \end{aligned}$

1. _____

2. _____

3. _____

4. _____

5. _____

6. _____

7. _____

8. _____

9. _____

10. _____

11. _____

12. _____

Copyright © 1979, Philippines copyright 1979, by Addison-Wesley Publishing Company, Inc. All rights reserved.

13. _____

14. _____

15. _____

16. _____

17. _____

18. _____

19. _____

20. _____

13.
$$
\begin{array}{l}
3x^5 - 6x^4 + 3x^3 \qquad\qquad - 1 \\
\qquad\quad 6x^4 - 4x^3 + 6x^2 \\
3x^5 \qquad\qquad + 2x^3 \\
\qquad\quad - 6x^4 \qquad\quad - 7x^2 \\
-5x^5 \qquad\qquad + 3x^3 \qquad\quad + 2 \\
\hline
\end{array}
$$

14.
$$
\begin{array}{l}
4x^5 - 3x^4 + 2x^3 \qquad\qquad - 2 \\
\qquad\quad 6x^4 + 5x^3 + 3x^2 \\
5x^5 \qquad\qquad + 4x^3 \\
\qquad\quad - 6x^4 \qquad\quad - 5x^2 \\
-3x^5 \qquad\qquad + 2x^3 \qquad\quad + 5 \\
\hline
\end{array}
$$

15.
$$
\begin{array}{l}
\qquad - x^3 + 6x^2 + 3x + 5 \\
x^4 \qquad\quad - 3x^2 \qquad\quad + 2 \\
\qquad\qquad\qquad\quad - 5x + 3 \\
6x^4 \qquad\quad + 4x^2 \qquad\quad - 1 \\
\qquad - x^3 \qquad\qquad + 6x \\
\hline
\end{array}
$$

16.
$$
\begin{array}{l}
\qquad - 2x^3 + 3x^2 + 5x + 3 \\
x^4 \qquad\qquad - 5x^2 \qquad\quad + 1 \\
\qquad\qquad\qquad\quad - 7x + 4 \\
4x^4 \qquad\qquad + 6x^2 \qquad\quad - 2 \\
\qquad - x^3 \qquad\qquad + 5x \\
\hline
\end{array}
$$

17.
$$
\begin{array}{l}
\qquad - 3x^4 + 6x^3 - 6x^2 + 5x + 1 \\
5x^5 \qquad\qquad - 3x^3 \qquad\quad - 5x \\
\qquad\quad 4x^4 + 7x^3 \qquad\qquad + 3x + 1 \\
-2x^5 \qquad\qquad\qquad + 7x^2 \qquad - 8 \\
\hline
\end{array}
$$

18.
$$
\begin{array}{l}
\qquad - 5x^4 + 4x^3 - 7x^2 + 3x + 2 \\
3x^5 \qquad\qquad - 7x^3 \qquad\quad - 6x \\
\qquad\quad 3x^4 + 5x^3 \qquad\qquad + 5x - 3 \\
-5x^5 \qquad\qquad\qquad + 10x^2 \qquad + 4 \\
\hline
\end{array}
$$

19.
$$
\begin{array}{l}
0.15x^4 + 0.10x^3 - 0.9x^2 \\
\qquad\quad - 0.01x^3 + 0.01x^2 + x \\
1.25x^4 \qquad\qquad + 0.11x^2 \qquad + 0.01 \\
\qquad\quad 0.27x^3 \qquad\qquad\qquad + 0.99 \\
-0.35x^4 \qquad\qquad + 15x^2 \qquad - 0.03 \\
\hline
\end{array}
$$

20.
$$
\begin{array}{l}
0.05x^4 + 0.12x^3 - 0.5x^2 \\
\qquad\quad - 0.02x^3 + 0.02x^2 + 2x \\
1.5x^4 \qquad\qquad + 0.01x^2 \qquad + 0.15 \\
\qquad\quad 0.25x^3 \qquad\qquad\qquad + 0.85 \\
-0.25x^4 \qquad\qquad + 10x^2 \qquad - 0.04 \\
\hline
\end{array}
$$

21. Add $6x^4 - 3x^2 + 9x - 2$ and $7x^4 + 2x^2 - 3$
and $-3x^3 + 6x + 1$.

22. Add $5x^4 - 2x^2 + 3x - 3$ and $8x^4 + 4x^2 - 5$
and $-5x^3 + 7x + 2$.

23. Add $3x^3 - 6x^2 + 2x - 1$ and $6x^2 - 3x + 1$ and $-2x + 1$
and $4x^4 - 7x^2 + 5$.

24. Add $5x^3 - 3x^2 + 4x - 2$ and $8x^2 - 4x + 1$ and $-3x + 4$
and $6x^4 - 3x^2 + 8$.

25. Add $-4x^5 + 6x^3 - 2x^2 + x$ and $3x^5 - 6x^4 - 2x^2 + \dfrac{1}{2}$

and $2x^2 - \dfrac{3}{2}x - \dfrac{3}{4}$.

ANSWERS

21. _____

22. _____

23. _____

24. _____

25. _____

Copyright © 1979, Philippines copyright 1979, by Addison-Wesley Publishing Company, Inc. All rights reserved.

ANSWERS

●● Subtract the second polynomial from the first.

26. $3x^2 - 6x + 1$ $\qquad$ $6x^2 + 8x - 3$

26. _____

27. $5x^4 + 6x^3 - 9x^2 + 1$ $\qquad$ $-6x^3 + 8x + 9$

27. _____

28. $5x^5 + 6x^2 - 3x + 6$ $\qquad$ $6x^3 + 7x^2 - 8x - 9$

28. _____

29. $3x^4 + 6x^2 + 8x - 1$ $\qquad$ $4x^5 - 6x^4 - 8x - 7$

29. _____

30. $6x^5 + 3x^2 - 7x + 2$ $\qquad$ $10x^5 + 6x^3 - 5x^2 - 2x + 4$

30. _____

Subtract.

31. _____

31. $\boxed{\text{▦}}$ $(345.099x^3 - 6.178x) - (-224.508x^3 + 8.99x)$

4.6 MULTIPLICATION OF POLYNOMIALS

⬤ MULTIPLYING MONOMIALS

To multiply two monomials, we multiply the coefficients and then use properties of exponents. We use parentheses to show multiplication. $(3x)(4x)$ means $(3x) \cdot (4x)$.

Examples Multiply.

1. $(\boxed{3}\,x)(\boxed{4}\,x) = (\boxed{3 \cdot 4})(x \cdot x)$ Multiplying the coefficients

            $= 12x^2$ Simplifying

2. $(3x)(-x) = (3x)(-1x)$

            $= (3)(-1)(x \cdot x)$

            $= -3x^2$

3. $(-7x^5)(4x^3) = (-7 \cdot 4)(x^5 \cdot x^3)$

            $= -28x^{5+3}$

            $= -28x^8$ Adding exponents and simplifying

After some practice you can do this mentally. Multiply the coefficients and add the exponents. Write only the answer.

DO EXERCISES 1–8.

⬤⬤ MULTIPLYING A MONOMIAL AND A BINOMIAL

Multiplications are based on the distributive laws.

Example 4 Multiply: $2x$ and $5x + 3$.

    $(\boxed{2x})(5x + 3) = (\boxed{2x})(5x) + (\boxed{2x})(3)$ Using a distributive law

            $= 10x^2 + 6x$ Multiplying the monomials

DO EXERCISES 9 AND 10.

⬤⬤⬤ MULTIPLYING TWO BINOMIALS

To multiply two binomials, we use the distributive laws more than once.

Example 5 Multiply: $x + 5$ and $x + 4$.

    $(\boxed{x + 5})(x + 4) = (\boxed{x + 5})x + (\boxed{x + 5})4$ Using a distributive law

                (a)   (b)

We now use a distributive law with parts (a) and (b).

OBJECTIVES

After finishing Section 4.6, you should be able to:

⬤ Multiply two monomials.

⬤⬤ Multiply a monomial and a binomial.

⬤⬤⬤ Multiply two binomials.

⬤⬤ Multiply a binomial and a trinomial.

⬤⬤⬤ Multiply any polynomials.

Multiply.

1. $3x$ and -5

2. $-x$ and x

3. $-x$ and $-x$

4. $-x^2$ and x^3

5. $3x^5$ and $4x^2$

6. $4x^5$ and $-2x^6$

7. $-7y^4$ and $-y$

8. $7x^5$ and 0

Multiply.

9. $4x$ and $2x + 4$

10. $3x^2$ and $-5x + 2$

Multiply.

11. $x + 8$ and $x + 5$

12. $(x + 5)(x - 4)$

Multiply.

13. $5x + 3$ and $x - 4$

14. $(2x - 3)(3x - 5)$

a) $(x + 5)\,x = x \cdot x + 5 \cdot x$ Distributive law

   $= x^2 + 5x$ Multiplying the monomials

b) $(x + 5)\,4 = x \cdot 4 + 5 \cdot 4$ Distributive law

   $= 4x + 20$ Multiplying the monomials

Now we replace parts (a) and (b) in the original expression with their answers and collect like terms:

$$(x + 5)(x + 4) = (x^2 + 5x) + (4x + 20)$$
$$= x^2 + 9x + 20.$$

DO EXERCISES 11 AND 12.

Example 6   Multiply: $4x + 3$ and $x - 2$.

$$(\,4x + 3\,)(x - 2) = (\,4x + 3\,)x + (\,4x + 3\,)(-2)$$
$$\qquad\qquad\qquad\quad (a)\qquad\qquad (b)$$

Now consider parts (a) and (b).

a) $(4x + 3)\,x = (4x)\,x + (3)\,x$

   $= 4x^2 + 3x$

b) $(4x + 3)(\,-2\,) = (4x)(\,-2\,) + (3)(\,-2\,)$

   $= -8x - 6$

Now we replace parts (a) and (b) in the original expression and collect like terms:

$$(4x + 3)(x - 2) = (4x^2 + 3x) + (-8x - 6)$$
$$= 4x^2 - 5x - 6.$$

DO EXERCISES 13 AND 14.

▪▪ MULTIPLYING A BINOMIAL AND A TRINOMIAL

Example 7   Multiply: $(x^2 + 2x - 3)(x^2 + 4)$.

$$(\,x^2 + 2x - 3\,)(x^2 + 4) = (\,x^2 + 2x - 3\,)(x^2) + (\,x^2 + 2x - 3\,)(4)$$
$$\qquad\qquad\qquad\qquad\qquad (a)\qquad\qquad\qquad\quad (b)$$

Consider parts (a) and (b).

a) $(x^2 + 2x - 3)\,x^2 = (x^2)(\,x^2\,) + 2x(\,x^2\,) - 3(\,x^2\,)$

   $= x^4 + 2x^3 - 3x^2$

b) $(x^2 + 2x - 3)\,4 = (x^2)\,4 + (2x)\,4 + (-3)\,4$

   $= 4x^2 + 8x - 12$

Now we replace parts (a) and (b) in the original expression and collect like terms:

$$(x^2 + 2x - 3)(x^2 + 4) = (x^4 + 2x^3 - 3x^2) + (4x^2 + 8x - 12)$$
$$= x^4 + 2x^3 + x^2 + 8x - 12.$$

DO EXERCISES 15 AND 16.

⠿ MULTIPLYING ANY POLYNOMIALS

Perhaps you have discovered the following.

> To multiply two polynomials, multiply each term of one by every term of the other. Then add the results.

We may use columns for long multiplications. We multiply each term at the top by every term at the bottom. Then we add.

Example 8 Multiply: $(4x^2 - 2x + 3)(x + 2)$.

$$
\begin{array}{l}
4x^2 - 2x\ + 3 \\
\underline{\quad x\ + 2} \\
4x^3 - 2x^2 + 3x \qquad \text{Multiplying the top row by } x \\
\underline{\qquad\quad 8x^2 - 4x + 6} \qquad \text{Multiplying the top row by 2} \\
4x^3 + 6x^2 -\ \ x + 6 \qquad \text{Adding}
\end{array}
$$

Example 9 Multiply: $(5x^3 - 3x + 4)(-2x^2 - 3)$.

$$
\begin{array}{l}
\quad 5x^3 - 3x + 4 \\
\underline{-2x^2 - 3} \\
-10x^5 \qquad\ + 6x^3 - 8x^2 \qquad \text{Multiplying by } -2x^2 \\
\underline{\qquad\qquad - 15x^3 \qquad\quad + 9x - 12} \qquad \text{Multiplying by } -3 \\
-10x^5 \qquad\quad - 9x^3 - 8x^2 + 9x - 12 \qquad \text{Adding}
\end{array}
$$

We left space for "missing terms." This helps in adding.

DO EXERCISES 17 AND 18.

Example 10 Multiply: $(2x^2 + 3x - 4)(2x^2 - x + 3)$.

$$
\begin{array}{l}
2x^2 + 3x\ - 4 \\
\underline{2x^2 -\ \ x\ + 3} \\
4x^4 + 6x^3 - 8x^2 \qquad\qquad\quad \text{Multiplying by } 2x^2 \\
\quad\ - 2x^3 - 3x^2 +\ 4x \qquad\ \text{Multiplying by } -x \\
\underline{\qquad\qquad\quad 6x^2 +\ 9x - 12} \qquad \text{Multiplying by 3} \\
4x^4 + 4x^3 - 5x^2 + 13x - 12 \qquad \text{Adding}
\end{array}
$$

DO EXERCISE 19.

Multiply.

15. $(x^2 + 3x - 4)(x^2 + 5)$

16. $(2x^3 - 2x + 5)(3x^2 - 7)$

Multiply.

17. $3x^2 - 2x + 4$
   $\underline{\quad x + 5\qquad}$

18. $-5x^2 + 4x + 2$
   $\underline{\ -4x^2 - 8\qquad}$

Multiply.

19. $3x^2 - 2x - 5$
   $\underline{\ 2x^2 +\ x - 2}$

SOMETHING EXTRA
CALCULATOR CORNER

1. *A number pattern.* Find each of the following using your calculator. Look for a pattern.

   1^3

   $1^3 + 2^3$

   $1^3 + 2^3 + 3^3$

   $1^3 + 2^3 + 3^3 + 4^3$

   Use the pattern to find each of the following without using your calculator.

   $1^3 + 2^3 + 3^3 + 4^3 + 5^3$

   $1^3 + 2^3 + 3^3 + 4^3 + 5^3 + 6^3$

2. Complete this table. Find decimal notation, rounding to four decimal places.

n	2^n	2^{-n}
1		
2		
3		
4		
5		
6		
7		

NAME

CLASS

Do Problems

EXERCISE SET 4.6

• Multiply.

1. $3x$ and -4

$-12X$

2. $4x$ and 5

$20X$

3. $6x^2$ and 7

$42x^2$

4. $-4x$ and -3

$12X$

5. $-5x$ and -6

6. $3x^2$ and -7

7. x^2 and $-2x$

8. $-x^3$ and $-x$

9. x^4 and x^2

10. $-x^5$ and x^3

11. $3x^4$ and $2x^2$

12. $-\dfrac{1}{5}x^3$ and $-\dfrac{1}{3}x$

13. $-4x^4$ and 0

14. $7x^5$ and x^5

15. $-0.1x^6$ and $0.2x^4$

• • Multiply.

16. $3x$ and $-x + 5$

17. $2x$ and $4x - 6$

18. $4x^2$ and $3x + 6$

19. $-6x^2$ and $x^2 + x$

20. $3x^2$ and $6x^4 + 8x^3$

21. $4x^4$ and $x^2 - 6x$

• • • Multiply.

22. $(x + 6)(x + 3)$

23. $(x + 6)(-x + 2)$

24. $(x + 3)(x - 3)$

ANSWERS

1. _____

2. _____

3. _____

4. _____

5. _____

6. _____

7. _____

8. _____

9. _____

10. _____

11. _____

12. _____

13. _____

14. _____

15. _____

16. _____

17. _____

18. _____

19. _____

20. _____

21. _____

22. _____

23. _____

24. _____

Copyright © 1979, Philippines copyright 1979, by Addison-Wesley Publishing Company, Inc. All rights reserved.

ANSWERS

25. _____

26. _____

27. _____

28. _____

29. _____

30. _____

31. _____

32. _____

33. _____

34. _____

35. _____

36. _____

37. _____

38. _____

39. _____

40. _____

25. $(x - 5)(2x - 5)$

26. $(2x + 5)(2x + 5)$

27. $(3x - 5)(3x + 5)$

28. $(3x + 1)(3x + 1)$

29. $\left(2x - \dfrac{1}{2}\right)\left(x + \dfrac{3}{2}\right)$

30. $(2x + 0.1)(3x - 0.1)$

Multiply.

31. $(x^2 + 6x + 1)(x' + 1)$

$x^3 + x^2$
$\quad 6x^2 + 6x$
$\qquad 1x + 1$
$\overline{x^3 + 7x^2 + 7x + 1}$

32. $(2x^3 + 6x' + 1)(2x' + 1)$

$4x^4 + 2x^3$
$\quad + 12x^2$
$\qquad + 6x'$
$\qquad + 2x'$
$\overline{4x^4 + 2x^3 + 12x^2 + 8x + 1}$

33. $(-5x^2 - 7x + 3)(2x + 1)$

34. $(2x^3 - 5x + 6)(2x + 2)$

Multiply.

35. $(3x^2 - 6x + 2)(x^2 - 3)$

36. $(x^2 + 6x - 1)(-3x^2 + 2)$

37. $(2t^2 - t - 4)(3t^2 + 2t - 1)$

38. $(3a^2 - 5a + 2)(2a^2 - 3a + 4)$

39. $(x^3 + x^2 + x + 1)(x - 1)$

40. $(x + 2)(x^2 - 2x + 4)$

4.7   SPECIAL PRODUCTS OF POLYNOMIALS

● ● PRODUCT OF A MONOMIAL AND ANY POLYNOMIAL

There is a quick way to multiply a monomial and any polynomial. Use the distributive law mentally, multiplying every term inside by the monomial. Just write the answer.

Example 1   Multiply: $5x(2x^2 - 3x + 4)$.

$$5x(2x^2 - 3x + 4) = 10x^3 - 15x^2 + 20x$$

DO EXERCISES 1 AND 2.

● ● PRODUCTS OF TWO BINOMIALS

To multiply two binomials, we multiply each term of one by every term of the other. We can do it like this.

$$(a + b)(c + d) = a \cdot c + a \cdot d + b \cdot c + b \cdot d$$

1. Multiply First terms: $a \cdot c$
2. Multiply Outside terms: $a \cdot d$
3. Multiply Inside terms: $b \cdot c$
4. Multiply Last terms: $b \cdot d$
   ↓
   FOIL   This will help you remember the rule.

Example 2   Multiply.

$$\begin{matrix} & \text{F} & \text{O} & \text{I} & \text{L} \\ (x + 8)(x^2 + 5) = & x^3 & + 5x & + 8x^2 & + 40 \end{matrix}$$

Often we can collect like terms after we multiply.

Examples   Multiply.

3. $(x + 6)(x - 6) = x^2 - 6x + 6x - 36$ Using FOIL
   $= x^2 - 36$ Collecting like terms
4. $(x + 3)(x - 2) = x^2 - 2x + 3x - 6$
   $= x^2 + x - 6$
5. $(x^3 + 5)(x^3 - 5) = x^6 - 5x^3 + 5x^3 - 25$
   $= x^6 - 25$
6. $(x^3 + 5)(x^3 - 2) = x^6 - 2x^3 + 5x^3 - 10$
   $= x^6 + 3x^3 - 10$

DO EXERCISES 3–9.

OBJECTIVES

After finishing Section 4.7, you should be able to:

● ● Multiply a monomial and a polynomial mentally.

● ● Multiply two binomials mentally.

Multiply. Just write the answer.

1. $4x(2x^2 - 3x + 4)$

2. $2y^3(5y^3 + 4y^2 - 5y)$

Multiply mentally. Just write the answer.

3. $(x + 3)(x + 4)$

4. $(x + 3)(x - 5)$

5. $(2x + 1)(x + 4)$

6. $(2x^2 - 3)(x - 2)$

7. $(6x^2 + 5)(2x^3 + 1)$

8. $(y^3 + 7)(y^3 - 7)$

9. $(2x^4 + x^2)(-x^3 + x)$

NAME _____   CLASS _____

ANSWERS

EXERCISE SET 4.7

⬤ Multiply. Just write the answer.

1. $4x(x + 1)$ **2.** $3x(x + 2)$ **3.** $-3x(x - 1)$

4. $-5x(-x - 1)$ **5.** $x^2(x^3 + 1)$ **6.** $-2x^3(x^2 - 1)$

7. $3x(2x^2 - 6x + 1)$ **8.** $-4x(2x^3 - 6x^2 - 5x + 1)$

⬤⬤ Multiply. Just write the answer.

9. $(x + 1)(x^2 + 3)$

$x^3 + 3x + 1x^2 + 3$

10. $(x^2 - 3)(x - 1)$

11. $(x^3 + 2)(x + 1)$ **12.** $(x^4 + 2)(x + 12)$

13. $(x + 2)(x - 3)$

$x^2 - x - 6$

14. $(x + 2)(x + 2)$

$x^2 + 4x + 4$

15. $(3x + 2)(3x + 3)$

$9x^2 + 15x + 6$

16. $(4x + 1)(2x + 2)$

$8x^2 + 10x + 2$

17. $(5x - 6)(x + 2)$

$5x^2 + 4x - 12$

18. $(x - 8)(x + 8)$

$x^2 - 64$

19. $(3x - 1)(3x + 1)$

$9x^2 - 1$

20. $(2x + 3)(2x + 3)$

$4x^2 + 12x + 9$

21. $(4x - 2)(x - 1)$

$4x^2 - 6x + 2$

22. $(2x - 1)(3x + 1)$

$6x^2 - 1x - 1$

23. $\left(x - \dfrac{1}{4}\right)\left(x + \dfrac{1}{4}\right)$

$1x^2 - \dfrac{1}{16}$

24. $\left(x + \dfrac{3}{4}\right)\left(x + \dfrac{3}{4}\right)$

$1x^2 + \dfrac{6}{8}x + \dfrac{9}{16}$

1.	
2.	
3.	
4.	
5.	
6.	
7.	
8.	
9.	
10.	
11.	
12.	
13.	
14.	
15.	
16.	
17.	
18.	
19.	
20.	
21.	
22.	
23.	
24.	

Copyright © 1979, Philippines copyright 1979, by Addison-Wesley Publishing Company, Inc. All rights reserved.

ANSWERS

25. _____

26. _____

27. _____

28. _____

29. _____

30. _____

31. _____

32. _____

33. _____

34. _____

35. _____

36. _____

37. _____

38. _____

39. _____

40. _____

41. _____

42. _____

43. _____

44. _____

45. _____

46. _____

47. _____

48. _____

25. $(x - 0.1)(x + 0.1)$ **26.** $(3x^2 + 1)(x + 1)$

27. $(2x^2 + 6)(x + 1)$ **28.** $(2x^2 + 3)(2x - 1)$

29. $(-2x + 1)(x + 6)$ **30.** $(3x + 4)(2x - 4)$

31. $(x + 7)(x + 7)$ **32.** $(2x + 5)(2x + 5)$

$x^2 + 14x + 49$ $4x^2 + 20 + 25$

33. $(1 + 2x)(1 - 3x)$ **34.** $(-3x - 2)(x + 1)$

35. $(x^2 + 3)(x^3 - 1)$ **36.** $(x^4 - 3)(2x + 1)$

37. $(x^2 - 2)(x - 1)$ **38.** $(x^3 + 2)(x - 3)$

39. $(3x^2 - 2)(x^4 - 2)$ **40.** $(x^{10} + 3)(x^{10} - 3)$

                                        $x^{20} - 9$

41. $(3x^5 + 2)(2x^2 + 6)$ **42.** $(1 - 2x)(1 + 3x^2)$

43. $(8x^3 + 1)(x^3 + 8)$ **44.** $(4 - 2x)(5 - 2x^2)$

45. $(4x^2 + 3)(x - 3)$ **46.** $(7x - 2)(2x - 7)$

47. $(4x^4 + x^2)(x^2 + x)$ **48.** $(5x^6 + 3x^3)(2x^6 + 2x^3)$

4.8   MORE SPECIAL PRODUCTS

● MULTIPLYING SUMS AND DIFFERENCES OF TWO EXPRESSIONS

Look for a pattern.

a) $(x + 2)(x - 2) = x^2 - 2x + 2x - 4$
$$= x^2 - 4$$
b) $(3x - 5)(3x + 5) = 9x^2 + 15x - 15x - 25$
$$= 9x^2 - 25$$

DO EXERCISES 1 AND 2.

Perhaps you discovered the following.

> **The product of the sum and difference of two expressions is the difference of their squares:**
> $$(A + B)(A - B) = A^2 - B^2.$$

Examples   Multiply.

1. $(x + 4)(x - 4) = x^2 - 4^2$ Squaring x, squaring 4, and writing a minus sign between the results

$$= x^2 - 16$$ Simplifying

2. $(2w + 5)(2w - 5) = (2w)^2 - 5^2$
$$= 4w^2 - 25$$

3. $(3x^2 - 7)(3x^2 + 7) = (3x^2)^2 - 7^2$
$$= 9x^4 - 49$$

4. $(-4x - 10)(-4x + 10) = (-4x)^2 - 10^2$
$$= 16x^2 - 100$$

DO EXERCISES 3–6.

●● SQUARING BINOMIALS

In this special product we multiply a binomial by itself. This is also called "squaring a binomial." Look for a pattern.

a) $(x + 3)(x + 3) = x^2 + 3x + 3x + 9 = x^2 + 6x + 9$
b) $(3x + 5)(3x + 5) = 9x^2 + 15x + 15x + 25 = 9x^2 + 30x + 25$

OBJECTIVES

After finishing Section 4.8, you should be able to:

● Multiply the sum and difference of two expressions mentally.
●● Square a binomial mentally.
●●● Find special products, such as those above and those in Section 4.7, mentally, when they are mixed together.

Multiply.

1. $(x + 5)(x - 5)$

2. $(2x - 3)(2x + 3)$

Multiply.

3. $(x + 2)(x - 2)$

4. $(x - 7)(x + 7)$

5. $(3t + 5)(3t - 5)$

6. $(2x^3 - 1)(2x^3 + 1)$

ANSWERS ON PAGE A–10

Multiply.

7. $(x + 8)(x + 8)$

8. $(x - 5)(x - 5)$

Multiply.

9. $(x + 2)^2$

10. $(a - 4)^2$

11. $(2x + 5)^2$

12. $(4x^2 - 3x)^2$

13. $(y + 9)(y + 9)$

14. $(3x^2 - 5)(3x^2 - 5)$

ANSWERS ON PAGE A–10

c) $(x - 3)(x - 3) = x^2 - 3x - 3x + 9 = x^2 - 6x + 9$

d) $(3x - 5)(3x - 5) = 9x^2 - 15x - 15x + 25 = 9x^2 - 30x + 25$

DO EXERCISES 7 AND 8.

Perhaps you discovered a quick way to square a binomial.

> *To square a binomial:* **Square the first term. Multiply the two terms and double. Square the last term. Then add.**
> $$(A + B)^2 = A^2 + 2 \cdot A \cdot B + B^2$$
> $$(A - B)^2 = A^2 - 2 \cdot A \cdot B + B^2$$

Examples   Multiply.

5. $(x + 3)^2 = x^2 + 2 \cdot x \cdot 3 + 3^2$
$$= x^2 + 6x + 9$$

6. $(t - 5)^2 = t^2 - 2 \cdot t \cdot 5 + 5^2$
$$= t^2 - 10t + 25$$

7. $(2x + 7)^2 = (2x)^2 + 2 \cdot 2x \cdot 7 + 7^2$
$$= 4x^2 + 28x + 49$$

8. $(3x^2 - 5x)^2 = (3x^2)^2 - 2 \cdot 3x^2 \cdot 5x + (5x)^2$
$$= 9x^4 - 30x^3 + 25x^2$$

DO EXERCISES 9–14.

●●● MULTIPLICATIONS OF VARIOUS TYPES

Now that we have considered how to multiply quickly certain kinds of polynomials, let us try several kinds mixed together so we can learn to sort them out. When you multiply, first see what kind of multiplication you have. Then use the best method. The methods you have used so far are as follows.

> 1. $(A + B)(A + B) = (A + B)^2 = A^2 + 2 \cdot A \cdot B + B^2$
> 2. $(A - B)(A - B) = (A - B)^2 = A^2 - 2 \cdot A \cdot B + B^2$
> 3. $(A - B)(A + B) = A^2 - B^2$
> 4. **FOIL**
> 5. **The product of a monomial and any polynomial. Multiply each term of the polynomial by the monomial.**

Example 9   Multiply.

$(x + 3)(x - 3) = x^2 - 9$ Using method 3 (the product of the sum and difference of two expressions)

Example 10   Multiply.

$(t + 7)(t - 5) = t^2 + 2t - 35$ Using method 4 (the product of two binomials, but neither the square of a binomial nor the product of the sum and difference of two expressions)

Example 11   Multiply.

$(x + 7)(x + 7) = x^2 + 14x + 49$ Using method 1 (the square of a binomial sum)

Example 12   Multiply.

$2x^3(9x^2 + x - 7) = 18x^5 + 2x^4 - 14x^3$ Using method 5 (the product of a monomial and a trinomial; multiply each term of the trinomial by the monomial)

Example 13   Multiply.

$(3x^2 - 7x)^2 = 9x^4 - 42x^3 + 49x^2$ Using method 2 (the square of a binomial difference)

Example 14   Multiply.

$\left(3x + \dfrac{1}{4}\right)^2 = 9x^2 + \dfrac{3}{2}x + \dfrac{1}{16}$ Using method 1 (the square of a binomial sum. To get the middle term, we multiply $3x$ by $\frac{1}{4}$ and double.)

DO EXERCISES 15–20.

Multiply.

15. $(x + 5)(x + 6)$

16. $(t - 4)(t + 4)$

17. $4x^2(-2x^3 + 5x^2 + 10)$

18. $(9x^2 + 1)^2$

19. $(2a - 5)(2a + 8)$

20. $\left(2x - \dfrac{1}{2}\right)^2$

ANSWERS ON PAGE A–10

SOMETHING EXTRA

POLYNOMIALS AND MULTIPLICATION

We can use special products to multiply.

Example 1   Find 99^2.

$$
\begin{aligned}
99^2 &= (100 - 1)^2 \\
&= 100^2 - 2 \cdot 100 \cdot 1 + 1^2 \qquad \text{Using } (A - B)^2 = A^2 - 2AB + B^2 \\
&= 10{,}000 - 200 + 1 \\
&= 9801
\end{aligned}
$$

Example 2   Find $98 \cdot 102$.

$$
\begin{aligned}
98 \cdot 102 &= (100 - 2)(100 + 2) \\
&= 100^2 - 2^2 \qquad \text{Using } (A - B)(A + B) = A^2 - B^2 \\
&= 10{,}000 - 4 \\
&= 9996
\end{aligned}
$$

EXERCISES

Use a special product to find each of the following. Show your work.

1. 98^2 2. 97^2 3. $97 \cdot 103$ 4. $96 \cdot 104$

5. 101^2 6. 203^2 7. 999^2 8. $198 \cdot 202$

NAME CLASS

EXERCISE SET 4.8

⬤ Multiply mentally.

1. $(x + 4)(x - 4)$ **2.** $(x + 1)(x - 1)$

3. $(2x + 1)(2x - 1)$ **4.** $(x^2 + 1)(x^2 - 1)$

5. $(5m - 2)(5m + 2)$ **6.** $(3x^4 + 2)(3x^4 - 2)$

7. $(2x^2 + 3)(2x^2 - 3)$ **8.** $(6x^5 - 5)(6x^5 + 5)$

9. $(3x^4 - 4)(3x^4 + 4)$ **10.** $(t^2 - 0.2)(t^2 + 0.2)$

11. $(x^6 - x^2)(x^6 + x^2)$ **12.** $(2x^3 - 0.3)(2x^3 + 0.3)$

13. $(x^4 + 3x)(x^4 - 3x)$ **14.** $\left(\dfrac{3}{4} + 2x^3\right)\left(\dfrac{3}{4} - 2x^3\right)$

15. $(x^{12} - 3)(x^{12} + 3)$ **16.** $(12 - 3x^2)(12 + 3x^2)$

17. $(2x^8 + 3)(2x^8 - 3)$ **18.** $\left(x - \dfrac{2}{3}\right)\left(x + \dfrac{2}{3}\right)$

⬤⬤ Multiply mentally.

19. $(x + 2)^2$ **20.** $(2x - 1)^2$ **21.** $(3x^2 + 1)^2$

22. $\left(3x + \dfrac{3}{4}\right)^2$ **23.** $\left(x - \dfrac{1}{2}\right)^2$ **24.** $\left(2x - \dfrac{1}{5}\right)^2$

25. $(3 + x)^2$ **26.** $(x^3 - 1)^2$ **27.** $(x^2 + 1)^2$

1. _____
2. _____
3. _____
4. _____
5. _____
6. _____
7. _____
8. _____
9. _____
10. _____
11. _____
12. _____
13. _____
14. _____
15. _____
16. _____
17. _____
18. _____
19. _____
20. _____
21. _____
22. _____
23. _____
24. _____
25. _____
26. _____
27. _____

Copyright © 1979, Philippines copyright 1979, by Addison-Wesley Publishing Company, Inc. All rights reserved.

ANSWERS

28. _____

29. _____

30. _____

31. _____

32. _____

33. _____

34. _____

35. _____

36. _____

37. _____

38. _____

39. _____

40. _____

41. _____

42. _____

43. _____

44. _____

45. _____

46. _____

47. _____

48. _____

49. _____

28. $(8x - x^2)^2$ 29. $(2 - 3x^4)^2$ 30. $(6x^3 - 2)^2$

31. $(5 + 6x^2)^2$ 32. $(3x^2 - x)^2$

●●● Multiply mentally.

33. $(3 - 2x^3)^2$ 34. $(x - 4x^3)^2$

35. $4x(x^2 + 6x - 3)$ 36. $8x(-x^5 + 6x^2 + 9)$

37. $\left(2x^2 - \dfrac{1}{2}\right)\left(2x^2 - \dfrac{1}{2}\right)$ 38. $(-x^2 + 1)^2$

39. $(-1 + 3p)(1 + 3p)$ 40. $(-3x + 2)(3x + 2)$

41. $3t^2(5t^3 - t^2 + t)$ 42. $-6x^2(x^3 + 8x - 9)$

43. $(6x^4 + 4)^2$ 44. $(8a + 5)^2$

45. $(3x + 2)(4x^2 + 5)$ 46. $(2x^2 - 7)(3x^2 + 9)$

47. $(8 - 6x^4)^2$ 48. $\left(\dfrac{1}{5}x^2 + 9\right)\left(\dfrac{3}{5}x^2 - 7\right)$

_____

49. ▦ Multiply: $(67.58x + 3.225)^2$.

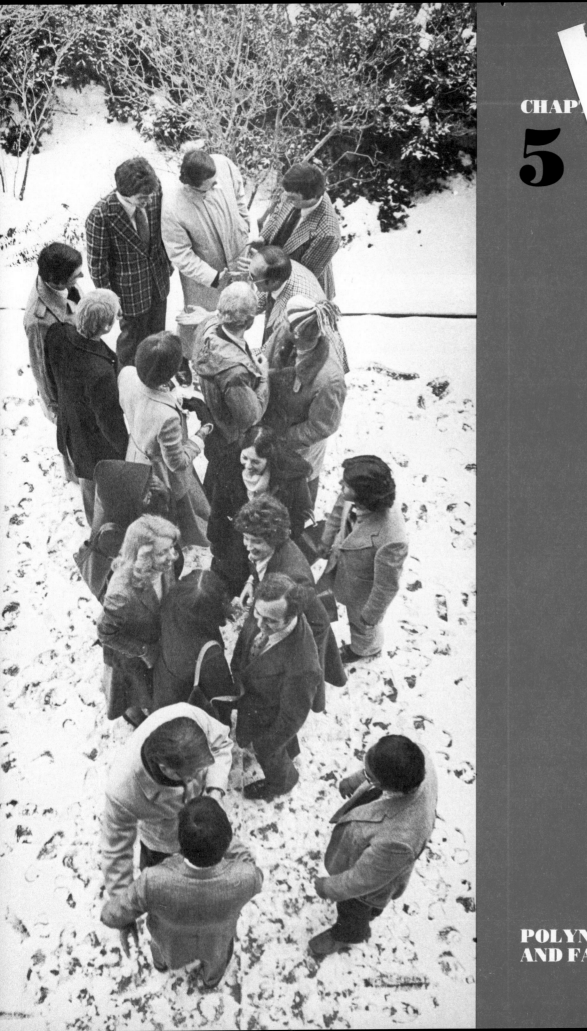

CHAPTER

5

POLYNOMIALS
AND FACTORING

READINESS CHECK: SKILLS FOR CHAPTER 5

1. Factor: $4y + 28 + 12z$.   **2.** Factor: $8x - 32$.

3. Solve: $4x + 9 = 17$.   **4.** Solve: $(4x + 3)(x - 7) = 0$.

Multiply.

5. $(-6x^8)(2x^5)$ **6.** $8x(2x^2 - 6x + 1)$

7. $(x + 6)(x - 4)$ **8.** $(7w + 6)(4w - 1)$

9. $(t - 9)^2$ **10.** $(5x + 3)^2$

11. $(p + 4)(p - 4)$

OBJECTIVES

**After finishing Section 5.1, you
should be able to:**

- ◼•◼ **Factor monomials.**
- ◼••◼ **Factor polynomials when the
terms have a common factor.**
- ◼•••◼ **Factor certain expressions with
four terms by grouping.**

1. a) Multiply: $(3x)(4x)$.

   $12x^2$

   b) Factor: $12x^2$.

   $(3x)(4x)$

2. a) Multiply: $(2x)(8x^2)$.

   $16x^3$

   b) Factor: $16x^3$.

   $(2x)(8x^2)$

Find three factorizations of each
monomial.

3. $8x^4$
   $(2 \cdot 4)x^4$
   $(2x^2)(4x^2)$
   $(2x^4)(4)$

4. $21x^2$
   $(3 \cdot 7)x^2$
   $(3x)(7x)$
   $(3x^2)(7)$

5. $6x^5$
   $(3 \cdot 2)x^5$
   $(3x^5)(2)$
   $(3)(2x^5)$

ANSWERS ON PAGE A–11

5.1 FACTORING POLYNOMIALS

Factoring is the reverse of multiplication. To factor quickly, we study
the quick methods of multiplication.

◼•◼ FACTORING MONOMIALS

To factor a monomial we find two monomials whose product is that
monomial. Compare.

	Multiplying	*Factoring*
a)	$(4x)(5x) = 20x^2$	$20x^2 = (4x)(5x)$
b)	$(2x)(10x) = 20x^2$	$20x^2 = (2x)(10x)$
c)	$(-4x)(-5x) = 20x^2$	$20x^2 = (-4x)(-5x)$
d)	$(x)(20x) = 20x^2$	$20x^2 = (x)(20x)$

The monomial $20x^2$ thus has many factorizations. There are still other
ways to factor $20x^2$.

DO EXERCISES 1 AND 2.

To factor a monomial, factor the coefficient first. Then shift some of
the letters to one factor and some to the other.

Example 1   Find three factorizations of $15x^3$.

a) $15x^3 = (3 \cdot 5)x^3$
       $= (3x)(5x^2)$

b) $15x^3 = (3 \cdot 5)x^3$
       $= (3x^2)(5x)$

c) $15x^3 = (-1) \cdot (-15)x^3$
       $= (-x)(-15x^2)$

DO EXERCISES 3–5.

●● FACTORING WHEN TERMS HAVE A COMMON FACTOR

To multiply a monomial and a polynomial with more than one term, we multiply each term by the monomial. To factor, we do the reverse.

Compare.

Multiply   *Factor*

$$5\,(x + 3) = \boxed{5}\,x + \boxed{5}\cdot 3 \qquad 5x + 15 = \boxed{5}\cdot x + \boxed{5}\cdot 3$$

$$= 5x + 15 \qquad\qquad\qquad = \boxed{5}\,(x + 3)$$

$$\boxed{3x}\,(x^2 + 2x - 4) \qquad 3x^3 + 6x^2 - 12x$$

$$= \boxed{3x}\cdot x^2 + \boxed{3x}\cdot 2x \qquad = \boxed{3x}\cdot x^2 + \boxed{3x}\cdot 2x$$

$$+\ \boxed{3x}\,(-4) \qquad\qquad +\ \boxed{3x}\cdot(-4)$$

$$= 3x^3 + 6x^2 - 12x \qquad = \boxed{3x}\,(x^2 + 2x - 4)$$

DO EXERCISES 6 AND 7.

We are finding a factor common to all the terms. There may not always be one other than 1. When there is, we usually use the factor with the largest possible coefficient and the largest exponent. In this way we "factor completely."

Example 2   Factor.

$$3x^2 + 6 = \boxed{3}\cdot x^2 + \boxed{3}\cdot 2 = \boxed{3}\,(x^2 + 2)$$

Example 3   Factor.

$$5x^4 + 20x^3 = \boxed{5x^3}\cdot x + \boxed{5x^3}\cdot 4 = \boxed{5x^3}\,(x + 4)$$

Example 4   Factor.

$$16x^3 + 20x^2 = (\,\boxed{4x^2}\,)(4x) + \boxed{4x^2}\cdot 5 = \boxed{4x^2}\,(4x + 5)$$

Example 5   Factor.

$$15x^5 - 12x^4 + 27x^3 - 3x^2$$

$$= (\,\boxed{3x^2}\,)(5x^3) - (\,\boxed{3x^2}\,)(4x^2) + (\,\boxed{3x^2}\,)(9x) - (\,\boxed{3x^2}\,)(1)$$

$$= \boxed{3x^2}\,(5x^3 - 4x^2 + 9x - 1)$$

If you can spot the common factor without factoring each term, you should just write the answer.

DO EXERCISES 8–12.

6. a) Multiply: $3(x + 2)$.

$$3x + 6$$

  b) Factor: $3x + 6$.

$$3(x + 2)$$

7. a) Multiply: $2x(x^2 + 5x + 4)$.

$$2x^3 + 10x^2 + 8x$$

  b) Factor: $2x^3 + 10x^2 + 8x$.

$$2x(x^2 + 5x + 4)$$

Factor.

8. $x^2 + 3x$

$$x(x + 3)$$

9. $3x^6 - 5x^3 + 2x^2$

$$x^2(3x^4 - 5x + 2)$$

10. $9x^4 - 15x^3 + 3x^2$

$$3x^2(3x^2 - 5x + 1)$$

11. $\dfrac{3}{4}x^3 + \dfrac{5}{4}x^2 + \dfrac{7}{4}x + \dfrac{1}{4}$

$$\frac{1}{4}(3x^3 + 5x^2 + 7x + 1)$$

12. $35x^7 - 49x^6 + 14x^5 - 63x^3$

$$7x^3(5x^4 - 7x^3 + 2x^2 - 9)$$

ANSWERS ON PAGE A–11

Factor.

13. $x^2 + 5x + 2x + 10$

$(2x^2 + 5x) + (2x + 10)$

$x(2x + 5) 2(4x + 5)$

$x(x + 5)\ 2(x + 5)$

$(x + 2)(x + 5)$

14. $x^2 - 4x + 3x - 12$

$(x^2 - 4x) + (3x - 12)$

$x(x - 4) + 3(x - 4)$

$(x + 3)(x - 4)$

15. $2x^2 + 8x - 3x - 12$

$(2x^2 + 8x) - (3x \cdot 12)$

$2x(x + 4) - 3(x - 4)$

$(2x - 3)(x + 4)$

16. $16x^2 + 20x - 12x - 15$

$(16x^2 + 12x) - (12x - 15)$

$4x(4x + 5) - 3(4x \cdot 5)$

$(4x - 3)(4x - 5)$

●●● FACTORING BY GROUPING

Consider

$$x^2 + 3x + 4x + 12.$$

There is no common factor other than 1. But we can factor $x^2 + 3x$ and $4x + 12$:

$$x^2 + 3x = x(x + 3);$$
$$4x + 12 = 4(x + 3).$$

Then

$$x^2 + 3x + 4x + 12 = x(\ x + 3\ ) + 4(\ x + 3\ ).$$

Note the common *binomial* factor $x + 3$. We can use the distributive law again like this:

$$x(\ x + 3\ ) + 4(\ x + 3\ ) = (x + 4)(\ x + 3\ ).$$

Examples   Factor.

6. $x^2 + 7x + 2x + 14 = (x^2 + 7x) + (2x + 14)$
$$= x(\ x + 7\ ) + 2(\ x + 7\ )$$
$$= (x + 2)(\ x + 7\ )$$

7. $x^2 + 3x - x - 3 = (x^2 + 3x) + (-x - 3)$

$$= x(x + 3) - 1(x + 3) \qquad \text{Factoring } -1 \text{ out of the second binomial}$$

$$= (x - 1)(x + 3)$$

8. $5x^2 - 10x + 2x - 4 = (5x^2 - 10x) + (2x - 4)$
$$= 5x(x - 2) + 2(x - 2)$$
$$= (5x + 2)(x - 2)$$

9. $2x^2 + 12x - 3x - 18 = (2x^2 + 12x) + (-3x - 18)$
$$= 2x(x + 6) - 3(x + 6)$$
$$= (2x - 3)(x + 6)$$

This method is called *factoring by grouping*. Not all expressions with four terms can be factored by this method.

DO EXERCISES 13–16.

NAME   CLASS   ANSWERS

EXERCISE SET 5.1

| ● | Find three factorizations of each monomial.

1. $6x^3$ **2.** $9x^4$ **3.** $-9x^5$

$(3\cdot2)x^3$ $(3\cdot3)x^4$ $(-3\cdot3)x^5$

$(3x^3)(2)$ $(3x^4)(3)$ $(-3x^5)(3)$

$(2x^3)(3)$ $(3x^2)(3x^2)$ $(-3)(3x^5)$

4. $-12x^6$ **5.** $24x^4$ **6.** $15x^5$

$(-2\cdot6)x^6$ $(3\cdot8)x^4$ $(3\cdot5)x^5$

$(-2x^6)(6)$ $(3x^2)(8x^2)$ $(3x^5)(5)$

$(-6x^6)(2)$ $(3x^4)(8)$ $(3)(5x^5)$

| ●● | Factor.

7. $x^2 - 4x$ **8.** $2x^2 + 6x$

$X(x-4)$ $2x(x+3)$

9. $x^3 + 6x^2$ **10.** $2x^2 + 2x - 8$

$x2(x+3)$ $2(x^2+x-4)$

11. $8x^4 - 24x^2$ **12.** $x^5 + x^4 + x^3 - x^2$

$8x^2(x^2-3)$ $x^2(x^3+x^2+x-1)$

13. $17x^5 + 34x^3 + 51x$ **14.** $6x^4 - 10x^3 + 3x^2$

$17x(x^4+2x^2+3)$ $x^2(6x^2-10x+3)$

15. $10x^3 + 25x^2 + 15x - 20$ **16.** $16x^4 - 24x^3 + 32x^2 + 64x$

$(10x^3+25x^2)+(15x-20)$ $8x(2x^3-3x^2+4x+8)$

$5x^2(2x+5)\ 5(3x-4)$

$5(2x^3+5x^2+3x-4)$

| ANSWERS |
| 1. |
| 2. |
| 3. |
| 4. |
| 5. |
| 6. |
| 7. |
| 8. |
| 9. |
| 10. |
| 11. |
| 12. |
| 13. |
| 14. |
| 15. |
| 16. |

Copyright © 1979, Philippines copyright 1979, by Addison-Wesley Publishing Company, Inc. All rights reserved.

ANSWERS

17. _____

18. _____

19. _____

20. _____

21. _____

22. _____

23. _____

24. _____

25. _____

26. _____

27. _____

28. _____

29. _____

30. _____

31. _____

32. _____

17. $\dfrac{5}{3}x^6 + \dfrac{4}{3}x^5 + \dfrac{1}{3}x^4 + \dfrac{1}{3}x^3$

$$\frac{1}{3}x^3\left(5x^3 + 4x^2 + 1x + 1\right)$$

18. $\dfrac{5}{7}x^7 + \dfrac{3}{7}x^5 - \dfrac{6}{7}x^3 - \dfrac{1}{7}x$

$$\frac{x}{7}\left(5x^6 + 3x^4 - 6x^2 - 1\right)$$

● ● ●   Factor.

19. $y^2 + 4y + y + 4$

$(y^2 + 4y) + 1(y + 4)$

$y(y + 4) + 1(y + 4)$

$(y + 1)(y + 4)$

20. $x^2 + 5x + 2x + 10$

$(x^2 + 5x) + (2x + 10)$

$x(x + 5)\, 2(x + 5)$

$(x + 2)(x + 5)$

21. $x^2 - 4x - x + 4$

$(x^2 - 4x) - (x + 4)$

$x(x - 4) - 1(x + 4)$

$(x - 1)(x - 4)$

22. $a^2 + 5a - 2a - 10$

$(a^2 + 5a) - (2a - 10)$

$a(a + 5) - 2(a + 5)$

$(a - 2)(a + 5)$

23. $6x^2 + 4x + 9x + 6$

$(6x^2 + 4x) + (9x + 6)$

$2x(3x + 2) + 3(3x + 2)$

$(2x + 3)(3x + 2)$

24. $3x^2 - 2x + 3x - 2$

25. $3x^2 - 4x - 12x + 16$

$(3x^2 - 4x) - (12x + 16)$

$x(3x - 4) - 4(3x + 4)$

$(x - 4)(3x - 4)$

26. $24 - 18y - 20y + 15y^2$

$(24 - 18y) - (20y + 15y^2)$

$6(4 - 3y) - 5y(4 - 3y)$

$(6 - 5y)(4 - 3y)$

27. $35x^2 - 40x + 21x - 24$

$(35x^2 - 40x) + (21x - 24)$

$5x(7x - 8) + 3(7x - 8)$

$(5x + 3)(7x - 8)$

28. $8x^2 - 6x - 28x + 21$

$(8x^2 - 6x) - (28x + 21)$

$2x(4x - 3) - 7(4x - 3)$

$(2x - 7)(4x - 3)$

29. $4x^2 + 6x - 6x - 9$

$(4x^2 + 6x) - (6x - 9)$

$2x(2x + 3) - 3(2x + 3)$

$(2x - 3)(2x + 3)$

30. $2x^4 - 6x^2 - 5x^2 + 15$

$2x^2(x^2 - 3) - 5(x^2 - 3)$

$(2x^2 - 5)(x^2 - 3)$

31. $2x^4 + 6x^2 + 5x^2 + 15$

$(2x^4 + 6x^2) + (5x^2 + 15)$

$2x^2(x^2 + 3) + 5(x^2 + 3)$

$(2x^2 + 5)(x^2 + 3)$

32. $4x^4 - 6x^2 - 6x^2 + 9$

$(4x^2 - 6x^2) - (6x^2 + 9)$

$2x^2(2 \cdot 3) - 3(2x^2 + 3)$

5.2 FACTORING TRINOMIALS

◼•◼ FACTORING TRINOMIALS OF THE TYPE $x^2 + px + q$

Consider this multiplication:

$$(x + 2)(x + 5) = x^2 + \underbrace{5x + 2x}_{} + 10$$

$$= x^2 + \quad 7x \quad + 10.$$

The coefficient 7 is $2 + 5$ and the 10 is $2 \cdot 5$. In general,

$$(x + a)(x + b) = x^2 + (a + b)x + ab.$$

To factor we can use this equation in reverse:

$$x^2 + (a + b)x + ab = (x + a)(x + b).$$

To factor

$$x^2 + 5x + 6,$$

we think of pairs of integers whose product is 6 and whose sum is 5.

Pairs of factors	Sum of factors
1, 6	7
−1, −6	−7
−2, −3	−5
2, 3	5

Thus, the numbers are 2 and 3. Then

$$x^2 + 5x + 6 = (x + 2)(x + 3).$$

You can check factoring by multiplying. Try to work mentally.

Example 1　Factor:　$x^2 - 8x + 12$.

We look for two numbers whose product is 12 and whose sum is -8. They are -2 and -6:

$$x^2 - 8x + 12 = (x - 2)(x - 6).$$

DO EXERCISES 1–3.

◼•◼ FACTORING TRINOMIALS OF THE TYPE $ax^2 + bx + c$ (METHOD 1)

Suppose the coefficient of the first term is not 1. Consider a multiplication.

OBJECTIVES

After finishing Section 5.2, you should be able to:

◼•◼ Factor trinomials of the type $x^2 + px + q$ by examining the last coefficient, q.

◼•◼ Factor trinomials of the type $ax^2 + bx + c$ by (Method 1) examining the first and last coefficients, a and c, or by (Method 2) examining the product of the first and last coefficients, ac.

Factor.

1. $x^2 + 7x + 12$

$(x + 4)(x + 3)$

$x^2 + 3x + 4x + 12$

$x^2 + 7x + 12$

2. $x^2 - 12x + 35$

$(x - 7)(x - 5)$

$x^2 - 5x - 7x + 35$

$x^2 - 12x + 35$

3. $x^2 - x - 2$

$(x + 2)(x + 1)$

$x^2 + 1x + 2x - 2$

$x^2 - x - 2$

Factor.

4. $6x^2 + 7x + 2$

$$12 \qquad -\frac{2}{3}$$

$(2x+2)(3x+1)$

$3x^2 + 2x \cdot 6x$
$8x$

$(6x \quad 1)(2x \quad 2)$

$(6x^2 + 6x \cdot 1x + 2)$

$6x^2 + 7x + 2$

$$\begin{array}{cccc} \mathrm{F} & \mathrm{O} & \mathrm{I} & \mathrm{L} \\ \end{array}$$
$$(2x + 5)(3x + 4) = 6x^2 + 8x + 15x + 20$$
$$= 6x^2 + \quad 23x \quad + 20$$

F	O + I	L
$2 \cdot 3$	$2 \cdot 4 + 5 \cdot 3$	$5 \cdot 4$

To factor $ax^2 + bx + c$, we look for two binomials like this

$$(\_x + \_)(\_x + \_),$$

where products of numbers in the blanks are as follows.

> 1. The numbers in the *first* blanks of each binomial have product a.
> 2. The *outside* product and the *inside* product add up to b.
> 3. The numbers in the *last* blanks of each binomial have product c.

We always look first for a common factor. If there is one, we remove that common factor before proceeding. This happens in Example 3.

Example 2 Factor: $3x^2 + 5x + 2$.

a) We first look for a factor common to all the terms. There is none, other than 1.

b) Next we look for two numbers whose product is 3. These are

$$1, 3 \quad \text{and} \quad -1, -3.$$

We have these possibilities:

$$(x + \_)(3x + \_) \quad \text{or} \quad (-x + \_)(-3x + \_).$$

Now we look for numbers whose product is 2. These are

$$1, 2 \quad \text{and} \quad -1, -2.$$

Here are some possibilities for factorizations. There are eight, but we have not listed all of them here.

$(x + 1)(3x + 2)$	$(-x + 1)(-3x + 2)$
$(x - 1)(3x - 2)$	$(-x - 1)(-3x - 2)$

When we multiply, we must get $3x^2 + 5x + 2$. When we multiply, we find that the shaded expressions are factorizations. We choose the one in which the first coefficients are positive. Thus, the factorization is

$$(x + 1)(3x + 2).$$

DO EXERCISE 4.

Example 3   Factor: $8x^2 + 8x - 6$.

a) First look for a factor common to all three terms. The number 2 is a common factor, so we factor it out:

$$\boxed{2}\,(4x^2 + 4x - 3).$$

b) Now we factor the trinomial $4x^2 + 4x - 3$. We look for pairs of numbers whose product is 4. These are

$$4, 1 \quad \text{and} \quad 2, 2. \qquad \text{Both positive}$$

We then have these possibilities:

$$(4x + \underline{\ \ })(x + \underline{\ \ }) \quad \text{and} \quad (2x + \underline{\ \ })(2x + \underline{\ \ }).$$

Next we look for pairs of numbers whose product is -3. They are

$$3, -1 \quad \text{and} \quad -3, 1.$$

Then we have these possibilities for factorizations:

$$(4x + 3)(x - 1), \qquad \boxed{(2x + 3)(2x - 1)},$$
$$(4x - 1)(x + 3), \qquad (2x - 3)(2x + 1).$$
$$(4x + 1)(x - 3),$$
$$(4x - 3)(x + 1),$$

We usually do not write all of these. We multiply until we find the factors that give the product $4x^2 + 4x - 3$. We find that the factorization is

$$(2x + 3)(2x - 1).$$

c) But don't forget the common factor. We must supply it to get a factorization of the original polynomial.

$$8x^2 + 8x - 6 = \boxed{2(2x + 3)(2x - 1)}$$

When we factor trinomials we must use trial and error. However, as you practice you will find that you can make better and better guesses. Don't forget, when factoring any polynomials, always look first for a common factor.

DO EXERCISES 5–8.

Factor.

5. $6x^2 + 15x + 9$

$3(2x^2 + 5x + 3)$

$3(1x + 3)(2x + 1)$

$2x^2 - 1x + 6x + 3$

$3(2x^2 - 1x)($

6. $2x^2 + 4x - 6$

$2(x^2 + 4x - 3)$

$2(x\ 1)(x\ 2)$ 2 3

$x^2\ 2x - 3x\ 6$ Fac

7. $4x^2 + 2x - 6$

$2(2x^2 + x - 3)$

$2(1x\ 3)(2x\ 1)$

$2x^2\ 1x\ 6x\ 3$

$2(1x - 1)(2x + 3)$

$2x^2\ 3x\ 2x - 3$

8. $6x^2 - 5x + 1$

$(3x - 1)(2x - 1)$

$(6x^2 - 3x - 2x + 1)$

$(3x - 1)(2x - 1)$

ANSWERS ON PAGE A–11

Factor.

9. $2x^2 - x - 15$

(handwritten work:)

$(2x+6)(2x+5)$

$2x^2 - 15x - 12x \; 15$

$2x^2 - x - 15$

$(1x+3)(2x+5)$

$2x^2 + 5x \; 6x - 15$

10. $12x^2 - 17x - 5$

$(x^2 \quad)(x \quad)$

$(3x-5)(4x+1)$

$12x \quad 3x \cdot 20x \; 5$

$(3x-5)(4x+1)$

FACTORING TRINOMIALS OF THE TYPE $ax^2 + bx + c$ (METHOD 2)

Another way to factor $ax^2 + bx + c$ is as follows.

> **a)** First look for a common factor.
> **b)** Multiply the first and last coefficients, a and c.
> **c)** Try to factor the product ac so that the sum of the factors is b.
> **d)** Write the middle term, bx, as a sum.
> **e)** Then factor by grouping.

Example 4   Factor: $3x^2 - 10x - 8$.

a) First look for a common factor. There is none (other than 1).

b) Multiply the first and last coefficients, 3 and -8:

$3(-8) = -24.$

c) Try to factor -24 so that the sum of the factors is -10.

Some pairs of factors	Sums of factors
$4, -6$	-2
$-4, \;\; 6$	2
$12, -2$	10
$-12, \;\; 2$	-10

The desired factors are -12 and 2.

d) Write $-10x$ as a sum using the results of part (c):

$-10x = -12x + 2x.$

e) Factor by grouping:

$$3x^2 - 10x - 8 = 3x^2 - 12x + 2x - 8 \qquad \text{Substituting} \\ -12x + 2x \text{ for} \\ -10x, \text{ from (d)}$$

$$= (3x^2 - 12x) + (2x - 8)$$

$$= 3x(\;x - 4\;) + 2(\;x - 4\;)$$

$$= (3x + 2)(\;x - 4\;).$$

DO EXERCISES 9 AND 10.

Example 5   Factor: $15x^2 - 22x + 8$.

a) First look for a common factor. There is none (other than 1).

b) Multiply the first and last coefficients, 15 and 8:

$15 \cdot 8 = 120.$

c) Try to factor 120 so that the sum of the factors is -22.

ANSWERS ON PAGE A—11

Some pairs of factors	Sum of factors
30, 4	34
−30, −4	−34
3, 40	43
−3, −40	−43
6, 20	26
−6, −20	−26
10, 12	22
−10, −12	−22

The desired factors are −10 and −12.

d) Write $-22x$ as a sum using the results of part (c):

$$-22x = -10x - 12x.$$

e) Factor by grouping:

$$15x^2 - 22x + 8 = 15x^2 - 10x - 12x + 8 \qquad \text{Substituting } -10x - 12x \text{ for } -22x, \text{ from (d)}$$

$$= (15x^2 - 10x) + (-12x + 8)$$

$$= 5x(\; 3x - 2\;) - 4(\; 3x - 2\;)$$

$$= (5x - 4)(\; 3x - 2\;).$$

DO EXERCISES 11 AND 12.

This method of factoring is based on the following:

$$(ax + b)(cx + d) = acx^2 + adx + bcx + bd$$

$$= \underbrace{acx^2} + (ad + bc)x + \underbrace{bd}$$

$$(ac)(bd) = (ad)(bc)$$

We multiply the outside coefficients and factor them in such a way that we can write the middle term as a sum.

There are trinomials that are not factorable. An example is

$$x^2 - x + 5.$$

Which method should you use? The answer is simple. Use the one that works best for you!

Factor.

11. $3x^2 - 19x + 20$

$(1x - 5)(3x - 4)$

$3x^2 - 4x - 15x + 20$

$(1x - 5)(3x - 4)$

12. $20x^2 - 46x + 24$

$2(10x^2 - 23x + 12)$

$(2x - 3)(5x - 4)$

$5x^2 - 8x - 15x + 12$

$2(2x - 3)(5x - 4)$

NAME

CLASS

EXERCISE SET 5.2 - All

● Factor.

1. $x^2 + 8x + 15$

$(x+3)(x+5)$

$x^2 + 5x + 3x + 15$

2. $x^2 + 5x + 6$

$(x+2)(x+3)$

$x^2 + 3x + 2x + 6$

$x^2 + 5x + 6$

3. $x^2 - 2x - 15$

$(x+3)(x-5)$

$x^2 - 5x + 3x - 15$

4. $x^2 - 10x + 25$

$(x-5)(x-5)$

$x^2 - 5x - 5x + 25$

5. $x^2 + 7x + 12$

$(x+4)(x+3)$

$x^2 + 3x + 4x + 12$

6. $x^2 + x - 42$

$(x+7)(x-6)$

$x^2 + 7x + 6x - 42$

$x^2 + x - 42$

7. $x^2 + 2x - 15$

$(x+5)(x-3)$

$x^2 - 3x + 5x - 15$

8. $x^2 + 8x + 12$

$(x+6)(x+2)$

$x^2 + 2x + 6x + 12$

$x^2 + 8x + 12$

9. $y^2 + 9y + 8$

$(y+1)(y+8)$

$y^2 + 8y + 1y + 8$

10. $x^2 - 6x + 9$

$(x-3)(x-3)$

$x^2 - 3x - 3x + 9$

11. $x^4 + 5x^2 + 6$

$(x^2+2)(x^2+3)$

$x^4 + 3x^2 + 2x^2 + 3$

12. $y^2 + 16y + 64$

$(y+8)(y+8)$

$y^2 + 8y + 8y + 64$

13. $x^2 + 3x - 28$

$(x \quad)(x \quad)$

14. $x^2 - 11x + 10$

15. $15 + 8x + x^2$

16. $6 + 5x + x^2$

17. $a^2 - 12a + 11$

18. $c^2 - 10c + 21$

19. $x^2 - \dfrac{2}{25}x + \dfrac{1}{25}$

$(x \quad)(x \quad)$

x^2

20. $x^2 + \dfrac{2}{3}x + \dfrac{1}{9}$

21. $y^2 - 0.2y - 0.08$

22. $t^2 - 0.3t - 0.10$

1. _____

2. _____

3. _____

4. _____

5. _____

6. _____

7. _____

8. _____

9. _____

10. _____

11. _____

12. _____

13. _____

14. _____

15. _____

16. _____

17. _____

18. _____

19. _____

20. _____

21. _____

22. _____

Copyright © 1979, Philippines copyright 1979, by Addison-Wesley Publishing Company, Inc. All rights reserved.

ANSWERS

23. _____

24. _____

25. _____

26. _____

27. _____

28. _____

29. _____

30. _____

31. _____

32. _____

33. _____

34. _____

35. _____

36. _____

37. _____

38. _____

39. _____

40. _____

41. _____

42. _____

43. _____

44. _____

● ● Factor. Use Method 1 or Method 2.

23. $3x^2 + 4x + 1$

$(3x+1)(1x+1) \overset{3}{;}$

$3x^2 \; 3x + 1x + 1$

24. $6x^2 + 13x + 6$

25. $12x^2 + 28x - 24$

$(3x - 2)(4x+12)$

$12x^2 \cancel{6x} \; 48$

$\begin{array}{cc} 3 & 26 \\ 4 & 12x \\ 2 & 4 \\ 6 & 6 \end{array}$

26. $6x^2 + 33x + 15$

27. $2x^2 - x - 1$

28. $15x^2 - 19x + 6$

29. $9x^2 + 18x - 16$

30. $14x^2 + 35x + 14$

31. $15x^2 - 25x - 10$

32. $18x^2 - 3x - 10$

33. $12x^2 + 31x + 20$

34. $15x^2 + 19x - 10$

35. $14x^2 + 19x - 3$

36. $35x^2 + 34x + 8$

37. $9x^4 + 18x^2 + 8$

38. $6 - 13x + 6x^2$

39. $9x^2 - 42x + 49$

40. $15x^4 - 19x^2 + 6$

41. $6x^3 + 4x^2 - 10x$

42. $18x^3 - 21x^2 - 9x$

43. $x^2 + 3x - 7$

44. $x^2 + 11x + 12$

5.3 SQUARES OF BINOMIALS

◖•◗ RECOGNIZING SQUARES OF BINOMIALS

Some trinomials are squares of binomials; for example,

$$x^2 + 10x + 25 = (x + 5)^2.$$

A trinomial like this is called the *square of a binomial*, or a *trinomial square*.

The equations for squaring a binomial can be used in reverse to factor trinomial squares:

$$A^2 + 2 \cdot A \cdot B + B^2 = (A + B)^2;$$
$$A^2 - 2 \cdot A \cdot B + B^2 = (A - B)^2.$$

Use the following to help recognize squares of binomials.

a) Two of the terms must be squares (A^2 and B^2), such as

   4, x^2, $25x^4$, $16t^2$.

b) There must be no minus sign before A^2 or B^2.

c) If we multiply A and B (the square roots of these expressions) and double the result, we get the remaining term $2 \cdot A \cdot B$, or its additive inverse, $-2 \cdot A \cdot B$.

Example 1   Is $x^2 + 6x + 9$ the square of a binomial?

a) x^2 and 9 are squares.

b) There is no minus sign before x^2 or 9.

c) If we multiply the square roots, x and 3, and double the product, we get the remaining term: $\boxed{2}$ $\cdot 3 \cdot x = 6x$.

Thus, $x^2 + 6x + 9$ is the square of a binomial.

Example 2   Is $x^2 + 6x + 11$ the square of a binomial?

The answer is *no*, because only one term is a square.

Example 3   Is $16x^2 - 56x + 49$ a trinomial square?

a) $16x^2$ and 49 are squares.

b) There is no minus sign before $16x^2$ or 49.

c) If we multiply the square roots, $4x$ and 7, and double the product, we get the additive inverse of the remaining term: $\boxed{2}$ $\cdot 4x \cdot 7 = 56x$.

Thus, $16x^2 - 56x + 49$ is a trinomial square.

DO EXERCISE 1.

OBJECTIVES

After finishing Section 5.3, you should be able to:

◖•◗ Recognize squares of binomials (also called trinomial squares).

◖••◗ Factor squares of binomials (or trinomial squares).

◖•••◗ Recognize differences of two squares.

▐▌ Factor differences of squares.

▐▌▌ Factor polynomials completely.

1. Which of the following are squares of binomials?

   a) $x^2 + 8x + 16$

   b) $x^2 - 10x + 25$

   c) $x^2 - 12x + 4$

   d) $4x^2 + 20x + 25$

   e) $5x^2 - 14x + 16$

   f) $16x^2 + 40x + 25$

   g) $x^2 + 6x - 9$

   h) $25x^2 - 30x - 9$

ANSWERS ON PAGE A–12

Factor.

2. $x^2 + 2x + 1$

$(x + 1)^2$

3. $x^2 - 2x + 1$

$(x - 1)^2$

4. $x^2 + 4x + 4$

$(x + 2)^2$

5. $25x^2 - 70x + 49$

$(5x - 7)^2$

6. $16x^2 - 56x + 49$

$(4x - 7^2)$

7. Which of the following are differences of squares?

a) $x^2 - 25$

b) $x^2 - 24$ *no*

c) $x^2 + 36$ *no*

d) $4x^2 - 15$ *no*

e) $16x^4 - 49$ *yes*

f) $9x^6 - 1$

● ● FACTORING SQUARES OF BINOMIALS

To factor squares of binomials, we use the following equations:

$$A^2 + 2 \cdot A \cdot B + B^2 = (A + B)^2;$$
$$A^2 - 2 \cdot A \cdot B + B^2 = (A - B)^2.$$

We use the square roots of the squared terms and the sign of the remaining term.

Example 4 Factor: $x^2 + 6x + 9$.

$$x^2 + 6x + 9 = x^2 + 2 \cdot 3 \cdot x + 3^2$$
$$= (x + 3)^2$$

$(x+3)(x+3)^2$

$2x^2 + 3x + 3x$

Example 5 Factor: $x^2 - 14x + 49$.

$$x^2 - 14x + 49 = x^2 - 2 \cdot 7 \cdot x + 7^2$$
$$= (x - 7)^2$$

Example 6 Factor: $16x^2 - 40x + 25$.

$$16x^2 - 40x + 25 = (4x)^2 - 2 \cdot 4x \cdot 5 + 5^2$$
$$= (4x - 5)^2$$

DO EXERCISES 2–6.

● ● ● RECOGNIZING DIFFERENCES OF TWO SQUARES

We can use the following equation to recognize a difference of squares:

$$A^2 - B^2 = (A - B)(A + B).$$

For a binomial to be a difference of squares:

a) There must be two expressions, both squares, such as

$$4x^2, \quad 9, \quad 25t^4, \quad 1, \quad x^6.$$

b) There must be a minus sign between them.

Example 7 Is $9x^2 - 64$ a difference of squares?

a) The first expression is a square: $9x^2 = (3x)^2$.
The second expression is a square: $64 = (8)^2$.

b) There is a minus sign between them.

So we have a difference of squares.

DO EXERCISE 7.

⠰⠿ FACTORING DIFFERENCES OF SQUARES

To factor a difference of squares we can use the following equation:

$$A^2 - B^2 = (A - B)(A + B).$$

We find the square roots of the expressions A^2 and B^2. We write a plus sign one time and a minus sign one time.

Example 8 Factor: $x^2 - 4$.

$$x^2 - 4 = \boxed{x}^2 - \boxed{2}^2 = (\boxed{x} - \boxed{2})(\boxed{x} + \boxed{2})$$

Example 9 Factor: $18x^2 - 50x^6$.

Always look first for a common factor to all terms. This time there is one.

$$18x^2 - 50x^6 = \boxed{2x^2}\,(9 - 25x^4)$$
$$= 2x^2[3^2 - (5x^2)^2]$$
$$= 2x^2(3 - 5x^2)(3 + 5x^2)$$

Example 10 Factor: $49x^4 - 9x^6$.

$$49x^4 - 9x^6 = \boxed{x^4}\,(49 - 9x^2) = x^4(7 - 3x)(7 + 3x)$$

DO EXERCISES 8–10.

⠰⠿ FACTORING COMPLETELY

If a factor can still be factored, you should factor it. When no factor can be factored further, we say that we have *factored completely*.

Example 11 Factor: $1 - 16x^{12}$.

$$1 - 16x^{12} = (1 - 4x^6)(1 + 4x^6) \qquad \text{Factoring a difference of squares}$$

$$= (1 - 2x^3)(1 + 2x^3)(1 + 4x^6) \qquad \begin{array}{l}\text{Factoring the first}\\\text{factor (again a dif-}\\\text{ference of squares)}\end{array}$$

Caution: A difference of squares does not factor like this:

$$A^2 - B^2 = (A - B)(A - B).$$

A sum of two squares cannot be factored! In particular, it cannot be factored like this:

$$A^2 + B^2 = (A + B)(A + B).$$

Factor.

8. $x^2 - 9$

$(x - 3)^2$

$(x - 3)(x + 3)$

9. $x^2 - 64$

$(x - 8)^2$

$(x + 8)(x - 8)$

10. $5 - 20x^6$
(*Hint:* $1 = 1^2$, $x^6 = (x^3)^2$.)

$5(1 - 4x^6)$

$5(1 - 2x^3)(1 + 2x^3)$

ANSWERS ON PAGE A–12

Factor and check by multiplying.
(Always factor completely, even though
such directions are not given.)

11. $81x^4 - 1$

$(9x^2 - 1)(9x^2 + 1)$

$(3x^2 - 1)(3x^2 + 1)(9x^2 + 1)$

12. $49x^4 - 25x^{10}$

0

$x^4(7x - 5x^3)(7 + 5x^3)$

Some other hints about factoring:
1. **Always look first for a common factor.**
2. **Always factor completely, even if the directions do not say so.**
3. **Never try to factor a sum of squares.**

You can always check by multiplying to see if you get the original polynomial.

DO EXERCISES 11 AND 12.

SOMETHING EXTRA
CALCULATOR CORNER: A NUMBER PATTERN

1. Calculate each of the following. Look for a pattern.

$(20 \cdot 20) - (21 \cdot 19)$

$(34 \cdot 34) - (35 \cdot 33)$

$(1999 \cdot 1999) - (2000 \cdot 1998)$

$(5.8 \times 5.8) - (6.8 \times 4.8)$

$[0.7 \times 0.7] - [(1.7)(-0.3)]$

$(999 \cdot 999) - (1000)(998)$

2. Use the pattern to find the following without using a calculator.

$(3778 \cdot 3778) - (3779 \cdot 3777)$

$(14.78 \times 14.78) - (15.78 \times 13.78)$

$[0.375 \times 0.375] - [(1.375) \times (-0.625)]$

3. Find an equation that describes the pattern above. Verify it.

NAME _____

CLASS _____

EXERCISE SET 5.3

● Which are squares of binomials? Write "yes" or "no."

1. $x^2 - 14x + 49$
  Yes

2. $x^2 - 16x + 64$
  No

3. $x^2 + 16x - 64$

4. $8x^2 + 40x + 25$

●● Factor. Remember to look first for a common factor.

5. $x^2 + 6x + 9$
  $(x + 3)(x + 3)$

6. $y^2 - 6y + 9$
  $(y + 3)(y - 3)$
  $y^2 - 3y + 9$

7. $2x^2 - 4x + 2$ 4
  $(1x + 2)(2x - 1)$
  $2x^2$ $1x + 2$

8. $2y^2 - 40y + 200$
  $2(y^2 - 20y + 100)$
  $(y - 10)(y - 10)$
  $y^2 - 10y 10y + 100$

10. $144x + 24x^2 + x^3$

9. $x^3 - 18x^2 + 81x$
  $x(x^2 - 18x + 81)$
  $(x - 9)(x - 9)$
  $x^2 - 9x - 9x + 81$

11. $y^2 - 12y + 36$
  $(y - 6)(y - 6)$
  $y^2 + 6y - 6y + 36$

12. $81 + 18x + x^2$

13. $64 - 16y + y^2$

14. $a^4 + 14a^2 + 49$

15. $12x^2 + 36x + 27$

16. $2x^2 - 44x + 242$

17. $x^6 - 16x^3 + 64$

18. $y^6 + 26y^3 + 169$

19. $49 - 42x + 9x^2$

20. $20x^2 + 100x + 125$

21. $5y^4 + 10y^2 + 5$

22. $16x^8 - 8x^4 + 1$

●●● Which are differences of squares? Write "yes" or "no."

23. $x^2 - 35$
  No

24. $x^2 + 36$
  yes

25. $16x^2 - 25$
  yes

26. $36t^2 - 1$
  yes

1. _____
2. _____
3. _____
4. _____
5. _____
6. _____
7. _____
8. _____
9. _____
10. _____
11. _____
12. _____
13. _____
14. _____
15. _____
16. _____
17. _____
18. _____
19. _____
20. _____
21. _____
22. _____
23. _____
24. _____
25. _____
26. _____

Copyright © 1979, Philippines copyright 1979, by Addison-Wesley Publishing Company, Inc. All rights reserved.

ANSWERS

27. _____

28. _____

29. _____

30. _____

31. _____

32. _____

33. _____

34. _____

35. _____

36. _____

37. _____

38. _____

39. _____

40. _____

41. _____

42. _____

43. _____

44. _____

45. _____

46. _____

47. _____

48. _____

49. _____

50. _____

51. _____

52. _____

53. _____

■■ Factor. Remember to look first for a common factor.

27. $x^2 - 4$

28. $x^2 - 36$

29. $8x^2 - 98$

30. $98x^4 - 2$

31. $4x^2 - 25$

32. $100a^2 - 9$

33. $16x - 81x^3$

34. $36b - 49b^3$

35. $16x^2 - 25x^4$

36. $x^{16} - 9x^2$

37. $49x^4 - 81$

38. $25a^2 - 9$

39. $a^{12} - 4a^2$

40. $81y^6 - 25$

41. $121a^8 - 100$

42. $x^6 - \dfrac{1}{9}$

■■ Factor completely.

43. $4x^4 - 64$

44. $5x^4 - 80$

45. $1 - y^8$

46. $x^{12} - 16$

47. $x^8 - 81$

48. $t^8 - 1$

49. $16 - t^4$

50. $1 - a^4$

51. $36a^2 - 15a + \dfrac{25}{16}$

52. $\dfrac{1}{81}x^6 - \dfrac{8}{27}x^3 + \dfrac{16}{9}$

53. Use a factoring formula to find $76^2 - 24^2$ without squaring either term.

5.4 SOLVING EQUATIONS AND PROBLEMS BY FACTORING

● SOLVING EQUATIONS BY FACTORING

Let's recall the principle of zero products.

A product is 0 if and only if at least one of the factors is 0.

Example 1 Solve $x^2 + 5x + 6 = 0$.

We first factor the polynomial. Then we use the principle of zero products.

$$x^2 + 5x + 6 = 0$$
$$(x + 2)(x + 3) = 0 \qquad \text{Factoring}$$
$$x + 2 = 0 \quad \text{or} \quad x + 3 = 0 \qquad \text{Using the principle of zero products}$$
$$x = -2 \quad \text{or} \qquad x = -3$$

Check:

$x^2 + 5x + 6 = 0$		$x^2 + 5x + 6 = 0$	
$(-2)^2 + 5(-2) + 6$	0	$(-3)^2 + 5(-3) + 6$	0
$4 - 10 + 6$		$9 - 15 + 6$	
$-6 + 6$		$-6 + 6$	
0		0	

The solutions are -2 and -3.

DO EXERCISE 1.

You *must* have 0 on one side before you use the principle of zero products.

Example 2 Solve $x^2 - 8x = -16$.

We first add 16 to get 0 on one side.

$$x^2 - 8x + 16 = 0 \qquad \text{Adding 16}$$
$$(x - 4)(x - 4) = 0 \qquad \text{Factoring}$$
$$x - 4 = 0 \quad \text{or} \quad x - 4 = 0 \qquad \text{Using the principle of zero products}$$
$$x = 4 \quad \text{or} \qquad x = 4$$

There is only one solution, 4. The check is left to the student.

DO EXERCISES 2 AND 3.

Example 3 Solve $x^2 + 5x = 0$.

$$x(x + 5) = 0 \qquad \text{Factoring out a common factor}$$
$$x = 0 \quad \text{or} \quad x + 5 = 0$$
$$x = 0 \quad \text{or} \qquad x = -5$$

The solutions are 0 and -5. The check is left to the student.

DO EXERCISES 4 AND 5.

OBJECTIVES

After finishing Section 5.4, you should be able to:

● Solve certain equations by factoring.

●● Solve applied problems involving equations that can be solved by factoring.

Solve.

1. $x^2 - x - 6 = 0$

Solve.

2. $x^2 - 3x = 28$

3. $x^2 + 9 = 6x$

Solve.

4. $x^2 - 4x = 0$

5. $x^2 - 16 = 0$

ANSWERS ON PAGE A–12

Translate to equations. Then solve and check.

6. One more than a number times one less than the number is 24.

7. Seven less than a number times eight less than the number is 0.

Translate to an equation. Then solve and check.

8. The square of a number minus the number is 20.

●● **APPLIED PROBLEMS**

Recall that to solve a problem, we

> 1. **Translate the problem to an equation. (Very often a drawing helps.)**
> 2. **Solve the equation.**
> 3. **Check the answers in the problem.**

Example 4   Solve this problem.

One more than a number times one less than that number is 8.

$$(x + 1) \qquad \qquad (x - 1) \qquad = 8$$

Translating

Solve: $(x + 1)(x - 1) = 8$

$$x^2 - 1 = 8 \qquad \text{Multiplying}$$
$$x^2 - 1 - 8 = 0 \qquad \text{Adding } -8 \text{ to get 0 on one side}$$
$$x^2 - 9 = 0$$
$$(x - 3)(x + 3) = 0 \qquad \text{Factoring}$$
$$x - 3 = 0 \quad \text{or} \quad x + 3 = 0 \qquad \text{Using the principle of zero products}$$
$$x = 3 \quad \text{or} \qquad x = -3$$

Check for 3:   One more than 3 (this is 4) times one less than 3 (this is 2) is 8. Thus, 3 checks.

Check for −3:   Left to the student.

There are two such numbers, 3 and −3.

DO EXERCISES 6 AND 7.

Example 5   Solve this problem.

The square of a number minus twice the number is 48.

$$x^2 \qquad \qquad - \qquad \qquad 2x \qquad = 48 \qquad \text{Translating}$$

$$x^2 - 2x = 48$$
$$x^2 - 2x - 48 = 0$$
$$(x - 8)(x + 6) = 0$$
$$x - 8 = 0 \quad \text{or} \quad x + 6 = 0 \qquad \text{Using the principle of zero products}$$
$$x = 8 \quad \text{or} \qquad x = -6$$

There are two such numbers, 8 and −6. They both check.

DO EXERCISE 8.

Sometimes it helps to reword a problem before translating.

Example 6   The height of a triangle is 7 cm more than the base. The area of the triangle is 30 cm². Find the height and the base. (Area is $\frac{1}{2} \cdot$ base $\cdot$ height.)

We first make a drawing.

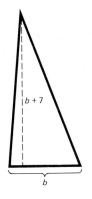

$\frac{1}{2}$ times the base times the base plus 7 is 30   Rewording

$\frac{1}{2} \quad \cdot \quad\quad b \quad\quad \cdot \quad\quad (b + 7) \quad\quad = 30$ Translating

$$\frac{1}{2} \cdot b \cdot (b + 7) = 30$$

$$\frac{1}{2}(b^2 + 7b) = 30 \quad\quad \text{Multiplying}$$

$$b^2 + 7b = 60 \quad\quad \text{Multiplying by 2}$$

$$b^2 + 7b - 60 = 0 \quad\quad \text{Adding} -60$$

$$(b + 12)(b - 5) = 0 \quad\quad \text{Factoring}$$

$b + 12 = 0 \quad$ or $\quad b - 5 = 0 \quad$ Using principle of zero products

$b = -12 \quad$ or $\quad\quad b = 5$

The solutions of the equation are -12 and 5. The base of a triangle cannot have a negative length. Thus, the base is 5 cm. The height is 7 cm more than the base, so the height is 12 cm. These numbers check.

DO EXERCISE 9.

Example 7   In a sports league of n teams in which each team plays every other team twice, the total number N of games to be played is given by

$$N = n^2 - n.$$

(a) A slow-pitch softball league has 17 teams. What is the total number of games to be played? **(b)** A basketball league plays a total of 90 games. How many teams are in the league?

a) We substitute 17 for n:

$$N = n^2 - n = 17^2 - 17 = 289 - 17 = 272.$$

---

9. The width of a rectangle is 2 cm less than the length. The area is 15 cm². Find the length and width.

10. Use $N = n^2 - n$ for the following.

a) A volleyball league has 19 teams. What is the total number of games to be played?

b) We substitute 90 for N and solve for n:

$$
\begin{aligned}
n^2 - n &= 90 && \text{Substituting 90 for } N \\
n^2 - n - 90 &= 0 && \text{Adding } -90 \text{ to get 0 on one side} \\
(n - 10)(n + 9) &= 0 && \text{Factoring} \\
n - 10 = 0 \quad &\text{or} \quad n + 9 = 0 && \text{Principle of zero products} \\
n = 10 \quad &\text{or} \qquad\quad n = -9
\end{aligned}
$$

Since the number of teams cannot be negative, -9 cannot be a solution. But 10 checks, so there are 10 teams in the league.

DO EXERCISE 10.

b) A slow-pitch softball league plays a total of 72 games. How many teams are in the league?

SOMETHING EXTRA
CALCULATOR CORNER: NESTED MULTIPLICATION

To evaluate polynomials with a calculator, it helps to first use factoring to change the polynomial to *nested form*, as shown in the example below.

Example   Transform to nested form and evaluate for $x = 5.3$:

$$
\begin{aligned}
2x^4 + 4x^3 - x^2 + 7x - 9 &= x\{2x^3 + 4x^2 - x + 7\} - 9 \\
&= x\{x[2x^2 + 4x - 1] + 7\} - 9 \\
&= x\{x[x(2x + 4) - 1] + 7\} - 9.
\end{aligned}
$$

To evaluate $x\{x[x(2x + 4) - 1] + 7\} - 9$, start from the inside and work out. For $x = 5.3$, you enter 5.3, multiply by 2, add 4, multiply by 5.3, subtract 1, multiply by 5.3, add 7, multiply by 5.3, and subtract 9. The answer is 2173.6142.

EXERCISES

Transform each polynomial to nested form. Then evaluate for $x = 5.3$, $x = 10.6$, and $x = -23$. Round to four decimal places.

1. $x^5 + 3x^4 - x^3 + x^2 - x + 9$

2. $5x^4 - 17x^3 + 2x^2 - x + 11$

3. $2x^4 - 3x^3 + 5x^2 - 2x + 18$

4. $-2x^5 + 4x^4 + 8x^3 - 4x^2 - 3x + 24$

NAME   CLASS   ANSWERS

EXERCISE SET 5.4

■ Solve.

1. $x^2 + 6x + 5 = 0$ *Check*

1. _____

2. $x^2 + 7x + 6 = 0$

2. _____

3. $x^2 - 5x = 0$

$x(x-5) = 0$

3. _____

4. $4x^2 - 9 = 0$

4. _____

5. $x^2 + 6x + 9 = 0$

5. _____

6. $3t^2 + t = 2$

6. _____

7. $6y^2 - 4y - 10 = 0$

7. _____

ANSWERS

8. $5x^2 = 6x$

Check

8. _____

9. $3x^2 = 7x + 20$

$3x^2 = 7x + 20$

$-7x + 3x^2 = 20$

$-4x^3 = 20$

9. _____

10. $0 = 2y^2 + 12y + 10$

10. _____

11. $-5x = -12x^2 + 2$

11. _____

12. $14 = x(x - 5)$

12. _____

13. $0 = -3x + 5x^2$

13. _____

14. $x^2 - 5x = 18 + 2x$

14. _____

■● Solve.

15. If you subtract a number from four times its square, the result is three. Find the number.

16. Eight more than the square of a number is six times the number. Find the number.

17. The product of two consecutive integers is 182. Find the integers.

18. The product of two consecutive even integers is 168. Find the integers.

19. The product of two consecutive odd integers is 195. Find the integers.

$$x(x+2)$$
$$x^2 + 2x = 195$$
$$x^2 + 2x - 195 = 0$$
$$(x \quad)(x \quad)$$

20. The length of a rectangle is 4 m greater than the width. The area of the rectangle is 96 m². Find the length and width.

Copyright © 1979. Philippines copyright 1979. by Addison-Wesley Publishing Company. Inc. All rights reserved.

21. The area of a square is 5 more than the perimeter. Find the length of a side.

22. The base of a triangle is 10 cm greater than the height. The area is 28 cm². Find the height and base.

21. _____

23. If the sides of a square are lengthened by 3 m, the area becomes 81 m². Find the length of a side of the original square.

22. _____

23. _____

24. The sum of the squares of two consecutive odd positive integers is 74. Find the integers.

24. _____

Use $N = n^2 - n$ for Exercises 25–28.

25. A slow-pitch softball league has 23 teams. What is the total number of games to be played?

25. _____

26. A basketball league has 14 teams. What is the total number of games to be played?

26. _____

27. A slow-pitch softball league plays a total of 132 games. How many teams are in the league?

27. _____

28. A basketball league plays a total of 240 games. How many teams are in the league?

28. _____

Copyright © 1979, Philippines copyright 1979, by Addison-Wesley Publishing Company, Inc. All rights reserved.

The number N of possible handshakes within a group of n people is given by

$$N = \frac{1}{2}(n^2 - n).$$

29. At a meeting there are 40 people. How many handshakes are possible?

29. _____

30. At a party there are 100 people. How many handshakes are possible?

30. _____

31. Everyone shook hands at a party. There were 190 handshakes in all. How many were at the party? (*Hint:* Multiply on both sides by 2.)

31. _____

32. Everyone shook hands at a meeting. There were 300 handshakes in all. How many were at the meeting?

32. _____

5.5 COMPLETING THE SQUARE

FACTORING DIFFERENCES OF SQUARES

A difference of squares can have more than two terms. This can happen if one of the squares is a trinomial.

Example 1 Factor: $x^2 + 2x + 1 - 25$.

Since $x^2 + 2x + 1 = (x + 1)^2$, we have

$$x^2 + 2x + 1 - 25 = (\boxed{x + 1})^2 - 5^2 = (\boxed{x + 1} - 5)(\boxed{x + 1} + 5)$$

$$= (x - 4)(x + 6).$$

Example 2 Factor: $x^2 + 6x + 9 - 16$.

$$x^2 + 6x + 9 - 16 = (\boxed{x + 3})^2 - 4^2$$

$$= (\boxed{x + 3} - 4)(\boxed{x + 3} + 4)$$

$$= (x - 1)(x + 7)$$

DO EXERCISES 1–4.

MAKING $x^2 + bx$ THE SQUARE OF A BINOMIAL

This is the square of a binomial:

$$x^2 + 10x + 25.$$

It is $(x + 5)^2$. We could find the 25 from $10x$, by taking half the coefficient of x and squaring it.

Example 3 What must be added to $x^2 + 8x$ to make it the square of a binomial?

$x^2 + 8x$
$\downarrow$
$\dfrac{8}{2} = 4$ Taking half the x-coefficient
$\downarrow$
$4^2 = 16$ Squaring
$\downarrow$
$x^2 + 8x + 16$ Adding

Thus, 16 must be added to $x^2 + 8x$ in order to make it the square of a binomial:

$$x^2 + 8x + 16 = (x + 4)^2.$$

OBJECTIVES

After finishing Section 5.5, you should be able to:

:•: Factor differences of squares when there are more than two terms.

:•:•: Make an expression $x^2 + bx$ the square of a binomial.

:•••: Factor trinomials by completing the square.

Factor.

1. $x^2 + 2x + 1 - 16$

2. $x^2 - 2x + 1 - 64$

3. $x^2 + 8x + 16 - 9$

4. $x^2 + 16x + 64 - 100$

ANSWERS ON PAGE A–12

What must be added to each to make it the square of a binomial?

5. $x^2 + 12x$

6. $x^2 - 6x$

7. $x^2 + 14x$

8. $x^2 - 22x$

Factor by completing the square.

9. $x^2 + 2x - 3$

10. $x^2 + 6x + 8$

11. $x^2 - 8x - 9$

Example 4   What must be added to $x^2 - 12x$ to make it the square of a binomial?

We take half of $\boxed{-12}$, obtaining $\boxed{-6}$. We square: $(\boxed{-6})^2 = 36$. Thus, 36 must be added to $x^2 - 12x$ to make it the square of a binomial:

$$x^2 - 12x + \boxed{36} = (x - 6)^2.$$

DO EXERCISES 5–8.

■■■   FACTORING BY COMPLETING THE SQUARE

Example 5   Factor: $x^2 - 6x - 16$.

This trinomial is not a square. The third term would have to be 9. To see this we take half the coefficient of x and square it. We add zero to the trinomial, naming it $\boxed{9 - 9}$.

$$
\begin{aligned}
x^2 - 6x - 16 &= x^2 - 6x + (\boxed{9 - 9}) - 16 && \text{Adding } 9 - 9 \\
&= (x^2 - 6x + 9) - 9 - 16 \\
&= (x - 3)^2 - 25 \\
&= (\boxed{x - 3})^2 - 5^2 \\
&= (\boxed{x - 3} - 5)(\boxed{x - 3} + 5) && \text{Factoring} \\
&= (x - 8)(x + 2)
\end{aligned}
$$

This is factoring by *completing the square*.

Example 6   Factor $x^2 + 8x + 7$ by completing the square.

The third term would have to be 16 (half of 8 squared) for the trinomial to be a square. We add zero, naming it $\boxed{16 - 16}$.

$$
\begin{aligned}
x^2 + 8x + 7 &= x^2 + 8x + (\boxed{16 - 16}) + 7 && \text{Adding } 16 - 16 \\
&= (x^2 + 8x + 16) - 16 + 7 \\
&= (x + 4)^2 - 9 \\
&= (\boxed{x + 4})^2 - 3^2 \\
&= (\boxed{x + 4} - 3)(\boxed{x + 4} + 3) && \text{Factoring} \\
&= (x + 1)(x + 7)
\end{aligned}
$$

Although you can factor most trinomials without completing the square, it is important that you learn to do it this way. Completing the square is an important technique, used in several different ways.

DO EXERCISES 9–11.

NAME   CLASS

EXERCISE SET 5.5

▪ Factor.

1. $x^2 + 12x + 36 - 49$ **2.** $x^2 + 2x + 1 - 9$ **3.** $x^2 + 14x + 49 - 16$

4. $x^2 - 16x + 64 - 100$ **5.** $x^2 - 18x + 81 - 1$ **6.** $x^2 + 14x + 49 - 121$

7. $x^4 + 20x^2 + 100 - 4$ **8.** $x^2 - 16x + 64 - 36$ **9.** $3x^6 - 6x^3 + 27x^3 - 54$

10. $2x^2 + 4x + 2 - \dfrac{2}{9}$ **11.** $a^4 - 6a^2 + 9 - 1$ **12.** $4x^4 + 12x^2 + 9 - 144$

▪▪ What must be added to each expression to make it the square of a binomial?

13. $x^2 + 4x$ **14.** $x^2 - 8x$ **15.** $x^2 - 14x$

16. $t^2 + 22t$ **17.** $d^2 + 20d$ **18.** $y^2 - 40y$

1. _____
2. _____
3. _____
4. _____
5. _____
6. _____
7. _____
8. _____
9. _____
10. _____
11. _____
12. _____
13. _____
14. _____
15. _____
16. _____
17. _____
18. _____

Copyright © 1979, Philippines copyright 1979, by Addison-Wesley Publishing Company, Inc. All rights reserved.

ANSWERS

19. _____

20. _____

21. _____

22. _____

23. _____

24. _____

25. _____

26. _____

27. _____

28. _____

29. _____

30. _____

31. _____

32. _____

●●● Factor by completing the square.

19. $x^2 + 6x + 8$ **20.** $y^2 + 8y + 15$

21. $x^2 - 8x + 12$ **22.** $y^2 + 10y + 16$

23. $2x^2 - 16x - 256$ **24.** $2x^2 - 28x + 26$

25. $a^2 - 24a - 25$ **26.** $c^2 - 8c + 15$

27. $10b^2 - 50b - 840$ **28.** $10t^2 - 200t + 640$

29. $x^2 - 14x - 72$ **30.** $r^2 - 22r + 21$

_____

31. ▦ Factor: $(x + 0.1)^2 - x^2$.   **32.** ▦ Factor by completing the square: $x^2 + 10.44x - 1201.9552$.

SYSTEMS OF
EQUATIONS
AND GRAPHS

READINESS CHECK: SKILLS FOR CHAPTER 6

Solve and check.

1. $\frac{5}{2} + y = \frac{1}{3}$ **2.** $-2x + 9 = -11$ **3.** $-8x + 3x = 25$

Evaluate each polynomial for $x = 4$.

4. $9 - 5x$ **5.** $3x^2 - x + 8$

6.1 GRAPHS AND EQUATIONS

POINTS AND ORDERED PAIRS

On a number line each point is the graph of a number. On a plane each point is the graph of a number pair. We use two perpendicular number lines called *axes*. They cross at a point called the *origin*. It is labeled 0. The arrows show the positive directions.

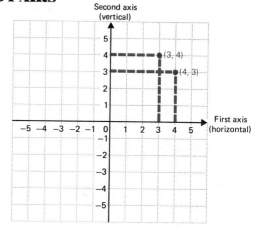

◦ PLOTTING POINTS

Note that (4,3) and (3,4) give different points. They are called *ordered pairs* of numbers because it makes a difference which number comes first.

Example 1   Plot the point $(-3,4)$.

The first number, -3, is negative. We go -3 units in the first direction (3 units to the left). The second number, 4, is positive. We go 4 units in the second direction (up).

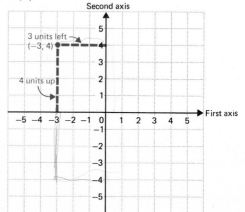

OBJECTIVES

After finishing Section 6.1, you should be able to:

◦ Plot points associated with ordered pairs of numbers.

◦◦ Determine the quadrant in which a point lies.

◦◦◦ Find coordinates of a point on a graph.

◦◦ Determine whether an ordered pair of numbers is a solution of an equation with two variables.

Plot these points on the graph below.

1. $(4,5)$ **2.** $(5,4)$

3. $(-2,5)$ **4.** $(-3,-4)$

5. $(5,-3)$ **6.** $(-2,-1)$

7. $(0,-3)$ **8.** $(2,0)$

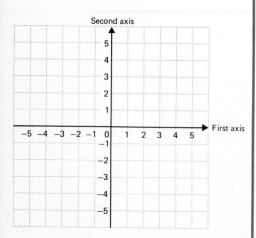

The numbers in an ordered pair are called *coordinates*. In $(-3,4)$, the *first coordinate* is -3 and the *second coordinate** is 4.

◼◼ QUADRANTS

This drawing shows some points and their coordinates. In region I (the first *quadrant*) both coordinates of any point are positive. In region II (the second quadrant) the first coordinate is negative and the second positive, and so on.

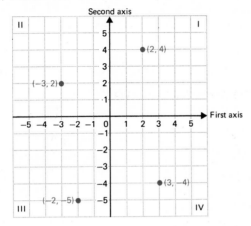

DO EXERCISES 9–14.

◼◼◼ FINDING COORDINATES

To find coordinates of a point, we see how far to the right or left of zero it is located and how far up or down.

Example 2   Find the coordinates of point A.

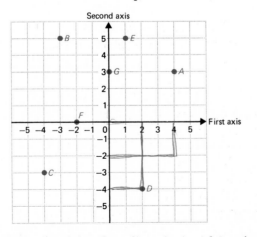

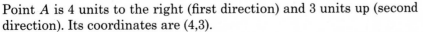

Point A is 4 units to the right (first direction) and 3 units up (second direction). Its coordinates are (4,3).

DO EXERCISE 15.

---

* The first coordinate of a point is sometimes called its *abscissa*, and the second coordinate its *ordinate*.

9. What can you say about the coordinates of a point in the third quadrant?

10. What can you say about the coordinates of a point in the fourth quadrant?

In which quadrant is each point located?

11. $(5,3)$

12. $(-6,-4)$

13. $(10,-14)$

14. $(-13,9)$

15. Find the coordinates of points B, C, D, E, F, and G in the drawing of Example 2.

16. Determine whether (2,3) is a solution of $y = 2x + 3$.

▄▄ SOLUTIONS OF EQUATIONS

An equation with two variables has *pairs* of numbers for solutions. We usually take the variables in alphabetical order. Then we get *ordered* pairs for solutions.

Example 3   Determine whether (3,7) is a solution of $y = 2x + 1$.

$$y = 2x + 1$$

7	$2 \cdot 3 + 1$
	$6 + 1$
	7

We substitute 3 for x and 7 for y (alphabetical order of variables)

The equation becomes true. (3,7) is a solution.

Example 4   Determine whether $(-2,3)$ is a solution of $2t = 4s - 8$.

$$2t = 4s - 8$$

$2 \cdot 3$	$4(-2) - 8$
6	$-8 - 8$
	-16

We substitute -2 for s and 3 for t.

The equation becomes false. $(-2,3)$ is not a solution.

17. Determine whether $(-2,4)$ is a solution of $4q - 3p = 22$.

DO EXERCISES 16 AND 17.

SOMETHING EXTRA

AN APPLICATION: COORDINATES

Three-dimensional objects can also be coordinatized. 0° latitude is the equator. 0° longitude is a line from the North Pole to the South Pole through France and Spain. In the drawing below, the hurricane Clara is at a point about 260 miles northwest of Bermuda near latitude 36.0 North, longitude 69.0 West.

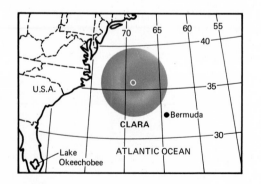

EXERCISES

1. Approximate the latitude and longitude of Bermuda.
2. Approximate the latitude and longitude of Lake Okeechobee.

NAME   CLASS   ANSWERS

EXERCISE SET 6.1

1. Plot these points.

    $(2,5)$ $(-1,3)$ $(3,-2)$ $(-2,-4)$

    $(0,4)$ $(0,-5)$ $(5,0)$ $(-5,0)$

2. Plot these points.

    $(4,4)$ $(-2,4)$ $(5,-3)$ $(-5,-5)$

    $(0,4)$ $(0,-4)$ $(3,0)$ $(-4,0)$

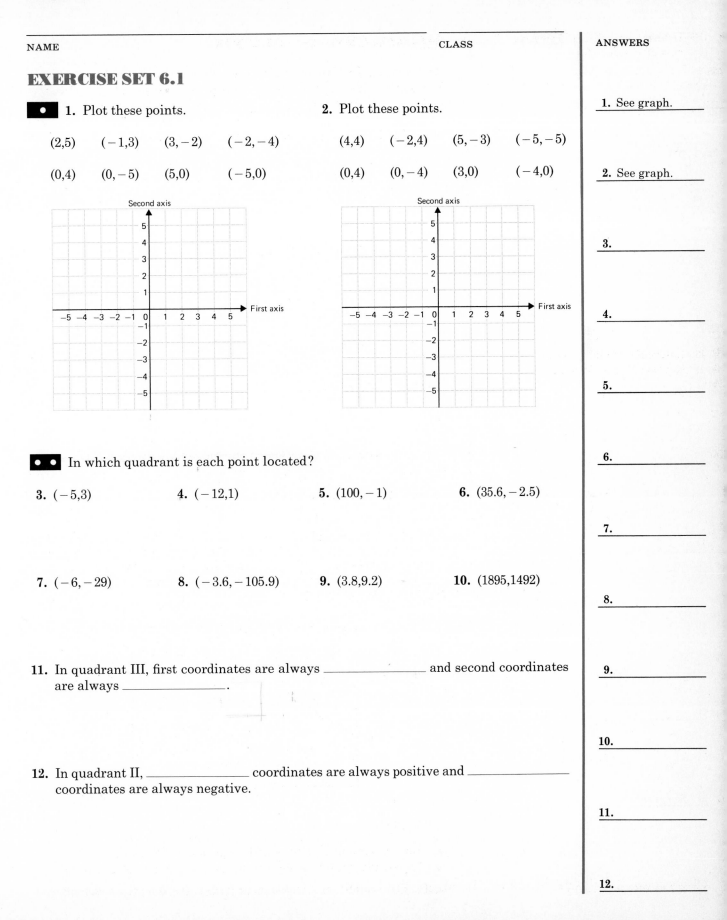

In which quadrant is each point located?

3. $(-5,3)$ **4.** $(-12,1)$ **5.** $(100,-1)$ **6.** $(35.6,-2.5)$

7. $(-6,-29)$ **8.** $(-3.6,-105.9)$ **9.** $(3.8,9.2)$ **10.** $(1895,1492)$

11. In quadrant III, first coordinates are always _____ and second coordinates are always _____ .

12. In quadrant II, _____ coordinates are always positive and _____ coordinates are always negative.

ANSWERS

1. See graph.

2. See graph.

3. _____

4. _____

5. _____

6. _____

7. _____

8. _____

9. _____

10. _____

11. _____

12. _____

Copyright © 1979, Philippines copyright 1979, by Addison-Wesley Publishing Company, Inc. All rights reserved.

ANSWERS

13. ●●● **13.** Find the coordinates of points A, B, C, D, and E.

13. _____

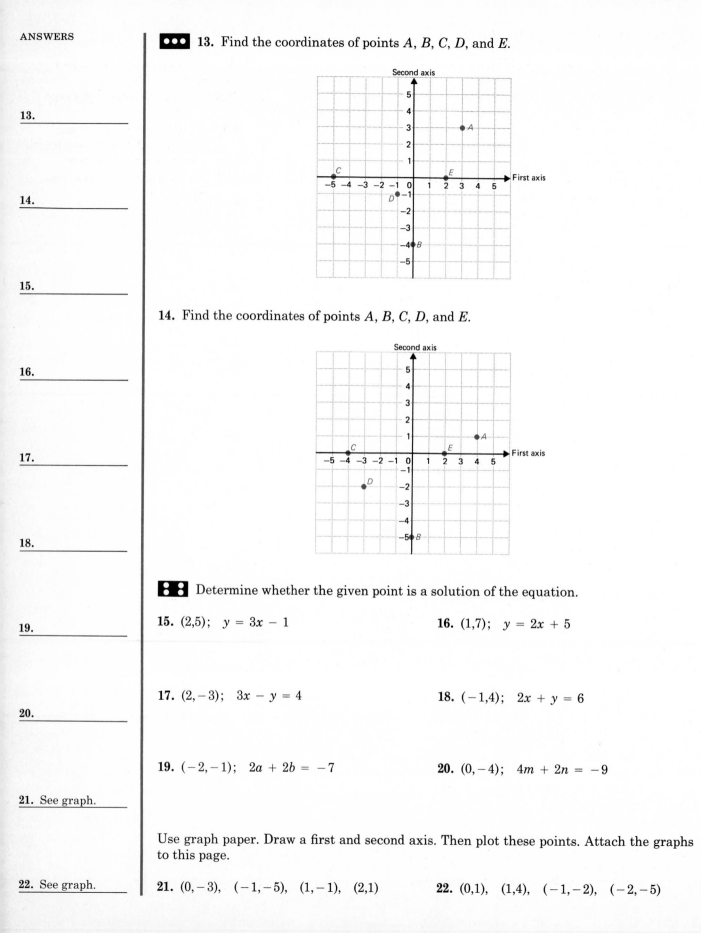

14. _____

14. Find the coordinates of points A, B, C, D, and E.

15. _____

16. _____

17. _____

18. _____

▐▌▐▌ Determine whether the given point is a solution of the equation.

15. $(2,5)$; $y = 3x - 1$ **16.** $(1,7)$; $y = 2x + 5$

19. _____

17. $(2,-3)$; $3x - y = 4$ **18.** $(-1,4)$; $2x + y = 6$

20. _____

19. $(-2,-1)$; $2a + 2b = -7$ **20.** $(0,-4)$; $4m + 2n = -9$

21. See graph. _____

Use graph paper. Draw a first and second axis. Then plot these points. Attach the graphs to this page.

22. See graph. _____

21. $(0,-3)$, $(-1,-5)$, $(1,-1)$, $(2,1)$ **22.** $(0,1)$, $(1,4)$, $(-1,-2)$, $(-2,-5)$

6.2 GRAPHING EQUATIONS

● GRAPHING EQUATIONS $y = mx$

To *graph* an equation means to make a drawing of its solutions. If an equation has a graph that is a line, we can graph it by plotting a few points and then drawing a line through them.

Example 1   Graph $y = x$.

We shall use alphabetical order. Thus, the first axis will be the x-axis and the second axis will be the y-axis. Next, we find some solutions of the equation. In this case it is easy. Here are a few:

$$(0,0), \quad (1,1), \quad (5,5), \quad (-2,-2), \quad (-4,-4).$$

Now we plot these points. We can see that if we were to plot a million solutions, the dots we draw would resemble a solid line. We see the pattern, so we can draw the line with a ruler. The line is the graph of the equation $y = x$. We label the line $y = x$ on the graph paper.

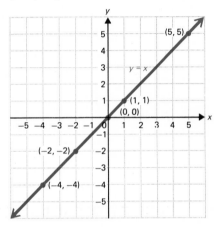

Example 2   Graph $y = 2x$.

We find some ordered pairs that are solutions, keeping the results in a table. We choose *any* number for x and then determine y by substitution. Suppose we choose 0 for x. Then

$$y = 2x = 2 \cdot 0 = 0.$$

We get a solution, the ordered pair $(0,0)$. Suppose we choose 3 for x. Then

$$y = 2x = 2 \cdot 3 = 6.$$

We get a solution, the ordered pair $(3,6)$. Continuing in this manner we get a table like the one shown below.

x	y
0	0
3	6
1	2
-2	-4
-3	-6

In this case, since $y = 2x$, we get y by doubling x.

After finishing Section 6.2, you should be able to:

● Graph equations of the type $y = mx$.

●● Given an equation $y = mx$, determine the slope and use it in graphing the equation.

●●● Graph equations of the type $y = mx + b$.

Graph.

1. $y = 3x$

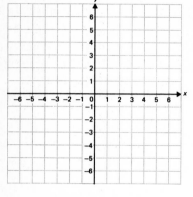

2. $y = \frac{1}{2}x$

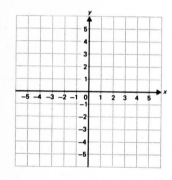

3. $y = -x$ (or $-1 \cdot x$)

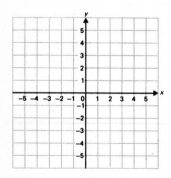

4. $y = -2x$

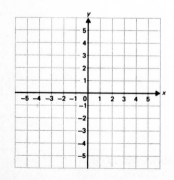

Now we plot these points. If we had enough of them, they would make a line. We draw it with a ruler and label it $y = 2x$.

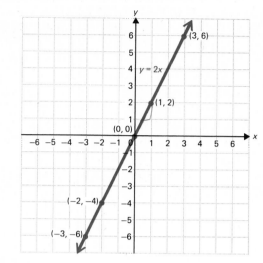

DO EXERCISES 1–4.

●● SLOPE

Every equation of the type $y = mx$ has a graph that is a straight line. It contains the origin $(0,0)$. The number m, called the *slope*, tells us how the line slants.

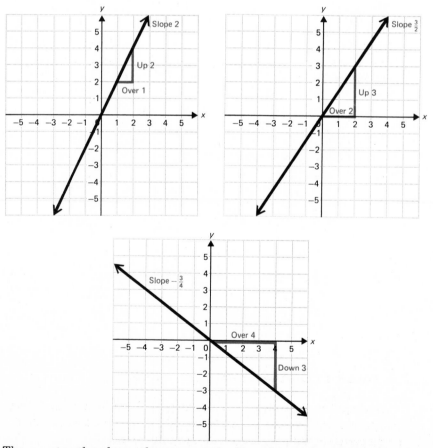

The greater the slope, the steeper the line slants, from left to right. If the slope is negative, the line slants down from left to right.

Example 3 Determine the slope of $y = \frac{2}{3}x$ and draw its graph.

The *slope* is $\frac{2}{3}$. This means that the graph slants up 2 units for every 3 across. This gives us a point (3,2), besides (0,0). We plot these points and draw a line through them. We could also have found these solutions by substituting 3 for x and 0 for x.

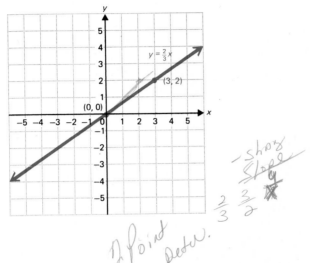

DO EXERCISE 5.

Example 4 Determine the slope of $y = -\frac{5}{3}x$ and draw its graph.

The *slope* is

$$-\frac{5}{3}, \quad \text{or} \quad \frac{-5}{3}.$$

Since the slope is negative, we know that the graph slants down 5 units for every 3 across. This gives us a point $(3, -5)$, besides (0,0). We plot these points and draw a line through them.

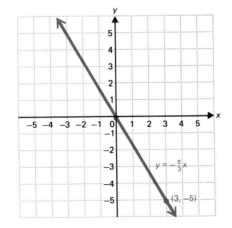

DO EXERCISE 6.

⬤⬤⬤ GRAPHS OF EQUATIONS $y = mx + b$

We know that the graph of any equation $y = mx$ is a straight line through the origin, with slope m. What will happen if we add a number b on the right side to get an equation $y = mx + b$?

5. Determine the slope of $y = \frac{3}{4}x$ and use it to draw a graph of the equation.

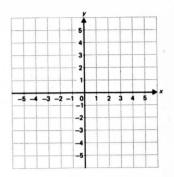

6. Determine the slope of $y = -\frac{4}{5}x$ and use it to draw a graph of the equation.

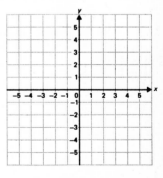

7. Graph $y = x + 3$ and compare it with $y = x$.

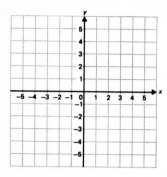

8. Graph $y = x - 1$ and compare it with $y = x$.

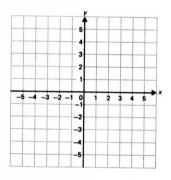

9. Graph $y = 2x + 3$ and compare it with $y = 2x$.

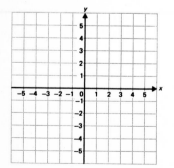

Example 5   Graph $y = x + 2$ and compare it with $y = x$.

We first make a table of values.

x	y (or $x + 2$)
0	2
1	3
-1	1
2	4
-2	0
3	5

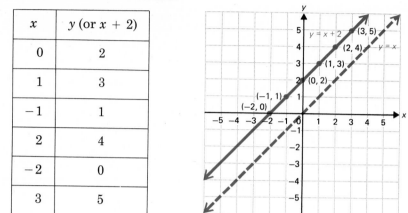

We plot these points. If we had enough of them they would make a line. We draw this line with a ruler and label it $y = x + 2$. The graph of $y = x$ is drawn for comparison. Note that the graph of $y = x + 2$ looks just like the graph of $y = x$, but it is moved up 2 units.

DO EXERCISES 7 AND 8.

Example 6   Graph $y = 2x - 3$ and compare it with $y = 2x$.

We first make a table of values.

x	y (or $2x - 3$)
0	-3
1	-1
2	1
-1	-5

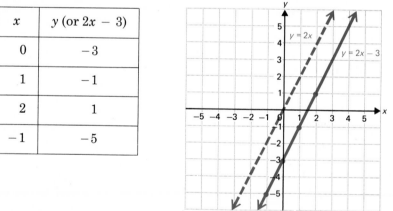

We draw the graph of $y = 2x - 3$. It looks just like the graph of $y = 2x$, but it is moved down 3 units.

DO EXERCISE 9.

The graph of $y = mx$ goes through the origin (0,0). The graph of any equation $y = mx + b$ is also a line. It is parallel to $y = mx$, but moved up or down. It goes through the point (0,b). This point is called the *y-intercept*. We may also refer to the number b as the y-intercept. The number m is still called the *slope*. It tells us how steeply the line slants.

Example 7 Graph $y = \frac{2}{5}x + 4$.

We know that the graph goes through the point (0,4) and has a slope of $\frac{2}{5}$. We locate the y-intercept, (0,4). If we go *over* (to the right) 5 units and *up* 2 units (because the slope is $\frac{2}{5}$), we get another point on the graph, (5,6). We could also go over 10 and up 4, or over 15 and up 6. We mark these points and draw the line.

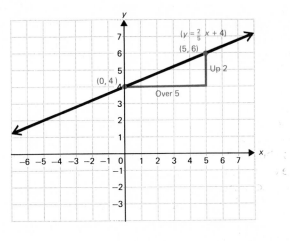

> **Any equation $y = mx + b$ has a graph that is a straight line. It goes through the point $(0,b)$, the y-intercept, and has slope m.**

Example 8 Graph $y = -\frac{3}{4}x - 2$.

We locate the y-intercept, $(0, -2)$. The slope is $-\frac{3}{4}$, so we go over 4 and down 3 and mark a point $(4, -5)$. Then we draw the line.

It is a good idea to find at least one other point as a check. If we choose -4 for x, we find by substitution that

$$y = -\frac{3}{4} \cdot (-4) - 2 = 1.$$

We plot the point $(-4,1)$ and find that it is on the graph.

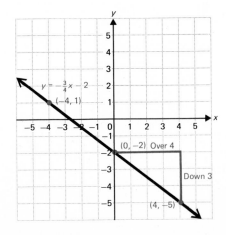

DO EXERCISES 10–13.

Determine the slope and y-intercept and then graph. Be sure to find a third point as a check.

10. $y = \frac{3}{5}x + 2$

11. $y = \frac{3}{5}x - 2$

12. $y = -\frac{3}{5}x - 1$

13. $y = -\frac{3}{5}x + 4$

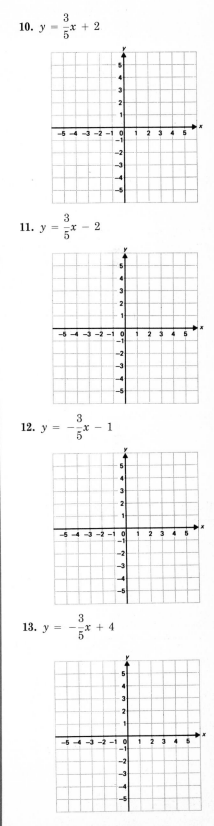

SOMETHING EXTRA

AN APPLICATION

The equation

$$P = 0.792t + 23$$

can be used to estimate the average percentage P of field goal completions by college basketball players t years from 1940. 1980 is 40 years from 1940 so the average percentage of field goal completions in 1980 is given by

$$P = 0.792(40) + 23$$
$$P = 54.68\%.$$

EXERCISES

Find the average percentage of field goal completions in

1. 1984.　　　　**2.** 1990.　　　　**3.** 1995.　　　　**4.** 2000.

EXERCISE SET 6.2

 Graph.

1. $y = 4x$

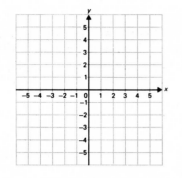

2. $y = 2x$

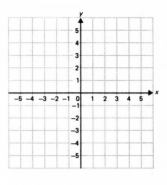

3. $y = -2x$

4. $y = -4x$

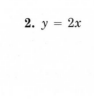 Determine the slope of each and use it to graph the equation.

5. $y = \dfrac{1}{3}x$

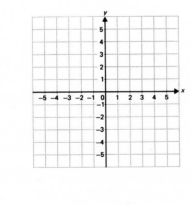

6. $y = \dfrac{1}{4}x$

7. $y = -\dfrac{3}{2}x$

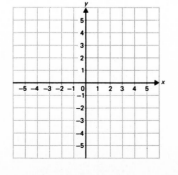

8. $y = -\dfrac{5}{4}x$

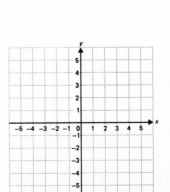

Copyright © 1979, Philippines copyright 1979, by Addison-Wesley Publishing Company, Inc. All rights reserved.

●●● Graph.

9. $y = x + 1$

10. $y = -x + 1$

11. $y = 2x + 2$

12. $y = 3x - 2$

13. $y = \dfrac{1}{3}x - 1$

14. $y = \dfrac{1}{2}x + 1$

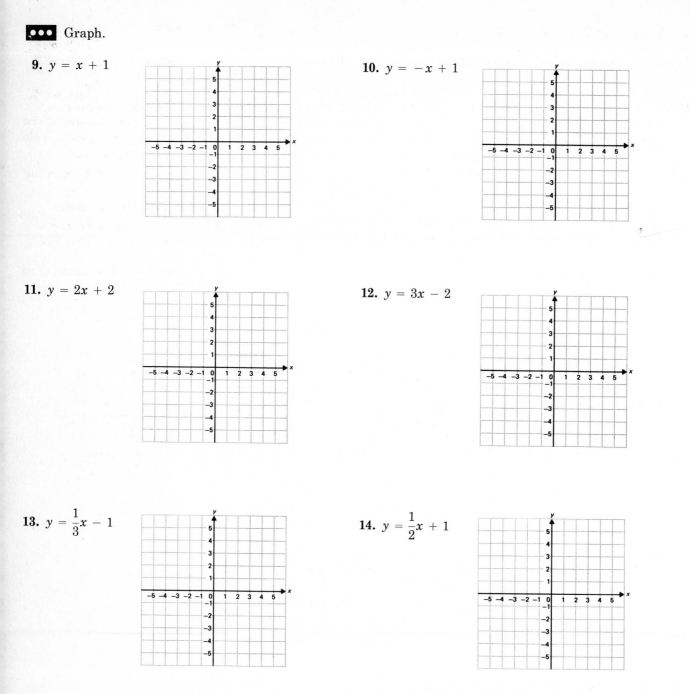

Use graph paper. Draw and label x- and y-axes. Graph these equations. Attach graphs to this page.

15. $y = -x - 3$

16. $y = -x - 2$

17. $y = \dfrac{5}{2}x + 3$

18. $y = \dfrac{5}{3}x - 2$

19. $y = -\dfrac{5}{2}x - 2$

20. $y = -\dfrac{5}{3}x + 3$

6.3 LINEAR EQUATIONS

● OTHER EQUATIONS HAVING LINES FOR GRAPHS

We know that any equation $y = mx + b$ has a straight-line graph. There are other equations that have straight-line graphs. If we solve an equation for y and get something like $y = mx + b$, we know that the graph is a straight line.

Example 1 Graph the equation $4x + 3y = 12$.

We solve the equation for y:

$$4x + 3y = 12$$
$$3y = -4x + 12 \qquad \text{Adding } -4x$$
$$y = \frac{1}{3}(-4x + 12) \qquad \text{Multiplying by } \frac{1}{3}$$
$$y = -\frac{4}{3}x + \frac{12}{3}$$
$$y = -\frac{4}{3}x + 4$$

From this equation we see that the slope is $-\frac{4}{3}$ and the y-intercept is $(0,4)$. We proceed as before to draw the graph.

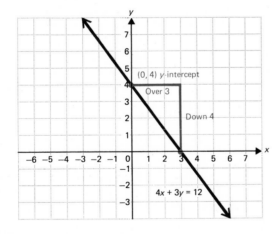

DO EXERCISE 1.

●● GRAPHING USING INTERCEPTS

There is another way to graph equations such as the one in Example 1. We find the points at which the line crosses the axes. These are called *intercepts*.

Example 2 Graph $4x + 3y = 12$.

To find the x-intercept, let $y = 0$. Then

$$4x + 3 \cdot 0 = 12$$
$$4x = 12$$
$$x = 3.$$

Thus, $(3,0)$ is the x-intercept.

OBJECTIVES

After finishing Section 6.3, you should be able to:

● Given certain equations, solve for y to put them in the form $y = mx + b$. Then determine the slope and y-intercept and graph.

●● Graph using intercepts.

●●● Graph equations with a missing variable.

⊟ Recognize linear equations.

⊟⊟ Determine whether graphs of linear equations are parallel.

1. Solve for y. Determine the slope and y-intercept and then graph.

$$2x - 3y = 6$$

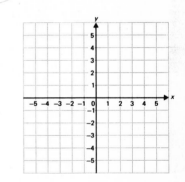

ANSWER ON PAGE A–17

Graph using intercepts.

2. $2x + 3y = 6$

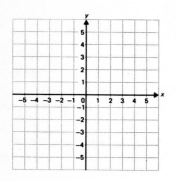

Note that this amounts to covering up the y-term and looking at the rest of the equation.

To find the y-intercept, let $x = 0$. Then

$$4 \cdot 0 + 3y = 12$$
$$3y = 12$$
$$y = 4.$$

Thus, $(0,4)$ is the y-intercept.

We plot these points and draw the line. A third point should be used as a check.

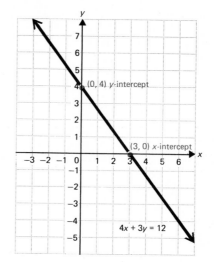

DO EXERCISES 2 AND 3.

3. $3y - 4x = 12$

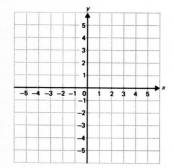

●●●   EQUATIONS WITH A MISSING VARIABLE

Consider the equation $y = 3$. We can think of it as $y = 0 \cdot x + 3$. No matter what number we choose for x, we find that y is 3.

Example 3   Graph $y = 3$.

Any ordered pair $(x,3)$ is a solution. So the line is parallel to the x-axis with y-intercept $(0,3)$.

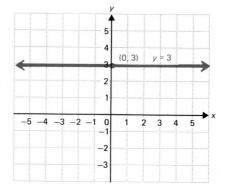

Example 4   Graph $x = -4$.

Any ordered pair $(-4, y)$ is a solution. So the line is parallel to the y-axis with x-intercept $(-4, 0)$.

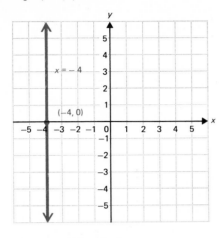

The graph of $y = b$ is a horizontal line. The graph of $x = a$ is a vertical line.

DO EXERCISES 4–7.

:: LINEAR EQUATIONS

Equations whose graphs are straight lines are called *linear equations*. An equation is linear if the variables occur to the first power only. There must be no products of variables or variables in denominators.

Example 5   Which of the following equations are linear?

a) $3x + 5 = 3y$

b) $xy = 2$

c) $5p^3 = 17q$

d) $5x = -4$

e) $9s - 15t = t$

f) $x = \dfrac{3}{y}$

Equations (a), (d), and (e) are linear equations.

In linear equations the variables occur to the first power only. Thus they are also called *first-degree equations*.

DO EXERCISE 8 (see top of p. 240).

:: PARALLEL LINES

Parallel lines are lines on the same plane that do not meet. Graphs of two linear equations meet (intersect) or they may be parallel.

Graph.

4. $x = 5$

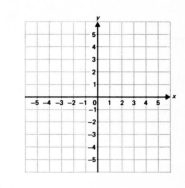

5. $y = -2$

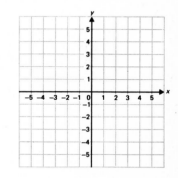

6. $x = 0$

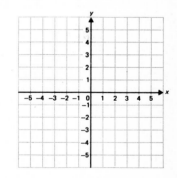

7. $x = -3$

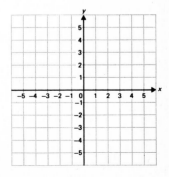

ANSWERS ON PAGE A–17

8. Which of these equations are linear (first degree)?

    a) $3y + 4x = 7$

    b) $5y = 10$

    c) $4p^2 + q = 5$

    d) $3x^2y = 4$

    e) $s = 3 + \dfrac{1}{t}$

    f) $4x + 3y = 2y + 5$

Without graphing, tell whether these lines are parallel.

9. $y + 4 = x$
    $x - y = -2$

10. $y + 1 = 2x$
    $3x - y = -5$

Example 6   The graphs of $3x - y = -1$ and $2y = -3x - 2$ intersect.

We solve each equation for y:

    $y = 3x + 1$;

    $y = -\dfrac{3}{2}x - 1$.

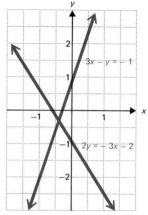

The slopes are different. The lines are not parallel.

Example 7   The graphs of $y - 3x = 5$ and $3x - y = 2$ are parallel.

Solve each equation for y:

    $y = 3x + 5$;
    $y = 3x - 2$.

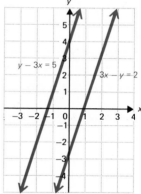

The slopes are the same, but the y-intercepts are different. The lines are parallel.

> To tell without graphing whether lines are parallel, solve the equations for y. Parallel lines have the same slope, but different y-intercepts.

DO EXERCISES 9 AND 10.

NAME   CLASS   ANSWERS

EXERCISE SET 6.3

● Solve for y. Determine the slope and y-intercept. Then graph.

1. $5x - 3y = 15$ **2.** $2x - 4y = 8$

1. _____

2. _____

3. $4x + 2y = 8$ **4.** $3x + 5y = 15$

3. _____

4. _____

●● Find the intercepts of each equation. Then graph.

5. $x - 1 = y$ **6.** $x - 3 = y$

5. _____

6. _____

7. $2x - 1 = y$ **8.** $3x - 2 = y$

7. _____

8. _____

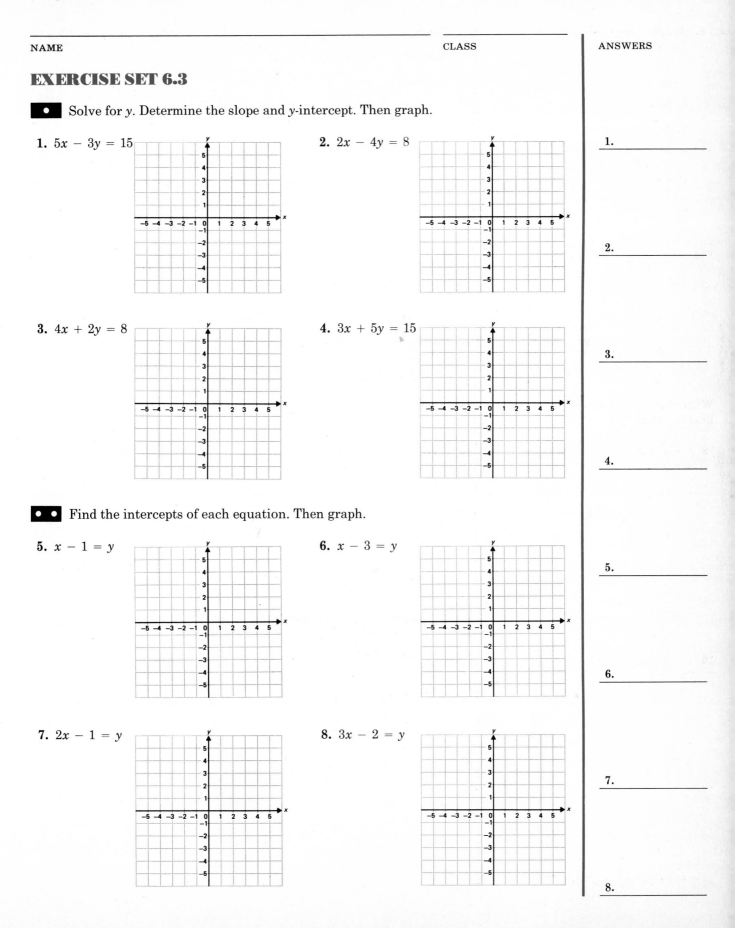

Copyright © 1979, Philippines copyright 1979, by Addison-Wesley Publishing Company, Inc. All rights reserved.

9. _____

10. _____

11. _____

12. _____

13. _____

14. _____

15. See graph. _____

16. See graph. _____

17. See graph. _____

18. See graph. _____

19. See graph. _____

20. See graph. _____

21. See graph. _____

22. See graph. _____

23. See graph. _____

24. See graph. _____

25. _____

26. _____

27. _____

28. _____

29. _____

30. _____

31. _____

32. _____

33. _____

34. _____

35. _____

36. _____

37. _____

38. _____

Use graph paper. Find the intercepts of each equation. Then graph. Attach graphs to this page.

9. $4x - 3y = 12$ **10.** $2x - 5y = 10$ **11.** $7x + 2y = 6$

12. $3x + 3y = 5$ **13.** $y = -4 - 4x$ **14.** $y = -3 - 3x$

●●● Graph.

15. $x = -2$

16. $x = -1$

17. $y = 2$

18. $y = 4$

Use graph paper. Graph each of the following. Attach graphs to this page.

19. $x = 2$ **20.** $x = 3$ **21.** $y = 0$

22. $y = -1$ **23.** $x = \dfrac{3}{2}$ **24.** $x = -\dfrac{5}{2}$

⋮⋮ Determine whether each equation is linear. Write "yes" or "no."

25. $3x^2 = 2x + 1$ **26.** $y = 4x - 6$ **27.** $x = 7$

28. $2x - 5y = 4$ **29.** $2s - 3 = \dfrac{1}{t}$ **30.** $3 = x^2 + y$

31. $2p^2 + q^2 = 9$ **32.** $p + 3q = 16$

⋮⋮⋮ Without graphing, tell which pairs of lines are parallel.

33. $x - y = 6$ **34.** $x - y = 8$ **35.** $2x - 3y = 4$
$\quad\ x - 7 = y$ $\quad\ x - 9 = y$ $\quad\ 7 - 6y = -4x$

36. $3x - 4y = 9$ **37.** $2x - y = 8$ **38.** $3x - 4y = 9$
$\quad\ 6 - 8y = -6x$ $\quad\ 2x + y = 3$ $\quad\ 3x + 2y = 0$

6.4   TRANSLATING PROBLEMS TO EQUATIONS

▪ ● The first and often hardest part of solving a problem is translating it to mathematical language. Translating becomes easier in many cases if we translate to more than one equation. When we do this, we use more than one variable.

Example 1   The sum of two numbers is 15. One number is four times the other. Find the numbers.

There are two statements in this problem. We translate the first one.

$$\underbrace{\text{The sum of two numbers}}_{x + y} \underbrace{\text{is}}_{=} \underbrace{15.}_{15}$$

We have used x and y for the numbers. Now we translate the second statement, remembering to use x and y.

$$\underbrace{\text{One number}}_{y} \underbrace{\text{is}}_{=} \underbrace{\text{four}}_{4} \underbrace{\text{times}}_{\cdot} \underbrace{\text{the other.}}_{x}$$

For the second statement we could have also translated to $x = 4y$. This would also have been correct. The problem has been translated to a pair or *system of equations*. We list what the variables represent and then list the equations.

Let $x =$ one number and $y =$ the other number.

$$x + y = 15$$
$$y = 4x$$

DO EXERCISE 1.

Example 2   Badger Rent-a-Car rents compact cars at a daily rate of $13.95 plus 10¢ per mile. Thirsty Rent-a-Car rents compact cars at a daily rate of $12.95 plus 12¢ per mile. For what mileage is the cost the same?

We translate the first statement, using $0.10 for 10¢.

$$\underbrace{13.95}_{13.95} \underbrace{\text{plus}}_{+} \underbrace{10¢}_{0.10} \underbrace{\text{times}}_{\cdot} \underbrace{\text{the number of miles driven}}_{m} \underbrace{\text{is}}_{=} \underbrace{\text{cost.}}_{c}$$

We have let m represent the mileage and c the cost. We translate the second statement, but again it helps to reword it first.

$$\underbrace{12.95}_{12.95} \underbrace{\text{plus}}_{+} \underbrace{12¢}_{0.12} \underbrace{\text{times}}_{\cdot} \underbrace{\text{the number of miles driven}}_{m} \underbrace{\text{is}}_{=} \underbrace{\text{cost.}}_{c}$$

We have now translated to a system of equations. Let $m =$ mileage and $c =$ cost.

$$13.95 + 0.10m = c$$
$$12.95 + 0.12m = c$$

DO EXERCISE 2.

ANSWERS ON PAGE A–19

OBJECTIVE

After finishing Section 6.4, you should be able to:

▪ ● Translate problems to systems of equations.

Translate to a system of equations. Do not attempt to solve. Save for later use.

1. The sum of two numbers is 115. The difference is 21. Find the numbers.

Translate to a system of equations. Do not attempt to solve. Save for later use.

2. Acme Rent-a-Car rents a station wagon at a daily rate of $21.95 plus 23¢ per mile. Speedo Rentzit rents a wagon for $24.95 plus 19¢ per mile. For what mileage is the cost the same?

Translate to a system of equations. Do not attempt to solve. Draw a picture if helpful. Save for later use.

3. The perimeter of a rectangle is 76 cm. The length is 17 cm more than the width. Find the length and width.

Making a drawing is often helpful. Whenever it makes sense to do so you should make a drawing before trying to translate.

Example 3 The perimeter of a rectangle is 90 cm. The length is 20 cm greater than the width. Find the length and width.

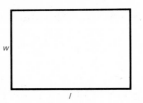

From the drawing we see that the perimeter (the distance around) of the rectangle is $l + l + w + w$, or $2l + 2w$. We translate the first statement.

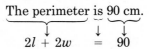

$$2l + 2w = 90$$

We translate the second statement.

The length is 20 cm greater than the width.

$$l = 20 + w$$

We have translated to a system of equations. Let w = width and l = length.

$$2w + 2l = 90$$
$$l = 20 + w$$

DO EXERCISE 3.

NAME   CLASS   ANSWERS

EXERCISE SET 6.4

▮•▮   Translate to a system of equations. Do not attempt to solve. Save for later use.

1. The sum of two numbers is 58. The difference is 16. Find the numbers.

        1. _____

2. The sum of two numbers is 26.4. One number is five times the other. Find the numbers.

        2. _____

3. The perimeter of a rectangle is 400 m. The width is 40 m less than the length. Find the length and width.

        3. _____

4. The perimeter of a rectangle is 76 cm. The width is 17 cm less than the length. Find the length and width.

        4. _____

5. Badger Rent-a-Car rents a compact car at a daily rate of $13.95 plus 10¢ per mile. Hartz Rent-a-Car rents a compact car at a daily rate of $14.95 plus 17¢ per mile. For what mileage is the cost the same?

        5. _____

6. Badger rents a basic car at a daily rate of $15.95 plus 10¢ per mile. Hartz rents a basic car at a daily rate of $15.95 plus 19¢ per mile. For what mileage is the cost the same?

        6. _____

7. The difference between two numbers is 16. Three times the larger number is seven times the smaller. What are the numbers?

        7. _____

8. The difference between two numbers is 18. Twice the smaller number plus three times the larger is 74. What are the numbers?

        8. _____

Copyright © 1979 Philippines copyright 1979 by Addison Wesley Publishing Co., Inc. All rights reserved.

ANSWERS

9. Two angles are supplementary. One is 8° more than 3 times the other. Find the angles. (Supplementary angles are angles whose sum is 180°.)

9. _____

10. Two angles are supplementary. One is 30° more than 2 times the other. Find the angles.

10. _____

11. Two angles are complementary. Their difference is 34°. Find the angles. (Complementary angles are angles whose sum is 90°.)

11. _____

12. Two angles are complementary. One angle is 42° more than $\frac{1}{2}$ the other. Find the angles.

12. _____

13. In a vineyard a vintner uses 820 hectares to plant Chardonnay and Riesling grapes. The vintner knows that profits will be greatest by planting 140 hectares more of Chardonnay than Riesling. How many hectares of each grape should be planted?

13. _____

14. The Hayburner Horse Farm allots 650 hectares to plant hay and oats. The owners know that their needs are best met if they plant 180 hectares more of hay than oats. How many hectares of each should they plant?

14. _____

15. The Skiddo Tire Shop has average monthly expenses of $6400 and pays the salespeople a 3% commission. The shop begins to make a profit when costs equal sales. Find the amount of sales at which the shop begins to profit.

15. _____

16. The Knitnest Yarn Store has average monthly expenses of $5200. The salespeople receive a 4% commission. The store begins to make a profit when costs equal sales. Find the amount of sales at which the store begins to profit.

16. _____

6.5 SYSTEMS OF EQUATIONS

■● IDENTIFYING SOLUTIONS

OBJECTIVES

After finishing Section 6.5, you should be able to:
■● Determine whether an ordered pair is a solution of a system of equations.
■●● Solve systems of equations by graphing.

A set of equations such as

$$x + y = 8$$
$$2x - y = 1$$

is called a *system of equations*. A solution of a system of two equations is an ordered pair that makes both equations true. Consider the system listed above. Look at their graphs. Which points (ordered pairs) satisfy *both* equations? The graph shows that there is only one. It is the point P where the graphs cross. This point looks as if its coordinates are (3,5). We check:

$$\frac{x + y = 8}{3 + 5 \mid 8}$$
$$8 \mid$$

$$\frac{2x - y = 1}{2 \cdot 3 - 5 \mid 1}$$
$$6 - 5 \mid$$
$$1 \mid$$

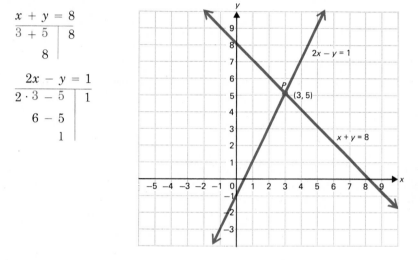

There is just one solution of the system of equations. It is (3,5). In other words, $x = 3$ and $y = 5$.

Example 1 Determine whether (1,2) is a solution of the system

$$y = x + 1$$
$$2x + y = 4.$$

$$\frac{y = x + 1}{2 \mid 1 + 1} \qquad \frac{2x + y = 4}{2 \cdot 1 + 2 \mid 4}$$
$$2 \mid 2 \qquad\qquad 2 + 2 \mid$$
$$4 \mid$$

(1,2) is a solution of the system.

Example 2 Determine whether $(-3,2)$ is a solution of the system

$$a + b = -1$$
$$b + 3a = 4.$$

$$\frac{a + b = -1}{-3 + 2 \mid -1} \qquad \frac{b + 3a = 4}{2 + 3(-3) \mid 4}$$
$$-1 \mid \qquad\qquad 2 - 9 \mid$$
$$-7 \mid$$

Determine whether the given ordered pair is a solution of the system of equations.

1. $(2, -3)$; $x = 2y + 8$

$\qquad\qquad 2x + y = 1$

2. $(20, 40)$; $a = \dfrac{1}{2}b$

$\qquad\qquad b - a = 60$

Solve by graphing.

3. $2x + y = 1$

$\quad\, x = 2y + 8$

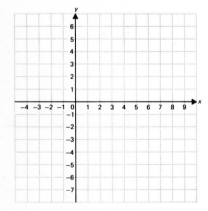

4. $3x - 2y = -4$

$\quad\, 2y - 3x = -2$

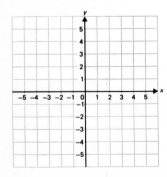

$(-3, 2)$ is not a solution of $b + 3a = 4$. Thus, it is not a solution of the system.

DO EXERCISES 1 AND 2.

● ●   SOLVING SYSTEMS BY GRAPHING

> To solve a system of equations by graphing, we graph both equations and find coordinates of the point(s) of intersection. Then we check. If the lines are parallel, there is no solution.

Example 3   Solve by graphing: $x + 2y = 7$

$\qquad\qquad\qquad\qquad\qquad\quad x = y + 4.$

We graph the equations. Point P looks as if it has coordinates $(5, 1)$.

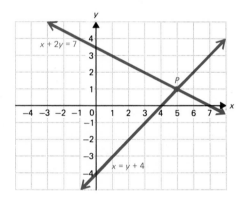

We check.

$$
\begin{array}{c|c}
x + 2y = 7 & \\
\hline
5 + 2 \cdot 1 & 7 \\
5 + 2 & \\
7 & \\
\end{array}
\qquad
\begin{array}{c|c}
x = y + 4 \\
\hline
5 & 1 + 4 \\
5 & 5 \\
\end{array}
$$

The solution is $(5, 1)$.

DO EXERCISES 3 AND 4.

EXERCISE SET 6.5

■ Determine whether the given ordered pair is a solution of the given system of equations. Remember to use alphabetical order of variables.

1. $(3,2)$; $2x + 3y = 12$
              $x - 4y = -5$

2. $(1,5)$; $5x - 2y = -5$
              $3x - 7y = -32$

1. _____

2. _____

3. $(-3,4)$; $2x = -y - 2$
               $y + 7x = 9$

4. $(3,-2)$; $3t - 2s = 0$
               $t + 2s = 15$

3. _____

4. _____

5. $(2,-2)$; $b + 2a = 2$
              $b - a = -4$

6. $(-1,-3)$; $3r + s = -6$
                $2r = 1 + s$

5. _____

6. _____

Copyright © 1979, Philippines copyright 1979, by Addison-Wesley Publishing Company, Inc. All rights reserved.

ANSWERS

●● Use graph paper. Solve by graphing. Check on this paper.

7. $x + 2y = 10$
$3x + 4y = 8$

Check: $x + 2y = 10$ $3x + 4y = 8$

7. _____

8. $u = v$
$4u = 2v - 6$

Check: $u = v$ $4u = 2v - 6$

8. _____

9. $8x - y = 29$
$2x + y = 11$

Check: $8x - y = 29$ $2x + y = 11$

9. _____

10. $4x - y = 10$
$3x + 5y = 19$

Check: $4x - y = 10$ $3x + 5y = 19$

10. _____

11. $a = 3b$
$3b - 6 = 2a$

Check: $a = 3b$ $3b - 6 = 2a$

11. _____

12. $x - 2y = 6$
$2x - 3y = 5$

Check: $x - 2y = 6$ $2x - 3y = 5$

12. _____

6.6 SOLVING BY SUBSTITUTION

● THE SUBSTITUTION METHOD

There are faster and more accurate methods of solving systems of equations than by graphing. One is called the *substitution method*.

Example 1   Solve the system: $x + y = 6$
$$x = y + 2.$$

The second equation says that x and $y + 2$ name the same thing. Thus in the first equation, we can substitute $y + 2$ for x:

$$x + y = 6$$
$$(y + 2) + y = 6. \qquad \text{Substituting } y + 2 \text{ for } x$$

This last equation has only one variable. We solve it:

$$(y + 2) + y = 6$$
$$2y + 2 = 6 \qquad \text{Collecting like terms}$$
$$2y = 4$$
$$y = 2.$$

We return to the original pair of equations. We substitute 2 for y in either of them. We use the first equation:

$$x + y = 6$$
$$x + 2 = 6 \qquad \text{Substituting } 2 \text{ for } y$$
$$x = 4.$$

The ordered pair (4,2) may be a solution. We check.

Check:

$x + y = 6$	$x = y + 2$
$4 + 2$ ∣ 6	4 ∣ $2 + 2$
6	4

Since (4,2) checks, we have the solution. We could also express the answer as $x = 4$, $y = 2$.

DO EXERCISE 1.

Example 2   Solve: $s = 13 - 3t$
$$s + t = 5.$$

We substitute $13 - 3t$ for s in the second equation:

$$s + t = 5$$
$$(13 - 3t) + t = 5. \qquad \text{Substituting } 13 - 3t \text{ for } s$$

Now we solve for t:

$$13 - 2t = 5 \qquad \text{Collecting like terms}$$
$$-2t = -8 \qquad \text{Adding } -13$$
$$t = \frac{-8}{-2}, \quad \text{or } 4 \qquad \text{Multiplying by } \frac{1}{-2}$$

OBJECTIVES

After finishing Section 6.6, you should be able to:

● Solve a system of two equations by the substitution method when one of them has a variable alone on one side.

●● Solve a system of two equations by the substitution method when neither equation has a variable alone on one side.

●●● Solve problems using the substitution method.

Solve by the substitution method. Do not graph.

1. $x + y = 5$
$x = y + 1$

Solve by the substitution method.

2. $a - b = 4$
$\quad b = 2 - a$

Next we substitute 4 for t in the second equation of the original system:

$$s + t = 5$$
$$s + 4 = 5 \qquad \text{Substituting 4 for } t$$
$$s = 1.$$

Check:
$$\begin{array}{c|c} s = 13 - 3t & s + t = 5 \\ \hline 1 & 13 - 3 \cdot 4 & 1 + 4 & 5 \\ & 13 - 12 & 5 \\ & 1 \end{array}$$

Since (1,4) checks, it is the solution.

DO EXERCISE 2.

●● SOLVING FOR THE VARIABLE FIRST

Sometimes neither equation of a pair has a variable alone on one side. Then we solve one equation for one of the variables and proceed as before.

Example 3   Solve: $x - 2y = 6$
$$\qquad\qquad\qquad 3x + 2y = 4.$$

We solve one equation for one variable. Since the coefficient of x is 1 in the first equation, it is easier to solve it for x:

$$x - 2y = 6$$
$$x = 6 + 2y.$$

We substitute $6 + 2y$ for x in the second equation of the original pair and solve:

$$3x + 2y = 4$$
$$3(6 + 2y) + 2y = 4 \qquad \text{Substituting } 6 + 2y \text{ for } x$$
$$18 + 6y + 2y = 4$$
$$18 + 8y = 4$$
$$8y = -14$$
$$y = \frac{-14}{8}, \quad \text{or} \quad -\frac{7}{4}.$$

We go back to either of the original equations and substitute $-\frac{7}{4}$ for y. It will be easier to solve for x in the first equation:

$$x - 2y = 6$$
$$x - 2\left(-\frac{7}{4}\right) = 6$$
$$x + \frac{7}{2} = 6$$
$$x = 6 - \frac{7}{2}$$
$$x = \frac{5}{2}$$

Check:

$$
\begin{array}{c|c}
x - 2y = 6 \\
\hline
\frac{5}{2} - 2(-\frac{7}{4}) & 6 \\
\frac{5}{2} + \frac{7}{2} & \\
\frac{12}{2} & \\
6 &
\end{array}
\qquad
\begin{array}{c|c}
3x + 2y = 4 \\
\hline
3 \cdot \frac{5}{2} + 2(-\frac{7}{4}) & 4 \\
\frac{15}{2} - \frac{7}{2} & \\
\frac{8}{2} & \\
4 &
\end{array}
$$

Since $(\frac{5}{2}, -\frac{7}{4})$ checks, it is the solution.

DO EXERCISE 3.

▄▄▄ SOLVING PROBLEMS

Let us solve a problem that we translated earlier.

Example 4   The sum of two numbers is 15. One number is four times the other. Find the numbers.

In Example 1 on p. 243, we translated this problem to the system

$$x + y = 15$$
$$y = 4x,$$

where we let $x =$ one number and $y =$ the other number. Now we solve. We substitute $4x$ for y in the first equation:

$$x + 4x = 15$$
$$5x = 15$$
$$x = 3 \qquad \text{Solving for } x$$

$$3 + y = 15 \qquad \text{Substituting 3 for } x \text{ in the first equation}$$
$$y = 12 \qquad \text{Solving for } y$$

We check in the original problem. The sum of 3 and 12 is 15. Four times 3 is 12. The numbers are 3 and 12.

DO EXERCISE 4.

Solve.

3. $x - 2y = 8$
   $2x + y = 8$

Translate to a system of equations. Then solve.

4. The sum of two numbers is 115. The difference is 21. Find the numbers. (We translated this problem in Margin Exercise 1, Section 6.4, p. 243.)

SOMETHING EXTRA

AN APPLICATION

DEPRECIATION: THE STRAIGHT-LINE METHOD (PART I)

A company buys a machine for $5200. The machine is expected to last for 8 years at which time its trade-in, or scrap, value will be $1300. Over its lifetime it depreciates $5200 − $1300, or $3900. If the company figures the decline in value to be the same each year—that is, $\frac{1}{8}$, or 12.5% of $3900, which is $487.50—then they are using what is called *straight-line depreciation*. We see this in the graph below. We see that the book value after one year is $5200 − $487.50, or $4712.50. After two years it is $4712.50 − $487.50, or $4225. After 3 years it is $4225 − $487.50, or $3737.50, and so on. (For more on depreciation, see p. 308.)

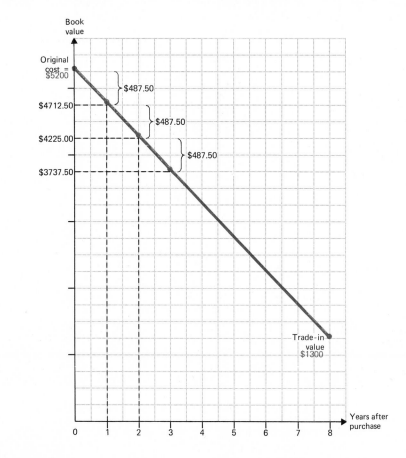

EXERCISE

Find the book values of the machine after each of the remaining years.

NAME   CLASS   ANSWERS

EXERCISE SET 6.6

▪ ○ Solve by the substitution method.

1. $x + y = 4$
$\quad y = 2x + 1$

2. $x + y = 10$
$\quad y = x + 8$

3. $y = x + 1$
$\quad 2x + y = 4$

4. $y = x - 6$
$\quad x + y = -2$

5. $y = 2x - 5$
$\quad 3y - x = 5$

6. $y = 2x + 1$
$\quad x + y = -2$

7. $x = -2y$
$\quad x + 4y = 2$

8. $r = -3s$
$\quad r + 4s = 10$

▪ ▪ Solve by the substitution method. Get one variable alone first.

9. $s + t = -4$
$\quad s - t = 2$

10. $x - y = 6$
$\quad x + y = -2$

11. $y - 2x = -6$
$\quad 2y - x = 5$

12. $x - y = 5$
$\quad x + 2y = 7$

1. _____

2. _____

3. _____

4. _____

5. _____

6. _____

7. _____

8. _____

9. _____

10. _____

11. _____

12. _____

Copyright © 1979, Philippines copyright 1979, by Addison-Wesley Publishing Company, Inc. All rights reserved.

ANSWERS

13. $2x + 3y = -2$ **14.** $x + 2y = 10$ **15.** $x - y = -3$ **16.** $3b + 2a = 2$
       $2x - y = 9$ $3x + 4y = 8$ $2x + 3y = -6$ $-2b + a = 8$

13. _____

14. _____

17. $r - 2s = 0$ **18.** $y - 2x = 0$
       $4r - 3s = 15$ $3x + 7y = 17$

15. _____

16. _____

●●● Solve.

19. The sum of two numbers is 58. The difference is 16. Find the numbers.

17. _____

18. _____

20. The sum of two numbers is 26.4. One number is five times the other. Find the numbers.

19. _____

20. _____

Solve by the substitution method.

21. ▦ $y - 2.35x = -5.97$
       $2.14y - x = 4.88$

21. _____

6.7 THE ADDITION METHOD

● SOLVING BY THE ADDITION METHOD

Another method of solving systems of equations is called the *addition method*.

Example 1   Solve: $x + y = 5$
                          $x - y = 1.$

We will use the addition principle for equations. According to the second equation, $x - y$ and 1 are the same thing. Thus, we can add $x - y$ to the left side of the first equation and 1 to the right side:

$$\begin{array}{r} x + y = 5 \\ x - y = 1 \\ \hline 2x + 0y = 6. \end{array}$$

We have made one variable "disappear." We have an equation with just one variable:

$$2x = 6.$$

We solve for x: $x = 3$. Next substitute 3 for x in either of the original equations:

$$\begin{array}{ll} x + y = 5 & \\ 3 + y = 5 & \text{Substituting 3 for } x \text{ in the first equation} \\ y = 2 & \text{Solving for } y \end{array}$$

Check:

$x + y = 5$		$x - y = 1$	
$3 + 2$	5	$3 - 2$	1
	5		1

Since (3,2) checks, it is the solution.

DO EXERCISES 1 AND 2.

●● USING THE MULTIPLICATION PRINCIPLE FIRST

The addition method allows us to eliminate a variable. We may need to multiply by -1 to make this happen.

Example 2   Solve: $2x + 3y = 8$
                          $x + 3y = 7.$

If we add, we do not eliminate a variable. However, if the $3y$ were $-3y$ in one equation we could. We multiply on both sides of the second equation by -1 and then add:

$$\begin{array}{rl} 2x + 3y = 8 & \\ -x - 3y = -7 & \text{Multiplying by } -1 \\ \hline x \phantom{{}- 3y} = 1. & \text{Adding} \end{array}$$

After finishing Section 6.7, you should be able to:

● Solve a system of two equations using the addition method when no multiplication is necessary.

●● Solve a system of two equations using the addition method when the multiplication principle must be used.

●●● Solve problems using the addition method.

Solve using the addition method.

1. $x + y = 5$
     $2x - y = 4$

2. $3x - 3y = 6$
     $3x + 3y = 0$

Solve. Multiply one equation by -1 first.

3. $5x + 3y = 17$
$\quad 5x - 2y = -3$

Now we substitute 1 for x in one of the original equations:

$$x + 3y = 7$$
$$1 + 3y = 7 \qquad \text{Substituting 1 for } x \text{ in the second equation}$$
$$3y = 6$$
$$y = 2 \qquad \text{Solving for } y$$

Check:

$$\begin{array}{c|c} 2x + 3y = 8 & x + 3y = 7 \\ \hline 2 \cdot 1 + 3 \cdot 2 \;\big|\; 8 & 1 + 3 \cdot 2 \;\big|\; 7 \\ 2 + 6 & 1 + 6 \\ 8 & 7 \end{array}$$

Since (1,2) checks, it is the solution.

DO EXERCISE 3.

In Example 2 we used the multiplication principle, multiplying by -1. We often need to multiply by something other than -1.

Example 3 Solve: $3x + 6y = -6$
$\qquad\qquad\qquad 5x - 2y = 14.$

This time we multiply by 3 on both sides of the second equation. Then we add:

$$\begin{array}{rll} 3x + 6y &= -6 & \\ 15x - 6y &= 42 & \text{Multiplying by 3} \\ \hline 18x &= 36 & \text{Adding} \\ x &= 2 & \text{Solving for } x \end{array}$$

Solve.

4. $4a + 7b = 11$
$\quad 2a + 3b = 5$

We go back to the first equation and substitute 2 for x:

$$3 \cdot 2 + 6y = -6 \qquad \text{Substituting}$$
$$6 + 6y = -6$$
$$6y = -12$$
$$y = -2 \qquad \text{Solving for } y$$

Check:

$$\begin{array}{c|c} 3x + 6y = -6 & 5x - 2y = 14 \\ \hline 3 \cdot 2 + 6 \cdot (-2) \;\big|\; -6 & 5 \cdot 2 - 2 \cdot (-2) \;\big|\; 14 \\ 6 + (-12) & 10 - (-4) \\ -6 & 14 \end{array}$$

Since $(2, -2)$ checks, it is the solution.

DO EXERCISE 4.

Example 4   Solve: $3x + 5y = 6$
$5x + 3y = 4.$

We use the multiplication principle with both equations:

$3x + 5y = 6$

$5x + 3y = 4$

$15x + 25y = 30$ Multiply on both sides of the first equation by 5

$\underline{-15x - 9y = -12}$ Multiply on both sides of the second equation by -3

$ 16y = 18$ Adding

$ y = \dfrac{18}{16}, \quad \text{or} \ \dfrac{9}{8}.$

We substitute $\dfrac{9}{8}$ for y in one of the original equations:

$3x + 5y = 6$

$3x + 5 \cdot \dfrac{9}{8} = 6$ Substituting $\dfrac{9}{8}$ for y in the first equation

$3x + \dfrac{45}{8} = 6$

$3x = 6 - \dfrac{45}{8}$

$3x = \dfrac{48}{8} - \dfrac{45}{8}$

$3x = \dfrac{3}{8}$

$x = \dfrac{3}{8} \cdot \dfrac{1}{3}, \quad \text{or} \ \dfrac{1}{8}$ Solving for x

Check:

$3x + 5y = 6$		$5x + 3y = 4$	
$3 \cdot \frac{1}{8} + 5 \cdot \frac{9}{8}$	6	$5 \cdot \frac{1}{8} + 3 \cdot \frac{9}{8}$	4
$\frac{3}{8} + \frac{45}{8}$		$\frac{5}{8} + \frac{27}{8}$	
$\frac{48}{8}$		$\frac{32}{8}$	
6		4	

The solution is $(\frac{1}{8}, \frac{9}{8})$.

DO EXERCISES 5 AND 6.

Solve.

5. $5x + 3y = 2$
$ 3x + 5y = -2$

6. $2x + 3y = 1$
$ 3x - 2y = 5$

Solve.

7. $2x + y = 15$
$4x + 2y = 23$

8. The perimeter of a rectangle is 76 cm. The length is 17 cm more than the width. Find the length and width.

Some systems have no solution.

Example 5   Solve: $y - 3x = 2$
$y - 3x = 1.$

We multiply by -1 on both sides of the second equation:

$$
\begin{array}{rl}
y - 3x = & 2 \\
-y + 3x = & -1 \qquad \text{Multiplying by } -1 \\
\hline
0 = & 1. \qquad \text{Adding}
\end{array}
$$

We obtain a false equation $0 = 1$, so there is no solution. The graphs of the equations are parallel lines. They do not intersect.

DO EXERCISE 7.

∷ SOLVING PROBLEMS

Let us solve a problem that we translated earlier. This time we use the addition method.

Example 6   The perimeter of a rectangle is 90 cm. The length is 20 cm greater than the width. Find the length and width.

In Example 3 on p. 244, we translated this problem to the system

$$2l + 2w = 90$$
$$l = 20 + w,$$

where we have let l = length and w = width. We solve, first adding $-w$ on both sides of $l = 20 + w$:

$$
\begin{array}{ll}
2l + 2w = 90 & 2l + 2w = 90 \\
l - w = 20 & \underline{2l - 2w = 40} \qquad \text{Multiplying by 2} \\
& 4l = 130 \qquad \text{Adding} \\
& l = 32.5
\end{array}
$$

$$
\begin{array}{ll}
32.5 - w = 20 & \text{Substituting in the equation } l - w = 20 \\
-w = -12.5 & \\
w = 12.5 &
\end{array}
$$

We leave the check to the student. The length is 32.5 cm and the width is 12.5 cm.

DO EXERCISE 8.

NAME   CLASS   ANSWERS

EXERCISE SET 6.7

▪ ▫   Solve using the addition method.

1. $x + y = 10$
   $x - y = 8$

2. $x - y = 7$
   $x + y = 3$

3. $x + y = 8$
   $2x - y = 7$

4. $x + y = 6$
   $3x - y = -2$

5. $3a + 4b = 7$
   $a - 4b = 5$

6. $7c + 5d = 18$
   $c - 5d = -2$

7. $8x - 5y = -9$
   $3x + 5y = -2$

8. $7a - 3b = -12$
    $-4a + 3b = -3$

▪ ▪   Solve using the addition method. Use the multiplication principle first.

9. $-x - y = 8$
   $2x - y = -1$

10. $x + y = -7$
    $3x + y = -9$

11. $x + 3y = 19$
    $x + 3y = -1$

1. _____

2. _____

3. _____

4. _____

5. _____

6. _____

7. _____

8. _____

9. _____

10. _____

11. _____

Copyright © 1979, Philippines copyright 1979, by Addison-Wesley Publishing Company, Inc. All rights reserved.

ANSWERS

12. $4x - y = 1$
    $4x - y = 7$

13. $3x - 2y = 10$
    $5x + 3y = 4$

14. $2p + 5q = 9$
    $3p - 2q = 4$

12. _____

13. _____

15. $2a + 3b = -1$
    $3a + 5b = -2$

16. $5x - 2y = 0$
    $2x - 3y = -11$

17. $0.3x + 0.2y = 0$
    $x + 0.5y = -0.5$

14. _____

15. _____

18. $0.4x + 0.1y = 0.1$
    $0.6x + 0.2y = 0.3$

19. $\dfrac{3}{4}x + \dfrac{1}{3}y = 8$

    $\dfrac{1}{2}x - \dfrac{5}{6}y = -1$

20. $\dfrac{2}{3}r - \dfrac{1}{5}s = 2$

    $\dfrac{4}{3}r + 4s = -4$

16. _____

17. _____

●●● Solve.

21. The perimeter of a rectangle is 400 m. The width is 40 m less than the length. Find the length and width.

18. _____

19. _____

22. The perimeter of a rectangle is 76 cm. The width is 17 cm less than the length. Find the length and width.

20. _____

21. _____

22. _____

Solve.

23. ▦ $4.05x + 2.53y = 1.23$
        $1.12x + 1.43y = 2.81$

23. _____

6.8 SOLVING PROBLEMS

When solving problems involving systems, any method, substitution or addition, can be used.

■● **Example 1**   Badger Rent-a-Car rents compact cars at a daily rate of $13.95 plus 10¢ per mile. Thirsty Rent-a-Car rents compact cars at a daily rate of $12.95 plus 12¢ per mile. For what mileage is the cost the same?

(See Example 2, p. 243.) If we let m = mileage and c = cost, we see that the cost for Badger will be

$$13.95 + 0.10m = c.$$

The cost for Thirsty will be

$$12.95 + 0.12m = c.$$

We solve the system

$$13.95 + 0.10m = c$$
$$12.95 + 0.12m = c.$$

We use substitution since there is a variable alone on one side of an equation—in fact, both:

$$13.95 + 0.10m = 12.95 + 0.12m$$
$$1.00 + 0.10m = 0.12m \qquad \text{Adding } -12.95$$
$$1.00 = 0.02m \qquad \text{Adding } -0.10m$$

$$\frac{1.00}{0.02} = m$$

$$50 = m.$$

Thus, if the cars are driven 50 miles, the cost will be the same.

DO EXERCISE 1.

Example 2   Howie Doon is 21 years older than Izzi Retyrd. In six years Howie will be twice as old as Izzi. How old are they now?

We translate the first statement.

$$\underbrace{\text{Howie's age}}_{x} \; \underbrace{\text{is}}_{=} \; \underbrace{21}_{21} \; \underbrace{\text{more than}}_{+} \; \underbrace{\text{Izzi's age.}}_{y} \qquad \begin{array}{l}\text{Rewording}\\[4pt]\text{Translating}\end{array}$$

Of course, x represents Howie's age and y represents Izzi's age. We translate the second statement.

$$\underbrace{\text{Howie's age in six years}}_{x+6} \; \underbrace{\text{will be}}_{=} \; \underbrace{\text{twice}}_{2 \,\cdot} \; \underbrace{\text{Izzi's age in six years.}}_{(y+6)} \qquad \begin{array}{l}\text{Rewording}\\[4pt]\text{Translating}\end{array}$$

We now have translated to a system of equations:

$$x = 21 + y$$
$$x + 6 = 2(y + 6).$$

1. Acme rents a station wagon at a daily rate of $21.95 plus 23¢ per mile. Speedo Rentzit rents a wagon for $24.95 plus 19¢ per mile. For what mileage is the cost the same? (See Exercise 2, p. 243.)

2. Person A is 26 years older than Person B. In five years, A will be twice as old as B. How old are they now?

We use the addition method. We first add $-y$ on both sides of the first equation:

$$x - y = 21.$$

We also simplify the second equation:

$$x + 6 = 2y + 12$$
$$x - 2y = 6.$$

We solve the system

$$x - y = 21$$
$$x - 2y = 6.$$

We multiply on both sides of the second equation by -1 and add:

$$
\begin{array}{ll}
x - \ y = \ \ 21 & \\
\underline{-x + 2y = \ -6} & \text{Multiplying by } -1 \\
\ \ \ \ \ \ \ y = \ \ 15. & \text{Adding}
\end{array}
$$

We find x by substituting 15 for y in $x - y = 21$:

$$x - 15 = 21$$
$$x = 36.$$

We check in the original problem. Howie's age is 36, which is 21 more than 15, Izzi's age. In six years when Howie will be 42 and Izzi 21, Howie's age will be twice Izzi's age. The answer is that Howie is 36 and Izzi is 15.

DO EXERCISE 2.

NAME   CLASS

EXERCISE SET 6.8

■ Translate to systems of equations and solve. Many problems have already been translated in Exercise Set 6.4, pp. 245–246.

1. Badger rents an intermediate-size car at a daily rate of $14.95 plus 15¢ per mile. Another company rents an intermediate-size car for $17.95 plus 10¢ per mile. For what mileage is the cost the same?

1. _____

2. Badger rents a basic car at a daily rate of $15.95 plus 12¢ per mile. Another company rents a basic car for $16.95 plus 10¢ per mile. For what mileage is the cost the same?

2. _____

3. Sammy Tary is twice as old as his daughter. In four years Sammy's age will be three times what his daughter's age was six years ago. How old is each at present?

3. _____

4. Ann Shent is eighteen years older than her son. She was three times as old one year ago. How old is each?

4. _____

Copyright © 1979, Philippines copyright 1979, by Addison-Wesley Publishing Company, Inc. All rights reserved.

ANSWERS

5. The difference between two numbers is 16. Three times the larger number is seven times the smaller. What are the numbers?

5. _____

6. The difference between two numbers is 18. Twice the smaller number plus three times the larger is 74. What are the numbers?

6. _____

7. Two angles are supplementary. One is 8° more than 3 times the other. Find the angles. (Supplementary angles are angles whose sum is 180°.)

7. _____

8. Two angles are supplementary. One is 30° more than 2 times the other. Find the angles.

8. _____

9. Two angles are complementary. Their difference is 34°. Find the angles. (Complementary angles are angles whose sum is 90°.)

10. Two angles are complementary. One angle is 42° more than $\frac{1}{2}$ the other. Find the angles.

9. _____

11. In a vineyard a vintner uses 820 hectares to plant Chardonnay and Riesling grapes. The vintner knows that profits will be greatest by planting 140 hectares more of Chardonnay than Riesling. How many hectares of each grape should be planted?

10. _____

12. The Hayburner Horse Farm allots 650 hectares to plant hay and oats. The owners know that their needs are best met if they plant 180 hectares more of hay than oats. How many hectares of each should they plant?

11. _____

12. _____

Copyright © 1979, Philippines copyright 1979, by Addison-Wesley Publishing Company, Inc. All rights reserved.

ANSWERS

13. The Skiddo Tire Shop has average monthly expenses of $6400 and pays the salespeople a 3% commission. The shop begins to make a profit when costs equal sales. Find the amount of sales at which the shop begins to profit.

13. _____

14. The Knitnest Yarn Store has average monthly expenses of $5200. The salespeople receive a 4% commission. The store begins to make a profit when costs equal sales. Find the amount of sales at which the store begins to profit.

14. _____

15. Several ancient Chinese books included problems that can be solved by translating to systems of equations. *Arithmetical Rules in Nine Sections* is a book of 246 problems compiled by a Chinese mathematician, Chang Tsang, who died in 152 B.C. One of the problems is: Suppose there are a number of rabbits and pheasants confined in a cage. In all there are 35 heads and 94 feet. How many rabbits and how many pheasants are there? Solve the problem.

15. _____

6.9 MOTION PROBLEMS

■ Many problems deal with distance, time, and speed. A basic formula is:

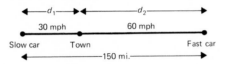

Speed $= \dfrac{\text{distance}}{\text{time}}, \qquad r = \dfrac{d}{t}.$

From $r = d/t$ we can also get the formula $d = rt$ by multiplying by t.

Example 1   Two cars leave town at the same time going in opposite directions. One of them travels 60 mph and the other 30 mph. In how many hours will they be 150 miles apart?

We first make a drawing: Note that d_1 is the distance for the first car, d_2 the distance for the second car.

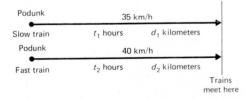

From the drawing we see that $d_1 + d_2 = 150$. Remember that $d = rt$ and substitute rt for d:

$$r_1 t_1 + r_2 t_2 = 150.$$

Note that we use r_1 and r_2 for the two speeds and t_1 and t_2 for the two times. Since the cars travel the same amount of time, t_1 and t_2 are the same, and we can just use t for time.

$$r_1 t + r_2 t = 150$$
$$30t + 60t = 150$$

Now we can solve this equation for t. We get

$$t = \frac{5}{3}, \quad \text{or } 1\frac{2}{3} \text{ hours. This checks.}$$

DO EXERCISES 1 AND 2.

Example 2   A train leaves Podunk traveling east at 35 kilometers per hour (km/h). An hour later another train leaves Podunk on a parallel track at 40 km/h. How far from Podunk will the trains meet?

First make a drawing.

From the drawing we see that the distances are the same,

$$d_1 = d_2.$$

So we just use d for distance.

OBJECTIVE

After finishing Section 6.9, you should be able to:

■ Solve motion problems using the formula $d = rt$.

1. Two cars leave town at the same time in opposite directions. One travels 48 mph and the other 60 mph. How far apart will they be 3 hours later? (*Hint:* The times are the same. Be SURE to make a drawing.)

2. Two cars leave town at the same time in the same direction. One travels 35 mph and the other 40 mph. In how many hours will they be 15 miles apart? (*Hint:* The times are the same. Be SURE to make a drawing.)

ANSWERS ON PAGE A–20

3. A car leaves Hereford traveling north at 56 km/h. Another car leaves Hereford one hour later traveling north at 84 km/h. How far from Hereford will the second car overtake the first? (*Hint:* The cars travel the same distance.)

The slow train travels 1 hour longer than the other, so

$$t_1 = t_2 + 1.$$

We can organize in a chart.

	r	t	d
Slow train	35	$t_2 + 1$	d
Fast train	40	t_2	d

Using $d = rt$ in each row of the chart, we get an equation. We get the system of equations

$$35(t_2 + 1) = d$$
$$40t_2 = d.$$

The solution is $t_2 = 7$ and $d = 280$. Thus, the distance is 280 km. This checks in the problem.

DO EXERCISE 3.

Example 3 A motorboat took 3 hr to make a downstream trip with a 6-km/h current. The return trip against the same current took 5 hr. Find the speed of the boat in still water.

First, we make a drawing. The drawing shows that the distances are the same,

$$d_1 = d_2,$$

so we just use d for distance.

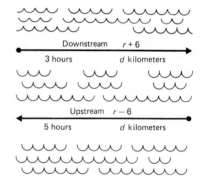

4. An airplane flew for 5 hours with a 25-km/h tail wind. The return flight against the same wind took 6 hours. Find the speed of the airplane in still air. (*Hint:* The distance is the same both ways. The speeds are $r + 25$ and $r - 25$, where r is the speed in still air.)

Let r = the speed of the boat in still water. Then, when traveling downstream, the speed of the boat is $r + 6$. When traveling upstream, the speed of the boat is $r - 6$. We can organize in a chart.

	r	t	d
Downstream	$r + 6$	3	d
Upstream	$r - 6$	5	d

Using $d = rt$ in each row of the chart, we get an equation. We get the system of equations

$$(r + 6) \cdot 3 = d$$
$$(r - 6) \cdot 5 = d$$

or

$$3r + 18 = d$$
$$5r - 30 = d.$$

The solution is $d = 90$ and $r = 24$. Thus, the speed in still water is 24 km/h. This checks in the problem.

ANSWERS ON PAGE A–20

DO EXERCISE 4.

NAME   CLASS   ANSWERS

EXERCISE SET 6.9

●   Solve.

1. Two cars leave town at the same time going in opposite directions. One travels 55 mph and the other travels 48 mph. In how many hours will they be 206 miles apart?

2. Two cars leave town at the same time going in opposite directions. One travels 44 mph and the other travels 55 mph. In how many hours will they be 297 miles apart?

3. Two cars leave town at the same time going in the same direction. One travels 30 mph and the other travels 46 mph. In how many hours will they be 72 miles apart?

4. Two cars leave town at the same time going in the same direction. One travels 32 mph and the other travels 47 mph. In how many hours will they be 69 miles apart?

1. _____

2. _____

3. _____

4. _____

Copyright © 1979, Philippines copyright 1979, by Addison-Wesley Publishing Company, Inc. All rights reserved.

ANSWERS

5. A train leaves a station and travels east at 72 km/h. Three hours later a second train leaves on a parallel track and travels east at 120 km/h. When will it overtake the first train?

5. _____

6. A private airplane leaves an airport and flies due south at 192 km/h. Two hours later a jet leaves the same airport and flies due south at 960 km/h. When will the jet overtake the private plane?

6. _____

7. A canoeist paddled for 4 hours with a 6-km/h current to reach a campsite. The return trip against the same current took 10 hours. Find the speed of the canoe in still water.

7. _____

8. An airplane flew for 4 hours with a 20-km/h tail wind. The return flight against the same wind took 5 hours. Find the speed of the plane in still air.

8. _____

9. It takes a passenger train 2 hours less time than it takes a freight train to make the trip from Central City to Clear Creek. The passenger train averages 96 km/h while the freight train averages 64 km/h. How far is it from Central City to Clear Creek?

9. _____

10. It takes a small jet plane 4 hours less time than it takes a propeller-driven plane to travel from Glen Rock to Oakville. The jet plane averages 637 km/h while the propeller plane averages 273 km/h. How far is it from Glen Rock to Oakville?

10. _____

11. An airplane took 2 hours to fly 600 km against a head wind. The return trip with the wind took $1\frac{2}{3}$ hours. Find the speed of the plane in still air.

11. _____

12. It took 3 hours to row a boat 18 km against the current. The return trip with the current took $1\frac{1}{2}$ hours. Find the speed of the rowboat in still water.

12. _____

Copyright © 1979, Philippines copyright 1979, by Addison-Wesley Publishing Company, Inc. All rights reserved.

13. A motorcycle breaks down and the rider has to walk the rest of the way to work. The motorcycle was being driven at 45 mph and the rider walks at a speed of 6 mph. The distance from home to work is 25 miles and the total time for the trip was 2 hours. How far did the motorcycle go before it broke down?

13. _____

14. A student walks and jogs to college each day. The student averages 5 mph walking and 9 mph jogging. The distance from home to college is 8 miles and the student makes the trip in 1 hour. How far does the student jog?

14. _____

15. ▦ An airplane flew for 4.23 hours with a 25.5-km/h tail wind. The return flight against the same wind took 4.97 hours. Find the speed of the plane in still air.

15. _____

6.10 COIN AND MIXTURE PROBLEMS

■• **Example 1**   A student has some nickels and dimes. The value of the coins is $1.65. There are 12 more nickels than dimes. How many of each kind of coin are there?

We will use d to represent the number of dimes and n to represent the number of nickels. We have one equation at once:

$$d + 12 = n.$$

The value of the nickels, in cents, is $5n$, since each is worth 5¢. The value of the dimes, in cents, is $10d$, since each is worth 10¢. The total value is given as $1.65. Since we have the values of the nickels and dimes *in cents*, we must use cents for the total value. This is 165. Now we have another equation:

$$10d + 5n = 165.$$

Thus we have a system of equations to solve:

$$d + 12 = n$$
$$10d + 5n = 165.$$

The solution of this system is $d = 7$ and $n = 19$. This checks, so the student has 7 dimes and 19 nickels.

DO EXERCISE 1.

Certain other types of problems are very much like coin problems. Although they are not about coins, they are solved in the same way.

Example 2   There were 411 people at a play. Admission for adults was $1.00 and for children $0.75. The receipts were $395.75. How many adults and how many children attended?

We will use a for the number of adults and c for the number of children. Since the total number is 411, we have this equation:

$$a + c = 411.$$

The amount paid in by the adults is $1 \cdot a$, since each paid $1.00. This amount is *in dollars*. In cents, the amount is $100a$.

The amount paid in by the children is $0.75c$, in dollars, or $75 \cdot c$ in cents. If we use cents instead of dollars we can avoid decimal points. Then we have this equation.

$$100a + 75c = 39575,$$

since the total amount taken in is 39575, *in cents*.

We solve the system of equations, and find that $a = 350$ and $c = 61$. This checks, so 350 adults and 61 children attended.

DO EXERCISE 2.

1. On a table are 20 coins, quarters and dimes. Their value is $3.05. How many of each are there?

2. There were 166 paid admissions to a game. The price was $2.10 for adults and $0.75 for children. The amount taken in was $293.25. How many adults and how many children attended?

3. One solution is 50% alcohol and a second is 70% alcohol. How much of each should be used to make 30 liters of a solution that is 55% alcohol?

Example 3   A chemist has one solution that is 80% acid and another that is 30% acid. What is needed is 200 liters of a solution that is 62% acid. The chemist will prepare it by mixing the two solutions on hand. How much of each should be used?

Suppose the chemist uses x liters of the first solution and y liters of the second. Since the total is to be 200 liters, we have

$$x + y = 200.$$

The *amount* of acid in the new mixture is to be 62% of 200 liters, or 124 liters. The amounts of acid from the two solutions are 80%x and 30%y. Thus,

$$80\%x + 30\%y = 124, \quad \text{or} \quad 0.8x + 0.3y = 124.$$

We eliminate the decimals by multiplying on both sides by 10:

$$\boxed{10} \, (0.8x + 0.3y) = \boxed{10} \cdot 124$$
$$8x + 3y \quad = 1240.$$

We solve this system of equations:

$$x + \; y = \;\; 200$$
$$8x + 3y = 1240.$$

The solution is $x = 128$ and $y = 72$.

Check: The sum of 128 and 72 is 200. Now 80% of 128 is 102.4 and 30% of 72 is 21.6. These add up to 124. Thus, the chemist should use 128 liters of the stronger acid and 72 liters of the other.

DO EXERCISE 3.

4. Grass seed A is worth $1.00 per pound and seed B is worth $1.35 per pound. How much of each would you use to make 50 lb of a mixture worth $1.14 per pound?

Example 4   A grocer wishes to mix some candy worth 45¢ per pound and some worth 80¢ per pound to make 350 pounds of a mixture worth 65¢ per pound. How much of each should be used?

We will use x and y for the amounts. Then

$$x + y = 350.$$

Our second equation will come from the values. The value of the first candy, in cents, is $45x$ (x pounds at 45¢ per pound). The value of the second is $80y$, and the value of the mixture is 350×65. Thus we have

$$45x + 80y = 350 \times 65.$$

Solving the system of equations, we get $x = 150$ lb and $y = 200$ lb. This checks.

DO EXERCISE 4.

NAME   CLASS   ANSWERS

EXERCISE SET 6.10

⬛⚫ Solve.

1. A collection of dimes and quarters is worth $15.25. There are 103 coins in all. How many of each are there?

1. _____

2. A collection of quarters and nickels is worth $1.25. There are 13 coins in all. How many of each are there?

2. _____

3. A collection of quarters and half dollars is worth $20. There are 51 coins in all. How many of each are there?

3. _____

4. A collection of half dollars and nickels is worth $13.40. There are 34 coins in all. How many of each are there?

4. _____

5. A collection of nickels and dimes is worth $25. There are three times as many nickels as dimes. How many of each are there?

5. _____

6. A collection of nickels and dimes is worth $2.90. There are 19 more nickels than dimes. How many of each are there?

6. _____

ANSWERS

7. There were 429 people at a play. Admission was $1 for adults and 75¢ for children. The receipts were $372.50. How many adults and how many children attended?

8. The attendance at a school concert was 578. Admission was $2 for adults and $1.50 for children. The receipts were $985. How many adults and how many children attended?

9. There were 200 tickets sold for a college basketball game. Tickets for students were $0.50 and for adults were $0.75. The total amount of money collected was $132.50. How many of each type of ticket were sold?

10. There were 203 tickets sold for a wrestling match. For activity card holders the price was $1.25 and for non-card holders the price was $2. The total amount of money collected was $310. How many of each type of ticket were sold?

11. Solution A is 50% acid and solution B is 80% acid. How much of each should be used to make 100 grams of a solution that is 68% acid?

12. Solution A is 30% alcohol and solution B is 75% alcohol. How much of each should be used to make 100 liters of a solution that is 50% alcohol?

7. _____

8. _____

9. _____

10. _____

11. _____

12. _____

13. Farmer Jones has 100 liters of milk that is 4.6% butterfat. How much skim milk (no butterfat) should be mixed with it to make milk that is 3.2% butterfat?

14. A tank contains 8000 liters of a solution that is 40% acid. How much water should be added to make a solution that is 30% acid?

13. _____

15. A solution containing 30% insecticide is to be mixed with a solution containing 50% insecticide to make 200 liters of a solution containing 42% insecticide. How much of each solution should be used?

14. _____

16. A solution containing 28% fungicide is to be mixed with a solution containing 40% fungicide to make 300 liters of a solution containing 36% fungicide. How much of each solution should be used?

15. _____

16. _____

Copyright © 1979, Philippines copyright 1979, by Addison-Wesley Publishing Company, Inc. All rights reserved.

ANSWERS

17. A candy mix sells for $2.20 per kilogram. It contains chocolates worth $1.80 per kilogram and other candy worth $3.00 per kilogram. How much of each are in 15 kg of the mixture?

17. _____

18. The "Nuthouse" has 10 kg of mixed cashews and pecans worth $8.40 per kilogram. Cashews alone sell for $8.00 per kilogram and pecans sell for $9.00 per kilogram. How many kilograms of each are in the mixture?

18. _____

19. A coffee shop mixes Brazilian coffee worth $5 per kilogram with Turkish coffee worth $8 per kilogram. The mixture is to sell for $7 per kilogram. How much of each type of coffee should be used to make a mixture of 300 kg?

19. _____

20. To make a weed and feed mixture, The Green Thumb Garden Shop mixes fertilizer worth $4 per kilogram with a weed killer worth $8 per kilogram. The mixture will cost $6 per kilogram. How much of each should be used to prepare a mixture of 500 kg?

20. _____

6.11 BUSINESS AND ECONOMIC APPLICATIONS (Optional)

● BREAK-EVEN ANALYSIS

When a company manufactures a product, it invests money (this is its *cost*). When it sells the product, it takes in money (called *revenue*). Hopefully it makes a profit. Profit is computed as follows.

> **Profit = Revenue – Cost**
> **or**
> $P \;=\; R \;-\; C$

If R is greater than C, the company makes a profit. If C is greater than R, the company has a loss. If $R = C$, the company breaks even.

There are two kinds of costs. First, there are costs like rent, insurance, buying machinery, maintenance, and so on. These must be paid whether a product is produced or not. These are called *fixed costs*. Then, when a product is being produced there are costs for labor, materials, marketing, and so on. These are called *variable costs*, because they vary according to the amount and kind of product being produced. Adding these together, we find the *total cost* of producing a product.

Example 1   Buzzo Electronics, Inc., is planning to make a new kind of calculator. Fixed costs the first year will be $80,000, and it will cost $20 to produce each calculator (variable costs). Each calculator will sell for $36. Determine the profit and break-even point.

a) We know that $P = R - C$, and that

$C = $ (Fixed costs) plus (Variable costs),

$C = 80,000 + 20x$,   where x is the number of calculators produced, and

$R = 36x$. $36 times the number of calculators produced

So we have

$P = R - C$
$P = 36x - (80,000 + 20x)$,   or
$P = 36x - 80,000 - 20x$
$P = 16x - 80,000$.

b) The break-even point occurs when revenue equals total cost. So we set $R = C$ and solve:

$36x = 20x + 80,000$
$16x = 80,000$
$\;\;x = 5000.$

The firm will need to produce and sell 5000 calculators to break even.

ANSWERS ON PAGE A–21

OBJECTIVES

After finishing Section 6.11, you should be able to:

● Given information on revenue and cost, find profit and the break-even point.

●● Given supply and demand curves, find the equilibrium point.

1. Suppose, in Example 1, that

$C = 28x + 100,000$

and

$R = 45x.$

a) Find the profit.

b) Find the break-even point.

It is interesting to graph the equations in a problem like this.

C = (Fixed costs) plus (Variable costs)

$C = 80,000 + 20x$ ①
$R = 36x$ ② C and R are both in dollars

Equation 1 has an intercept on the $-axis of 80,000 and has a slope of 20.

Equation 2 has a graph that goes through the origin, with a slope of 36.

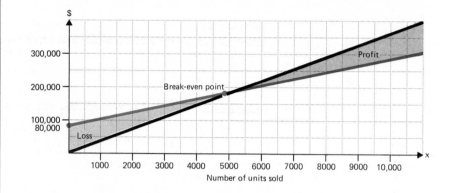

Profit occurs when the revenue is greater than the cost. Loss occurs when the revenue is less than cost. The break-even point is the point at which graphs cross.

●● SUPPLY AND DEMAND

DEMAND VARIES WITH PRICE

As the price of sugar varied over a period of years, the amount sold varied. The table and graph that follow show how the amount *demanded* by consumers went down as the price went up.

Demand schedule	
Price p (per bag)	Quantity q (number of 5-lb bags) in millions
$5	5
4	10
3	15
2	20
1	25

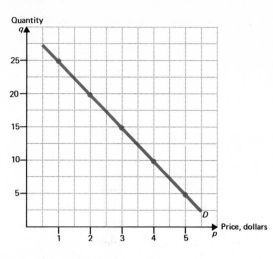

SUPPLY VARIES WITH PRICE

As the price of sugar varied, the amount available also varied. The table and graph below show how the amount of sugar on the market (the *supply*) went up as the price went up.

	Supply schedule	
Price p *(per bag)*	*Quantity q* *(number of 5-lb bags)* *in millions*	
$2	5	
2.50	10	
3	15	
3.50	20	
4	25	

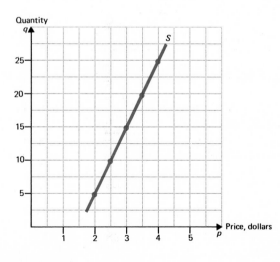

THE EQUILIBRIUM POINT

Now let us look at these graphs together. As supply increases, demand decreases. As supply decreases, demand increases. The point of intersection is called the *equilibrium point*. At this price, what the seller will supply is the same amount that the consumer will buy.

2. Demand and supply are given by

$$D = 5700 - 40p,$$

and

$$S = 325 + 3p.$$

Find the equilibrium point.

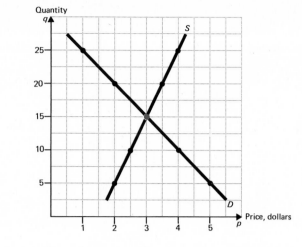

Example 2   Find the equilibrium point when demand and supply are given by

$$D = 1000 - 60p \quad \text{and} \quad S = 200 + 4p$$

a) We first set $D = S$ and solve:

$$1000 - 60p = 200 + 4p$$
$$1000 - 64p = 200 \qquad \text{Adding } -4p$$
$$-64p = -800 \qquad \text{Adding } -1000$$
$$p = \frac{-800}{-64}$$
$$p = 12.5$$

Thus $p_E = \$12.50$ per unit.

b) To find q_E we substitute p_E into either D or S. We use S:

$$200 + 4p = 200 + 4(12.5) = 200 + 50 = 250.$$

Thus, the equilibrium quantity is 250 units, and the equilibrium point is ($12.50, 250).

DO EXERCISE 2.

EXERCISE SET 6.11

■ ● ■   For each of the following, (a) find the profit; (b) find the break-even point.

1. $C = 45x + 600,000$
    $R = 65x$

2. $C = 25x + 360,000$
    $R = 70x$

3. $C = 10x + 120,000$
    $R = 60x$

4. $C = 30x + 49,500$
    $R = 85x$

5. $C = 20x + 10,000$
    $R = 100x$

6. $C = 40x + 22,500$
    $R = 85x$

■ ● ● ■   Find the equilibrium point.

7. $D = 2000 - 60p$
    $S = 460 + 94p$

8. $D = 1000 - 10p$
    $S = 250 + 5p$

9. $D = 760 - 13p$
    $S = 430 + 2p$

1. a) _____

   b) _____

2. a) _____

   b) _____

3. a) _____

   b) _____

4. a) _____

   b) _____

5. a) _____

   b) _____

6. a) _____

   b) _____

7. _____

8. _____

9. _____

Copyright © 1979, Philippines copyright 1979, by Addison-Wesley Publishing Company, Inc. All rights reserved.

ANSWERS

10. $D = 800 - 43p$
$S = 210 + 16p$

11. $D = 7500 - 25p$
$S = 6000 + 5p$

12. $D = 8800 - 30p$
$S = 7000 + 15p$

10. _____

11. _____

12. _____

13. a) _____

b) _____

c) _____

d) _____

e) _____

14. a) _____

b) _____

c) _____

d) _____

e) _____

13. A clock manufacturer is planning a new type of wall clock. For the first year, the fixed costs for setting up the production line are $22,500. Variable costs for producing each clock are estimated to be $40. The sales department projects that 3000 clocks can be sold during the first year. The revenue from each clock is $85.

a) Find the total cost C of producing x clocks.

b) Find the total revenue R from the sale of x clocks.

c) Find the total profit P.

d) What profit or loss will the company realize if expected sales of 3000 clocks occur?

e) Find the break-even point.

14. A clothing firm is planning a new line of pantsuits. For the first year, the fixed costs for setting up the production line are $10,000. Variable costs for producing each suit are $20. The sales department projects that 2000 suits can be sold during the first year. The revenue from each suit is to be $100.

a) Find the total cost C of producing x suits.

b) Find the total revenue R from the sale of x suits.

c) Find the total profit P.

d) What profit or loss will the company realize if expected sales of 2000 suits occur?

e) Find the break-even point.

NAME | CLASS | ANSWERS

TEST OR REVIEW—CHAPTER 6

If you miss an item, review the indicated section and objective.

Graph. Determine the slope and y-intercept.

[6.2, ●●●]   **1.** $y = -\dfrac{2}{5}x + 3$

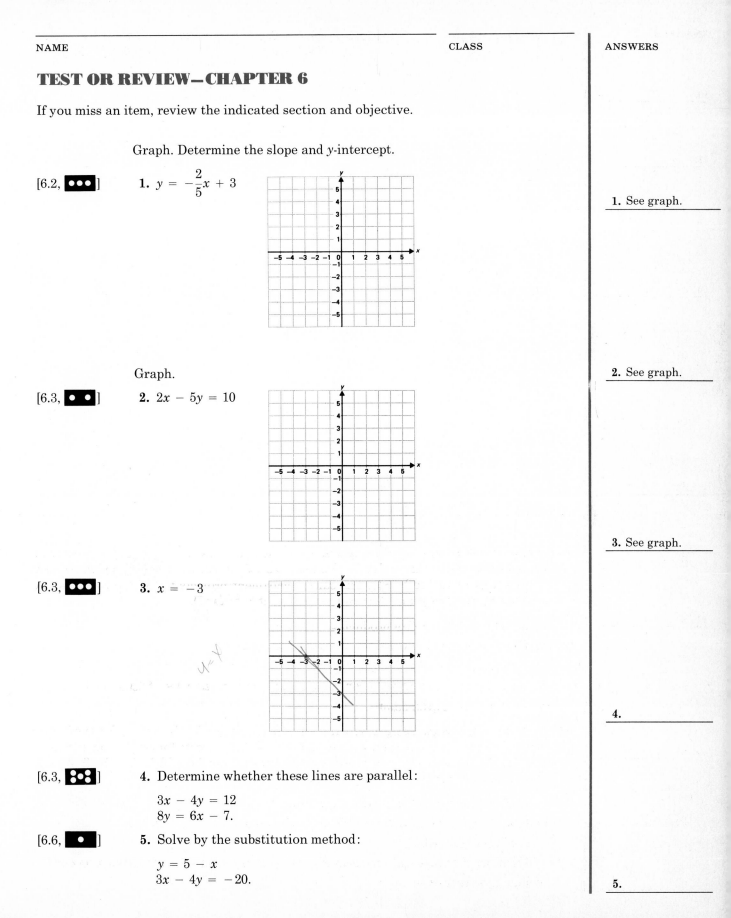

1. See graph.

Graph.

[6.3, ●●]   **2.** $2x - 5y = 10$

2. See graph.

[6.3, ●●●]   **3.** $x = -3$

3. See graph.

[6.3, ●●●●]   **4.** Determine whether these lines are parallel:

$3x - 4y = 12$
$8y = 6x - 7.$

4.

[6.6, ●]   **5.** Solve by the substitution method:

$y = 5 - x$
$3x - 4y = -20.$

5.

Copyright © 1979, Philippines copyright 1979, by Addison-Wesley Publishing Company, Inc. All rights reserved.

Solve by the addition method.

[6.7, ■]   **6.** $x + 2y = 9$
$3x - 2y = -5$

6. _____

[6.7, ■ ■]   **7.** $2x + 3y = 8$
$5x + 2y = -2$

7. _____

Solve.

[6.8, ■]   **8.** The sum of two numbers is 2. The difference is 8. Find the numbers.

$$X + Y = 2 \qquad 5 + Y = 2$$
$$\underline{X - Y = 8} \qquad Y = -3$$
$$2X = 10 \qquad 5 - 3 = 2 \checkmark$$
$$X = 5$$

[6.9, ■]   **9.** An airplane flew for 4 hours with a 15-km/h tail wind. The return flight against the same wind took 5 hours. Find the speed of the airplane in still air.

$$(X + 15)4 = (X - 15)5$$
$$4X + 60 = 5X - 75$$
$$-X = -135$$
$$X = 135$$

8. _____

[6.10, ■]   **10.** There were 508 people at an organ recital. Orchestra seats cost $5.00 per person with balcony seats costing $3.00. The total receipts were $2118. Find the number of orchestra and the number of balcony seats sold.

$$X + Y = 508 \qquad -3X - 3Y = -1524 \qquad 247 + Y = 508$$
$$\underline{5X + 3Y = 2118} \qquad \underline{5X + 3Y = 2118} \qquad Y = 261$$
$$-3[X + Y = 508] \qquad 2X = 594$$
$$\underline{5X + 3Y = 2118} \qquad X = 297$$

9. _____

[6.10, ■]   **11.** Solution A is 30% alcohol and solution B is 60% alcohol. How much of each is needed to make 80 liters of a solution that is 45% alcohol?

$$A + B = 80 \qquad -30[A + B = 80]$$
$$30A + 60B = 45(80) \qquad 30A + 60B = 3600$$
$$\underline{A + B = 80} \qquad -30A - 30B = -2400$$
$$30A + 60B = 3600 \qquad \underline{30A + 60B = 3600}$$

10. _____

[6.11, ■]   **12.** Find the break-even point.

$$C = 35x + 280,000$$
$$R = 75x$$

$$30B = 1200$$
$$B = 40$$
$$A + 40 = 80$$
$$A = 40$$
$$40 + 40 = 80$$
$$80 = 80 \checkmark$$

11. _____

12. _____

POLYNOMIALS
IN SEVERAL
VARIABLES

READINESS CHECK: SKILLS FOR CHAPTER 7

1. Evaluate for $x = -2$:

$x^3 - x^2 + 3x + 3$.

2. Collect like terms:

$6x^3 - 5x^4 - 3x^3 + 8x^4$.

3. Add:

$(4x^2 - 2x + 2) + (-3x^2 + 5x - 7)$.

4. Subtract:

$(4x^2 - 2x + 2) - (-3x^2 + 5x - 7)$.

Multiply.

5. $(3x^2 + 4x - 2)(x - 5)$

6. $(x + 2)(x - 3)$

7. $(3t + 1)(3t - 1)$

8. $(p + 4)^2$

Factor.

9. $x^2 - 7x + 2x - 14$

10. $3x^2 - 5x - 2$

11. $9t^2 - 6t + 1$

12. $x^5 - 16x^3$

13. Solve:

$3(x + 2) - 4 = 3 - 4(x - 5)$

14. Solve $d = rt$, for r.

7.1 POLYNOMIALS IN SEVERAL VARIABLES*

Most of the polynomials you have studied so far have had only one variable. A *polynomial in several variables* is an expression like those you have already seen, but we allow that there can be more than one variable. Here are some examples:

$$3x + xy^2 + 5y + 4, \quad 8xy^2z - 2x^3z - 13x^4y^2 + 5.$$

▫ EVALUATING POLYNOMIALS

Example 1 Evaluate the polynomial

$$4 + 3x + xy^2 + 8x^3y^3$$

for $x = -2$ and $y = 5$.

We replace x by -2 and y by 5:

$$4 + 3(-2) + (-2) \cdot 5^2 + 8(-2)^3 \cdot 5^3 = 4 - 6 - 50 - 8000 = -8052.$$

DO EXERCISES 1 AND 2.

Example 2 (*The magic number*) The Boston Red Sox are leading the New York Yankees for the Eastern Division championship of the American League. The magic number is 8. This means that any com-

---

OBJECTIVES

After finishing Section 7.1, you should be able to:

▫ Evaluate a polynomial in several variables for given values of the variables.

▫▫ Identify the coefficients and the degrees of the terms of a polynomial, and degrees of polynomials.

▫▫▫ Collect the like terms of a polynomial.

▫▫ Arrange a polynomial in ascending or descending order.

1. Evaluate the polynomial

$4 + 3x + xy^2 + 8x^3y^3$

for $x = 2$ and $y = -5$.

2. Evaluate the polynomial

$8xy^2 - 2x^3z - 13x^4y^2 + 5$

for $x = -1$, $y = 3$, and $z = 4$.

---

* *To the instructor:* This chapter is prerequisite only to Chapter 8. If time is short and that chapter is to be skipped, then this chapter might also be skipped.

bination of Red Sox wins and Yankee losses that totals 8 will ensure the championship for the Red Sox. The magic number is given by the polynomial

$$G - P - L + 1,$$

where G is the number of games in the season, P is the number of games the leading team has played, and L is the number of games ahead in the loss column.

Given the situation shown in the table and assuming a 162-game season, what is the magic number for the Philadelphia Phillies?

EASTERN DIVISION				
	W	L	Pct.	GB
Philadelphia	77	40	.658	—
Pittsburgh	65	53	.551	12½
New York	61	60	.504	18
Chicago	55	67	.451	24½
St. Louis	51	65	.440	25½
Montreal	41	73	.360	34½

We evaluate the polynomial for $G = 162$, $P = 77 + 40$, or 117, and $L = 53 - 40$, or 13:

$$162 - 117 - 13 + 1 = 33.$$

DO EXERCISE 3.

●● COEFFICIENTS AND DEGREES

The *degree* of a term is the sum of the exponents of the variables. The *degree of a polynomial* is the degree of the term of highest degree.

Example 3 Identify the coefficient and degree of each term of

$$9x^2y^3 - 14xy^2z^3 + xy + 4y + 5x^2 + 7.$$

Term	Coefficient	Degree	
$9x^2y^3$	9	5	
$-14xy^2z^3$	-14	6	
xy	1	2	
$4y$	4	1	Think: $4y = 4y^1$
$5x^2$	5	2	
7	7	0	Think: $7 = 7x^0$

Example 4 What is the degree of $5x^3y + 9xy^4 - 8x^3y^3$?

The term of highest degree is $-8x^3y^3$. Its degree is 6. The degree of the polynomial is 6.

DO EXERCISES 4 AND 5.

3. Given the situation below, what is the magic number for the Cincinnati Reds? Assume $G = 162$.

WESTERN DIVISION				
	W	L	Pct.	GB
Cincinnati	77	44	.636	—
Los Angeles	65	54	.546	11
San Diego	60	64	.484	18½
Houston	59	64	.480	19
Atlanta	56	65	.463	21
San Francisco	52	70	.426	25½

4. Identify the coefficient of each term.

$$-3xy^2 + 3x^2y - 2y^3 + xy + 2$$

5. Identify the degree of each term and the degree of the polynomial.

$$4xy^2 + 7x^2y^3z^2 - 5x + 2y + 4$$

Collect like terms.

6. $4x^2y + 3xy - 2x^2y$

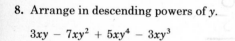

$2x^2y + 3xy$

7. $-3pq - 5pqr^3 + 8pq + 5pqr^3 + 4$

$5pq + 4$

8. Arrange in descending powers of y.

$3xy - 7xy^2 + 5xy^4 - 3xy^3$

9. Arrange in ascending powers of x.

$2x^2yz + 5xy^2z + 5x^3yz^2 - 2$

••• COLLECTING LIKE TERMS

Like terms (or *similar terms*) have exactly the same variables with exactly the same exponents. For example,

> $3x^2y^3$ and $-7x^2y^3$ are like terms;
> $9x^4z^7$ and $12x^4z^7$ are like terms.

But

> $13xy^2$ and $-2x^2y$ are *not* like terms;
> $3xyz^2$ and $4xy$ are *not* like terms.

Collecting like terms is based on the distributive law.

Examples   Collect like terms.

5. $5x^2y + 3xy^2 - 5x^2y - xy^2 = (5 - 5)x^2y + (3 - 1)xy^2 = 2xy^2$

6. $3xy - 5xy^2 + 3xy^2 + 9xy = -2xy^2 + 12xy$

7. $4ab^2 - 7a^2b^2 + 9a^2b^2 - 4a^2b = 4ab^2 + 2a^2b^2 - 4a^2b$

8. $3pq + 5pqr^3 - 8pq - 5pqr^3 - 4 = -5pq - 4$

DO EXERCISES 6 AND 7.

▪▪ ASCENDING AND DESCENDING ORDER

We usually arrange polynomials in one variable in descending order and sometimes in ascending order. For polynomials in several variables we choose one of the variables and arrange the terms with respect to it.

Examples

9. This polynomial is arranged in descending powers of x:

> $3x^5y - 5x^3 + 7xy + y^4 + 2.$

10. This polynomial is arranged in ascending powers of y:

> $3 + 13xy - x^3y^2 + 4xy^3.$

11. This polynomial is arranged in ascending powers of p:

> $3pq - 2p^2qr + 7p^3q^5r^2 + 5p^4q^2r^4 - 3p^5r^2.$

12. This polynomial is arranged in descending powers of t:

> $5s^5t^7 + 2s^6t^4 - 7s^6t^2 + 4s^3t + 6s^4.$

DO EXERCISES 8 AND 9.

NAME   CLASS   ANSWERS

EXERCISE SET 7.1

◦ Evaluate each polynomial for $x = 3$ and $y = -2$.

1. $x^2 - y^2 + xy$ **2.** $x^2 + y^2 - xy$

Evaluate each polynomial for $x = 2$, $y = -3$, and $z = -1$.

3. $xyz^2 + z$ **4.** $xy - xz + yz$

An amount of money P is invested at interest rate r. In 3 years it will grow to an amount given by the polynomial

$$P + 3rP + 3r^2P + r^3P.$$

5. Evaluate the polynomial for $P = 10,000$ and $r = 0.08$ to find the amount to which $10,000 will grow at 8% interest for 3 years.

6. Evaluate the polynomial for $P = 10,000$ and $r = 0.07$ to find the amount to which $10,000 will grow at 7% interest for 3 years.

The area of a right circular cylinder is given by the polynomial

$$2\pi rh + 2\pi r^2,$$

where h is the height and r is the radius of the base.

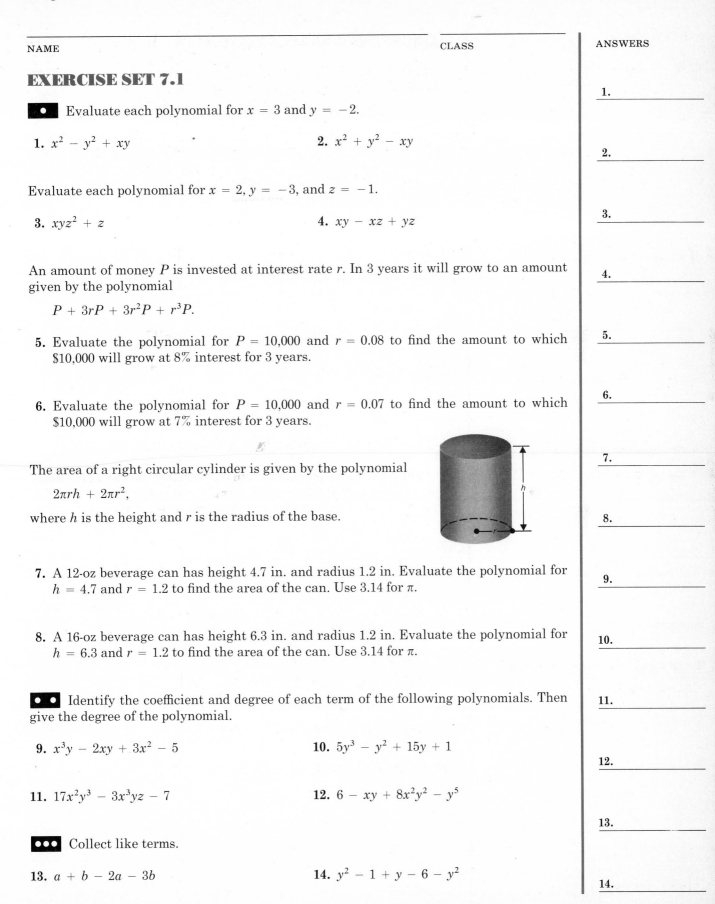

7. A 12-oz beverage can has height 4.7 in. and radius 1.2 in. Evaluate the polynomial for $h = 4.7$ and $r = 1.2$ to find the area of the can. Use 3.14 for π.

8. A 16-oz beverage can has height 6.3 in. and radius 1.2 in. Evaluate the polynomial for $h = 6.3$ and $r = 1.2$ to find the area of the can. Use 3.14 for π.

◦◦ Identify the coefficient and degree of each term of the following polynomials. Then give the degree of the polynomial.

9. $x^3y - 2xy + 3x^2 - 5$ **10.** $5y^3 - y^2 + 15y + 1$

11. $17x^2y^3 - 3x^3yz - 7$ **12.** $6 - xy + 8x^2y^2 - y^5$

◦◦◦ Collect like terms.

13. $a + b - 2a - 3b$ **14.** $y^2 - 1 + y - 6 - y^2$

ANSWERS

1. _____

2. _____

3. _____

4. _____

5. _____

6. _____

7. _____

8. _____

9. _____

10. _____

11. _____

12. _____

13. _____

14. _____

Copyright © 1979, Philippines copyright 1979, by Addison-Wesley Publishing Company, Inc. All rights reserved.

15. _____

16. _____

17. _____

18. _____

19. _____

20. _____

21. _____

22. _____

23. _____

24. _____

25. _____

26. _____

27. _____

28. _____

29. _____

30. _____

31. _____

32. _____

15. $3x^2y - 2xy^2 + x^2$ **16.** $m^3 + 2m^2n - 3m^2 + 3mn^2$

17. $2u^2v - 3uv^2 + 6u^2v - 2uv^2$

18. $3x^2 + 6xy + 3y^2 - 5x^2 - 10xy - 5y^2$

19. $6au + 3av - 14au + 7av$

20. $3x^2y - 2z^2y + 3xy^2 + 5z^2y$

21. $a^3 - a^2b + ab^2 + a^2b - ab^2$

22. $x^3 + x^2y + xy^2 - x^2y - xy^2 - y^3$

■■ Arrange in descending powers of y.

23. $-xy^2 - y^3 + 5x^2y$

24. $2xy^2 + 7xy + 4y^3$

25. $-xy^2 + x^3 - y^3$

26. $4y^3 - x^2y^2 - 2x^2y + xy^4 - y^5$

Arrange in ascending powers of m.

27. $5m^2n - 3mn^3 + 4n^2 - m^4n$

28. $5m^3n - 8mn^2 + 7m^2n - 3m^4n^3$

Arrange in descending powers of the second variable (using alphabetical order). Collect like terms if possible.

29. $x^2 + 3y^2 + 2xy$

30. $4n^2 - 3mn + m^2 - mn + n^2 + 5m^2$

31. $5uv - 8uv^2 + 7u^2v - 3uv^2$

32. $6r^2s + 11 + 3r^2s - 5r + 3$

7.2   OPERATIONS WITH POLYNOMIALS

Calculations with polynomials in several variables are very much like those for polynomials in one variable. Additional practice is needed, however, and you should strive for as much speed as possible without losing accuracy.

▪•▪ ADDITION

To add polynomials in several variables, we collect like terms.

Example 1   Add: $(3ax^2 + 4bx - 2) + (2ax^2 + 5bx + 4)$.

We can use columns. We write one polynomial under the other, keeping like terms in columns.

$$3ax^2 + 4bx - 2$$
$$\underline{2ax^2 + 5bx + 4}$$
$$5ax^2 + 9bx + 2$$

Usually you should not use columns. Try to write only the answer.

Example 2   Add: $(5xy^2 - 4x^2y + 5x^3 + 2) + (3xy^2 - 2x^2y + 3x^3y - 5)$.

We look for like terms. The like terms are $5xy^2$ and $3xy^2$, $-4x^2y$ and $-2x^2y$, and 2 and -5. We collect these. There are no more like terms. The answer is

$$8xy^2 - 6x^2y + 5x^3 + 3x^3y - 3.$$

DO EXERCISES 1 AND 2.

▪•▪• SUBTRACTION

We subtract a polynomial by adding its inverse. The additive inverse of a polynomial is found by replacing each coefficient by its additive inverse.

Example 3   Subtract.

$$(4x^2y + x^3y^2 + 3x^2y^3 + 6y) - (4x^2y - 6x^3y^2 + x^2y^2 - 5y)$$

$$= 4x^2y + x^3y^2 + 3x^2y^3 + 6y - 4x^2y$$
$$+ 6x^3y^2 - x^2y^2 + 5y \qquad \text{Taking the inverse}$$

$$= 7x^3y^2 + 3x^2y^3 - x^2y^2 + 11y \qquad \text{Adding (Try to write just the answer!)}$$

You can use columns as shown below, but you should avoid the use of columns wherever possible. Work for speed, with accuracy.

$$4x^2y + \ x^3y^2 + 3x^2y^3 \qquad\quad + 6y$$
$$\underline{4x^2y - 6x^3y^2 \qquad\quad + x^2y^2 - 5y} \qquad \text{Mentally change signs}$$
$$7x^3y^2 + 3x^2y^3 - x^2y^2 + 11y \qquad\quad \text{and add}$$

DO EXERCISES 3 AND 4.

OBJECTIVES

After finishing Section 7.2, you should be able to:

▪•▪ Add polynomials.

▪•▪• Subtract polynomials.

▪•▪•▪ Multiply polynomials.

Add.

1. $(13x^3y + 3x^2y - 5y)$
    $+ (x^3y + 4x^2y - 3xy + 3y)$

2. $(-5p^2q^4 + 2p^2q^2 + 3q)$
    $+ (6pq^2 + 3p^2q + 5)$

Subtract.

3. $(-4s^4t + s^3t^2 + 2s^2t^3)$
    $- (4s^4t - 5s^3t^2 + s^2t^2)$

4. $(-5p^4q + 5p^3q^2 - 3p^2q^3 - 7q^4)$
    $- (4p^4q - 5p^3q^2 + p^2q^3 + 2q^4)$

Multiply.

5. $(x^2y^3 + 2x)(x^3y^2 + 3x)$

6. $(p^4q - 2p^3q^2 + 3q^3)(p + 2q)$

●●● MULTIPLICATION

Multiplication of polynomials is based on the distributive laws. Recall that this means we can multiply each term of one by every term of the other. For most polynomials in several variables having three or more terms, you will probably want to use columns.

Example 4 Multiply: $(3x^2y - 2xy + 3y)(xy + 2y)$.

It may help to write one polynomial under the other.

$$
\begin{array}{l}
3x^2y - 2xy \quad + 3y \\
\underline{\quad xy + 2y} \\
3x^3y^2 - 2x^2y^2 + 3xy^2 \qquad \text{Multiplying by } xy \\
\underline{\qquad\quad 6x^2y^2 - 4xy^2 + 6y^2} \qquad \text{Multiplying by } 2y \\
3x^3y^2 + 4x^2y^2 - \quad xy^2 + 6y^2 \qquad \text{Adding}
\end{array}
$$

DO EXERCISES 5 AND 6.

NAME

CLASS

ANSWERS

EXERCISE SET 7.2

● Add.

1. $(2x^2 - xy + y^2) + (-x^2 - 3xy + 2y^2)$

2. $(2z - z^2 + 5) + (z^2 - 3z + 1)$

3. $(r - 2s + 3) + (2r + s) + (s + 4)$

4. $(b^3a^2 - 2b^2a^3 + 3ba + 4) + (b^2a^3 - 4b^3a^2 + 2ba - 1)$

5. $(2x^2 - 3xy + y^2) + (-4x^2 - 6xy - y^2) + (x^2 + xy - y^2)$

●● Subtract.

6. $(x^3 - y^3) - (-2x^3 + x^2y - xy^2 + 2y^3)$

7. $(xy - ab) - (xy - 3ab)$

8. $(3y^4x^2 + 2y^3x - 3y) - (2y^4x^2 + 2y^3x - 4y - 2x)$

9. $(-2a + 7b - c) - (-3b + 4c - 8d)$

10. Find the sum of $2a + b$ and $3a - 4b$. Then subtract $5a + 2b$.

●●● Multiply.

11. $(3z - u)(2z + 3u)$

12. $(a - b)(a^2 + b^2 + 2ab)$

13. $(a^2b - 2)(a^2b - 5)$

14. $(xy + 7)(xy - 4)$

15. $(a^2 + a - 1)(a^2 - y + 1)$

16. $(tx + r)(vx + s)$

1. _____

2. _____

3. _____

4. _____

5. _____

6. _____

7. _____

8. _____

9. _____

10. _____

11. _____

12. _____

13. _____

14. _____

15. _____

16. _____

Copyright © 1979, Philippines copyright 1979, by Addison-Wesley Publishing Company, Inc. All rights reserved.

ANSWERS

17. _____

18. _____

19. _____

20. _____

21. _____

22. _____

23. _____

24. _____

25. _____

26. _____

27. _____

28. _____

29. _____

30. _____

17. $(a^3 + bc)(a^3 - bc)$

18. $(m^2 + n^2 - mn)(m^2 + mn + n^2)$

19. $(y^4x + y^2 + 1)(y^2 + 1)$

20. $(a - b)(a^2 + ab + b^2)$

21. $(r + s)(r^2 + ry + s^2)$

22. $(a + 2b + 3c)(2a - b + 2c)$

Simplify.

23. $(xy + 1)(2xy - 3) + (xy + 1)(3xy - 1)$

24. $(2a + b)(3a - b) - (a - b)(2a - 3b)$

25. $(yz^2 + 2)(yz^2 - 2) + (2yz^2 + 1)(2yz^2 + 1)$

26. $(3a - d)(2a + 3d) - (4a + d)(2a - d)$

27. $(a - b)(a^3 + a^2b + ab^2 + b^3)$

28. $(c + d)(c^3 - c^2d + cd^2 - d^3)$

---

29. Express the colored area as a polynomial.

30. A square sandbox s ft on a side is placed on a square lawn x ft on a side. Express the area left over as a polynomial.

7.3 SPECIAL PRODUCTS OF POLYNOMIALS

● PRODUCTS OF TWO BINOMIALS

The methods of handling the special kinds of products you learned for polynomials in one variable are the same for polynomials in several variables.

Examples   Multiply.

$$\overset{\mathrm{F}\qquad\mathrm{O}\qquad\mathrm{I}\qquad\mathrm{L}}{}$$

1. $(x^2y + 2x)(xy^2 + y^2) = x^3y^3 + x^2y^3 + 2x^2y^2 + 2xy^2$

2. $(p + 5q)(2p - 3q) = 2p^2 - 3pq + 10pq - 15q^2$
$$= 2p^2 + 7pq - 15q^2$$

DO EXERCISES 1 AND 2.

●● SQUARES OF BINOMIALS

To square a binomial, square the first term. Multiply the two terms and double. Square the last term. Then add.

$$(A + B)^2 = A^2 + 2 \cdot A \cdot B + B^2$$
$$(A - B)^2 = A^2 - 2 \cdot A \cdot B + B^2$$

Examples   Multiply.

3. $(3x + 2y)^2 = (3x)^2 + 2(3x)(2y) + (2y)^2$
$$= 9x^2 + 12xy + 4y^2$$

4. $(2y^2 - 5x^2y)^2 = (2y^2)^2 - 2(2y^2)(5x^2y) + (5x^2y)^2$
$$= 4y^4 - 20x^2y^3 + 25x^4y^2$$

DO EXERCISES 3 AND 4.

●●● PRODUCTS OF SUMS AND DIFFERENCES

The product of the sum and difference of two expressions is the difference of their squares:

$$(A + B)(A - B) = A^2 - B^2.$$

OBJECTIVES

After finishing Section 7.3, you should be able to:

● Multiply two binomials mentally.

●● Square a binomial mentally.

●●● Multiply the sum and difference of two expressions mentally.

Multiply.

1. $(3xy + 2x)(x^2 + 2xy^2)$

2. $(x - 3y)(2x - 5y)$

Multiply.

3. $(4x + 5y)^2$

4. $(3x^2 - 2xy^2)^2$

Multiply.

5. $(2xy^2 + 3x)(2xy^2 - 3x)$

6. $(3xy^2 + 4y)(-3xy^2 + 4y)$

Multiply.

7. $(3y + 4 - 3x)(3y + 4 + 3x)$

8. $(2a + 5b + c)(2a - 5b - c)$

Examples   Multiply.

5. $(3x^2y + 2y)(3x^2y - 2y) = (3x^2y)^2 - (2y)^2$
$$= 9x^4y^2 - 4y^2$$

6. $(-2x^3y^2 + 5t)(2x^3y^2 + 5t) = (5t - 2x^3y^2)(5t + 2x^3y^2)$
$$= (5t)^2 - (2x^3y^2)^2$$
$$= 25t^2 - 4x^6y^4$$

DO EXERCISES 5 AND 6.

One of the expressions may have more than one term.

Examples   Multiply.

7. $(\boxed{2x + 3} - 2y)(\boxed{2x + 3} + 2y) = (\boxed{2x + 3})^2 - (2y)^2$
$$= 4x^2 + 12x + 9 - 4y^2$$

8. $(4x + 3y + 2)(4x - 3y - 2) = (4x + \boxed{3y + 2})(4x - [\,\boxed{3y + 2}\,])$
$$= 16x^2 - (\boxed{3y + 2})^2$$
$$= 16x^2 - (9y^2 + 12y + 4)$$
$$= 16x^2 - 9y^2 - 12y - 4$$

DO EXERCISES 7 AND 8.

ANSWERS ON PAGE A-22

NAME   CLASS   ANSWERS

EXERCISE SET 7.3

•   Multiply.

1. $(3xy - 1)(4xy + 2)$ **2.** $(m^3n + 8)(m^3n - 6)$

3. $(3 - c^2d^2)(4 + c^2d^2)$ **4.** $(6x - 2y)(5x - 3y)$

5. $(m^2 - n^2)(m + n)$ **6.** $(pq + 0.2)(0.4pq - 0.1)$

7. $(x^5y^5 + xy)(x^4y^4 - xy)$ **8.** $(x - y^3)(x + 2y^3)$

••   Multiply.

9. $(x + h)^2$ **10.** $(3a + 2b)^2$

11. $(r^3t^2 - 4)^2$ **12.** $(3a^2b - b^2)^2$

13. $(p^4 + m^2n^2)^2$ **14.** $(ab + cd)^2$

15. $\left(2a^3 - \frac{1}{2}b^3\right)^2$ **16.** $-5x(x + 3y)^2$

17. $3a(a - 2b)^2$ **18.** $(a^2 + b + 2)^2$

•••   Multiply.

19. $(2a - b)(2a + b)$ **20.** $(x - y)(x + y)$

21. $(c^2 - d)(c^2 + d)$ **22.** $(p^3 - 5q)(p^3 + 5q)$

1. _____
2. _____
3. _____
4. _____
5. _____
6. _____
7. _____
8. _____
9. _____
10. _____
11. _____
12. _____
13. _____
14. _____
15. _____
16. _____
17. _____
18. _____
19. _____
20. _____
21. _____
22. _____

Copyright © 1979, Philippines copyright 1979, by Addison-Wesley Publishing Company, Inc. All rights reserved.

ANSWERS

23. $(ab + cd^2)(ab - cd^2)$ **24.** $(xy + pq)(xy - pq)$

23. _____

24. _____

25. $(x + y - 3)(x + y + 3)$ **26.** $(p + q + 4)(p + q - 4)$

25. _____

27. $[x + y + z][x - (y + z)]$ **28.** $[a + b + c][a - (b + c)]$

26. _____

27. _____

29. $(a + b + c)(a - b - c)$ **30.** $(3x + 2 - 5y)(3x + 2 + 5y)$

28. _____

29. _____

Perform the indicated operations and simplify.

30. _____

31. $(a - b)^2 - (a + b)(a - b)$ **32.** $(a - b)^2 + (a + b)^2$

31. _____

33. $(a + b)^2 - (a - b)^2$ **34.** $9(x - 1)^2 - 25(x + 1)^2$

32. _____

33. _____

35. $y^2 - (y - 2)^2$ **36.** $(c^2 - cd + d^2)(c^2 + cd + d^2)$

34. _____

35. _____

37. $(a - b)^2 - (b - a)$ **38.** $(4x - y)(y - 4x)$

36. _____

37. _____

38. _____

39. Find a formula for $(A + B)^3$.

39. _____

40. Use the formula from Exercise 39 to multiply out the compound interest formula for $t = 3$:

40. _____ $A = P(1 + r)^3$.

7.4 FACTORING

Factoring polynomials in several variables is quite similar to factoring polynomials in one variable.

▪ TERMS WITH COMMON FACTORS

Whenever you factor polynomials, look for common factors in the terms before trying any other kind of factoring.

Example 1 Factor.

$$20x^3y + 12x^2y = (\,4x^2y\,)(5x) + (\,4x^2y\,) \cdot 3$$
$$= 4x^2y\,(5x + 3)$$

Try to write the answer directly, as follows.

Example 2 Factor.

$$6x^2y - 21x^3y^2 + 3x^2y^3 = 3x^2y\,(2 - 7xy + y^2)$$

DO EXERCISES 1 AND 2.

▪▪ FACTORING BY GROUPING

Sometimes a common factor is itself a binomial.

Example 3 Factor.

$$(\,p + q\,)(x + 2) + (\,p + q\,)(x + y) = (\,p + q\,)[(x + 2) + (x + y)]$$
$$= (p + q)(2x + y + 2)$$

Sometimes pairs of terms have a common factor that can be removed.

Example 4 Factor.

$$px + py + qx + qy = p(\,x + y\,) + q(\,x + y\,)$$
$$= (p + q)(\,x + y\,)$$

DO EXERCISES 3 AND 4.

OBJECTIVES

After finishing Section 7.4, you should be able to:

▪ Factor polynomials when the terms have a common factor.

▪▪ Factor by grouping.

▪▪▪ Factor differences of squares.

▪▪▪▪ Factor polynomials completely.

Factor.

1. $x^4y^2 + 2x^3y + 3x^2y$

2. $10p^6q^2 - 4p^5q^3 + 2p^4q^4$

Factor.

3. $(a - b)(x + 5) + (a - b)(x + y^2)$

4. $ax^2 + ay + bx^2 + by$

Factor.

5. $9a^2 - 16x^4$

6. $25x^2y^4 - 4a^2$

Factor.

7. $5 - 5x^2y^6$

Factor completely.

8. $16y^2 - 81x^4y^2$

9. $3a^2 + 6ab + 3b^2 - 75$

●●● DIFFERENCES OF SQUARES

Differences of squares can be factored as before.

$$A^2 - B^2 = (A + B)(A - B)$$

Example 5 Factor.

$$36x^2 - 25y^6 = (6x)^2 - (5y^3)^2$$
$$= (6x + 5y^3)(6x - 5y^3)$$

DO EXERCISES 5 AND 6.

If a factor can be factored again, it should be factored.

Example 6 Factor.

$$32x^4y^4 - 50 = 2(16x^4y^4 - 25)$$
$$= 2(4x^2y^2 + 5)(4x^2y^2 - 5)$$

DO EXERCISE 7.

⣿ FACTORING COMPLETELY

Sometimes one or more of the factors can be factored.

Example 7 Factor completely.

$$5x^4 - 5y^4 = 5(x^4 - y^4)$$
$$= 5(x^2 + y^2)(x^2 - y^2)$$
$$= 5(x^2 + y^2)(x + y)(x - y)$$

Example 8 Factor completely.

$$2x^2 + 4ax + 2a^2 - 50 = 2(x^2 + 2ax + a^2 - 25)$$

$$= 2[(x + a)^2 - 25] \quad \text{Factoring first three terms}$$

$$= 2(x + a + 5)(x + a - 5) \quad \text{Factoring a difference of two squares}$$

DO EXERCISES 8 AND 9.

NAME   CLASS

EXERCISE SET 7.4

■● Factor.

1. $12n^2 + 24n^3$ **2.** $ax^2 + ay^2$

3. $9x^2y^2 - 36xy$ **4.** $x^2y - xy^2$

5. $2\pi rh + 2\pi r^2$ **6.** $10p^4q^4 + 35p^3q^3 + 10p^2q^2$

●● Factor.

7. $(a + b)(x - 3) + (a + b)(x + 4)$

8. $5c(a^3 + b) - (a^3 + b)$

9. $(x - 1)(x + 1) - y(x + 1)$ **10.** $x^2 + x + xy + y$

11. $n^2 + 2n + np + 2p$ **12.** $a^2 - 3a + ay - 3y$

13. $2x^2 - 4x + xz - 2z$ **14.** $6y^2 - 3y + 2py - p$

●●● Factor.

15. $x^2 - y^2$ **16.** $a^2 - h^2$

17. $a^2b^2 - 9$ **18.** $p^2q^2 - r^2$

19. $9x^4y^2 - b^2$ **20.** $36t^2 - 49p^2q^2$

ANSWERS

1. _____

2. _____

3. _____

4. _____

5. _____

6. _____

7. _____

8. _____

9. _____

10. _____

11. _____

12. _____

13. _____

14. _____

15. _____

16. _____

17. _____

18. _____

19. _____

20. _____

Copyright © 1979, Philippines copyright 1979, by Addison-Wesley Publishing Company, Inc. All rights reserved.

ANSWERS

21. _____

22. _____

23. _____

24. _____

25. _____

26. _____

27. _____

28. _____

29. _____

30. _____

31. _____

32. _____

33. _____

34. _____

35. _____

36. _____

37. _____

38. _____

39. _____

40. _____

21. $3x^2 - 48y^2$

22. $9s^4 - 9s^2$

23. $64z^2 - 25c^2d^2$

24. $5t^2 - 20m^2$

⚏ Factor completely.

25. $7p^4 - 7q^4$

26. $a^4b^4 - 16$

27. $81a^4 - b^4$

28. $1 - 16x^{12}y^{12}$

29. $18m^4 + 12m^3 + 2m^2$

30. $(a + b)^2 - c^2$

31. $(x - a)^2 - (y - b)^2$

32. $ay^2 - a - y^2 + 1$

33. $(b - 4)^2 - a^6$

34. $15x^2y - 20xy - 35y$

35. $9a^2 - (a + 3b)^2$

36. $(a - 3)^2 - (b - 3)^2$

37. $(s - 2)^2 - (t - 2)^2$

38. $c^2xy - c^3 - x^2y + cx$

---

39. Factor: $(x + h)^2 - x^2$.

40. Use the answer to Exercise 39 to find $(4.1)^2 - 4^2$ and $(4.01)^2 - 4^2$.

7.5 FACTORING TRINOMIALS

◖•◗ TRINOMIAL SQUARES

If a trinomial is the square of a binomial, then it is easy to factor. So, whenever you have a trinomial to factor, you should check to see whether it is the square of a binomial.

Example 1   Factor $25x^2 + 20xy + 4y^2$.

a) Find whether $25x^2 + 20xy + 4y^2$ is the square of a binomial.

The first term and the last term are squares:

$$25x^2 = (\,5x\,)^2 \quad \text{and} \quad 4y^2 = (\,2y\,)^2.$$

Twice the product of $5x$ and $2y$ should be the other term:

$$2 \cdot \boxed{5x} \cdot \boxed{2y} = 20xy.$$

Thus, the trinomial is the square of a binomial.

b) We factor by writing the square roots of the square terms and the sign of the other term:

$$25x^2 + 20xy + 4y^2 = (5x + 2y)^2.$$

We can check by squaring $5x + 2y$.

DO EXERCISES 1 AND 2.

◖•◗ TRINOMIALS THAT ARE NOT SQUARES

If a trinomial is not a square, we use trial and error.

Examples   Factor.

2. $p^2q^2 + 7pq + 12 = (pq)^2 + (3 + 4)pq + 3 \cdot 4$
$$ = (pq + 3)(pq + 4)$$

3. $8x^4 - 20x^2y - 12y^2 = 4(2x^4 - 5x^2y - 3y^2)$ Removing a
common factor

$$ = 4[(2x^2)(x^2) + (-6 + 1)x^2y + (-3y)y]$$
$$ = 4(2x^2 + y)(x^2 - 3y)$$

> **Remember always to look first for common factors. Remember, too, that some of the factors obtained may be factorable, and factor completely. Furthermore, not all trinomials can be factored.**

DO EXERCISES 3 AND 4.

OBJECTIVES

After finishing Section 7.5, you should be able to:

◖•◗ Factor trinomials that are squares of binomials.

◖•◗ Factor trinomials that are not squares.

Factor.

1. $x^4 + 2x^2y^2 + y^4$

2. $-4x^2 + 12xy - 9y^2$
(*Hint:* First factor out -1.)

Factor.

3. $x^2y^2 + 5xy + 4$

4. $2x^4y^6 + 6x^2y^3 - 20$

SOMETHING EXTRA
AN APPLICATION

DEPRECIATION: THE STRAIGHT-LINE METHOD (PART II)

Recall that on p. 254 we considered a machine being bought for $5200. It was expected to last for 8 years at which time the trade-in value would be $1300. Each year it depreciates $\frac{1}{8}$, or 12.5%, of $3900, which is $487.50.

The book values of the machine for successive years are shown in the depreciation schedule below.

Year	Rate of depreciation	Annual depreciation	Book value (V)	Total depreciation
0			$5200	
1	$\frac{1}{8}$ or 12.5%	$487.50	4712.50	$ 487.50
2	12.5%	487.50	4225.00	975.00
3	12.5%	487.50	3737.50	1462.50
4	12.5%	487.50	3250.00	1950.00
5	12.5%	487.50	2762.50	2437.50
6	12.5%	487.50	2275.00	2925.00
7	12.5%	487.50	1787.50	3412.50
8	12.5%	487.50	1300.00	3900.00

The book value V of an item n years after purchase is given by

$$V = C - n\left(\frac{C - S}{N}\right),$$

where C = the original cost of the item, N = the total number of years of useful life, and S = the trade-in or salvage value. For the machine above,

$$\text{the annual depreciation} = \frac{C - S}{N} = \frac{5200 - 1300}{8} = \$487.50;$$

$$\text{the rate of depreciation} = \frac{1}{N} = \frac{1}{8} = 12.5\%;$$

$$\text{the book value is given by } V = 5200 - n\left(\frac{5200 - 1300}{8}\right)$$

$$= 5200 - (487.50)n.$$

There are other depreciation methods where the decline in value is not the same each year.

EXERCISES

In each situation, find (a) a depreciation schedule and (b) a formula for V.

1. Cost = $8700
   Expected life = 5 years
   Salvage value = $1600

2. Cost = $9400
   Expected life = 10 years
   Salvage value = $2200

NAME   CLASS   ANSWERS

EXERCISE SET 7.5

◼ • Factor.

1. $x^2 - 2xy + y^2$ **2.** $a^2 - 4ab + 4b^2$

3. $9c^2 + 6cd + d^2$ **4.** $16x^2 + 24xy + 9y^2$

5. $49m^4 - 112m^2n + 64n^2$ **6.** $4x^2y^2 + 12xyz + 9z^2$

7. $y^4 + 10y^2z^2 + 25z^4$ **8.** $0.01x^4 - 0.1x^2y^2 + 0.25y^4$

9. $\frac{1}{4}a^2 + \frac{1}{3}ab + \frac{1}{9}b^2$ **10.** $4p^2q + pq^2 + 4p^3$

◼ ◼ Factor.

11. $a^2 - ab - 2b^2$ **12.** $3b^2 - 17ab - 6a^2$

13. $m^2 + 2mn - 360n^2$ **14.** $x^2y^2 + 8xy + 15$

15. $m^2n^2 - 4mn - 32$ **16.** $p^2q^2 + 7pq + 6$

17. $a^5b^2 + 3a^4b - 10a^3$ **18.** $m^2n^6 + 4mn^5 - 32n^4$

ANSWERS

1. _____

2. _____

3. _____

4. _____

5. _____

6. _____

7. _____

8. _____

9. _____

10. _____

11. _____

12. _____

13. _____

14. _____

15. _____

16. _____

17. _____

18. _____

Copyright © 1979, Philippines copyright 1979, by Addison-Wesley Publishing Company, Inc. All rights reserved

ANSWERS

19. _____

20. _____

21. _____

22. _____

23. _____

24. _____

25. _____

26. _____

27. _____

28. _____

29. _____

30. _____

31. _____

32. _____

33. _____

34. _____

35. _____

36. _____

19. $(r - s)^2 - 2(r - s) + 1$
(*Hint:* Think of this as
$\boxed{}^2 - 2\,\boxed{} + 1$.)

20. $4p^2 - q^2 - 6q - 9$
[*Hint:* Think of this as
$4p^2 - (q^2 + 6q + 9)$.]

21. $t^2 - 4t + 4 - s^2$

22. $r^2 + 6rs + 9s^2$

23. $-k^2 + 36kl - 324l^2$

24. $4a^2 - 4ab - 36 + b^2$

25. $1 - n^2 - 2nx - x^2$

26. $x^4 + x^3 + x^2$

27. $3b^2 - 17ab - 6a^2$

28. $6a^2 - 47ab - 63b^2$

29. $a^2c^2 - 4a^2 - b^2c^2 + 4b^2$

30. $-16a^2b - 10a^2br - a^2br^2$

31. $-p^2 + 20pt - 100t^2$

32. $n^2 + 23nr - 420r^2$

33. $a^2 - b^2 + (a - b)^2$

34. $5ac - 5ad + 5bc - 5bd$

35. $k^2t^2 - 9s^2 - 9t^2 + k^2s^2$

36. $42m^2n - 24mn^2 - 18n^3$

7.6 SOLVING EQUATIONS

Sometimes a letter represents a specific number, although we may not know what the number is. In this case it is called a *constant* rather than a variable. In this section we agree that letters near the end of the alphabet

x, y, z will be variables.

Letters at the beginning of the alphabet

a, b, c will be constants.

This is a fairly common agreement. Thus, when we solve equations with letters for constants, we solve for one of the letters x, y, or z.

OBJECTIVE

After finishing Section 7.6, you should be able to:

- Solve equations where some letters are used as constants.

Solve.

1. $ay - b = 3$

EQUATIONS WITH UNKNOWN CONSTANTS

We treat unknown constants in the same way as known ones. The laws of numbers still apply.

Example 1   Solve: $cx + b^2 = 2$.

By our agreement x is a variable and b and c are constants. Thus, we solve for x:

$cx = 2 - b^2$ Adding $-b^2$

$x = \dfrac{2 - b^2}{c}$.   Multiplying by $\dfrac{1}{c}$

We treated c and b as if they were known. In this sense they are like the 2 in the equation.

DO EXERCISE 1.

To solve an equation we try to get the variable alone on one side. If the variable appears in several terms, we usually collect like terms, or factor out the variable. We may need to multiply to remove parentheses.

Example 2   Solve: $cdy + 4d = 3cdy$.

We solve for y, the letter near the end of the alphabet:

$4d = 3cdy - cdy$ Getting the y-terms on one side

$4d = (3cd - cd)y$ Collecting like terms (or factoring out y)

$4d = 2cdy$

$\dfrac{4d}{2cd} = y$ Multiplying by $\dfrac{1}{2cd}$

$\dfrac{2}{c} = y$.   Simplifying

Solve.

2. $2abx - 6a = abx$

DO EXERCISE 2.

Solve.

3. $3(7 - 2y) + 7(2y + 1) = 36$

Solve.

4. $(x + a)(x + b) = x^2 + 2$

Solve.

5. $(x - a)(x + a) = x^2 - 2ax + b^2$

Example 3   Solve: $3(7 + 2x) = 30 + 7(x - 1)$.

$$21 + 6x = 30 + 7x - 7 \qquad \text{Multiplying to remove parentheses}$$

$$21 = 23 + 7x - 6x \qquad \text{Adding } -6x \text{ and collecting like terms (the 30 and } -7)$$

$$-2 = x \qquad \text{Adding } -23 \text{ and collecting } x\text{-terms}$$

DO EXERCISE 3.

In Examples 4 and 5 that follow, remember that x, y, and z are used as variables and a, b, and c are used as constants.

Example 4   Solve: $(x - a)(x + a) = x^2 - x$.

$$x^2 - a^2 = x^2 - x \qquad \text{Multiplying}$$

$$-a^2 = -x \qquad \text{Adding } -x^2$$

$$a^2 = x \qquad \text{Multiplying by } -1$$

DO EXERCISE 4.

Example 5   Solve: $(x + a)(x + 2b) = x^2 + a^2 + b^2$.

$$x^2 + 2bx + ax + 2ab = x^2 + a^2 + b^2 \qquad \text{Multiplying to remove parentheses}$$

$$2bx + ax + 2ab = a^2 + b^2 \qquad \text{Adding } -x^2$$

$$(2b + a)x + 2ab = a^2 + b^2 \qquad \text{Collecting } x\text{-terms (or factoring out } x)$$

$$(2b + a)x = a^2 + b^2 - 2ab \qquad \text{Adding } -2ab$$

$$x = \frac{a^2 - 2ab + b^2}{2b + a}, \qquad \text{Multiplying by } \frac{1}{2b + a}$$

$$\text{or } \frac{(a - b)^2}{2b + a} \qquad \text{Factoring numerator}$$

DO EXERCISE 5.

NAME   CLASS   ANSWERS

EXERCISE SET 7.6

■● Solve for x, y, or z.

1. $2a - 3x = 12$ **2.** $z - k = 3k$

3. $3b - 2y = 1$ **4.** $5a + 2x = 3b$

5. $3acy - 9a = acy$ **6.** $(z + c)(z - d) = z^2 + 5$

7. $ax - bx = a^2 - 2ab + b^2$ **8.** $a(x + b) = c$

9. $3(z + c) - 5(z + b) = 4(z - 3)$

10. $(2x - 5)(3x + 6) = 6x^2 - 8$

11. $b(x - b) = x - (2 - b)$ **12.** $3z - a = b + 11$

13. $a(3 + 2x) = 5 + x$ **14.** $24a - 3x = 5x$

15. $4(x - a) + 4a = 7(b + 3) - 21$

16. $-3(c + y) + 3c = 6(c - 2) + 12$

17. $5(2a - z) + 6 = -2(z - 5a)$ **18.** $x(a + 2) - 3 = b(x + 7)$

19. $bx - b^2 = 3x + b - 12$ **20.** $cy + dy = c^2 + 2cd + d^2$

ANSWERS

1.

2.

3.

4.

5.

6.

7.

8.

9.

10.

11.

12.

13.

14.

15.

16.

17.

18.

19.

20.

Copyright © 1979, Philippines copyright 1979, by Addison-Wesley Publishing Company, Inc. All rights reserved.

ANSWERS

21. $ax + b^2 = a^2 + bx$ **22.** $m(m - x) = n(n - x)$

21. _____

22. _____

23. $b^2(ay - 1) = a^2(1 - ay)$ **24.** $m^2x = m + 1$

23. _____

25. $3(x - a) = x + 3a$ **26.** $cy - 1 = c^2 + 2c - y$

24. _____

25. _____

27. $dx + x = (d + 1)(d - 1)$ **28.** $a^2x = a^6$

26. _____

27. _____

29. $rxt = c$ **30.** $mx + 4 = x + 4m^2$

28. _____

29. _____

31. $ax + bx = 3ax - c$ **32.** $b^2x = 4b$

30. _____

31. _____

33. $3(m - x) = x - 7m$ **34.** $2(3 - x) + b = b(x - 5)$

32. _____

33. _____

35. $(x - c)^2 = (x - d)^2$ **36.** $(y + a)^2 = (y - b)^2$

34. _____

35. _____

36. _____

7.7 FORMULAS

⦿ The use of formulas is important in many applications of mathematics. It is important to be able to solve a formula for a letter.

Example 1 *(Gravitational force)* The gravitational force f between planets of mass M and m, at a distance d from each other, is given by

$$f = \frac{kMm}{d^2},$$

where k is a constant. Solve for m.

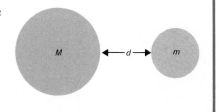

$fd^2 = kMm$ Multiplying by d^2

$\dfrac{fd^2}{kM} = m$ Multiplying by $\dfrac{1}{kM}$

DO EXERCISE 1.

It is difficult, if not impossible, to look at a formula and tell which symbols are variables and which are constants. In fact, in a formula, a certain letter may be a variable at times and a constant at others, depending on how the formula is used.

Example 2 *(The area of a trapezoid)* The area A of a trapezoid is half the product of the height h and the sum of the lengths b_1 and b_2 of the parallel sides:

$$A = \frac{1}{2}(b_1 + b_2)h.$$

Solve for b_2.

Note that b_1 and b_2 are different variables (or constants). Don't forget to write the subscripts (the 1 and 2).

$$2A = (b_1 + b_2)h \qquad \text{Multiplying by 2}$$
$$2A = b_1h + b_2h$$
$$2A - b_1h = b_2h \qquad \text{Adding } -b_1h$$
$$\frac{2A - b_1h}{h} = b_2 \qquad \text{Multiplying by } \frac{1}{h}$$

Each of the following is also a correct answer:

$$\frac{2A}{h} - b_1 = b_2 \quad \text{and} \quad \frac{1}{h}(2A - b_1h) = b_2.$$

DO EXERCISE 2.

In both of the examples, the letter for which we solved was on the right side of the equation. Ordinarily we put the letter for which we solve on the left. This is a matter of choice, since all equations are reversible.

OBJECTIVE

After finishing Section 7.7, you should be able to:

⦿ Solve a formula for a letter.

1. Solve $f = \dfrac{kMm}{d^2}$ for M.

2. Solve $V = \dfrac{1}{6}\pi h(h^2 + 3a^2)$ for a^2.

ANSWERS ON PAGE A–23

SOMETHING EXTRA

APPLICATIONS

The following are some further applications of polynomials in several variables.

The area A of a rectangle of length l and width w:

$$A = lw.$$

The volume V of a rectangular solid of length x, width y, and height h:

$$V = xyh.$$

The total revenue R from the sale of x units of one product at \$4 each and y units of another product at \$7 each:

$$R = 4x + 7y.$$

The amount A that P dollars will grow to at interest rate r, compounded annually for 3 years:

$$A = P + 3rP + 3r^2P + r^3P.$$

The height S, in feet, of an object t seconds after being given an upward velocity of v feet per second from an altitude h:

$$S = -16t^2 + vt + h.$$

The approximate length L of a pulley belt around pulleys whose centers are D units apart and whose circumferences are C_1 and C_2:

$$L = \frac{1}{2}C_1 + \frac{1}{2}C_2 + 2D.$$

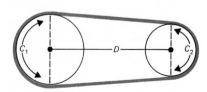

EXERCISE

Solve the pulley formula for D.

NAME   CLASS   ANSWERS

EXERCISE SET 7.7

● Solve for the variable indicated.

1. $S = 2\pi rh$; r

2. (*An interest formula*) $A = P(1 + rt)$; t

3. (*The area of a triangle*) $A = \frac{1}{2}bh$; b

4. $s = \frac{1}{2}gt^2$; g

5. $S = 180(n - 2)$; n

6. $S = \frac{n}{2}(a + l)$; a

7. $V = \frac{1}{3}k(B + b + 4M)$; b

8. $A = P + Prt$; P
    (*Hint:* Factor the right-hand side.)

9. $S(r - 1) = rl - a$; r

10. $T = mg - mf$; m
    (*Hint:* Factor the right-hand side.)

11. $A = \frac{1}{2}h(b_1 + b_2)$; h

12. (*The area of a right circular cylinder*)
    $S = 2\pi r(r + h)$; h

13. $r = \frac{v^2 pL}{a}$; a

14. $L = \frac{Mt - g}{t}$; M

ANSWERS

1. _____

2. _____

3. _____

4. _____

5. _____

6. _____

7. _____

8. _____

9. _____

10. _____

11. _____

12. _____

13. _____

14. _____

Copyright © 1979, Philippines copyright 1979, by Addison-Wesley Publishing Company, Inc. All rights reserved.

ANSWERS

15. _____

16. _____

17. _____

18. _____

19. _____

20. _____

21. _____

22. _____

23. _____

24. _____

25. _____

26. _____

27. _____

28. _____

29. _____

30. _____

15. $A = \frac{1}{2}h(b_1 + b_2)$; b_1

16. $l = a + (n - 1)d$; n

17. $A = \frac{\pi r^2 E}{180}$; E

18. $R = \frac{WL - x}{L}$; W

19. $V = -h(B + c + 4M)$; M

20. $W = I^2 R$; R

21. $y = \frac{v^2 pL}{a}$; L

22. $V = \frac{1}{3}bh$; b

23. $r = \frac{v^2 pL}{a}$; p

24. $P = 2(l + w)$; l

25. $\frac{a}{c} = n + bn$; n

(*Hint:* Factor the right-hand side.)

26. $C = \frac{Ka - b}{a}$; K

27. (*A temperature conversion formula*)

$C = \frac{5}{9}(F - 32)$; F

28. (*The volume of a sphere*)

$V = \frac{4}{3}\pi r^3$; π

29. $f = \frac{gm - t}{m}$; g

30. $S = \frac{rl - a}{r - l}$; a

NAME

SCORE

TEST OR REVIEW—CHAPTER 7

If you miss an item, review the indicated section and objective.

[7.1, ●●●]   **1.** Collect like terms:
$3x^2yz^3 - 2xy^2 + x^2yz^3 + x^2y^2 + 2xy^2$.

1. _____

[7.1, ⦂ ⦂]   **2.** Arrange in descending powers of a:
$b^3 + 3a^2b + 3ab^2 + a^3$.

2. _____

[7.2, ●]   **3.** Add: $(a^3b - 6a^2b^2 - ab^3 + 3) + (a^3b - 4a^2b^2 + 8)$

3. _____

[7.2, ●●]   **4.** Subtract: $(7x - 4y + 6z) - (3x - 7y + z)$.

4. _____

Multiply.

[7.2, ●●●]   **5.** $(y^3z^2 - 2yz - 3)(2yz - 1)$

5. _____

[7.3, ●]   **6.** $(a^2b - 3a)(ab^2 + 2a)$

6. _____

[7.3, ●●]   **7.** $(a^2 + 2b)^2$

7. _____

Copyright © 1979, Philippines copyright 1979, by Addison-Wesley Publishing Company, Inc. All rights reserved.

ANSWERS

[7.3, ●●●]   **8.** $(c^2 + d)(c^2 - d)$

8. _____

Factor.

[7.4, ●]   **9.** $p^4q - p^3q - pq$

9. _____

[7.4, ●●]   **10.** $2uw - 6ux - 3vx + vw$

10. _____

[7.4, ●●●]   **11.** $16x^2y^2 - 1$

11. _____

[7.4, ⦂⦂]   **12.** $16c^4 - d^4$

12. _____

[7.4, ⦂⦂]   **13.** $c^2 + 6c + 9 - 16x^2$

13. _____

[7.5, ●]   **14.** $9p^2 + 42pq + 49q^2$

14. _____

[7.5, ●●]   **15.** $35x^2 - 22xy + 3y^2$

15. _____

[7.6, ●]   **16.** Solve for x:
$$r(x - r) + s(x - s) = s(s + x) - (r + s).$$

16. _____

[7.7, ●]   **17.** Solve for P: $A - P = Prt$.

17. _____

**FRACTIONAL
EXPRESSIONS
AND EQUATIONS**

READINESS CHECK: SKILLS FOR CHAPTER 8

Multiply and simplify.

1. $\dfrac{1}{6} \cdot \dfrac{3}{8}$

2. $\dfrac{7}{9} \cdot \dfrac{9}{7}$

3. Find the reciprocal of $\dfrac{2}{5}$.

4. Divide and simplify: $\dfrac{2}{5} \div \dfrac{4}{15}$.

5. Add and simplify: $\dfrac{4}{8} + \dfrac{2}{5}$.

6. Subtract and simplify: $\dfrac{13}{15} - \dfrac{2}{45}$.

7. Subtract:

$(x^2 + 6x + 8) - (x^2 - 3x - 4)$.

8. Multiply:

$(x - 2)(x + 2)$.

Factor.

9. $x^2 + 3x + 2$

10. $x^2 - 6x + 9$

Solve.

11. $3t + 2t = 12$

12. $x^2 - 5x + 6 = 0$

OBJECTIVES

After finishing Section 8.1, you should be able to:

■□ Multiply fractional expressions.

■■□ Multiply a fractional expression by 1, using an expression such as A/A.

■■■ Simplify fractional expressions by factoring numerator and denominator and removing factors of 1.

■■ Multiply fractional expressions and simplify.

8.1 MULTIPLYING AND SIMPLIFYING

These are *fractional expressions*:

$$\frac{3}{4}, \quad \frac{5}{x + 2}, \quad \frac{x^2 + 3x - 10}{7x^2 - 4}.$$

A fractional expression is a quotient of two polynomials and, therefore, indicates division. For example,

$$\frac{3}{4} \quad \text{means} \quad 3 \div 4$$

and

$$\frac{x^2 + 3x - 10}{7x^2 - 4} \quad \text{means} \quad (x^2 + 3x - 10) \div (7x^2 - 4).$$

■□ MULTIPLYING

For fractional expressions, multiplication is done as in arithmetic.

> **To multiply two fractional expressions, multiply numerators and multiply denominators.**

Examples Multiply.

1. $\dfrac{x - 2}{3} \cdot \dfrac{x + 2}{x + 3} = \dfrac{(x - 2)(x + 2)}{3(x + 3)}$ Multiplying numerators and multiplying denominators

$$= \dfrac{x^2 - 4}{3x + 9}$$

2. $\dfrac{-2}{2y + 3} \cdot \dfrac{3}{y - 5} = \dfrac{-2 \cdot 3}{(2y + 3)(y - 5)}$

$= \dfrac{-6}{2y^2 - 7y - 15}$

DO EXERCISES 1 AND 2.

●● MULTIPLYING BY 1

Any fractional expression with the same numerator and denominator is a symbol for 1:

$\dfrac{x + 2}{x + 2} = 1, \quad \dfrac{3x^2 - 4}{3x^2 - 4} = 1, \quad \dfrac{-1}{-1} = 1.$

It should be noted that certain replacements are not sensible. For example, in

$\dfrac{x + 2}{x + 2}$

we should not substitute -2 for x. We would get 0 for the denominator.

We can multiply by 1 to get an equivalent expression.

Examples   Multiply.

3. $\dfrac{3x + 2}{x + 1} \cdot \dfrac{2x}{2x} = \dfrac{(3x + 2)2x}{(x + 1)2x}$

$= \dfrac{6x^2 + 4x}{2x^2 + 2x}$

4. $\dfrac{x + 2}{x - 1} \cdot \dfrac{x + 1}{x + 1} = \dfrac{(x + 2)(x + 1)}{(x - 1)(x + 1)}$

$= \dfrac{x^2 + 3x + 2}{x^2 - 1}$

5. $\dfrac{2 + x}{2 - x} \cdot \dfrac{-1}{-1} = \dfrac{(2 + x)(-1)}{(2 - x)(-1)}$

$= \dfrac{2(-1) + x(-1)}{2(-1) - x(-1)}$

$= \dfrac{-2 - x}{-2 + x}, \quad \text{or} \quad \dfrac{-2 - x}{x - 2}$

DO EXERCISES 3–5.

Multiply.

1. $\dfrac{x + 3}{5} \cdot \dfrac{x + 2}{x + 4}$

2. $\dfrac{-3}{2x + 1} \cdot \dfrac{4}{2x - 1}$

Multiply.

3. $\dfrac{2x + 1}{3x - 2} \cdot \dfrac{x}{x}$

4. $\dfrac{x + 1}{x - 2} \cdot \dfrac{x + 2}{x + 2}$

5. $\dfrac{x - 8}{x - y} \cdot \dfrac{-1}{-1}$

Simplify.

6. $\dfrac{5y}{y}$

ANSWERS ON PAGE A–24

7. $\dfrac{8x^2}{24x}$

●●● SIMPLIFYING FRACTIONAL EXPRESSIONS

To simplify, we can do the reverse of multiplying. We factor numerator and denominator and "remove" a factor of 1.

Example 6 Simplify.

$$\frac{3x}{x} = \frac{3 \cdot x}{1 \cdot x} \qquad \text{Factoring numerator and denominator}$$

$$= \frac{3}{1} \cdot \boxed{\frac{x}{x}} \qquad \text{Factoring the fractional expression}$$

$$= \frac{3}{1} \cdot 1 \qquad \frac{x}{x} = 1$$

$$= 3 \qquad \text{We "removed" a factor of 1.}$$

In this example we supplied a 1 in the denominator. This can always be done, but it is not necessary.

Example 7 Simplify.

$$\frac{7x^2}{14x} = \frac{7x \cdot x}{7x \cdot 2} \qquad \text{Factoring numerator and denominator}$$

$$= \boxed{\frac{7x}{7x}} \cdot \frac{x}{2} \qquad \text{Factoring the fractional expression}$$

$$= \frac{x}{2} \qquad \text{"Removing" a factor of 1}$$

DO EXERCISES 6 AND 7.

Examples Simplify.

8. $\dfrac{6a + 12}{7(a + 2)} = \dfrac{6(a + 2)}{7(a + 2)}$

$$= \frac{6}{7} \cdot \boxed{\frac{a + 2}{a + 2}}$$

$$= \frac{6}{7} \qquad \text{"Removing" the factor } \frac{a + 2}{a + 2}$$

9. $\dfrac{6x^2 + 4x}{2x^2 + 2x} = \dfrac{2x(3x + 2)}{2x(x + 1)} \qquad \text{Factoring numerator and denominator}$

$$= \boxed{\frac{2x}{2x}} \cdot \frac{3x + 2}{x + 1} \qquad \text{Factoring the fractional expression}$$

$$= \frac{3x + 2}{x + 1} \qquad \text{"Removing" a factor of 1}$$

10. $\dfrac{x^2 + 3x + 2}{x^2 - 1} = \dfrac{(x + 2)(x + 1)}{(x + 1)(x - 1)}$

$\qquad\qquad = \dfrac{x + 1}{x + 1} \cdot \dfrac{x + 2}{x - 1}$

$\qquad\qquad = \dfrac{x + 2}{x - 1}$

11. $\dfrac{5a + 15}{10} = \dfrac{5(a + 3)}{5 \cdot 2}$

$\qquad\qquad = \dfrac{5}{5} \cdot \dfrac{a + 3}{2}$

$\qquad\qquad = \dfrac{a + 3}{2}$

DO EXERCISES 8–11.

⣿ MULTIPLYING AND SIMPLIFYING

Example 12 Multiply and simplify.

$\dfrac{x^2 + 6x + 9}{x^2 - 4} \cdot \dfrac{x - 2}{x + 3}$

$\quad = \dfrac{(x^2 + 6x + 9)(x - 2)}{(x^2 - 4)(x + 3)}$ Multiplying numerators and also denominators

$\quad = \dfrac{(x + 3)(x + 3)(x - 2)}{(x + 2)(x - 2)(x + 3)}$ Factoring numerator and denominator

$\quad = \dfrac{(x + 3)(x - 2)}{(x + 3)(x - 2)} \cdot \dfrac{x + 3}{x + 2}$ Factoring the fractional expression

$\quad = \dfrac{x + 3}{x + 2}$ Simplifying by removing a factor of 1

Canceling is a shortcut for removing a factor of 1 in the preceding examples.

Simplify.

8. $\dfrac{2x^2 + x}{3x^2 + 2x}$

9. $\dfrac{x^2 - 1}{2x^2 - x - 1}$

10. $\dfrac{7x + 14}{7}$

11. $\dfrac{12y + 24}{48}$

Multiply and simplify.

12. $\dfrac{a^2 - 4a + 4}{a^2 - 9} \cdot \dfrac{a + 3}{a - 2}$

Example 13   Simplify.

$$\frac{x^2 + x - 2}{2x^2 - 3x + 1} = \frac{(x + 2)(x - 1)}{(x - 1)(2x - 1)} \qquad \text{Factoring numerator and denominator}$$

$$= \frac{(x + 2)(\cancel{x - 1})}{(\cancel{x - 1})(2x - 1)} \qquad \text{Canceling (removing a factor of 1)}$$

$$= \frac{x + 2}{2x - 1}$$

Note: **If you can't factor, you can't cancel!**

The use of canceling causes a great many errors because it is sometimes used mechanically without understanding. It should be done very cautiously, if at all.

DO EXERCISES 12 AND 13.

13. $\dfrac{x^2 - 25}{6} \cdot \dfrac{3}{x + 5}$

NAME   CLASS   ANSWERS

EXERCISE SET 8.1

⚫ Multiply.

1. $\dfrac{x-2}{x-5} \cdot \dfrac{x-2}{x+5}$

2. $\dfrac{x-1}{x+2} \cdot \dfrac{x+1}{x+2}$

3. $\dfrac{c-3d}{c+d} \cdot \dfrac{c+3d}{c-d}$

4. $\dfrac{a+2b}{a+b} \cdot \dfrac{a-2b}{a-b}$

⚫⚫ Multiply.

5. $\dfrac{2a-1}{2a-1} \cdot \dfrac{3a-1}{3a+2}$

6. $\dfrac{3x-2}{x+7} \cdot \dfrac{2x+5}{2x+5}$

⚫⚫⚫ Simplify.

7. $\dfrac{x(3x+2)(x+1)}{x(x+1)(3x-2)}$

8. $\dfrac{x(3x+4)(5x+7)}{x(3x-4)(5x+7)}$

9. $\dfrac{8a+8b}{8a-8b}$

10. $\dfrac{6x-6y}{6x+6y}$

11. $\dfrac{t^2-25}{t^2+t-20}$

12. $\dfrac{a^2-9}{a^2+5a+6}$

13. $\dfrac{2x^2+6x+4}{4x^2-12x-16}$

14. $\dfrac{x^2-3x-4}{2x^2+10x+8}$

1. _____

2. _____

3. _____

4. _____

5. _____

6. _____

7. _____

8. _____

9. _____

10. _____

11. _____

12. _____

13. _____

14. _____

Copyright © 1979, Philippines copyright 1979, by Addison-Wesley Publishing Company, Inc. All rights reserved.

ANSWERS

15. $\dfrac{a^2 - 10a + 21}{a^2 - 11a + 28}$

16. $\dfrac{x^2 - 3x - 18}{x^2 - 2x - 15}$

15. _____

16. _____

17. $\dfrac{6x + 12}{x^2 - x - 6}$

18. $\dfrac{5y + 5}{y^2 + 7y + 6}$

17. _____

18. _____

19. $\dfrac{a^2 + 1}{a + 1}$

20. $\dfrac{t^2 - 1}{t + 1}$

19. _____

20. _____

⦙⦙ Multiply and simplify.

21. $\dfrac{t^2}{t^2 - 4} \cdot \dfrac{t^2 - 5t + 6}{t^2 - 3t}$

22. $\dfrac{x^2 - 3x - 10}{(x - 2)^2} \cdot \dfrac{x - 2}{x - 5}$

21. _____

22. _____

23. $\dfrac{24a^2}{3(a^2 - 4a + 4)} \cdot \dfrac{3a - 6}{2a}$

24. $\dfrac{5v + 5}{v - 2} \cdot \dfrac{v^2 - 4v + 4}{v^2 - 1}$

23. _____

24. _____

25. $\dfrac{ab - b^2}{2a} \cdot \dfrac{2a + 2b}{a^2b - b^3}$

26. $\dfrac{c^2 - 6c}{c - 6} \cdot \dfrac{c + 3}{c}$

25. _____

26. _____

27. ▦ Multiply: $\dfrac{625}{x - 345.1} \cdot \dfrac{34.2}{x + 345.1}$.

27. _____

8.2 DIVISION AND RECIPROCALS

◖•◗ FINDING RECIPROCALS

Two expressions are reciprocals of each other if their product is 1. The reciprocal of a fractional expression is found by interchanging numerator and denominator.

Examples

1. The reciprocal of $\frac{2}{5}$ is $\frac{5}{2}$.

2. The reciprocal of $\frac{2x^2 - 3}{x + 4}$ is $\frac{x + 4}{2x^2 - 3}$.

3. The reciprocal of $x + 2$ is $\frac{1}{x + 2}$. $\left(\text{Think of } x + 2 \text{ as } \frac{x + 2}{1}.\right)$

DO EXERCISES 1–4.

◖••◗ DIVISION

To divide, we multiply by a reciprocal and simplify the result.

(To review the reason for this, see Chapter 1, p. 26.)

Examples   Divide.

4. $\dfrac{3}{4} \div \dfrac{2}{5} = \dfrac{3}{4} \cdot \boxed{\dfrac{5}{2}}$ Multiplying by the reciprocal

$= \dfrac{3 \cdot 5}{4 \cdot 2}$

$= \dfrac{15}{8}$

5. $\dfrac{x + 1}{x + 2} \div \dfrac{x - 1}{x + 3} = \dfrac{x + 1}{x + 2} \cdot \boxed{\dfrac{x + 3}{x - 1}}$ Multiplying by the reciprocal

$= \dfrac{x^2 + 4x + 3}{x^2 + x - 2}$

DO EXERCISES 5 AND 6.

OBJECTIVES

After finishing Section 8.2, you should be able to:

◖•◗ **Find the reciprocal of a fractional expression.**

◖••◗ **Divide fractional expressions and simplify.**

Find the reciprocal.

1. $\dfrac{7}{2}$

2. $\dfrac{x^2 + 5}{2x^3 - 1}$

3. $x - 5$

4. $\dfrac{1}{x^2 - 3}$

Divide.

5. $\dfrac{3}{5} \div \dfrac{7}{2}$

6. $\dfrac{x - 3}{x + 5} \div \dfrac{x + 5}{x - 2}$

ANSWERS ON PAGE A–24

Divide and simplify.

7. $\dfrac{x-3}{x+5} \div \dfrac{x+2}{x+5}$

Example 6   Divide and simplify.

$$\dfrac{x+1}{x^2-1} \div \dfrac{x+1}{x^2-2x+1}$$

$$= \dfrac{x+1}{x^2-1} \cdot \dfrac{x^2-2x+1}{x+1} \qquad \text{Multiplying by the reciprocal}$$

$$= \dfrac{(x+1)(x^2-2x+1)}{(x^2-1)(x+1)}$$

$$= \dfrac{(x+1)(x-1)(x-1)}{(x-1)(x+1)(x+1)} \qquad \begin{array}{l}\text{Factoring numerator and}\\\text{denominator}\end{array}$$

$$= \dfrac{(x+1)(x-1)}{(x+1)(x-1)} \cdot \dfrac{x-1}{x+1} \qquad \begin{array}{l}\text{Factoring the fractional}\\\text{expression}\end{array}$$

$$= \dfrac{x-1}{x+1} \qquad \text{Simplifying}$$

DO EXERCISES 7 AND 8.

8. $\dfrac{y^2-1}{y+1} \div \dfrac{y^2-2y+1}{y+1}$

EXERCISE SET 8.2

● Find the reciprocal.

1. $\dfrac{4}{x}$

2. $\dfrac{a + 3}{a - 1}$

3. $x^2 - y^2$

4. $\dfrac{1}{a + b}$

5. $\dfrac{x^2 + 2x - 5}{x^2 - 4x + 7}$

6. $\dfrac{x^2 - 3xy + y^2}{x^2 + 7xy - y^2}$

●● Divide and simplify.

7. $\dfrac{2}{5} \div \dfrac{4}{3}$

8. $\dfrac{5}{6} \div \dfrac{2}{3}$

9. $\dfrac{2}{x} \div \dfrac{8}{x}$

10. $\dfrac{x}{2} \div \dfrac{3}{x}$

11. $\dfrac{x^2}{y} \div \dfrac{x^3}{y^3}$

12. $\dfrac{a}{b^2} \div \dfrac{a^2}{b^3}$

13. $\dfrac{a + 2}{a - 3} \div \dfrac{a - 1}{a + 3}$

14. $\dfrac{y + 2}{4} \div \dfrac{y}{2}$

15. $\dfrac{x^2 - 1}{x} \div \dfrac{x + 1}{x - 1}$

16. $\dfrac{4y - 8}{y + 2} \div \dfrac{y - 2}{y^2 - 4}$

17. $\dfrac{x + 1}{6} \div \dfrac{x + 1}{3}$

18. $\dfrac{a}{a - b} \div \dfrac{b}{a - b}$

19. $\dfrac{x^2 - 9}{4x + 12} \div \dfrac{x - 3}{6}$

20. $\dfrac{c^2 + 3c}{c^2 + 2c - 3} \div \dfrac{c}{c + 1}$

ANSWERS

1. _____
2. _____
3. _____
4. _____
5. _____
6. _____
7. _____
8. _____
9. _____
10. _____
11. _____
12. _____
13. _____
14. _____
15. _____
16. _____
17. _____
18. _____
19. _____
20. _____

Copyright © 1979, Philippines copyright 1979, by Addison-Wesley Publishing Company, Inc. All rights reserved.

ANSWERS

21. $\dfrac{x+y}{x-y} \div \dfrac{x^2+y}{x^2-y^2}$

22. $\dfrac{x-b}{2x} \div \dfrac{x^2-b^2}{5x^2}$

21. _____

22. _____

23. $\dfrac{x^2-x-20}{x^2+7x+12} \div \dfrac{x^2-10x+25}{x^2+6x+9}$

24. $\dfrac{2y^2-7y+3}{2y^2+3y-2} \div \dfrac{6y^2-5y+1}{3y^2+5y-2}$

23. _____

25. $\dfrac{c^2+10c+21}{c^2-2c-15} \div (c^2+2c-35)$

26. $(1-z) \div \dfrac{1-z}{1+2z-z^2}$

24. _____

25. _____

27. $\dfrac{(t+5)^3}{(t-5)^3} \div \dfrac{(t+5)^2}{(t-5)^2}$

28. $\dfrac{(y-3)^3}{(y+3)^3} \div \dfrac{(y-3)^2}{(y+3)^2}$

26. _____

Perform the indicated operations and simplify.

29. $\left[\dfrac{r^2-4s^2}{r+2s} \div (r+2s) \right] \cdot \dfrac{2s}{r-2s}$

30. $\left[\dfrac{15-13a+2a^2}{4a^2-9} \cdot \dfrac{2a+1}{1-2a} \right] \div \dfrac{5-a}{2a-1}$

27. _____

28. _____

29. _____

31. $\left[\dfrac{d^2-d}{d^2-6d+8} \cdot \dfrac{d-2}{-d^2-5d} \right] \div \dfrac{5d}{d^2+d-20}$

30. _____

31. _____

8.3   ADDITION

◾◦ ADDITION WHEN DENOMINATORS ARE THE SAME

Addition is done as in arithmetic.

> **When denominators are the same, we add the numerators and keep the denominator.**

Examples   Add.

1. $\dfrac{x}{x+1} + \dfrac{2}{x+1} = \dfrac{x+2}{x+1}$

2. $\dfrac{2x^2 + 3x - 7}{2x+1} + \dfrac{x^2 + x - 8}{2x+1} = \dfrac{(2x^2 + 3x - 7) + (x^2 + x - 8)}{2x+1}$

$$= \dfrac{3x^2 + 4x - 15}{2x+1}$$

DO EXERCISES 1–3.

◦◦ ADDITION WHEN DENOMINATORS ARE ADDITIVE INVERSES

When one denominator is the additive inverse of the other, we first multiply one expression by $-1/-1$.

Examples   Add.

3. $\dfrac{x}{2} + \dfrac{3}{-2} = \dfrac{x}{2} + \boxed{\dfrac{-1}{-1}} \cdot \dfrac{3}{-2}$ Multiplying by $\dfrac{-1}{-1}$

$$= \dfrac{x}{2} + \dfrac{(-1)3}{(-1)(-2)}$$

$$= \dfrac{x}{2} + \dfrac{-3}{2}$$ Denominators are now the same

$$= \dfrac{x + (-3)}{2} = \dfrac{x - 3}{2}$$

4. $\dfrac{3x+4}{x-2} + \dfrac{x-7}{2-x} = \dfrac{3x+4}{x-2} + \boxed{\dfrac{-1}{-1}} \cdot \dfrac{x-7}{2-x}$

$$= \dfrac{3x+4}{x-2} + \dfrac{-1(x-7)}{-1(2-x)} \quad \textit{Note:}\ -1(2-x) = -2 + x$$
$$\phantom{= \dfrac{3x+4}{x-2} + \dfrac{-1(x-7)}{-1(2-x)} \quad \textit{Note:}\ } = x - 2$$

$$= \dfrac{3x+4}{x-2} + \dfrac{7-x}{x-2}$$

$$= \dfrac{(3x+4) + (7-x)}{x-2} = \dfrac{2x+11}{x-2}$$

DO EXERCISES 4 AND 5.

OBJECTIVES

After finishing Section 8.3, you should be able to:

◾◦ Add fractional expressions having the same denominator.

◦◦ Add fractional expressions whose denominators are additive inverses of each other.

Add.

1. $\dfrac{5}{9} + \dfrac{2}{9}$

2. $\dfrac{3}{x-2} + \dfrac{x}{x-2}$

3. $\dfrac{4x+5}{x-1} + \dfrac{2x-1}{x-1}$

Add.

4. $\dfrac{x}{4} + \dfrac{5}{-4}$

5. $\dfrac{2x+1}{x-3} + \dfrac{x+2}{3-x}$

ANSWERS ON PAGE A–25

SOMETHING EXTRA
AN APPLICATION: HANDLING DIMENSION SYMBOLS (Part II)

We can treat dimension symbols much like numerals and variables, since correct results can thus be obtained.

Example 1 Compare

$$4 \text{ m} \cdot 3 \text{ m} = 4 \cdot 3 \cdot \text{m} \cdot \text{m} = 12 \text{ m}^2 \text{ (square meters)}$$

with

$$4x \cdot 3x = 4 \cdot 3 \cdot x \cdot x = 12x^2.$$

Example 2 Compare

$$5 \text{ persons} \cdot 8 \text{ hr} = 5 \cdot 8 \text{ person-hr} = 40 \text{ person-hr}$$

with

$$5x \cdot 8y = 5 \cdot 8 \cdot x \cdot y = 40xy.$$

Example 3 Compare

$$480 \text{ cm} \cdot \frac{1 \text{ m}}{100 \text{ cm}} = \frac{480}{100} \text{ cm} \cdot \frac{\text{m}}{\text{cm}} = 4.8 \cdot \frac{\text{cm}}{\text{cm}} \cdot \text{m} = 4.8 \text{ m}$$

with

$$480x \cdot \frac{y}{100x} = \frac{480}{100} x \cdot \frac{y}{x} = 4.8 \cdot \frac{x}{x} \cdot y = 4.8y.$$

In each example, dimension symbols are treated as though they are variables or numerals, and as though a symbol like "3 ft" represents a product of "3" by "ft." A symbol like km/hr (standard notation is "km/h") is treated as if it represents a division of kilometers by hours.

Any two measures can be "multiplied" or "divided." For example,

$$6 \text{ ft} \cdot 4 \text{ lb} = 24 \text{ ft-lb}, \qquad 3 \text{ km} \cdot 4 \text{ sec} = 12 \text{ km-sec},$$

$$\frac{8 \text{ grams}}{4 \text{ min}} = 2 \frac{\text{g}}{\text{min}}, \qquad \frac{3 \text{ in.} \cdot 8 \text{ days}}{6 \text{ lb}} = 4 \frac{\text{in.-day}}{\text{lb}}.$$

EXERCISES

Perform these calculations and simplify if possible. *Do not* make unit changes.

1. $12 \text{ ft} \cdot \dfrac{1 \text{ yd}}{3 \text{ ft}}$

2. $6 \text{ lb} \cdot \dfrac{16 \text{ oz}}{1 \text{ lb}}$

3. $9 \dfrac{\text{km}}{\text{hr}} \cdot 3 \text{ hr}$

4. $12 \dfrac{\text{m}}{\text{sec}} \cdot 5 \text{ sec}$

5. $3 \text{ cm} \cdot \dfrac{2 \text{ g}}{2 \text{ cm}}$

6. $\dfrac{9 \text{ mi}}{3 \text{ days}} \cdot 6 \text{ days}$

7. $2347 \text{ m} \cdot \dfrac{1 \text{ km}}{1000 \text{ m}}$

8. $55 \text{ cm} \cdot \dfrac{10 \text{ mm}}{1 \text{ cm}}$

9. $\dfrac{3 \text{ kg}}{5 \text{ m}} \cdot \dfrac{7 \text{ kg}}{6 \text{ m}}$

10. $\dfrac{2000 \text{ lb} \cdot (6 \text{ mi/hr})^2}{100 \text{ ft}}$

11. $\dfrac{7 \text{ m} \cdot 8 \text{ kg/sec}}{4 \text{ sec}}$

NAME   CLASS   ANSWERS

EXERCISE SET 8.3

| ● | Add. Simplify, if possible.

1. $\dfrac{5}{12} + \dfrac{7}{12}$ **2.** $\dfrac{3}{14} + \dfrac{5}{14}$ **3.** $\dfrac{1}{3+x} + \dfrac{5}{3+x}$

4. $\dfrac{4x+1}{6x+5} + \dfrac{3x-7}{5+6x}$ **5.** $\dfrac{x^2+7x}{x^2-5x} + \dfrac{x^2-4x}{x^2-5x}$ **6.** $\dfrac{a}{x+y} + \dfrac{b}{y+x}$

| ● ● | Add. Simplify, if possible.

7. $\dfrac{7}{8} + \dfrac{5}{-8}$ **8.** $\dfrac{11}{6} + \dfrac{5}{-6}$ **9.** $\dfrac{3}{t} + \dfrac{4}{-t}$

10. $\dfrac{5}{-a} + \dfrac{8}{a}$ **11.** $\dfrac{2x+7}{x-6} + \dfrac{3x}{6-x}$ **12.** $\dfrac{3x-2}{4x-3} + \dfrac{2x-5}{3-4x}$

13. $\dfrac{y^2}{y-3} + \dfrac{9}{3-y}$ **14.** $\dfrac{t^2}{t-2} + \dfrac{4}{2-t}$ **15.** $\dfrac{b-7}{b^2-16} + \dfrac{7-b}{16-b^2}$

16. $\dfrac{a-3}{a^2-25} + \dfrac{a-3}{25-a^2}$ **17.** $\dfrac{z}{(y+z)(y-z)} + \dfrac{y}{(z+y)(z-y)}$

ANSWERS

1. _____
2. _____
3. _____
4. _____
5. _____
6. _____
7. _____
8. _____
9. _____
10. _____
11. _____
12. _____
13. _____
14. _____
15. _____
16. _____
17. _____

Copyright © 1979, Philippines copyright 1979, by Addison-Wesley Publishing Company, Inc. All rights reserved.

ANSWERS

18. $\dfrac{a^2}{a-b} + \dfrac{b^2}{b-a}$

19. $\dfrac{x+3}{x-5} + \dfrac{2x-1}{5-x} + \dfrac{2(3x-1)}{x-5}$

20. $\dfrac{3(x-2)}{2x-3} + \dfrac{5(2x+1)}{2x-3} + \dfrac{3(x-1)}{3-2x}$

21. $\dfrac{2(4x+1)}{5x-7} + \dfrac{3(x-2)}{7-5x} + \dfrac{-10x-1}{5x-7}$

22. $\dfrac{5(x-2)}{3x-4} + \dfrac{2(x-3)}{4-3x} + \dfrac{3(5x+1)}{4-3x}$

23. $\dfrac{x+1}{(x+3)(x-3)} + \dfrac{4(x-3)}{(x-3)(x+3)} + \dfrac{(x-1)(x-3)}{(3-x)(x+3)}$

24. $\dfrac{2(x+5)}{(2x-3)(x-1)} + \dfrac{3x+4}{(2x-3)(1-x)} + \dfrac{x-5}{(3-2x)(x-1)} + \dfrac{2(x+3)}{(3-2x)(1-x)}$

18. _____

19. _____

20. _____

21. _____

22. _____

23. _____

24. _____

8.4   SUBTRACTION

SUBTRACTION WHEN DENOMINATORS ARE THE SAME

Subtraction is done as in arithmetic.

> **When denominators are the same, we subtract the numerators and keep the denominator.**

Example 1   Subtract.

$$\frac{3x}{x+2} - \frac{x-2}{x+2} = \frac{3x-(x-2)}{x+2}$$

The parentheses are important to make sure that you subtract the entire numerator.

$$= \frac{3x-x+2}{x+2}$$

$$= \frac{2x+2}{x+2}$$

DO EXERCISES 1 AND 2.

 SUBTRACTION WHEN DENOMINATORS ARE ADDITIVE INVERSES

When one denominator is the additive inverse of the other, we first multiply one expression by $-1/-1$.

Example 2   Subtract.

$$\frac{x}{5} - \frac{3x-4}{-5} = \frac{x}{5} - \frac{-1}{-1} \cdot \frac{3x-4}{-5}$$

$$= \frac{x}{5} - \frac{(-1)(3x-4)}{(-1)(-5)} = \frac{x}{5} - \frac{4-3x}{5}$$

$$= \frac{x-(4-3x)}{5} \qquad \text{Remember the parentheses!}$$

$$= \frac{x-4+3x}{5} = \frac{4x-4}{5}$$

Example 3   Subtract.

$$\frac{5y}{y-5} - \frac{2y-3}{5-y} = \frac{5y}{y-5} - \frac{-1}{-1} \cdot \frac{2y-3}{5-y}$$

$$= \frac{5y}{y-5} - \frac{(-1)(2y-3)}{(-1)(5-y)} = \frac{5y}{y-5} - \frac{3-2y}{y-5}$$

$$= \frac{5y-(3-2y)}{y-5} \qquad \text{Remember the parentheses!}$$

$$= \frac{5y-3+2y}{y-5} = \frac{7y-3}{y-5}$$

DO EXERCISES 3 AND 4.

OBJECTIVES

After finishing Section 8.4, you should be able to:

 Subtract fractional expressions having the same denominator.

Subtract fractional expressions whose denominators are additive inverses of each other.

Subtract.

1. $\dfrac{7}{11} - \dfrac{3}{11}$

2. $\dfrac{2x^2+3x-7}{2x+1} - \dfrac{x^2+x-8}{2x+1}$

Subtract.

3. $\dfrac{x}{3} - \dfrac{2x-1}{-3}$

4. $\dfrac{3x}{x-2} - \dfrac{x-3}{2-x}$

SOMETHING EXTRA
AN APPLICATION: HANDLING DIMENSION SYMBOLS (PART III)

Changes of unit can be achieved by substitutions. The table of measures on the inside front cover may be helpful.

Example 1 Change to inches: 25 yd.

$$25 \text{ yd} = 25 \cdot 1 \text{ yd}$$
$$= 25 \cdot 3 \text{ ft} \qquad \text{Substituting 3 ft for 1 yd}$$
$$= 25 \cdot 3 \cdot 1 \text{ ft}$$
$$= 25 \cdot 3 \cdot 12 \text{ in.} \qquad \text{Substituting 12 in. for 1 ft}$$
$$= 900 \text{ in.}$$

Example 2 Change to meters: 4 km.

$$4 \text{ km} = 4 \cdot 1 \text{ km}$$
$$= 4 \cdot 1000 \text{ m} \qquad \text{Substituting 1000 m for 1 km}$$
$$= 4000 \text{ m}$$

The notion of "multiplying by 1" can also be used to change units.

Example 3 Change to yd: 7.2 in.

$$7.2 \text{ in.} = 7.2 \text{ in.} \cdot \frac{1 \text{ ft}}{12 \text{ in.}} \cdot \frac{1 \text{ yd}}{3 \text{ ft}} \qquad \text{Both of these are equal to 1.}$$

$$= \frac{7.2}{12 \cdot 3} \cdot \frac{\text{in.}}{\text{in.}} \cdot \frac{\text{ft}}{\text{ft}} \cdot \text{yd} = 0.2 \text{ yd}$$

Example 4 Change to mm: 55 cm.

$$55 \text{ cm} = 55 \text{ cm} \cdot \frac{1 \text{ m}}{100 \text{ cm}} \cdot \frac{1000 \text{ mm}}{1 \text{ m}} = \frac{55 \cdot 1000}{100} \cdot \frac{\text{cm}}{\text{cm}} \cdot \frac{\text{m}}{\text{m}} \cdot \text{mm}$$
$$= 550 \text{ mm}$$

Example 5 Change to $\dfrac{\text{m}}{\text{sec}}$: $95 \dfrac{\text{km}}{\text{hr}}$.

$$95 \frac{\text{km}}{\text{hr}} = 95 \frac{\text{km}}{\text{hr}} \cdot \frac{1000 \text{ m}}{1 \text{ km}} \cdot \frac{1 \text{ hr}}{60 \text{ min}} \cdot \frac{1 \text{ min}}{60 \text{ sec}}$$

$$= \frac{95 \cdot 1000}{60 \cdot 60} \cdot \frac{\text{km}}{\text{km}} \cdot \frac{\text{hr}}{\text{hr}} \cdot \frac{\text{min}}{\text{min}} \cdot \frac{\text{m}}{\text{sec}} = 26.4 \frac{\text{m}}{\text{sec}}$$

Below is a shortcut for the procedure in Example 5. It can also be used in Examples 3 and 4.

$$95 \frac{\text{km}}{\text{hr}} = \overset{19}{95} \frac{\text{km}}{\text{hr}} \cdot \frac{1000 \text{ m}}{1 \text{ km}} \cdot \frac{1 \text{ hr}}{60 \text{ min}} \cdot \frac{1 \text{ min}}{60 \text{ sec}} = 26.4 \frac{\text{m}}{\text{sec}}$$

12

EXERCISES

Perform the following changes of unit, using substitution or multiplying by 1.

1. 72 in., change to ft

2. 17 hr, change to min

3. 2 days, change to sec

4. 360 sec, change to hr

5. $60 \dfrac{\text{kg}}{\text{m}}$, change to $\dfrac{\text{g}}{\text{cm}}$

6. $44 \dfrac{\text{ft}}{\text{sec}}$, change to $\dfrac{\text{mi}}{\text{hr}}$

7. 216 m², change to cm²

8. $60 \dfrac{\text{lb}}{\text{ft}^3}$, change to $\dfrac{\text{ton}}{\text{yd}^3}$

9. $\dfrac{\$36}{\text{day}}$, change to $\dfrac{\cancel{c}}{\text{hr}}$

10. 1440 person-hr, change to person-days

11. $186{,}000 \dfrac{\text{mi}}{\text{sec}}$ (speed of light), change to $\dfrac{\text{mi}}{\text{yr}}$. Let 365 days = 1 yr.

12. $1100 \dfrac{\text{ft}}{\text{sec}}$ (speed of sound), change to $\dfrac{\text{mi}}{\text{yr}}$. Let 365 days = 1 yr.

NAME   CLASS   ANSWERS

EXERCISE SET 8.4

◖●◗ Subtract. Simplify, if possible.

1. $\dfrac{7}{8} - \dfrac{3}{8}$

2. $\dfrac{5}{y} - \dfrac{7}{y}$

3. $\dfrac{x}{x-1} - \dfrac{1}{x-1}$

4. $\dfrac{x^2}{x+4} - \dfrac{16}{x+4}$

5. $\dfrac{x+1}{x^2-2x+1} - \dfrac{5-3x}{x^2-2x+1}$

6. $\dfrac{2x-3}{x^2+3x-4} - \dfrac{x-7}{x^2+3x-4}$

◖●●◗ Subtract. Simplify, if possible.

7. $\dfrac{11}{6} - \dfrac{5}{-6}$

8. $\dfrac{7}{8} - \dfrac{5}{-8}$

9. $\dfrac{5}{a} - \dfrac{8}{-a}$

10. $\dfrac{3}{t} - \dfrac{4}{-t}$

11. $\dfrac{x}{4} - \dfrac{3x-5}{-4}$

12. $\dfrac{2}{x-1} - \dfrac{2}{1-x}$

13. $\dfrac{3-x}{x-7} - \dfrac{2x-5}{7-x}$

14. $\dfrac{t^2}{t-2} - \dfrac{4}{2-t}$

15. $\dfrac{x-8}{x^2-16} - \dfrac{x-8}{16-x^2}$

ANSWERS

1. _____

2. _____

3. _____

4. _____

5. _____

6. _____

7. _____

8. _____

9. _____

10. _____

11. _____

12. _____

13. _____

14. _____

15. _____

ANSWERS

16. $\dfrac{x-2}{x^2-25} - \dfrac{6-x}{25-x^2}$

17. $\dfrac{4-x}{x-9} - \dfrac{3x-8}{9-x}$

18. $\dfrac{3-x}{x-7} - \dfrac{2x-5}{7-x}$

19. $\dfrac{2(x-1)}{2x-3} - \dfrac{3(x+2)}{2x-3} - \dfrac{x-1}{3-2x}$

20. $\dfrac{3(x-2)}{2x-3} - \dfrac{5(2x+1)}{2x-3} - \dfrac{3(x-1)}{3-2x}$

Perform the indicated operations and simplify.

21. $\dfrac{3(2x+5)}{x-1} - \dfrac{3(2x-3)}{1-x} + \dfrac{6x-1}{x-1}$

22. $\dfrac{2x-y}{x-y} + \dfrac{x-2y}{y-x} - \dfrac{3x-3y}{x-y}$

23. $\dfrac{x-y}{x^2-y^2} + \dfrac{x+y}{x^2-y^2} - \dfrac{2x}{x^2-y^2}$

24. $\dfrac{x+y}{2(x-y)} - \dfrac{2x-2y}{2(x-y)} + \dfrac{x-3y}{2(y-x)}$

25. $\dfrac{10}{2y-1} - \dfrac{6}{1-2y} + \dfrac{y}{2y-1} + \dfrac{y-4}{1-2y}$

26. $\dfrac{(x+3)(2x-1)}{(2x-3)(x-3)} - \dfrac{(x-3)(x+1)}{(3-x)(3-2x)} + \dfrac{(2x+1)(x+3)}{(3-2x)(x-3)}$

16. _____

17. _____

18. _____

19. _____

20. _____

21. _____

22. _____

23. _____

24. _____

25. _____

26. _____

8.5 LEAST COMMON MULTIPLES

◖•◗   LEAST COMMON MULTIPLES

To add when denominators are different we first find a common denominator. For example, to add $\frac{5}{12}$ and $\frac{7}{30}$ we first look for a common multiple of both 12 and 30. Any multiple will do, but we prefer the smallest such number, the *Least Common Multiple* (LCM). To find the LCM, we factor.

$$12 = 2 \cdot 2 \cdot 3$$
$$30 = 2 \cdot 3 \cdot 5$$

The LCM is the number that has 2 as a factor twice, 3 as a factor once, and 5 as a factor once:

$$\text{LCM} = 2 \cdot 2 \cdot 3 \cdot 5, \quad \text{or} \quad 60.$$

> **To find the LCM, we use each factor the greatest number of times that it appears in any one factorization.**

Example 1   Find the LCM of 24 and 36.

$$\left. \begin{array}{l} 24 = \boxed{2 \cdot 2 \cdot 2} \cdot 3 \\ 36 = 2 \cdot 2 \cdot \boxed{3 \cdot 3} \end{array} \right\} \quad \text{LCM} = \boxed{2 \cdot 2 \cdot 2} \cdot \boxed{3 \cdot 3}, \quad \text{or} \quad 72.$$

DO EXERCISES 1–4.

◖••◗   ADDING USING THE LCM

Let us finish adding $\frac{5}{12}$ and $\frac{7}{30}$.

$$\frac{5}{12} + \frac{7}{30} = \frac{5}{2 \cdot 2 \cdot 3} + \frac{7}{2 \cdot 3 \cdot 5}$$

The LCM is $2 \cdot 2 \cdot 3 \cdot 5$. To get the LCM in the first denominator we need a 5. To get the LCM in the second denominator we need another 2. We get these numbers by multiplying by 1.

$$\frac{5}{12} + \frac{7}{30} = \frac{5}{2 \cdot 2 \cdot 3} \cdot \boxed{\frac{5}{5}} + \frac{7}{2 \cdot 3 \cdot 5} \cdot \boxed{\frac{2}{2}} \qquad \text{Multiplying by 1}$$

$$= \frac{25}{2 \cdot 2 \cdot 3 \cdot 5} + \frac{14}{2 \cdot 3 \cdot 5 \cdot 2} \qquad \begin{array}{l}\text{Denominators are now} \\ \text{the LCM}\end{array}$$

$$= \frac{39}{2 \cdot 2 \cdot 3 \cdot 5} \qquad \begin{array}{l}\text{Adding the numerators and keeping} \\ \text{the LCM}\end{array}$$

$$= \frac{13}{20} \qquad \text{Simplifying}$$

Example 2   Add: $\frac{5}{12} + \frac{11}{18}$.

$$\left. \begin{array}{l} 12 = \boxed{2 \cdot 2} \cdot 3 \\ 18 = 2 \cdot \boxed{3 \cdot 3} \end{array} \right\} \quad \text{LCM} = \boxed{2 \cdot 2} \cdot \boxed{3 \cdot 3}, \quad \text{or} \quad 36.$$

OBJECTIVES

After finishing Section 8.5, you should be able to:

◖•◗ Find the LCM of several numbers by factoring.

◖••◗ Add fractions, first finding the LCM of the denominators.

◖•••◗ Find the LCM of algebraic expressions by factoring.

Find the LCM by factoring.

1. 16,   18

2. 6,   12

3. 2,   5

4. 24,   30,   20

Add, first finding the LCM of the denominators. Simplify, if possible.

5. $\dfrac{3}{16} + \dfrac{1}{18}$

6. $\dfrac{1}{6} + \dfrac{1}{12}$

7. $\dfrac{1}{2} + \dfrac{3}{5}$

8. $\dfrac{1}{24} + \dfrac{1}{30} + \dfrac{3}{20}$

Find the LCM.

9. $12xy^2, \quad 15x^3y$

10. $y^2 + 5y + 4, \quad y^2 + 2y + 1$

11. $t^2 + 16, \quad t - 2, \quad 7$

12. $x^2 + 2x + 1, \quad 3x - 3x^2, \quad x^2 - 1$

$$\frac{5}{12} + \frac{11}{18} = \frac{5}{2 \cdot 2 \cdot 3} \cdot \boxed{\frac{3}{3}} + \frac{11}{2 \cdot 3 \cdot 3} \cdot \boxed{\frac{2}{2}}$$

$$= \frac{15}{2 \cdot 2 \cdot 3 \cdot 3} + \frac{22}{2 \cdot 3 \cdot 3 \cdot 2} = \frac{37}{2 \cdot 2 \cdot 3 \cdot 3} = \frac{37}{36}$$

DO EXERCISES 5–8.

■■■ LCM'S OF ALGEBRAIC EXPRESSIONS

To find the LCM of two or more algebraic expressions, we factor them. Then we use each factor the greatest number of times it occurs in any one expression.

Example 3   Find the LCM of $12x$, $16y$, and $8xyz$.

$$\left.\begin{array}{l} 12x = 2 \cdot 2 \cdot \boxed{3 \cdot x} \\[4pt] 16y = \boxed{2 \cdot 2 \cdot 2 \cdot 2} \cdot \boxed{y} \\[4pt] 8xyz = 2 \cdot 2 \cdot 2 \cdot x \cdot y \cdot \boxed{z} \end{array}\right\} \quad \begin{array}{l} \text{LCM} = \boxed{2 \cdot 2 \cdot 2 \cdot 2} \cdot \boxed{3 \cdot x} \cdot \boxed{y} \cdot \boxed{z} \\[4pt] \qquad\; = 48xyz \end{array}$$

Example 4   Find the LCM of $x^2 + 5x - 6$ and $x^2 - 1$.

$$\left.\begin{array}{l} x^2 + 5x - 6 = \boxed{(x + 6)(x - 1)} \\[4pt] x^2 - 1 = \boxed{(x + 1)}\,\boxed{(x - 1)} \end{array}\right\} \quad \text{LCM} = \boxed{(x + 6)(x - 1)}\ \boxed{(x + 1)}$$

Example 5   Find the LCM of $x^2 + 4$, $x + 1$, and 5.

These expressions are not factorable, so the LCM is their product: $5(x^2 + 4)(x + 1)$.

We left answers in factored form. This is best for most purposes. If an LCM is multiplied by -1, we still consider the answer to be an LCM. For example:

the LCM of 2 and -3 is either 6 or -6;

the LCM of $x - 2$ and $x + 2$ is either $x^2 - 4$ or $4 - x^2$.

Example 6   Find the LCM of $x^2 - y^2$ and $2y - 2x$.

$$\left.\begin{array}{l} x^2 - y^2 = \boxed{(x + y)(x - y)} \\[4pt] 2y - 2x = \boxed{2}\,(y - x) \end{array}\right\} \quad \begin{array}{l} \text{LCM} = \boxed{2}\ \boxed{(x + y)(x - y)} \\[4pt] \text{or}\quad 2(x + y)(y - x) \end{array}$$

Example 7   Find the LCM of $x^2 - 4y^2$, $x^2 - 4xy + 4y^2$, and $x - 2y$.

$$\left.\begin{array}{l} x^2 - 4y^2 = (x - 2y)\,\boxed{(x + 2y)} \\[4pt] x^2 - 4xy + 4y^2 = \boxed{(x - 2y)(x - 2y)} \\[4pt] x - 2y = x - 2y \end{array}\right\} \quad \begin{array}{l} \text{LCM} = \\[4pt] \boxed{(x + 2y)}\ \boxed{(x - 2y)(x - 2y)} \\[4pt] = (x + 2y)(x - 2y)^2 \end{array}$$

DO EXERCISES 9–12.

NAME   CLASS

EXERCISE SET 8.5

● Find the LCM.

1. 12, 27   **2.** 10, 15   **3.** 8, 9

4. 12, 15   **5.** 6, 9, 21   **6.** 8, 36, 40

7. 24, 36, 40   **8.** 3, 4, 5   **9.** 28, 42, 60

●● Add, first finding the LCM of the denominators. Simplify, if possible.

10. $\dfrac{7}{24} + \dfrac{11}{18}$

11. $\dfrac{7}{60} + \dfrac{6}{75}$

12. $\dfrac{1}{6} + \dfrac{3}{40} + \dfrac{2}{75}$

13. $\dfrac{5}{24} + \dfrac{3}{20} + \dfrac{7}{30}$

14. $\dfrac{2}{15} + \dfrac{5}{9} + \dfrac{3}{20}$

15. $\dfrac{1}{20} + \dfrac{1}{30} + \dfrac{2}{45}$

1. _____

2. _____

3. _____

4. _____

5. _____

6. _____

7. _____

8. _____

9. _____

10. _____

11. _____

12. _____

13. _____

14. _____

15. _____

Copyright © 1979, Philippines copyright 1979, by Addison-Wesley Publishing Company, Inc. All rights reserved.

ANSWERS

● ● ● Find the LCM.

16. $6x^2, 12x^3$

17. $2a^2b, 8ab^2$

18. $2x^2, 6xy, 18y^2$

16. _____

17. _____

18. _____

19. c^2d, cd^2, c^3d

20. $2(y - 3), 6(3 - y)$

21. $4(x - 1), 8(1 - x)$

19. _____

20. _____

22. $x^2 - y^2, 2x + 2y, x^2 + 2xy + y^2$

23. $a + 1; (a - 1)^2; a^2 - 1$

21. _____

22. _____

24. $m^2 - 5m + 6; m^2 - 4m + 4$

25. $2 + 3k; 9k^2 - 4; 2 - 3k$

23. _____

24. _____

26. $10v^2 + 30v; -5v^2 - 35v - 60$

27. $9x^3 - 9x^2 - 18x; 6x^5 - 24x^4 + 24x^3$

25. _____

26. _____

27. _____

28. When is the LCM of two expressions the same as their product?

The planets Earth, Jupiter, Saturn, and Uranus revolve about the sun about once each 1, 12, 30, and 84 years, respectively.

28. _____

29. How often will Jupiter and Saturn appear in the same direction in the night sky as seen from Earth?

29. _____

30. How often will Jupiter, Saturn, and Uranus all appear in the same direction in the night sky as seen from Earth?

30. _____

8.6 ADDITION WITH DIFFERENT DENOMINATORS

● Now that we know how to find LCM's, we can add fractional expressions with different denominators. We first find the LCM of the denominators (the least common denominator) and then add.

Example 1 Add: $\dfrac{3}{x + 1} + \dfrac{5}{x - 1}$.

The denominators do not factor, so the LCM is their product. We multiply by 1 to get the LCM in each expression.

$$\frac{3}{x + 1} \cdot \frac{x - 1}{x - 1} + \frac{5}{x - 1} \cdot \frac{x + 1}{x + 1} = \frac{3(x - 1) + 5(x + 1)}{(x - 1)(x + 1)}$$

$$= \frac{3x - 3 + 5x + 5}{(x - 1)(x + 1)}$$

$$= \frac{8x + 2}{(x - 1)(x + 1)}$$

The numerator and denominator have no common factor (other than 1), so we cannot simplify. If we wish, we can multiply out the denominator to get $(8x + 2)/(x^2 - 1)$, but we usually leave answers in factored form.

DO EXERCISE 1.

Example 2 Add: $\dfrac{5}{x^2 + x} + \dfrac{4}{2x + 2}$.

First find the LCM of the denominators.

$$\left.\begin{array}{l} x^2 + x = \boxed{x(x + 1)} \\[2mm] 2x + 2 = \boxed{2}\,(x + 1) \end{array}\right\} \quad \text{LCM} = \boxed{2}\ \boxed{x(x + 1)}$$

Multiply by 1 to get the LCM in each expression. Then add and simplify.

$$\frac{5}{x(x + 1)} \cdot \frac{2}{2} + \frac{4}{2(x + 1)} \cdot \frac{x}{x} = \frac{10}{2x(x + 1)} + \frac{4x}{2x(x + 1)} \quad \text{Multiplying by 1}$$

$$= \frac{10 + 4x}{2x(x + 1)} \quad \text{Adding}$$

$$= \frac{2(5 + 2x)}{2x(x + 1)} \quad \text{Factoring numerator}$$

$$= \frac{2}{2} \cdot \frac{5 + 2x}{x(x + 1)} \quad \text{Factoring the fractional expression}$$

$$= \frac{5 + 2x}{x(x + 1)} \quad \text{or} \quad \frac{5 + 2x}{x^2 + x} \quad \text{Simplifying}$$

DO EXERCISES 2 AND 3.

ANSWERS ON PAGE A–26

OBJECTIVE

After finishing Section 8.6, you should be able to:

● Add fractional expressions with different denominators and simplify the result.

Add.

1. $\dfrac{x}{x - 2} + \dfrac{4}{x + 2}$

Add.

2. $\dfrac{3}{x^3 - x} + \dfrac{4}{x^2 + 2x + 1}$

3. $\dfrac{5}{x^2 + 17x + 16} + \dfrac{3}{x^2 + 9x + 8}$

Add.

4. $\dfrac{a-1}{a+2} + \dfrac{a+3}{a-9}$

Example 3   Add: $\dfrac{x+4}{x-2} + \dfrac{x-7}{x+5}$.

First, find the LCM of the denominators. It is just the product.

$$\text{LCM} = (x-2)(x+5)$$

Multiply by 1 to get the LCM in each expression. Then add and simplify.

$$\frac{x+4}{x-2} \cdot \frac{x+5}{x+5} + \frac{x-7}{x+5} \cdot \frac{x-2}{x-2} = \frac{(x+4)(x+5)}{(x-2)(x+5)} + \frac{(x-7)(x-2)}{(x-2)(x+5)}$$

$$= \frac{x^2+9x+20}{(x-2)(x+5)} + \frac{x^2-9x+14}{(x-2)(x+5)}$$

$$= \frac{x^2+9x+20+x^2-9x+14}{(x-2)(x+5)}$$

$$= \frac{2x^2+34}{(x-2)(x+5)}$$

DO EXERCISE 4.

NAME

CLASS

ANSWERS

EXERCISE SET 8.6

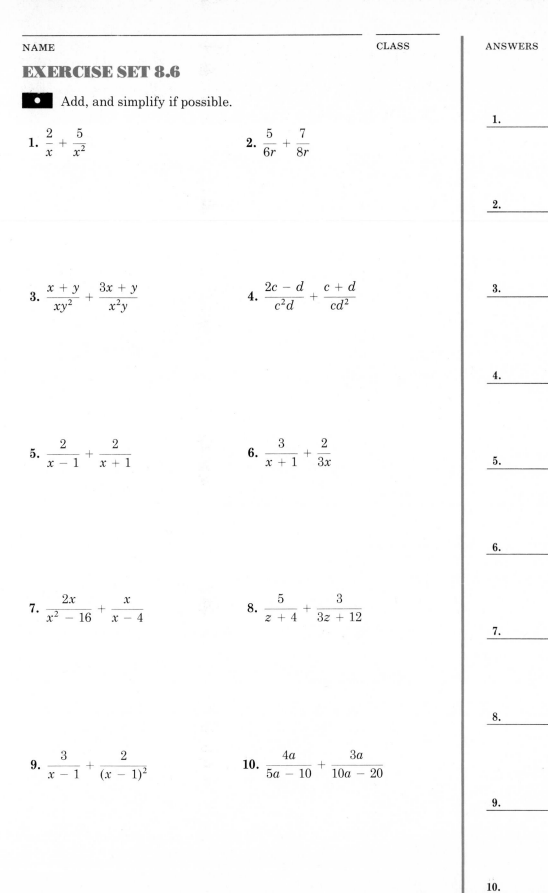

● Add, and simplify if possible.

1. $\dfrac{2}{x} + \dfrac{5}{x^2}$

2. $\dfrac{5}{6r} + \dfrac{7}{8r}$

3. $\dfrac{x+y}{xy^2} + \dfrac{3x+y}{x^2y}$

4. $\dfrac{2c-d}{c^2d} + \dfrac{c+d}{cd^2}$

5. $\dfrac{2}{x-1} + \dfrac{2}{x+1}$

6. $\dfrac{3}{x+1} + \dfrac{2}{3x}$

7. $\dfrac{2x}{x^2-16} + \dfrac{x}{x-4}$

8. $\dfrac{5}{z+4} + \dfrac{3}{3z+12}$

9. $\dfrac{3}{x-1} + \dfrac{2}{(x-1)^2}$

10. $\dfrac{4a}{5a-10} + \dfrac{3a}{10a-20}$

1. _____

2. _____

3. _____

4. _____

5. _____

6. _____

7. _____

8. _____

9. _____

10. _____

Copyright © 1979, Philippines copyright 1979, by Addison-Wesley Publishing Company, Inc. All rights reserved.

ANSWERS

11. $\dfrac{x}{x^2 + 2x + 1} + \dfrac{1}{x^2 + 5x + 4}$ 12. $\dfrac{7}{a^2 + a - 2} + \dfrac{5}{a^2 - 4a + 3}$

11. _____

12. _____

13. $\dfrac{x + 3}{x - 5} + \dfrac{x - 5}{x + 3}$ 14. $\dfrac{3x}{2y - 3} + \dfrac{2x}{3y - 2}$

13. _____

14. _____

15. $\dfrac{a}{a^2 - 1} + \dfrac{2a}{a^2 - a}$ 16. $\dfrac{3x + 2}{3x + 6} + \dfrac{x - 2}{x^2 - 4}$

15. _____

16. _____

17. $\dfrac{6}{x - y} + \dfrac{4x}{y^2 - x^2}$ 18. $\dfrac{a - 2}{3 - a} + \dfrac{4 - a^2}{a^2 - 9}$

17. _____

18. _____

19. $\dfrac{10}{x^2 + x - 6} + \dfrac{3x}{x^2 - 4x + 4}$ 20. $\dfrac{2}{z^2 - z - 6} + \dfrac{3}{z^2 - 9}$

19. _____

20. _____

8.7 SUBTRACTION WITH DIFFERENT DENOMINATORS

 SUBTRACTION WITH DIFFERENT DENOMINATORS

Subtraction is like addition, except that we subtract numerators.

Example 1 Subtract: $\dfrac{x + 2}{x - 4} - \dfrac{x + 1}{x + 4}$.

$$\text{LCM} = (x - 4)(x + 4)$$

$$\frac{x + 2}{x - 4} \cdot \frac{x + 4}{x + 4} - \frac{x + 1}{x + 4} \cdot \frac{x - 4}{x - 4}$$

$$= \frac{(x + 2)(x + 4)}{(x - 4)(x + 4)} - \frac{(x + 1)(x - 4)}{(x - 4)(x + 4)}$$

$$= \frac{x^2 + 6x + 8}{(x - 4)(x + 4)} - \frac{x^2 - 3x - 4}{(x - 4)(x + 4)}$$

$$= \frac{x^2 + 6x + 8 - (x^2 - 3x - 4)}{(x - 4)(x + 4)} \qquad \text{Subtracting numerators (don't forget parentheses)}$$

$$= \frac{x^2 + 6x + 8 - x^2 + 3x + 4}{(x - 4)(x + 4)}$$

$$= \frac{9x + 12}{(x - 4)(x + 4)}$$

$$= \frac{3(3x + 4)}{(x - 4)(x + 4)}$$

DO EXERCISE 1.

 SIMPLIFYING COMBINED ADDITIONS AND SUBTRACTIONS

Example 2 Simplify: $\dfrac{1}{x} - \dfrac{1}{x^2} + \dfrac{2}{x + 1}$.

$$\text{LCM} = x^2(x + 1)$$

$$\frac{1}{x} \cdot \frac{x(x + 1)}{x(x + 1)} - \frac{1}{x^2} \cdot \frac{x + 1}{x + 1} + \frac{2}{x + 1} \cdot \frac{x^2}{x^2}$$

$$= \frac{x(x + 1)}{x^2(x + 1)} - \frac{x + 1}{x^2(x + 1)} + \frac{2x^2}{x^2(x + 1)}$$

$$= \frac{x(x + 1) - (x + 1) + 2x^2}{x^2(x + 1)}$$

$$= \frac{x^2 + x - x - 1 + 2x^2}{x^2(x + 1)}$$

$$= \frac{3x^2 - 1}{x^2(x + 1)}$$

DO EXERCISE 2.

OBJECTIVES

After finishing Section 8.7, you should be able to:

 Subtract fractional expressions with different denominators.

 Simplify combined additions and subtractions of fractional expressions.

Subtract.

1. $\dfrac{x - 2}{3x} - \dfrac{2x - 1}{5x}$

Simplify.

2. $\dfrac{1}{x} - \dfrac{5}{3x} + \dfrac{2x}{x + 1}$

NAME	CLASS

EXERCISE SET 8.7

▪● Subtract, and simplify if possible.

1. $\dfrac{x-2}{6} - \dfrac{x+1}{3}$

2. $\dfrac{a+2}{2} - \dfrac{a-4}{4}$

3. $\dfrac{4z-9}{3z} - \dfrac{3z-8}{4z}$

4. $\dfrac{x-1}{4x} - \dfrac{2x+3}{x}$

5. $\dfrac{4x+2t}{3xt^2} - \dfrac{5x-3t}{x^2 t}$

6. $\dfrac{5x+3y}{2x^2 y} - \dfrac{3x+4y}{xy^2}$

7. $\dfrac{5}{x+5} - \dfrac{3}{x-5}$

8. $\dfrac{2z}{z-1} - \dfrac{3z}{z+1}$

9. $\dfrac{3}{2t^2-2t} - \dfrac{5}{2t-2}$

10. $\dfrac{8}{x^2-4} - \dfrac{3}{x+2}$

ANSWERS

1. _____

2. _____

3. _____

4. _____

5. _____

6. _____

7. _____

8. _____

9. _____

10. _____

Copyright © 1979, Philippines copyright 1979, by Addison-Wesley Publishing Company, Inc. All rights reserved.

ANSWERS

11. $\dfrac{2s}{t^2 - s^2} - \dfrac{s}{t - s}$

12. $\dfrac{3}{12 + x - x^2} - \dfrac{2}{x^2 - 9}$

11. _____

12. _____

● ● Simplify.

13. $\dfrac{4y}{y^2 - 1} - \dfrac{2}{y} - \dfrac{2}{y + 1}$

14. $\dfrac{x + 6}{4 - x^2} - \dfrac{x + 3}{x + 2} + \dfrac{x - 3}{2 - x}$

13. _____

14. _____

15. $\dfrac{2z}{1 - 2z} + \dfrac{3z}{2z + 1} - \dfrac{3}{4z^2 - 1}$

16. $\dfrac{1}{x + y} + \dfrac{1}{x - y} - \dfrac{2x}{x^2 - y^2}$

15. _____

16. _____

17. $\dfrac{5}{3 - 2x} + \dfrac{3}{2x - 3} - \dfrac{x - 3}{2x^2 - x - 3}$

18. $\dfrac{2r}{r^2 - s^2} + \dfrac{1}{r + s} - \dfrac{1}{r - s}$

17. _____

18. _____

19. $\dfrac{3}{2c - 1} - \dfrac{1}{c + 2} - \dfrac{5}{2c^2 + 3c - 2}$

20. $\dfrac{3y - 1}{2y^2 + y - 3} - \dfrac{2 - y}{y - 1}$

19. _____

20. _____

8.8 COMPLEX FRACTIONAL EXPRESSIONS

● A *complex fractional expression* is one that has a fractional expression in its numerator or its denominator, or both. Here are some examples:

$$\frac{1+\dfrac{2}{x}}{3}, \quad \frac{\dfrac{x+y}{2}}{\dfrac{2x}{x+1}}, \quad \frac{\dfrac{1}{3}+\dfrac{1}{5}}{\dfrac{2}{x}-\dfrac{x}{y}}.$$

> To simplify a complex fractional expression we first add or subtract, if necessary, to get a single fractional expression in both numerator and denominator. Then we divide, by multiplying by the reciprocal of the denominator.

Example 1   Simplify.

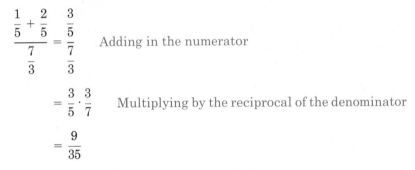

$$\frac{\dfrac{1}{5}+\dfrac{2}{5}}{\dfrac{7}{3}} = \frac{\dfrac{3}{5}}{\dfrac{7}{3}} \qquad \text{Adding in the numerator}$$

$$= \frac{3}{5}\cdot\frac{3}{7} \qquad \text{Multiplying by the reciprocal of the denominator}$$

$$= \frac{9}{35}$$

Example 2   Simplify.

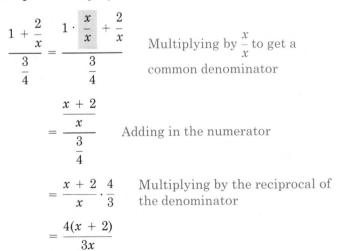

$$\frac{1+\dfrac{2}{x}}{\dfrac{3}{4}} = \frac{1\cdot\dfrac{x}{x}+\dfrac{2}{x}}{\dfrac{3}{4}} \qquad \text{Multiplying by } \dfrac{x}{x} \text{ to get a common denominator}$$

$$= \frac{\dfrac{x+2}{x}}{\dfrac{3}{4}} \qquad \text{Adding in the numerator}$$

$$= \frac{x+2}{x}\cdot\frac{4}{3} \qquad \text{Multiplying by the reciprocal of the denominator}$$

$$= \frac{4(x+2)}{3x}$$

DO EXERCISES 1 AND 2.

ANSWERS ON PAGE A–26

After finishing Section 8.8, you should be able to:

● **Simplify complex fractional expressions.**

Simplify.

1. $\dfrac{\dfrac{2}{7}+\dfrac{3}{7}}{\dfrac{3}{4}}$

2. $\dfrac{3+\dfrac{x}{2}}{\dfrac{5}{4}}$

Simplify.

3. $\dfrac{\dfrac{x}{2} + \dfrac{2x}{3}}{\dfrac{1}{x} - \dfrac{x}{2}}$

4. $\dfrac{b - \dfrac{x^2}{b}}{\dfrac{x}{b} + 1}$

Example 3   Simplify.

$$\frac{\dfrac{3}{x} + \dfrac{1}{2x}}{\dfrac{1}{3x} - \dfrac{3}{4x}} = \frac{\dfrac{3}{x} \cdot \boxed{\dfrac{2}{2}} + \dfrac{1}{2x}}{\dfrac{1}{3x} \cdot \boxed{\dfrac{4}{4}} - \dfrac{3}{4x} \cdot \boxed{\dfrac{3}{3}}}$$

$$= \frac{\dfrac{6}{2x} + \dfrac{1}{2x}}{\dfrac{4}{12x} - \dfrac{9}{12x}}$$

$$= \frac{\dfrac{7}{2x}}{\dfrac{-5}{12x}} \qquad \text{Adding in the numerator, subtracting in the denominator}$$

$$= \frac{7}{2x} \cdot \frac{12x}{-5} \qquad \text{Multiplying by the reciprocal of the denominator}$$

$$= \boxed{\frac{2x}{2x}} \cdot \frac{7 \cdot 6}{-5}$$

$$= \frac{42}{-5} = -\frac{42}{5} \qquad \text{Simplifying}$$

Example 4   Simplify.

$$\frac{1 - \dfrac{x}{a}}{a - \dfrac{x^2}{a}} = \frac{\boxed{\dfrac{a}{a}} - \dfrac{x}{a}}{a \cdot \boxed{\dfrac{a}{a}} - \dfrac{x^2}{a}} = \frac{\dfrac{a - x}{a}}{\dfrac{a^2 - x^2}{a}}$$

$$= \frac{a - x}{a} \cdot \frac{a}{a^2 - x^2}$$

$$= \boxed{\frac{a}{a}} \cdot \frac{a - x}{a^2 - x^2}$$

$$= \frac{a - x}{(a - x)(a + x)}$$

$$= \boxed{\frac{a - x}{a - x}} \cdot \frac{1}{a + x}$$

$$= \frac{1}{a + x}$$

DO EXERCISES 3 AND 4.

NAME   CLASS   ANSWERS

EXERCISE SET 8.8

⬤ Simplify.

1. $\dfrac{1 + \dfrac{9}{16}}{1 - \dfrac{3}{4}}$

2. $\dfrac{9 - \dfrac{1}{4}}{3 + \dfrac{1}{2}}$

3. $\dfrac{1 - \dfrac{3}{5}}{1 + \dfrac{1}{5}}$

4. $\dfrac{\dfrac{5}{27} - 5}{\dfrac{1}{3} + 1}$

5. $\dfrac{\dfrac{1}{x} + 3}{\dfrac{1}{x} - 5}$

6. $\dfrac{\dfrac{3}{s} + s}{\dfrac{s}{3} + s}$

7. $\dfrac{1 - \dfrac{x}{y}}{\dfrac{x - y}{y}}$

8. $\dfrac{a - 1}{a - \dfrac{1}{a}}$

9. $\dfrac{\dfrac{2}{y} + \dfrac{1}{2y}}{y + \dfrac{y}{2}}$

1. _____

2. _____

3. _____

4. _____

5. _____

6. _____

7. _____

8. _____

9. _____

Copyright © 1979, Philippines copyright 1979, by Addison-Wesley Publishing Company, Inc. All rights reserved.

ANSWERS

10. _____

11. _____

12. _____

13. _____

14. _____

15. _____

16. _____

17. _____

18. _____

10. $\dfrac{4 - \dfrac{1}{x^2}}{2 - \dfrac{1}{x}}$

11. $\dfrac{c + \dfrac{c}{d}}{1 + \dfrac{1}{d}}$

12. $\dfrac{2 - \dfrac{a}{b}}{2 - \dfrac{b}{a}}$

13. $\dfrac{\dfrac{1}{b} - \dfrac{1}{a}}{\dfrac{b - a}{5}}$

14. $\dfrac{2 - \dfrac{1}{x}}{\dfrac{2}{x}}$

15. $\dfrac{\dfrac{x}{x - y}}{\dfrac{x^2}{x^2 - y^2}}$

16. $\dfrac{\dfrac{x}{y} - \dfrac{y}{x}}{\dfrac{1}{y} + \dfrac{1}{x}}$

17. $\dfrac{x - 3 + \dfrac{2}{x}}{x - 4 + \dfrac{3}{x}}$

18. $\dfrac{1 + \dfrac{a}{b - a}}{\dfrac{a}{a + b} - 1}$

8.9 DIVISION OF POLYNOMIALS

▪ DIVISOR A MONOMIAL

Division can be shown by a fractional expression.

Example 1   Divide $x^3 + 10x^2 + 8x$ by $2x$.

We write a fractional expression to show division:

$$\frac{x^3 + 10x^2 + 8x}{2x}.$$

This is equivalent to

$$\frac{x^3}{2x} + \frac{10x^2}{2x} + \frac{8x}{2x}. \qquad \text{To see this, add and get the original expression.}$$

Next, we do the separate divisions:

$$\frac{x^3}{2x} + \frac{10x^2}{2x} + \frac{8x}{2x} = \frac{1}{2}x^2 + 5x + 4.$$

DO EXERCISES 1 AND 2.

Example 2   Divide and check: $(5y^2 - 2y + 4) \div 2$.

$$\frac{5y^2 - 2y + 4}{2} = \frac{5y^2}{2} - \frac{2y}{2} + \frac{4}{2} = \frac{5}{2}y^2 - y + 2$$

Check: $\dfrac{5}{2}y^2 - y + 2$

$$\frac{\qquad\qquad\quad 2}{5y^2 - 2y + 4} \qquad \begin{array}{l}\text{We multiply.}\\ \text{The answer checks.}\end{array}$$

Try to write only the answer.

> **To divide by a monomial, we can divide each term by the monomial.**

DO EXERCISES 3 AND 4.

▪▪ DIVISOR NOT A MONOMIAL

When the divisor is not a monomial, we use long division very much as we do in arithmetic.

Example 3   Divide $x^2 + 5x + 6$ by $x + 2$.

$$
\begin{array}{r}
x \\
x + 2 \overline{)\, x^2 + 5x + 6} \\
x^2 + 2x \\
\hline
3x
\end{array}
$$

Divide first term by first term, to get x.
Ignore the term 2.
Multiply x by divisor.
Subtract.

We now "bring down" the next term of the dividend, 6.

OBJECTIVES

After finishing Section 8.9, you should be able to:

▪ Divide a polynomial by a monomial and check the result.

▪▪ Divide a polynomial by a divisor that is not a monomial and, if there is a remainder, express the result in two ways.

Divide.

1. $\dfrac{2x^3 + 6x^2 + 4x}{2x}$

2. $(6x^2 + 3x - 2) \div 3$

Divide and check.

3. $(8x^2 - 3x + 1) \div 2$

4. $\dfrac{2x^4 - 3x^3 + 5x^2}{x^2}$

ANSWERS ON PAGE A–26

Divide and check.

5. $(x^2 + x - 6) \div (x + 3)$

$$
\begin{array}{r}
x + 3 \\
x + 2\overline{)x^2 + 5x + 6} \\
\underline{x^2 + 2x} \\
3x + 6 \\
\underline{3x + 6} \\
0
\end{array}
$$

Divide first term by first term to get 3.

Multiply 3 by divisor.

Subtract.

The quotient is $x + 3$.

To check, multiply the quotient by the divisor and add the remainder, if any, to see if you get the dividend:

$$(x + 2)(x + 3) = x^2 + 5x + 6. \quad \text{The division checks.}$$

Example 4   Divide and check: $(x^2 + 2x - 12) \div (x - 3)$.

$$
\begin{array}{r}
x + 5 \\
x - 3\overline{)x^2 + 2x - 12} \\
\underline{x^2 - 3x} \\
5x - 12 \\
\underline{5x - 15} \\
3
\end{array}
$$

Check:

$(x - 3)(x + 5) + 3 = x^2 + 2x - 15 + 3$

$ = x^2 + 2x - 12$

Quotient

Remainder

6. $x - 2\overline{)x^2 + 2x - 8}$

The quotient is $x + 5$ and the remainder is 3. We can write this $x + 5$, R 3.

The answer can also be given in this way: $x + 5 + \dfrac{3}{x - 3}$.

DO EXERCISES 5 AND 6.

Example 5   Divide: $(x^3 + 1) \div (x + 1)$.

$$
\begin{array}{r}
x^2 - x + 1 \\
x + 1\overline{)x^3 + 1} \\
\underline{x^3 + x^2} \\
- x^2 \\
\underline{- x^2 - x} \\
x + 1 \\
\underline{x + 1}
\end{array}
$$

Leave space for missing terms.

Divide and check. Express your answer in two ways.

7. $x + 3\overline{)x^2 + 7x + 10}$

The answer is $x^2 - x + 1$.

You need not write a 0 remainder.

Example 6   Divide: $(x^4 - 3x^2 + 1) \div (x - 4)$.

$$
\begin{array}{r}
x^3 + 4x^2 + 13x + 52 \\
x - 4\overline{)x^4 - 3x^2 + 1} \\
\underline{x^4 - 4x^3} \\
4x^3 - 3x^2 \\
\underline{4x^3 - 16x^2} \\
13x^2 \\
\underline{13x^2 - 52x} \\
52x + 1 \\
\underline{52x - 208} \\
209
\end{array}
$$

8. $(x^3 - 1) \div (x - 1)$

The answer can be expressed as $x^3 + 4x^2 + 13x + 52$, R 209, or

$$x^3 + 4x^2 + 13x + 52 + \dfrac{209}{x - 4}.$$

DO EXERCISES 7 AND 8.

EXERCISE SET 8.9

● Divide and check.

1. $\dfrac{u - 2u^2 - u^5}{u}$

2. $\dfrac{50x^5 - 7x^4 + x^2}{x}$

3. $(15t^3 + 24t^2 - 6t) \div 3t$

4. $(25x^3 + 15x^2 - 30x) \div 5x$

5. $\dfrac{20x^6 - 20x^4 - 5x^2}{-5x^2}$

6. $\dfrac{24x^6 + 32x^5 - 8x^2}{-8x^2}$

7. $\dfrac{4x^4y - 8x^6y^2 + 12x^8y^6}{4x^4y}$

8. $\dfrac{9r^2s^2 + 3r^2s - 6rs^2}{-3rs}$

●● Divide and check.

9. $(x^2 - 10x - 25) \div (x - 5)$

10. $(x^2 + 8x - 16) \div (x + 4)$

1. _____

2. _____

3. _____

4. _____

5. _____

6. _____

7. _____

8. _____

9. _____

10. _____

Copyright © 1979, Philippines copyright 1979, by Addison-Wesley Publishing Company, Inc. All rights reserved.

ANSWERS

11. $(x^2 + 4x + 4) \div (x + 2)$ 12. $(x^2 - 6x + 9) \div (x - 3)$

11. _____

12. _____

13. $(x^2 + 4x - 14) \div (x + 6)$ 14. $(x^2 + 5x - 9) \div (x - 2)$

13. _____

14. _____

15. $(x^5 - 1) \div (x - 1)$ 16. $(x^5 + 1) \div (x + 1)$

15. _____

16. _____

17. $(t^3 - t^2 + t - 1) \div (t - 1)$ 18. $(t^3 - t^2 + t - 1) \div (t + 1)$

17. _____

18. _____

19. $(x^6 - 13x^3 + 42) \div (x^3 - 7)$ 20. $(x^6 + 5x^3 - 24) \div (x^3 - 3)$

19. _____

20. _____

8.10 SOLVING FRACTIONAL EQUATIONS

● A *fractional equation* is an equation containing one or more fractional expressions. Here are some examples:

$$\frac{2}{3} + \frac{5}{6} = \frac{x}{9}, \quad x + \frac{6}{x} = -5, \quad \frac{x^2}{x-1} = \frac{1}{x-1}.$$

> **To solve a fractional equation, multiply on both sides by the LCM of all the denominators. This is called** *clearing of fractions.*

Example 1   Solve: $\dfrac{2}{3} + \dfrac{5}{6} = \dfrac{x}{9}$.

The LCM of all denominators is 18, or $2 \cdot 3 \cdot 3$. We multiply on both sides by this.

$$2 \cdot 3 \cdot 3 \left(\frac{2}{3} + \frac{5}{6} \right) = 2 \cdot 3 \cdot 3 \cdot \frac{x}{9}$$

$$2 \cdot 3 \cdot 3 \cdot \frac{2}{3} + 2 \cdot 3 \cdot 3 \cdot \frac{5}{6} = 2 \cdot 3 \cdot 3 \cdot \frac{x}{9} \quad \begin{array}{l}\text{Multiplying to} \\ \text{remove parentheses}\end{array}$$

$$2 \cdot 3 \cdot 2 + 3 \cdot 5 = 2 \cdot x \qquad \text{Simplifying}$$

$$12 + 15 = 2x$$

$$27 = 2x$$

$$\frac{27}{2} = x$$

All the denominators disappeared and the resulting equation was easier to solve.

> **When clearing an equation of fractions, be sure to multiply** *all* **terms in the equation by the LCM.**

Example 2   Solve: $\dfrac{1}{x} = \dfrac{1}{4-x}$.

The LCM is $x(4-x)$. We multiply on both sides by this.

$$x(4-x) \cdot \frac{1}{x} = x(4-x) \cdot \frac{1}{4-x}$$

$$4 - x = x \qquad \text{Simplifying}$$

$$4 = 2x$$

$$x = 2$$

Check: $\dfrac{1}{x} = \dfrac{1}{4-x}$

$$\begin{array}{c|c} \dfrac{1}{2} & \dfrac{1}{4-2} \\ \hline & \dfrac{1}{2} \end{array}$$

This checks, so 2 is the solution.

DO EXERCISES 1 AND 2.

OBJECTIVE

After finishing Section 8.10, you should be able to:

● Solve fractional equations.

Solve.

1. $\dfrac{3}{4} + \dfrac{5}{8} = \dfrac{x}{12}$

2. $\dfrac{1}{x} = \dfrac{1}{6-x}$

ANSWERS ON PAGE A-26

Solve.

3. $x + \dfrac{1}{x} = 2$

4. $\dfrac{x^2}{x + 2} = \dfrac{4}{x + 2}$

5. $\dfrac{1}{2x} + \dfrac{1}{x} = -12$

The next example shows how important it is to multiply all terms in an equation by the LCM.

Example 3 Solve: $x + \dfrac{6}{x} = -5$.

The LCM is x. We multiply by this.

$$x\left(x + \dfrac{6}{x}\right) = -5\,x \qquad \text{Multiplying on both sides by } x$$

$$x^2 + x \cdot \dfrac{6}{x} = -5x \qquad \begin{array}{l}\text{Note that \textbf{each term} on the}\\ \text{left is now multiplied by } x.\end{array}$$

$$x^2 + 6 = -5x$$

$$x^2 + 5x + 6 = 0$$

$$(x + 3)(x + 2) = 0 \qquad \text{Factoring}$$

$$x + 3 = 0 \ \text{ or } \ x + 2 = 0 \qquad \text{Principle of zero products}$$

$$x = -3 \ \text{ or } \ x = -2$$

Check:

$$\begin{array}{c|c} x + \dfrac{6}{x} = -5 & \\ \hline -3 + \dfrac{6}{-3} \ \Big| \ -5 \\ \hline -5 \ \Big| \end{array} \qquad \begin{array}{c|c} x + \dfrac{6}{x} = -5 & \\ \hline -2 + \dfrac{6}{-2} \ \Big| \ -5 \\ \hline -5 \ \Big| \end{array}$$

Both of these check, so there are two solutions, -3 and -2.

It is important *always* to check when solving fractional equations.

Example 4 Solve: $\dfrac{x^2}{x - 1} = \dfrac{1}{x - 1}$.

The LCM is $x - 1$. We multiply by this.

$$(x - 1) \cdot \dfrac{x^2}{x - 1} = (x - 1) \cdot \dfrac{1}{x - 1}$$

$$x^2 = 1$$

$$x^2 - 1 = 0$$

$$(x - 1)(x + 1) = 0 \qquad \text{Factoring}$$

$$x - 1 = 0 \ \text{ or } \ x + 1 = 0 \qquad \text{Principle of zero products}$$

$$x = 1 \ \text{ or } \qquad x = -1$$

Possible solutions are 1 and -1.

Check:

$$\begin{array}{c|c} \dfrac{x^2}{x - 1} = \dfrac{1}{x - 1} & \\ \hline \dfrac{1^2}{1 - 1} \ \Big| \ \dfrac{1}{1 - 1} \\ \hline \dfrac{1}{0} \ \Big| \ \dfrac{1}{0} \end{array} \qquad \begin{array}{c|c} \dfrac{x^2}{x - 1} = \dfrac{1}{x - 1} & \\ \hline \dfrac{(-1)^2}{-1 - 1} \ \Big| \ \dfrac{1}{-1 - 1} \\ \hline -\dfrac{1}{2} \ \Big| \ -\dfrac{1}{2} \end{array}$$

The number -1 is a solution, but 1 is not because it makes a denominator zero.

ANSWERS ON PAGE A–26

DO EXERCISES 3–5.

NAME　　　　　　　　　　　　　　　　　CLASS　　　　　　　　ANSWERS

EXERCISE SET 8.10

▪ ● Solve.

1. $\dfrac{5}{x} = \dfrac{6}{x} - \dfrac{1}{3}$

2. $\dfrac{5}{3x} + \dfrac{3}{x} = 1$

3. $\dfrac{x-7}{x+2} = \dfrac{1}{4}$

4. $\dfrac{a-2}{a+3} = \dfrac{3}{8}$

5. $\dfrac{2}{x+1} = \dfrac{1}{x-2}$

6. $\dfrac{5}{x-1} = \dfrac{3}{x+2}$

7. $\dfrac{x}{8} - \dfrac{x}{12} = \dfrac{1}{8}$

8. $\dfrac{x+1}{3} - \dfrac{x-1}{2} = 1$

9. $\dfrac{a-3}{3a+2} = \dfrac{1}{5}$

10. $\dfrac{2}{x+3} = \dfrac{5}{x}$

11. $\dfrac{x-1}{x-5} = \dfrac{4}{x-5}$

12. $\dfrac{x-7}{x-9} = \dfrac{2}{x-9}$

1. _____

2. _____

3. _____

4. _____

5. _____

6. _____

7. _____

8. _____

9. _____

10. _____

11. _____

12. _____

Copyright © 1979, Philippines copyright 1979, by Addison-Wesley Publishing Company, Inc. All rights reserved.

ANSWERS

13. $\dfrac{x-2}{x-3} = \dfrac{x-1}{x+1}$

14. $\dfrac{2b-3}{3b+2} = \dfrac{2b+1}{3b-2}$

13. _____

14. _____

15. $\dfrac{6x-2}{2x-1} = \dfrac{9x}{3x+1}$

16. $\dfrac{2a}{a+1} = 2 - \dfrac{5}{2a}$

15. _____

16. _____

17. $\dfrac{1}{x+3} + \dfrac{1}{x-3} = \dfrac{1}{x^2-9}$

18. $\dfrac{4}{x-3} + \dfrac{2x}{x^2-9} = \dfrac{1}{x+3}$

17. _____

18. _____

19. $\dfrac{x}{x+4} - \dfrac{4}{x-4} = \dfrac{x^2+16}{x^2-16}$

20. $\dfrac{5}{y-3} - \dfrac{30}{y^2-9} = 1$

19. _____

20. _____

8.11 APPLIED PROBLEMS AND FORMULAS

⬛● APPLIED PROBLEMS

We now solve applied problems using fractional equations.

Example 1   The reciprocal of 2 less than a certain number is twice the reciprocal of the number itself. What is the number?

First, translate to an equation. Let x = the number. Then 2 less than the number is $x - 2$, and the reciprocal of the number is $1/x$.

$$\underbrace{\left(\begin{array}{c}\text{Reciprocal of 2}\\\text{less than number}\end{array}\right)}_{\dfrac{1}{x-2}} \quad \overset{\text{is}}{\Big\downarrow} \quad \underbrace{\left(\begin{array}{c}\text{Twice the reciprocal}\\\text{of the number}\end{array}\right)}_{2\cdot\dfrac{1}{x}}$$

$$\frac{1}{x-2} = 2\cdot\frac{1}{x}$$

Now we solve.

$$\frac{1}{x-2} = \frac{2}{x} \qquad \text{The LCM is } x(x-2).$$

$$\frac{\boxed{x(x-2)}}{x-2} = \frac{2\boxed{x(x-2)}}{x} \qquad \text{Multiplying by LCM}$$

$$x = 2(x-2) \qquad \text{Simplifying}$$

$$x = 2x - 4$$

$$x = 4$$

Check: Go to the original problem. The number to be checked is 4. Two less than 4 is 2. The reciprocal of 2 is $\frac{1}{2}$. The reciprocal of the number itself is $\frac{1}{4}$. Now $\frac{1}{2}$ is twice $\frac{1}{4}$, so the conditions are satisfied. Thus, the solution is 4.

DO EXERCISE 1.

Example 2   One car travels 20 km/h faster than another. While one of them goes 240 km, the other goes 160 km. Find their speeds.

To translate, recall that distance, speed, and time are related by $r = d/t$. Also, note that both cars travel for the same time. Thus, $t_1 = t_2$. We solve $r = d/t$ for t and get $t = d/r$. Since $t_1 = t_2$, we have

$$\frac{d_1}{r_1} = \frac{d_2}{r_2}.$$

Now, $d_1 = 160$ and $d_2 = 240$. We note that $r_2 = r_1 + 20$. We can summarize in a table.

	r	t	d
First car	r_1	t_1	160
Second car	$r_1 + 20$	t_2	240

We get

$$\frac{240}{r_1 + 20} = \frac{160}{r_1}.$$

OBJECTIVES

After finishing Section 8.11, you should be able to:

⬛● Solve applied problems using fractional equations.

⬛●● Solve a formula for a given letter when fractional expressions occur in the formula.

Solve.

1. The reciprocal of two more than a number is three times the reciprocal of the number. Find the number.

ANSWER ON PAGE A–27

2. One car goes 10 km/h faster than another. While one car goes 120 km, the other goes 150 km. How fast is each car?

Solve:

$$LCM = r_1(r_1 + 20)$$

$$\frac{240 \; r_1(r_1 + 20)}{r_1 + 20} = \frac{160 \; r_1(r_1 + 20)}{r_1} \qquad \text{Multiplying by LCM}$$

$$240r_1 = 160(r_1 + 20)$$

$$240r_1 = 160r_1 + 3200$$

$$80r_1 = 3200$$

$$r_1 = 40 \text{ km/h}$$

$$r_2 = r_1 + 20 = 60 \text{ km/h.} \qquad \text{These check.}$$

DO EXERCISE 2.

Suppose it takes a person 4 hours to do a certain job. Then in 1 hour $\frac{1}{4}$ of the job is done.

> **If a job can be done in t hours (or days, or some other unit of time), then $1/t$ of it can be done in 1 hour (or day).**

Example 3   The head of a secretarial pool examines work records and finds that it takes Helen Huntinpeck 4 hr to type a certain report. It takes Willie Typitt 6 hr to type the same report. How long would it take them, working together, to type the same report?

It takes Helen 4 hr and Willie 6 hr. Then in 1 hr Helen types $\frac{1}{4}$ of the report and Willie types $\frac{1}{6}$ of the report. Working together, they can type

$$\frac{1}{4} + \frac{1}{6}$$

of the report in 1 hr. Let t = the time it takes to type the report, if they work together. Then in 1 hr they can type

$$\frac{1}{t}$$

of the report. Thus,

$$\frac{1}{4} + \frac{1}{6} = \frac{1}{t}.$$

Solve:

$$12t \cdot \left(\frac{1}{4} + \frac{1}{6}\right) = 12t \cdot \frac{1}{t} \qquad \text{The LCM is } 2 \cdot 2 \cdot 3t, \text{ or } 12t.$$

$$\frac{12t}{4} + \frac{12t}{6} = \frac{12t}{t}$$

$$3t + 2t = 12$$

$$5t = 12$$

$$t = \frac{12}{5}, \quad \text{or} \quad 2\frac{2}{5}.$$

Thus it takes $2\frac{2}{5}$ hr.

Note that this answer is less than the time it takes either person to do the job alone.

DO EXERCISE 3.

● ●   FORMULAS

We solve some formulas containing fractional expressions.

Example 4   Solve for P: $A/P = 1 + r$. This is an interest formula.

The LCM is P. We multiply by this.

$$P \cdot \frac{A}{P} = P(1 + r) \qquad \text{Multiplying by } P$$

$$A = P(1 + r) \qquad \text{Simplifying the left side}$$

$$\frac{A}{1 + r} = \frac{P(1 + r)}{1 + r} \qquad \text{Multiplying by } \frac{1}{1 + r}$$

$$\frac{A}{1 + r} = P \qquad \text{Simplifying}$$

Example 5   If one person can do a job in a hr and another can do the same job in b hr, then working together they can do the same job in t hr, where a, b, and t are related by

$$\frac{1}{a} + \frac{1}{b} = \frac{1}{t}.$$

Solve for t.

The LCM is abt. We multiply by this.

$$abt \cdot \left(\frac{1}{a} + \frac{1}{b}\right) = abt \cdot \frac{1}{t} \qquad \text{Multiplying by } abt$$

$$abt \cdot \frac{1}{a} + abt \cdot \frac{1}{b} = \frac{abt}{t}$$

$$\frac{abt}{a} + \frac{abt}{b} = \frac{abt}{t}$$

$$bt + at = ab \qquad \text{Simplifying}$$

$$(b + a)t = ab \qquad \text{Factoring out } t$$

$$t = \frac{ab}{b + a} \qquad \text{Multiplying by } \frac{1}{b + a}$$

This answer can be used to find solutions to problems such as Example 3:

$$t = \frac{4 \cdot 6}{6 + 4} = \frac{24}{10} = 2\frac{2}{5}.$$

DO EXERCISES 4 AND 5.

3. By checking work records, a contractor finds that it takes Red Bryck 6 hr to construct a wall of a certain size. It takes Lotta Mudd 8 hr to construct the same wall. How long would it take if they worked together?

4. Solve for p: $\dfrac{n}{p} = 2 - m$.

5. Solve for f: $\dfrac{1}{p} + \dfrac{1}{q} = \dfrac{1}{f}.$

(This is an optics formula.)

SOMETHING EXTRA
A NUMBER PATTERN

Show that each of the following is true by simplifying the left side of the equation. Look for a pattern.

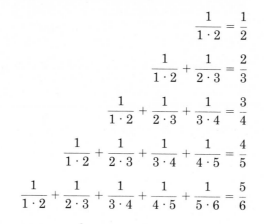

$$\frac{1}{1 \cdot 2} = \frac{1}{2}$$

$$\frac{1}{1 \cdot 2} + \frac{1}{2 \cdot 3} = \frac{2}{3}$$

$$\frac{1}{1 \cdot 2} + \frac{1}{2 \cdot 3} + \frac{1}{3 \cdot 4} = \frac{3}{4}$$

$$\frac{1}{1 \cdot 2} + \frac{1}{2 \cdot 3} + \frac{1}{3 \cdot 4} + \frac{1}{4 \cdot 5} = \frac{4}{5}$$

$$\frac{1}{1 \cdot 2} + \frac{1}{2 \cdot 3} + \frac{1}{3 \cdot 4} + \frac{1}{4 \cdot 5} + \frac{1}{5 \cdot 6} = \frac{5}{6}$$

EXERCISES

Use the pattern to find these sums without adding.

1. $\dfrac{1}{1 \cdot 2} + \dfrac{1}{2 \cdot 3} + \dfrac{1}{3 \cdot 4} + \dfrac{1}{4 \cdot 5} + \dfrac{1}{5 \cdot 6} + \dfrac{1}{6 \cdot 7} + \dfrac{1}{7 \cdot 8}$

2. $\dfrac{1}{1 \cdot 2} + \dfrac{1}{2 \cdot 3} + \dfrac{1}{3 \cdot 4} + \cdots + \dfrac{1}{100 \cdot 101}$ (The dots stand for the symbols that we did not write.)

3. $\dfrac{1}{1 \cdot 2} + \dfrac{1}{2 \cdot 3} + \dfrac{1}{3 \cdot 4} + \cdots + \dfrac{1}{n(n + 1)}$

NAME   CLASS   ANSWERS

EXERCISE SET 8.11

▟●▙ Solve.

1. The reciprocal of 4 plus the reciprocal of 5 is the reciprocal of what number?

2. The reciprocal of 3 plus the reciprocal of 8 is the reciprocal of what number?

3. One number is 5 more than another. The quotient of the larger divided by the smaller is $\frac{4}{3}$. Find the numbers.

4. One number is 4 more than another. The quotient of the larger divided by the smaller is $\frac{5}{2}$. Find the numbers.

5. One car travels 40 km/h faster than another. While one travels 150 km, the other goes 350 km. Find their speeds.

6. One car travels 30 km/h faster than another. While one goes 250 km, the other goes 400 km. Find their speeds.

7. A person traveled 120 mi in one direction. The return trip was accomplished at double the speed, and took 3 hr less time. Find the speed going.

8. After making a trip of 126 mi, a person found that the trip would have taken 1 hr less time by increasing the speed by 8 mph. What was the actual speed?

9. The speed of a freight train is 14 km/h slower than the speed of a passenger train. The freight train travels 330 km in the same time that it takes the passenger train to travel 400 km. Find the speed of each train.

10. The speed of a freight train is 15 km/h slower than the speed of a passenger train. The freight train travels 390 kilometers in the same time that it takes the passenger train to travel 480 kilometers. Find the speed of each train.

1. _____

2. _____

3. _____

4. _____

5. _____

6. _____

7. _____

8. _____

9. _____

10. _____

Copyright © 1979, Philippines copyright 1979, by Addison-Wesley Publishing Company, Inc. All rights reserved.

ANSWERS

11. _____

12. _____

13. _____

14. _____

15. _____

16. _____

17. _____

18. _____

19. _____

20. _____

21. _____

22. _____

23. _____

24. _____

11. It takes painter A 4 hours to paint a certain area of a house. It takes painter B 5 hours to do the same job. How long would it take them, working together, to do the painting job?

12. By checking work records a carpenter finds that worker A can build a certain type of garage in 12 hours. Worker B can do the same job in 16 hours. How long would it take if they worked together?

13. By checking work records a plumber finds that worker A can do a certain job in 12 hours. Worker B can do the same job in 9 hours. How long would it take if they worked together?

14. A tank can be filled in 18 hours by pipe A alone and 24 hours by pipe B alone. How long would it take to fill the tank if both pipes were working?

●● Solve each formula for the given letter.

15. $\dfrac{1}{p} + \dfrac{1}{q} = \dfrac{1}{f}$; p

16. $\dfrac{1}{a} + \dfrac{1}{b} = \dfrac{1}{t}$; b

17. $\dfrac{A}{P} = 1 + r$; A

18. $\dfrac{2A}{h} = a + b$; h

19. $\dfrac{1}{R} = \dfrac{1}{r_1} + \dfrac{1}{r_2}$; R

(An electricity formula)

20. $\dfrac{1}{R} = \dfrac{1}{r_1} + \dfrac{1}{r_2}$; r_1

21. $\dfrac{A}{B} = \dfrac{C}{D}$; D

22. $\dfrac{A}{B} = \dfrac{C}{D}$; C

23. $h_1 = q\left(1 + \dfrac{h_2}{p}\right)$; h_2

24. $S = \dfrac{a - ar^n}{1 - r}$; a

8.12   RATIO, PROPORTION, AND VARIATION

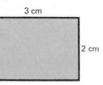 RATIO

The *ratio* of two quantities is their quotient. For example, in the rectangle, the ratio of width to length is

$$\frac{2 \text{ cm}}{3 \text{ cm}}, \quad \text{or} \quad \frac{2}{3}.$$

3 cm

2 cm

An older way to write this is 2:3, read "2 to 3." The ratio of two different kinds of measure is called a *rate*.

Example 1   Betty Cuthbert of Australia set a world record in the 60-meter dash of 7.2 seconds. What was her rate, or *speed*, in meters per second?

$$\frac{60 \text{ m}}{7.2 \text{ sec}}, \quad \text{or} \quad 8.3 \, \frac{\text{m}}{\text{sec}} \quad \text{(Rounded to the nearest tenth)}$$

DO EXERCISES 1–4.

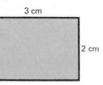 PROPORTIONS

In applied problems a single ratio is often expressed in two ways. For example, it takes 9 gallons of gas to drive 120 miles, and we wish to find how much will be required to go 550 miles. We can set up ratios:

$$\frac{9 \text{ gal}}{120 \text{ mi}} \quad \frac{x \text{ gal}}{550 \text{ mi}}.$$

Assuming that the car uses gas at the same rate throughout the trip, the ratios are the same.

$$\begin{array}{c} \text{Gas} \longrightarrow \\ \text{Miles} \longrightarrow \end{array} \frac{9}{120} = \frac{x}{550} \begin{array}{c} \longleftarrow \text{Gas} \\ \longleftarrow \text{Miles} \end{array}$$

To solve, we multiply by 550 to get x alone on one side:

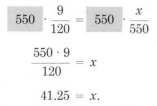

$$550 \cdot \frac{9}{120} = 550 \cdot \frac{x}{550}$$

$$\frac{550 \cdot 9}{120} = x$$

$$41.25 = x.$$

Thus, 41.25 gallons will be required. (Note that we could have multiplied by the LCM of 120 and 550, which is 6600, but in this case, that would have been more complicated.)

An equality of ratios, $A/B = C/D$, is called a *proportion*. The numbers named in a true proportion are said to be *proportional*.

After finishing Section 8.12, you should be able to:

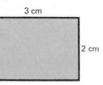 Find the ratio of one quantity to another.

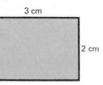 Solve proportion problems.

 Solve problems involving direct variation.

1. Find the ratio of 145 km to 2.5 liters (ℓ).

2. In a recent World Series, Fred Lynn of the Boston Red Sox got 7 hits in 25 times at bat. What was his rate, or batting average, in hits per times at bat?

3. Impulses in nerve fibers travel 310 km in 2.5 hours. What is the rate, or speed, in kilometers per hour?

4. A lake of area 550 square yards contains 1320 fish. What is the population density of the lake in fish per square yard?

5. Nolan Ryan, a pitcher for the California Angels, gave up 76 earned runs in 198 innings of pitching in a recent year. What was his earned run average?

Example 2   Randy Jones, a pitcher for the San Diego Padres, gave up 71 earned runs in 285 innings in a recent year. At this rate, how many runs did he give up every 9 innings?

$$\text{Earned runs each 9 innings} \longrightarrow \frac{A}{9} = \frac{71}{285} \begin{array}{l} \longleftarrow \text{Earned runs} \\ \longleftarrow \text{Innings pitched} \end{array}$$

Solve:
$$\boxed{9} \cdot \frac{A}{9} = \boxed{9} \cdot \frac{71}{285}$$

$$A = \frac{9 \cdot 71}{285}$$

$$A = 2.24 \quad \text{(Rounded to the nearest hundredth)}$$

A stands for the *earned run average*.

DO EXERCISES 5 AND 6.

Example 3   (*Estimating wildlife populations*) To determine the number of fish in a lake, a conservationist catches 225 fish, tags them, and throws them back into the lake. Later, 108 fish are caught. Fifteen of them are found to be tagged. Estimate how many fish are in the lake.

6. A sample of 184 light bulbs contained 6 defective bulbs. How many would you expect in 1288 bulbs?

Let F = the number of fish in the lake. Then translate to a proportion.

$$\begin{array}{l} \text{Fish tagged originally} \longrightarrow \frac{225}{F} = \frac{15}{108} \begin{array}{l} \longleftarrow \text{Tagged fish caught later} \\ \longleftarrow \text{Fish caught later} \end{array} \\ \text{Fish in lake} \longrightarrow \end{array}$$

This time we multiply by the LCM, which is $108F$:

$$108F \cdot \frac{225}{F} = 108F \cdot \frac{15}{108}$$

$$108 \cdot 225 = F \cdot 15$$

$$\frac{108 \cdot 225}{15} = F$$

$$1620 = F.$$

Thus, we estimate that there are about 1620 fish in the lake.

7. To determine the number of deer in a forest, a conservationist catches 612 deer, tags them, and lets them loose. Later 244 deer are caught. 72 of them are tagged. Estimate how many deer are in the forest.

DO EXERCISE 7.

◖◗◗ DIRECT VARIATION

A bicycle is traveling at 10 km/h. In 1 hour it goes 10 km. In 2 hours it goes 20 km. In 3 hours it goes 30 km, and so on. This gives rise to a set of pairs of numbers, all having the same ratio:

(1,10), (2,20), (3,30), (4,40),   and so on.

The ratio of distance to time is always $\frac{10}{1}$, or 10.

Whenever a situation gives rise to pairs of numbers in which the ratio is constant, we say that there is *direct variation*. Here the *distance varies directly as the time*.

$$\frac{d}{t} = 10 \quad \text{(a constant)}, \quad \text{or} \quad d = 10t.$$

> **Whenever a situation gives rise to a relation among variables $y = kx$, where k is a constant, we say that there is *direct variation*. The number k is called the *variation constant*.**

Example 4 Consumer specialists have determined that the amount F that a family spends on food varies directly as its income I. A family making $9600 a year will spend $2496 on food. How much will a family making $10,500 spend on food?

a) First find an equation of variation.

$F = kI$

$2496 = k \cdot 9600$ Substituting 2496 for F and 9600 for I

$\dfrac{2496}{9600} = k$

$0.26 = k$

The equation of variation is $F = 0.26I$.

b) Use the equation to find how much a family making $10,500 will spend on food.

$F = 0.26I$

$F = 0.26(10{,}500)$

$F = 2730$

The family will spend $2730 for food.

DO EXERCISES 8 AND 9.

We can solve the same kinds of problems with variation equations as with proportions. You should use the approach that makes the work easier.

8. The cost C of operating a TV varies directly as the number N of hours it is in operation. It costs $6.00 to operate a standard-size color TV continuously for 30 days. At this rate, how much would it cost to operate the TV for 1 day? 1 hour?

9. The number of servings S of meat that can be obtained from a turkey varies directly as its weight W. From a 15-lb turkey, one can get 20 servings of meat. How many servings can be obtained from an 8-lb turkey?

SOMETHING EXTRA

CALCULATOR CORNER: THE PRICE-EARNINGS RATIO

If a company has total earnings one year of $5,000,000 and there are 100,000 shares of stock issued, the earnings per share are $50. The price per share of IBM at one time was $263\frac{1}{8}$ and the earnings per share were $17.60. The *price-earnings ratio*, P/E, is the price of the stock divided by the earnings per share. For the IBM stock the price-earnings ratio, P/E, is given by

$$\frac{P}{E} = \frac{263\frac{1}{8}}{17.60}$$

$$= \frac{263.125}{17.60} \qquad \text{Converting to decimal notation}$$

$$\approx 15.0 \qquad \text{Dividing and rounding to the nearest tenth}$$

EXERCISES

Calculate the price-earnings ratio for each stock.

Stock	Price per share	Earnings
1. General Motors	$68\frac{5}{8}$	$11.50
2. K-Mart	31	2.60
3. United Airlines	$18\frac{5}{8}$	4.00
4. AT&T	62	6.60

NAME   CLASS   ANSWERS

EXERCISE SET 8.12

▪ Find the ratio of

1. 54 days, 6 days   **2.** 800 mi, 50 gallons

3. A black racer snake travels 4.6 kilometers in 2 hours. What is the speed in kilometers per hour?

4. Light travels 558,000 miles in 3 seconds. What is the speed in miles per second?

▪ ▪ Solve.

5. The coffee beans from 14 trees are required to produce 7.7 kilograms of coffee (this is the average that each person in the United States drinks each year). How many trees are required to produce 320 kilograms of coffee?

6. Last season a minor league baseball player got 240 hits in 600 times at bat. This season his ratio of hits to number of times at bat is the same. He batted 500 times. How many hits has he had?

7. A student traveled 234 kilometers in 14 days. At this same ratio, how far would the student travel in 42 days?

8. In a bread recipe, the ratio of milk to flour is $\frac{4}{3}$. If 5 cups of milk are used, how many cups of flour are used?

9. 10 cm^3 of a normal specimen of human blood contains 1.2 grams of hemoglobin. How many grams would 16 cm^3 of the same blood contain?

10. The winner of an election for class president won by a vote of 3 to 2, with 324 votes. How many votes did the loser get?

11. To determine the number of trout in a lake, a conservationist catches 112 trout, tags them, and throws them back into the lake. Later, 82 trout are caught; 32 of them are tagged. How many trout are in the lake?

12. To determine the number of deer in a game preserve, a conservationist catches 318 deer, tags them, and lets them loose. Later, 168 deer are caught; 56 of them are tagged. How many deer are in the preserve?

ANSWERS

1. _____

2. _____

3. _____

4. _____

5. _____

6. _____

7. _____

8. _____

9. _____

10. _____

11. _____

12. _____

Copyright © 1979, Philippines copyright 1979, by Addison-Wesley Publishing Company, Inc. All rights reserved.

ANSWERS

13. The ratio of the weight of an object on the moon to the weight of an object on earth is 0.16 to 1.

   a) How much would a 12-ton rocket weigh on the moon?

   b) How much would a 90-kilogram astronaut weigh on the moon?

13. _____

14. The ratio of the weight of an object on Mars to the weight of an object on earth is 0.4 to 1.

   a) How much would a 12-ton rocket weigh on Mars?

   b) How much would a 90-kilogram astronaut weigh on Mars?

14. _____

■■■ Solve.

15. The number B of bolts a machine can make varies directly as the time it operates. It can make 1000 bolts in 2 hours. How many can it make in 5 hours?

15. _____

16. A person's paycheck P varies directly as the number H of hours worked. For working 15 hours the pay is $48.75. Find the pay for 35 hours of work.

16. _____

17. The amount C that a family spends on car expenses varies directly as its income I. A family making $10,880 a year will spend $1632 a year for car expenses. How much will a family making $10,000 a year spend for car expenses?

17. _____

18. The number of kg of water W in a human body varies directly as the total body weight B. A person weighing 96 kg contains 64 kg of water. How many kilograms of water are in a person weighing 81 kg?

18. _____

19. The weight M of an object on Mars varies directly as its weight E on earth. A person who weighs 95 kg on earth weighs 38 kg on Mars. How much would an 80-kg person weigh on Mars?

19. _____

20. The number of servings S of meat that can be obtained from round steak varies directly as the weight W. From 9 kg of round steak one can get 70 servings of meat. How many servings can one get from 12 kg of round steak?

20. _____

NAME _____   SCORE _____   ANSWERS

TEST OR REVIEW—CHAPTER 8

If you miss an item, review the indicated section and objective.

[8.1, ●●●]   **1.** Simplify: $\dfrac{14x^2 - x - 3}{2x^2 - 7x + 3}$ · $= \dfrac{(7x+3)(2x+1)}{(2x-1)(x-3)}$

1. $\dfrac{7x+3}{x-3}$

[8.1, ●●]   **2.** Multiply and simplify: $\dfrac{a^2 - 36}{10a} \cdot \dfrac{2a}{a + 6}$

$\dfrac{(A+6)(A-6)}{\underset{5}{10 \cdot A}} \cdot \dfrac{2 \cdot A}{(A+6)}$

2. $\dfrac{A-6}{5}$

[8.2, ● ●]   **3.** Divide and simplify: $\dfrac{4x^4}{x^2 - 1} \div \dfrac{2x^3}{x^2 - 2x + 1}$

$\dfrac{2 \cdot 4 \cdot x \cdot x \cdot x \cdot x}{(x+1)(x-1)} \cdot \dfrac{(x-1)(x-1)}{2 \cdot x \cdot x \cdot x}$

3. $\dfrac{2x(x-1)}{x+1}$

Add.

[8.3, ● ●]   **4.** $\dfrac{x + 3}{x - 2} + \dfrac{x}{2 - x} = \dfrac{x+3}{x-2} + \dfrac{-x}{x-2}$

4. $\dfrac{3}{x-2}$

[8.6, ●]   **5.** $\dfrac{3}{3x - 9} + \dfrac{x - 2}{3 - x} = \dfrac{3}{3(x-3)} + \dfrac{x-2}{3-x} =$

$\dfrac{3}{3(x-3)} + \dfrac{2-x}{x-3} = \dfrac{3+6-3x}{3(x-3)} = \dfrac{9-3x}{3 \, LCD} =$

Subtract. $\dfrac{3(3-x)}{3(x-3)}$

5. -1

[8.4, ● ●]   **6.** $\dfrac{x + 3}{x - 2} - \dfrac{x}{2 - x} =$

$\dfrac{x+3}{x-2} - \dfrac{-x}{x-2} = \dfrac{x+3+x}{x-2} = \dfrac{2x+3}{x-2}$

6. $\dfrac{2x+3}{x-2}$

[8.7, ●]   **7.** $\dfrac{1}{x^2 - 25} - \dfrac{x - 5}{x^2 - 4x - 5} = \dfrac{1}{(x+5)(x-5)} - \dfrac{x-5}{(x-5)(x+1)} =$

$\dfrac{1}{(x+5)(x-5)(x+1)} - \dfrac{x-5}{(x+5)(x-5)(x+1)} =$

$\dfrac{1(x+1) - (x+5)(x-5)}{LCD} = \dfrac{1(x+1) - (x^2 - 25)}{LCD} =$

$x+1-x^2+25$

7. $\dfrac{-x^2 + x + 26}{(x+5)(x-5)(x+1)}$

Copyright © 1979, Philippines copyright 1979, by Addison-Wesley Publishing Company, Inc. All rights reserved.

ANSWERS

8. $\quad 2$

9. $2X^2 - 8X + 25\pi - 79$

10. $X = 1$

11.

12. $\dfrac{Er - er}{e} = R$

13. 100

[8.8, ▬●▬]   8. Simplify: $\dfrac{\dfrac{1}{a} - 2}{\dfrac{1}{2a} - 1}$.

$= \dfrac{\dfrac{1-2A}{A}}{\dfrac{1-2A}{2A}} = \dfrac{1-2A}{A} \div \dfrac{1-2A}{2A} =$

$\dfrac{1-2A}{A} \cdot \dfrac{2A}{1-2A} = 2$

[8.9, ●●]   9. Divide: $(2x^3 - 2x^2 + x - 4) \div (x + 3)$.

$\begin{array}{r} 2X^2 - 8X + 25 \\ x+3 \overline{)\, 2X^3 - 2X^2 + x - 4} \\ 2X^3 + 6X^2 \\ \hline -8X^2 + X \\ -8X^2 - 24X \\ \hline 25X - 4 \\ 25X + 75 \\ \hline -79 \end{array}$

[8.10, ▬●▬]   10. Solve: $\dfrac{15}{x} = \dfrac{45}{x+2}$.

$= \quad 15(X+2) = 45X$

$15X + 30 = 45X$

$30 = 30X$

$1 = X$

[8.11, ▬●▬]   Solve.

11. One vehicle travels 90 km in the same time a car traveling 10 km/h slower travels 60 km. Find the speed of each.

	R	T	D
I	R	T	90
II	R-10	T	60km

$\dfrac{90}{R} = \dfrac{60}{R-10}$

$90(R-10) = 60R$

$90R - 900 = 60R$

$30R = 900$

$R = 30$

$\dfrac{90}{30} = \dfrac{60}{20}$

$180 = 180$ ✓

[8.11, ●●]   12. Solve for R: $\dfrac{E}{e} = \dfrac{R+r}{r}$.

$\dfrac{E}{e} = \dfrac{R+r}{r}$

$Er = eR + er$

$Er - er = eR$

$\dfrac{Er - er}{e} = R$

[8.12, ●●●]   Solve.

13. A student's test score varies directly as the time T studied. A score of 80 is made after studying 4 hours. What score would be made after studying 5 hours?

$S = kT$

$80 = k(4)$

$20 = k$

$20(5) = 100$

**RADICAL
EXPRESSIONS
AND EQUATIONS**

1. What is the meaning of 5^2?

2. Multiply: $(-5)(-5)$.

3. Multiply: $\dfrac{4}{3} \cdot \dfrac{4}{3}$.

Simplify.

4. $\dfrac{18}{50}$

5. $\dfrac{81}{27}$

Write decimal notation.

6. $\dfrac{7}{8}$

7. $\dfrac{5}{11}$

8. Simplify: $|-8|$.

9. Multiply and simplify: $x^3 \cdot x^3$.

10. Factor: $x^3 - x^2$.

11. Factor: $x^2 + 2x + 1$.

12. Solve: $x + 1 = 2x - 5$.

9.1 RADICAL EXPRESSIONS

■ SQUARE ROOTS

The number c is a square root of a if $c^2 = a$.

Every positive number has two square roots. For example, the square roots of 25 are 5 and -5, because

$$5^2 = 25 \quad \text{and} \quad (-5)^2 = 25.$$

Zero has only one square root, 0 itself.

Example 1 Find the square roots of 81.

The square roots are 9 and -9, because $9^2 = 81$ and $(-9)^2 = 81$.

DO EXERCISES 1–3.

The symbol $\sqrt{}$ is called a *radical* symbol. We use it in naming the nonnegative square root of a number. We call $\sqrt{a}$ the *principal square root* of a.

Examples

2. Find $\sqrt{25}$.

$\quad \sqrt{25} = 5 \qquad$ Remember, $\sqrt{}$ means to take the principal, *nonnegative*, square root.

3. Find $\sqrt{225}$.

$\quad \sqrt{225} = 15$

4. Find $\sqrt{0}$.

$\quad \sqrt{0} = 0$

DO EXERCISES 4–6.

OBJECTIVES

After finishing Section 9.1, you should be able to:

■ Find principal square roots and their additive inverses for squares of whole numbers from 0^2 to 25^2.

■■ Identify radicands.

■■■ Identify meaningless radical expressions.

■■ Simplify radical expressions with a perfect square radicand.

Find the square roots of each number.

1. 36

2. 64

3. 225

Find.

4. $\sqrt{16}$

5. $\sqrt{49}$

6. $\sqrt{100}$

To name the negative square root of a number, we use $-\sqrt{}$

Examples

5. Find $-\sqrt{25}$

$$-\sqrt{25} = -5 \qquad \sqrt{25} = 5, \text{ so } -\sqrt{25} = -5$$

6. Find $-\sqrt{64}$

$$-\sqrt{64} = -8$$

DO EXERCISES 7–9

●● RADICAL EXPRESSIONS

When an expression is written under a radical, we have a *radical expression*. These are radical expressions:

$$\sqrt{14}, \quad \sqrt{x}, \quad \sqrt{x^2 + 4}, \quad \sqrt{\dfrac{x^2 - 5}{2}}.$$

The expression written under the radical is called the *radicand*.

Examples　Identify the radicand in each expression.

7. $\sqrt{x}$;　　The radicand is x.

8. $\sqrt{y^2 - 5}$;　　The radicand is $y^2 - 5$.

DO EXERCISES 10 AND 11.

●●● MEANINGLESS EXPRESSIONS

Negative numbers do not have square roots. This is because the square of any negative number is positive. For example,

$$(-5)^2 = 25, \quad (-11)^2 = 121.$$

Thus, the following expressions are meaningless:

$$\sqrt{-100}, \quad \sqrt{-49}, \quad -\sqrt{-3}.$$

Any number that makes a radicand negative is a nonsensible replacement.

DO EXERCISES 12–15.

Find.

7. $-\sqrt{16}$

8. $-\sqrt{49}$

9. $-\sqrt{169}$

Identify the radicand in each expression.

10. $\sqrt{45 + x}$

11. $\sqrt{\dfrac{x}{x + 2}}$

Is the expression meaningless? Write "yes" or "no."

12. $\sqrt{-25}$

13. $-\sqrt{25}$

14. $-\sqrt{-36}$

15. $-\sqrt{36}$

Simplify.

16. $\sqrt{(xy)^2}$

17. $\sqrt{x^2y^2}$

18. $\sqrt{(x-1)^2}$

19. $\sqrt{x^2 + 8x + 16}$

20. $\sqrt{25x^2}$

▪▪ PERFECT SQUARE RADICANDS

In the expression $\sqrt{x^2}$, the radicand is a perfect square. Since squares are never negative, all replacements are sensible.

Suppose $x = 3$. Then we have $\sqrt{3^2}$, which is $\sqrt{9}$, or 3.

Suppose $x = -3$. Then we have $\sqrt{(-3)^2}$, which is $\sqrt{9}$, or 3.

In any case, $\sqrt{x^2} = |x|$.

> **Any radical expression $\sqrt{A^2}$ can be simplified to $|A|$.**

Examples   Simplify.

9. $\sqrt{(3x)^2} = |3x|$ Don't forget the absolute values!

10. $\sqrt{a^2b^2} = \sqrt{(ab)^2} = |ab|$

11. $\sqrt{x^2 + 2x + 1} = \sqrt{(x+1)^2} = |x+1|$

We can sometimes simplify absolute value notation. In Example 9, $|3x|$ simplifies to $|3| \cdot |x|$ or $3|x|$. In Example 10, we can change $|ab|$ to $|a| \cdot |b|$ if we wish. The absolute value of a product is always the product of the absolute values.

> **For any numbers a and b, $|a \cdot b| = |a| \cdot |b|$.**

DO EXERCISES 16–20.

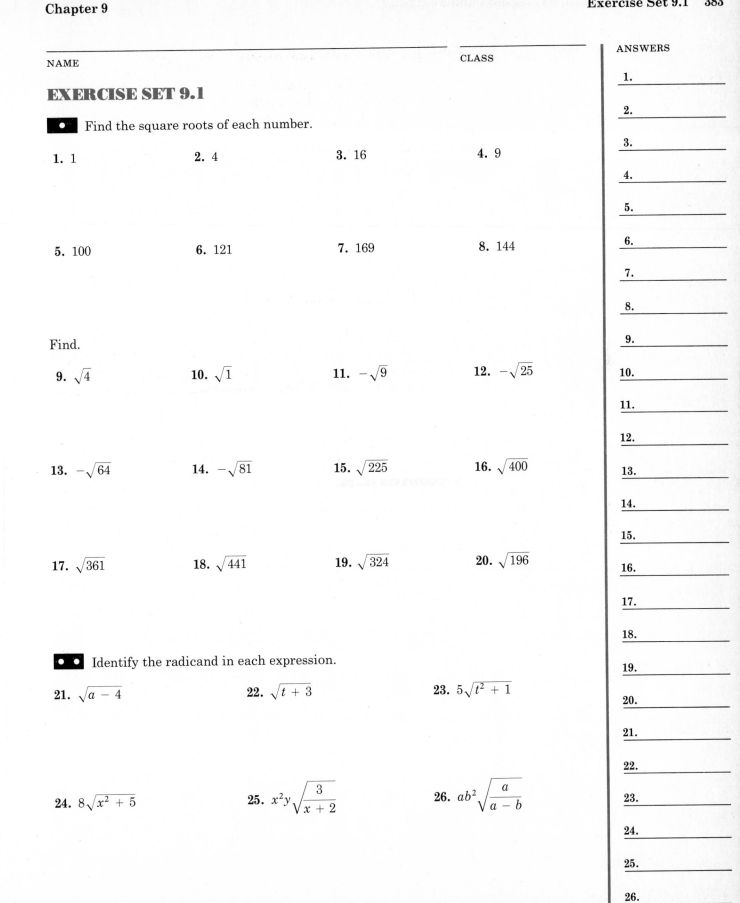

NAME

CLASS

EXERCISE SET 9.1

⬤ Find the square roots of each number.

1. 1
2. 4
3. 16
4. 9

5. 100
6. 121
7. 169
8. 144

Find.

9. $\sqrt{4}$
10. $\sqrt{1}$
11. $-\sqrt{9}$
12. $-\sqrt{25}$

13. $-\sqrt{64}$
14. $-\sqrt{81}$
15. $\sqrt{225}$
16. $\sqrt{400}$

17. $\sqrt{361}$
18. $\sqrt{441}$
19. $\sqrt{324}$
20. $\sqrt{196}$

⬤⬤ Identify the radicand in each expression.

21. $\sqrt{a-4}$
22. $\sqrt{t+3}$
23. $5\sqrt{t^2+1}$

24. $8\sqrt{x^2+5}$
25. $x^2 y \sqrt{\dfrac{3}{x+2}}$
26. $ab^2 \sqrt{\dfrac{a}{a-b}}$

ANSWERS

1. _____
2. _____
3. _____
4. _____
5. _____
6. _____
7. _____
8. _____
9. _____
10. _____
11. _____
12. _____
13. _____
14. _____
15. _____
16. _____
17. _____
18. _____
19. _____
20. _____
21. _____
22. _____
23. _____
24. _____
25. _____
26. _____

Copyright © 1979, Philippines copyright 1979, by Addison-Wesley Publishing Company, Inc. All rights reserved.

ANSWERS

27. _____

28. _____

29. _____

30. _____

31. _____

32. _____

33. _____

34. _____

35. _____

36. _____

37. _____

38. _____

39. _____

40. _____

41. _____

42. _____

43. _____

44. _____

45. _____

46. _____

47. _____

48. _____

●●● Which of these expressions are meaningless? Write "yes" or "no."

27. $\sqrt{-16}$ **28.** $\sqrt{-81}$ **29.** $-\sqrt{81}$ **30.** $-\sqrt{64}$

31. $-\sqrt{-49}$ **32.** $-\sqrt{-144}$ **33.** $\sqrt{(-3)^2}$ **34.** $\sqrt{(-5)^2}$

⬛⬛ Simplify.

35. $\sqrt{x^2}$ **36.** $\sqrt{t^2}$ **37.** $\sqrt{4a^2}$

38. $\sqrt{9x^2}$ **39.** $\sqrt{(-5)^2}$ **40.** $\sqrt{(-7)^2}$

41. $\sqrt{(-3b)^2}$ **42.** $\sqrt{(-4d)^2}$ **43.** $\sqrt{(x-7)^2}$

44. $\sqrt{(x+3)^2}$ **45.** $\sqrt{x^2+2x+1}$ **46.** $\sqrt{a^2-10a+25}$

47. $\sqrt{4x^2-20x+25}$ **48.** $\sqrt{9x^2+12x+4}$

9.2 IRRATIONAL NUMBERS AND REAL NUMBERS

◼•◼ IRRATIONAL NUMBERS

Recall that all rational numbers can be named by fractional notation

$$\frac{a}{b}, \quad \text{where } a \text{ and } b \text{ are integers and } b \neq 0.$$

Rational numbers can be named in other ways, such as with decimal notation, but they are all nameable with fractional notation. Suppose we try to find a rational number for $\sqrt{2}$. We look for a number a/b for which $a/b \cdot a/b = 2$. We can find rational numbers whose squares are quite close to 2:

$$\left(\frac{14}{10}\right)^2 = (1.4)^2 = 1.96,$$

$$\left(\frac{141}{100}\right)^2 = (1.41)^2 = 1.9881,$$

$$\left(\frac{1414}{1000}\right)^2 = (1.414)^2 = 1.999396,$$

$$\left(\frac{14142}{10000}\right)^2 = (1.4142)^2 = 1.99996164,$$

but we can never find one whose square is exactly 2. This can be proved but we will not do it here. Since $\sqrt{2}$ is not a rational number, we call it an *irrational* number.

> An *irrational* number is a number that cannot be named by fractional notation a/b, where a and b are integers and $b \neq 0$.

Unless a whole number is a perfect square, its square root is irrational.

Examples   Identify the rational numbers and the irrational numbers.

1. $\sqrt{3}$ is irrational.   3 is not a perfect square.
2. $\sqrt{25}$ is rational.   25 is a perfect square.
3. $\sqrt{35}$ is irrational.   35 is not a perfect square.

There are many irrational numbers. Many are *not* obtained by taking square roots. For example,

π is irrational,

and it is not a square root of any rational number.

DO EXERCISES 1–5.

◼•◼ DECIMAL NOTATION FOR IRRATIONAL NUMBERS

Decimal notation for a rational number either ends or repeats. For example,

$$\frac{1}{4} = 0.25, \quad \frac{1}{3} = 0.\boxed{3}\,33\ldots, \quad \frac{5}{11} = 0.\boxed{45}\,4545\ldots.$$

ends   3 repeats   45 repeats

OBJECTIVES

After finishing Section 9.2, you should be able to:

◼•◼ Identify rational numbers and irrational numbers when named with radical notation.

◼•◼ Identify rational numbers and irrational numbers when named by fractional notation and decimal notation.

◼•◼ Find the rational number halfway between two rational numbers.

◼◼ Approximate square roots using a table.

1. Define an irrational number.

Identify the rational numbers and the irrational numbers.

2. $\sqrt{5}$

3. $\sqrt{36}$

4. $\sqrt{32}$

5. $\sqrt{101}$

Identify the rational numbers
and the irrational numbers.

6. $\dfrac{-3}{5}$

7. $\dfrac{95}{37}$

8. 6.12

9. 0.353535 . . . (Numeral repeats)

10. 3.01001000100001 . . .
(Numeral does not repeat)

Find the number halfway between
each pair of numbers.

11. $\dfrac{3}{64}$ and $\dfrac{4}{64}$

12. -0.678 and -0.6782

13. 5.698 and 5.6999

14. Describe the set of real numbers.

Use Table 1 to approximate these
square roots.

15. $\sqrt{7}$

16. $\sqrt{72}$

The decimal numeral for an irrational number will never end and it will never repeat. The number π is an example:

$$\pi = 3.1415926535\ldots.$$

Decimal notation for π never ends and never repeats. Note that 3.14 and 22/7 are both rational approximations for π. Decimal notation for 22/7 is 3.142857 142857 . . . . It repeats. Here are some other examples of irrational numbers:

2.818118111811118111118 . . .   Numeral does not repeat
0.0350355035550355550 . . .   Numeral does not repeat

DO EXERCISES 6–10.

▪▪▪ REAL NUMBERS

The rational numbers are very close together. Yet no matter how close together two rational numbers are, we can find infinitely many rational numbers between them. By averaging, we can find the number *halfway* between.

Example 4 Find the number halfway between $\frac{1}{32}$ and $\frac{2}{32}$.

We average: $\dfrac{\frac{1}{32} + \frac{2}{32}}{2} = \dfrac{\frac{3}{32}}{2} = \dfrac{3}{64}.$ Add and divide by 2.

Example 5 Find the number halfway between -0.32 and -0.323.

$$\dfrac{(-0.32) + (-0.323)}{2} = \dfrac{-0.643}{2} = -0.3215$$

DO EXERCISES 11–13.

Although the rational numbers seem to fill up the number line, they do not. There are many points on the line for which there is no rational number. For these points, there are *irrational* numbers.

> The *real numbers* **consist of the rational numbers and the irrational numbers. There is a real number for each point on a number line.**

DO EXERCISE 14.

▪▪ APPROXIMATING IRRATIONAL NUMBERS

Most of the time we use rational numbers to approximate irrational numbers. Table 1 on the inside back cover contains rational approximations for square roots; for example,

$$\sqrt{10} \approx 3.162.$$ Rounded to three decimal places

The symbol $\approx$ means "is approximately equal to."

DO EXERCISES 15 AND 16.

NAME   CLASS

EXERCISE SET 9.2

■ Identify the rational numbers and the irrational numbers.

1. $\sqrt{2}$ **2.** $\sqrt{6}$ **3.** $-\sqrt{8}$ **4.** $-\sqrt{10}$

5. $\sqrt{49}$ **6.** $\sqrt{100}$ **7.** $\sqrt{98}$ **8.** $-\sqrt{12}$

■■ Identify the rational numbers and the irrational numbers.

9. $-\dfrac{2}{3}$ **10.** $\dfrac{136}{51}$ **11.** 23   **12.** 4.23

13. 0.424242 . . . (Numeral repeats)   **14.** 0.1565656 . . . (Numeral repeats)

15. 4.28228222822228 . . . (Numeral does not repeat)

16. 7.76776777677776 . . . (Numeral does not repeat)

17. -1 **18.** 0

19. -45.6919119111911119 . . . (Numeral does not repeat)

20. -63.03003000300003 . . . (Numeral does not repeat)

■■■ Find the number halfway between each pair of numbers.

21. $\dfrac{2}{7}, \dfrac{5}{7}$ **22.** $-\dfrac{3}{5}, -\dfrac{4}{5}$ **23.** $\dfrac{2}{3}, \dfrac{7}{10}$ **24.** $\dfrac{7}{8}, \dfrac{9}{10}$

ANSWERS

1. _____

2. _____

3. _____

4. _____

5. _____

6. _____

7. _____

8. _____

9. _____

10. _____

11. _____

12. _____

13. _____

14. _____

15. _____

16. _____

17. _____

18. _____

19. _____

20. _____

21. _____

22. _____

23. _____

24. _____

Copyright © 1979, Philippines copyright 1979, by Addison-Wesley Publishing Company, Inc. All rights reserved.

ANSWERS

25. _____

26. _____

27. _____

28. _____

29. _____

30. _____

31. _____

32. _____

33. _____

34. _____

35. _____

36. _____

37. _____

38. _____

39. _____

40. _____

41. _____

25. $6\frac{1}{3}, 6\frac{1}{4}$ **26.** $8\frac{1}{6}, 8\frac{1}{7}$ **27.** $-1.45, -1.452$ **28.** $6.374, 6.375$

29. $0.25, \dfrac{3}{8}$ **30.** $0.58, \dfrac{5}{8}$

⋮⋮ Use Table 1 to approximate these square roots.

31. $\sqrt{5}$ **32.** $\sqrt{43}$ **33.** $\sqrt{81}$ **34.** $\sqrt{87}$

35. $\sqrt{93}$ **36.** $\sqrt{17}$ **37.** $\sqrt{63}$ **38.** $\sqrt{50}$

Complete.

39. Every _____ number can be named as the quotient of two _____.

40. The real numbers consist of the rational numbers and the _____ numbers.

41. *Parking.* When a parking lot has attendants to park cars, it uses spaces for drivers to pull in and leave cars before they are taken to permanent parking stalls. The number N of such spaces needed is given by the formula

$$N = 2.5\sqrt{A},$$

where A is the average number of arrivals in peak hours. Find the number of spaces needed when the average number of arrivals in peak hours is 25, 36, 49, and 64.

9.3 MULTIPLICATION AND FACTORING

◻• MULTIPLICATION

To see how we can multiply with radical notation, look at the following examples.

Example 1   Simplify.

a) $\sqrt{9} \cdot \sqrt{4} = 3 \cdot 2 = 6$ This is a product of square roots.

b) $\sqrt{9 \cdot 4} = \sqrt{36} = 6$ This is the square root of a product.

Example 2   Simplify.

a) $\sqrt{4} \cdot \sqrt{25} = 2 \cdot 5 = 10$

b) $\sqrt{4 \cdot 25} = \sqrt{100} = 10$

Example 3

$\sqrt{-9}\sqrt{-4}$ Meaningless (negative radicands)

DO EXERCISE 1.

We can multiply radical expressions by multiplying the radicands.* However, we cannot allow any radicand to be negative.

> **For any nonnegative radicands A and B, $\sqrt{A} \cdot \sqrt{B} = \sqrt{A \cdot B}$. (The product of square roots, provided they exist, is the square root of the product of the radicands.)**

Examples   Multiply. Assume all radicands nonnegative.

4. $\sqrt{5}\sqrt{7} = \sqrt{5 \cdot 7} = \sqrt{35}$

5. $\sqrt{8}\sqrt{8} = \sqrt{8 \cdot 8} = \sqrt{64} = 8$

6. $\sqrt{4}\sqrt{5} = 2\sqrt{5}$ Here a simpler answer was obtained by not multiplying.

7. $\sqrt{\dfrac{2}{3}}\sqrt{\dfrac{4}{5}} = \sqrt{\dfrac{2}{3} \cdot \dfrac{4}{5}} = \sqrt{\dfrac{8}{15}}$

8. $\sqrt{2x}\sqrt{3x - 1} = \sqrt{2x(3x - 1)} = \sqrt{6x^2 - 2x}$

DO EXERCISES 2–5.

---

*A proof (optional).   We consider a product $\sqrt{A}\sqrt{B}$, where A and B are not negative. We square this product, to show that we get AB, and thus the product is the square root of AB (or $\sqrt{AB}$).

$$(\sqrt{A}\sqrt{B})^2 = (\sqrt{A}\sqrt{B})(\sqrt{A}\sqrt{B})$$
$$= (\sqrt{A}\sqrt{A})(\sqrt{B}\sqrt{B})$$
$$= AB$$

OBJECTIVES

After finishing Section 9.3, you should be able to:

◻• Multiply with radical notation.

◻◻ Factor radical expressions and where possible simplify.

◻◻◻ Approximate square roots not in Table 1 using factoring and the square root table.

1. Simplify.

   a) $\sqrt{4} \cdot \sqrt{16}$

   b) $\sqrt{4 \cdot 16}$

Multiply. Assume all radicands nonnegative.

2. $\sqrt{3}\sqrt{7}$

3. $\sqrt{5}\sqrt{5}$

4. $\sqrt{x}\sqrt{x + 1}$

5. $\sqrt{x + 1}\sqrt{x - 1}$

ANSWERS ON PAGE A–28

Factor. Simplify where possible.

6. $\sqrt{32}$

7. $\sqrt{x^2 - 81}$

8. $\sqrt{25x^2}$

9. $\sqrt{36m^2}$

10. $\sqrt{76}$

11. $\sqrt{x^2 - 1}$

12. $\sqrt{64t^2}$

13. $\sqrt{100a^2}$

Approximate these square roots. Round to three decimal places.

14. $\sqrt{275}$

15. $\sqrt{102}$

■■ FACTORING AND SIMPLIFYING

We know that for nonnegative radicands,

$$\sqrt{A}\sqrt{B} = \sqrt{AB}.$$

To factor radical expressions we can think of this equation in reverse:

$$\sqrt{AB} = \sqrt{A}\sqrt{B}.$$

In some cases we can simplify after factoring.

Examples   Factor and simplify.

9. $\sqrt{18} = \sqrt{9 \cdot 2}$ Factoring the radicand: 9 is a perfect square.

$\qquad = \sqrt{9} \cdot \sqrt{2}$ Factoring the radical expression

$\qquad = 3\sqrt{2}$ $3\sqrt{2}$ means $3 \cdot \sqrt{2}$

10. $\sqrt{25x} = \sqrt{25} \cdot \sqrt{x} = 5\sqrt{x}$

11. $\sqrt{x^2 - 4} = \sqrt{(x - 2)(x + 2)} = \sqrt{x - 2}\sqrt{x + 2}$

12. $\sqrt{36x^2} = \sqrt{36}\sqrt{x^2} = 6|x|$

DO EXERCISES 6–13.

■■■ APPROXIMATING SQUARE ROOTS

Table 1 goes only to 100. We can use it to find approximate square roots for other numbers. When possible, we factor out the largest perfect square.

Examples   Approximate these square roots.

13. $\sqrt{160} = \sqrt{16 \cdot 10}$ Factoring the radicand (make one factor a perfect square, if you can)

$\qquad = \sqrt{16}\sqrt{10}$ Factoring the radical expression

$\qquad = 4\sqrt{10}$

$\qquad \approx 4(3.162)$ From Table 1, $\sqrt{10} \approx 3.162$

$\qquad \approx 12.648$

14. $\sqrt{341} = \sqrt{11 \cdot 31}$ Factoring (There is no perfect square factor.)

$\qquad = \sqrt{11}\sqrt{31}$

$\qquad \approx 3.317 \times 5.568$ Table 1

$\qquad \approx 18.469$ Rounded to 3 decimal places

DO EXERCISES 14 AND 15.

ANSWERS ON PAGE A–28

EXERCISE SET 9.3

● Multiply.

1. $\sqrt{2}\sqrt{3}$

2. $\sqrt{3}\sqrt{5}$

3. $\sqrt{3}\sqrt{3}$

4. $\sqrt{6}\sqrt{6}$

5. $\sqrt{9}\sqrt{2}$

6. $\sqrt{16}\sqrt{3}$

7. $\sqrt{\dfrac{2}{5}}\sqrt{\dfrac{3}{4}}$

8. $\sqrt{\dfrac{4}{5}}\sqrt{\dfrac{15}{8}}$

9. $\sqrt{2}\sqrt{x}$

10. $\sqrt{3}\sqrt{t}$

11. $\sqrt{x}\sqrt{x-3}$

12. $\sqrt{5}\sqrt{2x-1}$

13. $\sqrt{x+2}\sqrt{x+1}$

14. $\sqrt{x+4}\sqrt{x-4}$

15. $\sqrt{x+y}\sqrt{x-y}$

16. $\sqrt{a-b}\sqrt{a-b}$

17. $\sqrt{-3}\sqrt{2x}$

18. $\sqrt{-5}\sqrt{4x}$

● ● Factor. Simplify where possible.

19. $\sqrt{12}$

20. $\sqrt{8}$

21. $\sqrt{75}$

22. $\sqrt{50}$

ANSWERS	
1.	
2.	
3.	
4.	
5.	
6.	
7.	
8.	
9.	
10.	
11.	
12.	
13.	
14.	
15.	
16.	
17.	
18.	
19.	
20.	
21.	
22.	

Copyright © 1979, Philippines copyright 1979, by Addison-Wesley Publishing Company, Inc. All rights reserved.

ANSWERS

23. _____

24. _____

25. _____

26. _____

27. _____

28. _____

29. _____

30. _____

31. _____

32. _____

33. _____

34. _____

35. _____

36. _____

37. _____

38. _____

39. _____

40. _____

41. _____

42. _____

43. _____

23. $\sqrt{200x}$

24. $\sqrt{300x}$

25. $\sqrt{16a^2}$

26. $\sqrt{64y^2}$

27. $\sqrt{49t^2}$

28. $\sqrt{81p^2}$

29. $\sqrt{x^3 - 2x^2}$

30. $\sqrt{t^3 + t^2}$

31. $\sqrt{2x^2 - 5x - 12}$

32. $\sqrt{x^2 - x - 2}$

33. $\sqrt{a^2 - b^2}$

34. $\sqrt{x^2 - 4}$

● ● ●   Approximate these square roots. Round to three decimal places.

35. $\sqrt{125}$

36. $\sqrt{180}$

37. $\sqrt{360}$

38. $\sqrt{105}$

39. $\sqrt{300}$

40. $\sqrt{143}$

41. $\sqrt{122}$

42. $\sqrt{2000}$

43. *Speed of a skidding car.* How do police determine the speed of a car after an accident? The formula

$$r = 2\sqrt{5L}$$

can be used to approximate the speed r, in mph, of a car that has left a skid mark of length L, in feet. What was the speed of a car that left skid marks of lengths 20 ft? 70 ft? 90 ft?

9.4 MULTIPLYING AND SIMPLIFYING

● SIMPLIFYING

To simplify radical expressions we usually try to factor out as many perfect square factors as possible. Compare

$$\sqrt{50} = \sqrt{10 \cdot 5} = \sqrt{10}\sqrt{5}$$

Not a perfect square

and

$$\sqrt{50} = \sqrt{25 \cdot 2} = \sqrt{25}\sqrt{2} = 5\sqrt{2}.$$

A perfect square

In the second case we factored out the perfect square 25. $5\sqrt{2}$ is usually considered simpler than $\sqrt{50}$ or $\sqrt{10}\sqrt{5}$.

Examples   Simplify.

1. $\sqrt{48} = \sqrt{16 \cdot 3} = \sqrt{16}\sqrt{3} = 4\sqrt{3}$

2. $\sqrt{20t^2} = \sqrt{4}\sqrt{t^2}\sqrt{5} = 2 \cdot |t| \cdot \sqrt{5}$

3. $\sqrt{18y^2} = \sqrt{9} \cdot \sqrt{y^2} \cdot \sqrt{2} = 3|y|\sqrt{2}$

4. $\sqrt{3x^2 + 6x + 3} = \sqrt{3(x^2 + 2x + 1)}$

$$= \sqrt{3}\sqrt{x^2 + 2x + 1}$$

$$= \sqrt{3}\sqrt{(x + 1)^2}$$

$$= \sqrt{3}\,|x + 1| \quad \text{Taking principal square root}$$

Remember the absolute values!

DO EXERCISES 1–3.

●● SQUARE ROOTS OF POWERS

We can take square roots of powers.

Examples   Simplify.

5. $\sqrt{x^6}$; $x^3 \cdot x^3 = x^6$,   so $\sqrt{x^6} = |x^3|$

6. $\sqrt{x^{10}}$; $x^5 \cdot x^5 = x^{10}$,   so $\sqrt{x^{10}} = |x^5|$

To take a square root of a power such as x^6, the exponent must be even. We then take half the exponent. We should remember to use absolute values.

Example 7   Simplify.

$$\sqrt{x^{12}} = |x^6| \qquad \frac{1}{2} \cdot 12 = 6$$

When odd powers occur, express the power in terms of the nearest even power. Then simplify the even power.

OBJECTIVES

After finishing Section 9.4, you should be able to:

● Simplify radical expressions by factoring out as many perfect powers as possible.

●● Simplify radical expressions where radicands are powers.

●●● Multiply and simplify radical expressions.

Simplify.

1. $\sqrt{32}$

2. $\sqrt{50h^2}$

3. $\sqrt{3x^2 - 6x + 3}$

Simplify.

4. $\sqrt{x^8}$

5. $\sqrt{(x + 2)^{14}}$

6. $\sqrt{x^{15}}$

Multiply and simplify.

7. $\sqrt{3}\sqrt{6}$

8. $\sqrt{2}\sqrt{50}$

Multiply and simplify.

9. $\sqrt{2x^3}\sqrt{8x^3y^4}$

10. $\sqrt{10xy^2}\sqrt{5x^2y^3}$

Example 8 Simplify.

$$\sqrt{x^9} = \sqrt{x^8 \cdot x}$$
$$= \sqrt{x^8}\sqrt{x}$$
$$= |x^4|\sqrt{x}$$

DO EXERCISES 4–6.

••• MULTIPLYING AND SIMPLIFYING

Sometimes we can simplify after multiplying.

Example 9 Multiply and simplify.

$$\sqrt{2}\sqrt{14} = \sqrt{2 \cdot 14} \qquad \text{Multiplying}$$
$$= \sqrt{2 \cdot 2 \cdot 7} \qquad \text{Factoring and looking for}$$
$$\qquad\qquad\qquad\qquad \text{perfect square factors}$$
$$= \sqrt{2 \cdot 2} \cdot \sqrt{7}$$
$$\qquad\qquad \text{Perfect square}$$
$$= 2\sqrt{7}$$

DO EXERCISES 7 AND 8.

Example 10 Multiply and simplify.

$$\sqrt{3x^2}\sqrt{9x^3} = \sqrt{3 \cdot 9x^5} \qquad \text{Multiplying}$$
$$= \sqrt{3 \cdot 9 \cdot x^4 \cdot x} \qquad \text{Factoring and looking for}$$
$$\qquad\qquad\qquad\qquad\qquad \text{perfect square factors}$$
$$= \sqrt{9 \cdot x^4 \cdot 3 \cdot x}$$
$$\qquad\qquad\qquad \text{Perfect squares}$$
$$= \sqrt{9}\sqrt{x^4}\sqrt{3x}$$
$$= 3|x^2|\sqrt{3x}$$

The absolute value signs could be omitted in this case because x^2 cannot be negative. However, it is not wrong to write them.

DO EXERCISES 9 AND 10.

NAME	CLASS	ANSWERS

EXERCISE SET 9.4

● Simplify.

1. $\sqrt{24}$ **2.** $\sqrt{12}$ **3.** $\sqrt{40}$

4. $\sqrt{200}$ **5.** $\sqrt{175}$ **6.** $\sqrt{243}$

7. $\sqrt{48x}$ **8.** $\sqrt{40m}$ **9.** $\sqrt{28x^2}$

10. $\sqrt{20x^2}$ **11.** $\sqrt{8x^2 + 8x + 2}$ **12.** $\sqrt{36y + 12y^2 + y^3}$

●● Simplify.

13. $\sqrt{t^6}$ **14.** $\sqrt{x^4}$ **15.** $\sqrt{x^5}$ **16.** $\sqrt{t^{19}}$

17. $\sqrt{(y - 2)^8}$ **18.** $\sqrt{4(x + 5)^{10}}$ **19.** $\sqrt{36m^3}$ **20.** $\sqrt{8a^5}$

1. _____

2. _____

3. _____

4. _____

5. _____

6. _____

7. _____

8. _____

9. _____

10. _____

11. _____

12. _____

13. _____

14. _____

15. _____

16. _____

17. _____

18. _____

19. _____

20. _____

Copyright © 1979, Philippines copyright 1979, by Addison-Wesley Publishing Company, Inc. All rights reserved.

21. $\sqrt{448x^6y^3}$ **22.** $\sqrt{243x^5y^4}$

21. _____

22. _____

23. _____

◯◯◯ Multiply and simplify.

23. $\sqrt{3}\sqrt{18}$ **24.** $\sqrt{5}\sqrt{10}$ **25.** $\sqrt{18}\sqrt{14}$

24. _____

25. _____

26. _____

26. $\sqrt{12}\sqrt{18}$ **27.** $\sqrt{10}\sqrt{10}$ **28.** $\sqrt{11}\sqrt{11}$

27. _____

28. _____

29. $\sqrt{5b}\sqrt{15b}$ **30.** $\sqrt{6a}\sqrt{18a}$ **31.** $\sqrt{ab}\sqrt{ac}$

29. _____

30. _____

31. _____

32. $\sqrt{xy}\sqrt{xz}$ **33.** $\sqrt{18x^2y^3}\sqrt{6xy^4}$ **34.** $\sqrt{12x^3y^2}\sqrt{8xy}$

32. _____

33. _____

35. $\sqrt{50ab}\sqrt{10a^2b^4}$ **36.** $\sqrt{18xy}\sqrt{14x^3y^2}$

34. _____

35. _____

36. _____

9.5 FRACTIONAL RADICANDS

▪ ○ ▪ PERFECT SQUARE RADICANDS

Some expressions with fractional radicands can be simplified by taking the square root of the numerator and denominator separately.

Examples   Simplify.

1. $\sqrt{\dfrac{25}{9}} = \dfrac{5}{3}$ because $\dfrac{5}{3} \cdot \dfrac{5}{3} = \dfrac{25}{9}$.

2. $\sqrt{\dfrac{1}{16}} = \dfrac{1}{4}$ because $\dfrac{1}{4} \cdot \dfrac{1}{4} = \dfrac{1}{16}$.

Sometimes a fractional expression can be simplified to one that has a perfect square numerator and denominator.

Example 3   Simplify.

$$\sqrt{\frac{18}{50}} = \sqrt{\frac{9 \cdot 2}{25 \cdot 2}} = \sqrt{\frac{9}{25} \cdot \boxed{\frac{2}{2}}}$$

$$= \sqrt{\frac{9}{25} \cdot \boxed{1}} = \sqrt{\frac{9}{25}} = \frac{3}{5} \quad \text{because} \quad \frac{3}{5} \cdot \frac{3}{5} = \frac{9}{25}.$$

Example 4   Simplify.

$$\sqrt{\frac{2560}{2890}} = \sqrt{\frac{256 \cdot 10}{289 \cdot 10}} = \sqrt{\frac{256}{289} \cdot \boxed{\frac{10}{10}}}$$

$$= \sqrt{\frac{256}{289} \cdot \boxed{1}} = \sqrt{\frac{256}{289}} = \frac{16}{17} \quad \text{because} \quad \frac{16}{17} \cdot \frac{16}{17} = \frac{256}{289}.$$

DO EXERCISES 1–5.

▪ ○ ○ ▪ RATIONALIZING DENOMINATORS

When neither the numerator nor denominator is a perfect square, we can simplify to an expression that has a whole-number radicand.

Example 5   Simplify: $\sqrt{\dfrac{2}{3}}$.

We multiply by 1, choosing 3/3 for 1. This makes the denominator a perfect square.

$$\sqrt{\frac{2}{3}} = \sqrt{\frac{2}{3} \cdot \boxed{\frac{3}{3}}} \qquad \text{Multiplying by 1}$$

$$= \sqrt{\frac{6}{9}}$$

$$= \sqrt{\frac{1}{9} \cdot 6} \qquad \text{Factoring to get a perfect square factor}$$

$$= \sqrt{\frac{1}{9}}\sqrt{6} = \frac{1}{3}\sqrt{6}$$

OBJECTIVES

After finishing Section 9.5, you should be able to:

▪ ○ ▪   Simplify radical expressions with fractional radicands, when numerator and denominator are perfect squares (or can be so simplified).

▪ ○ ○ ▪   Simplify radical expressions with fractional radicands when numerator and denominator are not perfect squares so that there will be only a whole-number radicand.

▪ ○ ○ ○ ▪   Use a table to approximate square roots of fractions.

Simplify.

1. $\sqrt{\dfrac{16}{9}}$

2. $\sqrt{\dfrac{1}{25}}$

3. $\sqrt{\dfrac{1}{9}}$

4. $\sqrt{\dfrac{18}{32}}$

5. $\sqrt{\dfrac{2250}{2560}}$

Rationalize the denominator.

6. $\sqrt{\dfrac{3}{5}}$

7. $\sqrt{\dfrac{5}{8}}$

(*Hint:* Multiply the radicand by $\frac{2}{2}$.)

Approximate to three decimal places.

8. $\sqrt{\dfrac{2}{7}}$

9. $\sqrt{\dfrac{5}{8}}$

We can always multiply by 1 to make a denominator a perfect square. This is called *rationalizing the denominator.*

Example 6   Rationalize the denominator: $\sqrt{\dfrac{5}{12}}$.

$$\sqrt{\frac{5}{12}} = \sqrt{\frac{5}{12} \cdot \frac{3}{3}} = \sqrt{\frac{15}{36}} = \sqrt{\frac{1}{36} \cdot 15} = \sqrt{\frac{1}{36}}\sqrt{15} = \frac{1}{6}\sqrt{15}$$

In Example 6, we did not multiply by 12/12. This would have given us a correct answer, but not the simplest. We chose 3/3 to make the denominator the smallest multiple of 12 that is a perfect square.

DO EXERCISES 6 AND 7.

■■■ APPROXIMATING SQUARE ROOTS OF FRACTIONS

Now we can use the square root table to find square roots of fractions.

Example 7   Approximate $\sqrt{\dfrac{3}{5}}$ to three decimal places.

a) $\sqrt{\dfrac{3}{5}} = \sqrt{\dfrac{3}{5} \cdot \dfrac{5}{5}} = \sqrt{\dfrac{15}{25}} = \sqrt{\dfrac{1}{25} \cdot 15} = \sqrt{\dfrac{1}{25}}\sqrt{15} = \dfrac{1}{5}\sqrt{15}$

b) From Table 1, $\sqrt{15} \approx 3.873$. We divide by 5.

$$\begin{array}{r} 0.7746 \\ 5\overline{)3.8730} \\ \underline{35} \\ 37 \\ \underline{35} \\ 23 \\ \underline{20} \\ 30 \end{array}$$

Then $\sqrt{\dfrac{3}{5}} \approx 0.775$ to three decimal places.

Example 8   Approximate $\sqrt{\dfrac{7}{18}}$ to three decimal places.

a) $\sqrt{\dfrac{7}{18}} = \sqrt{\dfrac{7}{18} \cdot \dfrac{2}{2}} = \sqrt{\dfrac{14}{36}} = \sqrt{\dfrac{1}{36} \cdot 14} = \sqrt{\dfrac{1}{36}} \cdot \sqrt{14} = \dfrac{1}{6}\sqrt{14}$

b) From Table 1, $\sqrt{14} \approx 3.742$.

We divide by 6 and get

$$\sqrt{\frac{7}{18}} \approx 0.624.\quad \text{Rounded to three decimal places}$$

DO EXERCISES 8 AND 9.

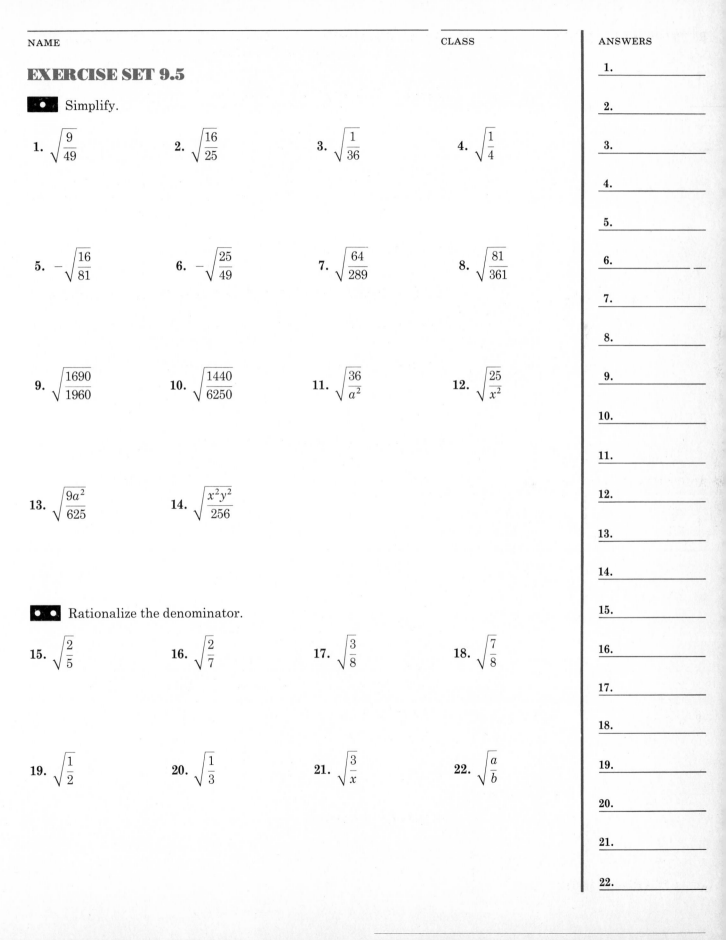

NAME

CLASS

EXERCISE SET 9.5

● ○ Simplify.

1. $\sqrt{\dfrac{9}{49}}$

2. $\sqrt{\dfrac{16}{25}}$

3. $\sqrt{\dfrac{1}{36}}$

4. $\sqrt{\dfrac{1}{4}}$

5. $-\sqrt{\dfrac{16}{81}}$

6. $-\sqrt{\dfrac{25}{49}}$

7. $\sqrt{\dfrac{64}{289}}$

8. $\sqrt{\dfrac{81}{361}}$

9. $\sqrt{\dfrac{1690}{1960}}$

10. $\sqrt{\dfrac{1440}{6250}}$

11. $\sqrt{\dfrac{36}{a^2}}$

12. $\sqrt{\dfrac{25}{x^2}}$

13. $\sqrt{\dfrac{9a^2}{625}}$

14. $\sqrt{\dfrac{x^2y^2}{256}}$

● ● Rationalize the denominator.

15. $\sqrt{\dfrac{2}{5}}$

16. $\sqrt{\dfrac{2}{7}}$

17. $\sqrt{\dfrac{3}{8}}$

18. $\sqrt{\dfrac{7}{8}}$

19. $\sqrt{\dfrac{1}{2}}$

20. $\sqrt{\dfrac{1}{3}}$

21. $\sqrt{\dfrac{3}{x}}$

22. $\sqrt{\dfrac{a}{b}}$

ANSWERS
1.
2.
3.
4.
5.
6.
7.
8.
9.
10.
11.
12.
13.
14.
15.
16.
17.
18.
19.
20.
21.
22.

Copyright © 1979, Philippines copyright 1979, by Addison-Wesley Publishing Company, Inc. All rights reserved.

●●● Approximate to three decimal places.

23. $\sqrt{\dfrac{3}{7}}$ **24.** $\sqrt{\dfrac{3}{2}}$ **25.** $\sqrt{\dfrac{1}{3}}$ **26.** $\sqrt{\dfrac{1}{5}}$

23. _____

24. _____

27. $\sqrt{\dfrac{7}{20}}$ **28.** $\sqrt{\dfrac{3}{20}}$ **29.** $\sqrt{\dfrac{12}{5}}$ **30.** $\sqrt{\dfrac{8}{3}}$

25. _____

26. _____

27. _____

31. *Pendulums.* The *period T* of a pendulum is the time it takes to make a move from one side to the other and back. A formula for the period is

$$T = 2\pi\sqrt{\dfrac{L}{32}},$$

where T is in seconds and L is in feet. Find the periods of pendulums of lengths 2 ft, 8 ft, 64 ft, and 100 ft. Use 3.14 for π.

28. _____

29. _____

30. _____

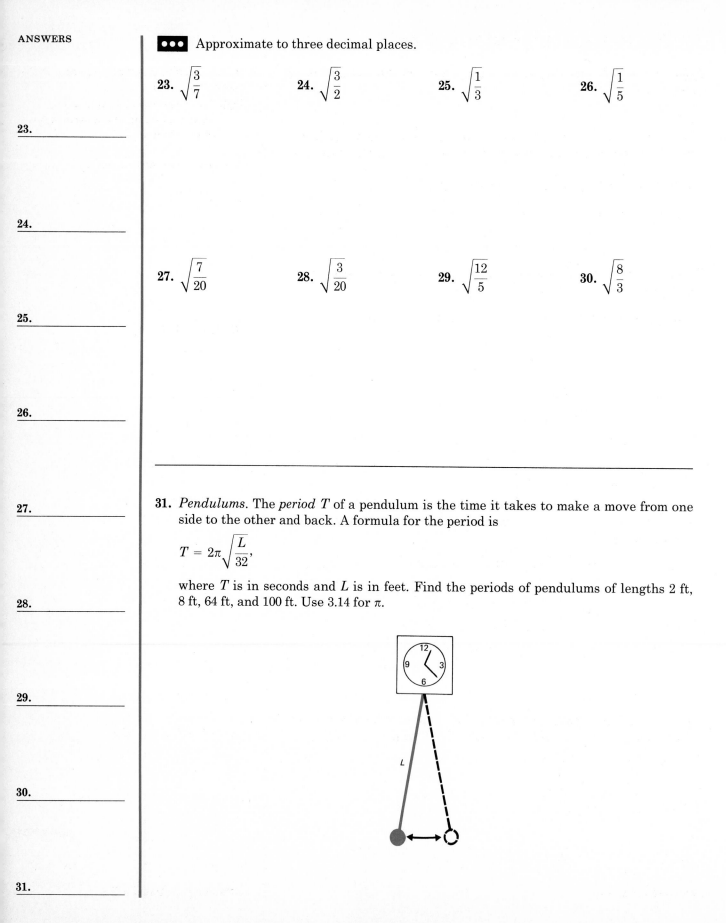

31. _____

9.6 ADDITION AND SUBTRACTION

■ We can add any two real numbers. The sum of 5 and $\sqrt{2}$ can be expressed as

$$5 + \sqrt{2}.$$

We cannot simplify this unless we use rational approximations. Sometimes, however, a sum can be simplified using the distributive laws.

Example 1 Add $3\sqrt{5}$ and $4\sqrt{5}$ and simplify.

$$3\sqrt{5} + 4\sqrt{5} = (3+4)\sqrt{5} \qquad \text{Using the distributive law to factor out } \sqrt{5}$$
$$= 7\sqrt{5}$$

To simplify like this, the radical expressions must be the same. Sometimes we can make them the same.

Example 2 Subtract $\sqrt{8}$ from $\sqrt{2}$ and simplify.

$$\sqrt{2} - \sqrt{8} = \sqrt{2} - \sqrt{4 \cdot 2} \qquad \text{Factoring 8}$$
$$= \sqrt{2} - \sqrt{4}\sqrt{2}$$
$$= \sqrt{2} - 2\sqrt{2}$$
$$= 1\sqrt{2} - 2\sqrt{2}$$
$$= (1-2)\sqrt{2} \qquad \text{Using the distributive law to factor out the common factor } \sqrt{2}$$
$$= -1 \cdot \sqrt{2}$$
$$= -\sqrt{2}$$

Example 3 Simplify.

$$\sqrt{x^3 - x^2} + \sqrt{4x - 4} = \sqrt{x^2(x-1)} + \sqrt{4(x-1)} \qquad \text{Factoring radicands}$$
$$= \sqrt{x^2}\sqrt{x-1} + \sqrt{4}\sqrt{x-1}$$
$$= |x|\sqrt{x-1} + 2\sqrt{x-1}$$
$$= (|x| + 2)\sqrt{x-1} \qquad \text{Using the distributive law to factor out the common factor } \sqrt{x-1}$$

Warning! Do not make the mistake of thinking that the sum of square roots is the square root of a sum. For example,

$$\sqrt{9} + \sqrt{16} = 3 + 4 = 7$$

but

$$\sqrt{9+16} = \sqrt{25} = 5.$$

In general,

$$\sqrt{a} + \sqrt{b} \neq \sqrt{a+b}.$$

After finishing Section 9.6, you should be able to:

■ Add or subtract with radical notation, using the distributive law to simplify.

Add or subtract and simplify.

1. $3\sqrt{2} + 9\sqrt{2}$

2. $8\sqrt{5} - 3\sqrt{5}$

3. $2\sqrt{10} - 7\sqrt{40}$

4. $\sqrt{24} + \sqrt{54}$

5. $\sqrt{9x+9} - \sqrt{4x+4}$

Simplify.

6. $\sqrt{2} + \sqrt{\dfrac{1}{2}}$

7. $\sqrt{\dfrac{5}{3}} - \sqrt{\dfrac{3}{5}}$

Sometimes eliminating fractional expressions will enable us to factor and then combine expressions.

Example 4 Simplify.

$$\sqrt{3} + \sqrt{\frac{1}{3}} = \sqrt{3} + \sqrt{\frac{1}{3} \cdot \frac{3}{3}} \quad \text{Multiplying by 1}$$

$$= \sqrt{3} + \sqrt{\frac{3}{9}}$$

$$= \sqrt{3} + \sqrt{\frac{1}{9} \cdot 3}$$

$$= \sqrt{3} + \sqrt{\frac{1}{9}}\sqrt{3}$$

$$= \sqrt{3} + \frac{1}{3}\sqrt{3}$$

$$= \left(1 + \frac{1}{3}\right)\sqrt{3} \quad \text{Factoring and simplifying}$$

$$= \frac{4}{3}\sqrt{3}$$

DO EXERCISES 6 AND 7.

SOMETHING EXTRA

AN APPLICATION: WIND CHILL TEMPERATURE

Wind speed affects the actual temperature, making a person feel colder. The *wind chill temperature* is what the temperature would have to be with no wind to give the same chilling effect. A formula for finding the wind chill temperature, T_w, is given by

$$T_w = 91.4 - \frac{(10.45 + 6.68\sqrt{v} - 0.447v)(457 - 5T)}{110},$$

where T is the actual temperature as given by a thermometer, in degrees Fahrenheit, and v is the wind speed in mph.

EXERCISES

You will need a calculator. You can get the square roots from Table 1. Find the wind chill temperature in each case. Round to the nearest one degree.

1. $T = 30°F$, $v = 25$ mph

2. $T = 20°F$, $v = 20$ mph

3. $T = 20°F$, $v = 40$ mph

4. $T = -10°F$, $v = 30$ mph

NAME   CLASS   ANSWERS

EXERCISE SET 9.6

▶ Simplify.

1. $3\sqrt{2} + 4\sqrt{2}$

2. $7\sqrt{5} + 3\sqrt{5}$

3. $6\sqrt{a} - 14\sqrt{a}$

4. $10\sqrt{x} - 13\sqrt{x}$

5. $3\sqrt{12} + 2\sqrt{3}$

6. $5\sqrt{8} + 15\sqrt{2}$

7. $\sqrt{27} - 2\sqrt{3}$

8. $\sqrt{45} - \sqrt{20}$

9. $\sqrt{72} + \sqrt{98}$

10. $\sqrt{45} + \sqrt{80}$

11. $3\sqrt{18} - 2\sqrt{32} - 5\sqrt{50}$

12. $2\sqrt{12} + 4\sqrt{27} - 5\sqrt{48}$

1. _____

2. _____

3. _____

4. _____

5. _____

6. _____

7. _____

8. _____

9. _____

1C. _____

11. _____

12. _____

ANSWERS

13. $\sqrt{4x} + \sqrt{81x^3}$

14. $\sqrt{27} - \sqrt{12x^2}$

13. _____

14. _____

15. $\sqrt{8x + 8} + \sqrt{2x + 2}$

16. $\sqrt{x^5 - x^2} + \sqrt{9x^3 - 9}$

15. _____

16. _____

17. $3x\sqrt{y^3x} - x\sqrt{yx^3} + y\sqrt{y^3x}$

18. $4a\sqrt{a^2b} + a\sqrt{a^2b^3} - 5\sqrt{b^3}$

17. _____

18. _____

19. $\sqrt{3} - \sqrt{\dfrac{1}{3}}$

20. $\sqrt{2} - \sqrt{\dfrac{1}{2}}$

21. $5\sqrt{2} + 3\sqrt{\dfrac{1}{2}}$

19. _____

20. _____

21. _____

22. $\sqrt{\dfrac{2}{3}} - \sqrt{\dfrac{1}{6}}$

23. $\sqrt{\dfrac{1}{12}} - \sqrt{\dfrac{1}{27}}$

24. $\sqrt{\dfrac{2}{3}} + \sqrt{\dfrac{3}{2}}$

22. _____

23. _____

24. _____

25. Can you find any pairs of numbers a and b, for which
$$\sqrt{a} + \sqrt{b} = \sqrt{a + b}?$$
If so, name them.

25. _____

9.7   DIVISION

◼•◻ DIVISION

To multiply with radical notation, we multiply radicands. To divide, we divide the radicands.

> **For any nonnegative radicands A and B,**
> $$\frac{\sqrt{A}}{\sqrt{B}} = \sqrt{\frac{A}{B}}.$$
> **(The quotient of square roots, provided they exist, is the square root of the quotient of the radicands.)**

Example 1

$$\frac{\sqrt{27}}{\sqrt{3}} = \sqrt{\frac{27}{3}} = \sqrt{9} = 3$$

Example 2

$$\frac{\sqrt{7}}{\sqrt{14}} = \sqrt{\frac{7}{14}} = \sqrt{\frac{1}{2}} = \sqrt{\frac{1}{2} \cdot \frac{2}{2}} = \sqrt{\frac{2}{4}} = \sqrt{\frac{1}{4} \cdot 2} = \sqrt{\frac{1}{4}} \cdot \sqrt{2} = \frac{1}{2}\sqrt{2}$$

Example 3

$$\frac{\sqrt{30a^3}}{\sqrt{6a^2}} = \sqrt{\frac{30a^3}{6a^2}} = \sqrt{5a}$$

DO EXERCISES 1–3.

◼•◻◻ RATIONALIZING DENOMINATORS

Expressions with radicals are considered simpler if there are no radicals in denominators. We can simplify by multiplying by 1, but this time we do it a bit differently.

Example 4   Rationalize the denominator.

$$\frac{\sqrt{2}}{\sqrt{3}} = \frac{\sqrt{2}}{\sqrt{3}} \cdot \boxed{\frac{\sqrt{3}}{\sqrt{3}}} = \frac{\sqrt{2} \cdot \sqrt{3}}{\sqrt{3} \cdot \sqrt{3}} = \frac{\sqrt{6}}{3}, \quad \text{or} \quad \frac{1}{3}\sqrt{6}$$

Example 5   Rationalize the denominator.

$$\frac{\sqrt{49a^5}}{\sqrt{12}} = \frac{\sqrt{49a^5}}{\sqrt{12}} \cdot \boxed{\frac{\sqrt{3}}{\sqrt{3}}} = \frac{\sqrt{49a^5}\sqrt{3}}{\sqrt{12}\sqrt{3}} = \frac{\sqrt{49a^5}\sqrt{3}}{\sqrt{36}}$$

$$= \frac{\sqrt{49a^4 \cdot 3a}}{\sqrt{36}} = \frac{7|a^2|\sqrt{3a}}{6}$$

DO EXERCISES 4 AND 5.

OBJECTIVES

After finishing Section 9.7, you should be able to:

◼•◻ **Divide with radical notation.**

◼•◻◻ **Rationalize denominators.**

Divide and simplify.

1. $\dfrac{\sqrt{5}}{\sqrt{45}}$

2. $\dfrac{\sqrt{2}}{\sqrt{6}}$

3. $\dfrac{\sqrt{42x^5}}{\sqrt{7x^2}}$

Rationalize the denominator.

4. $\dfrac{\sqrt{5}}{\sqrt{7}}$

5. $\dfrac{\sqrt{x}}{\sqrt{y}}$

ANSWERS ON PAGE A–29

SOMETHING EXTRA
CALCULATOR CORNER: FINDING SQUARE ROOTS ON A CALCULATOR

Some calculators have a square root key. If yours does not, you can use Table 1 if the number occurs there, or you can use the following method.

> 1. **Make a guess.**
> 2. **Square the guess on a calculator.**
> 3. **Adjust the guess.**

Example   Approximate $\sqrt{27}$ to three decimal places using a calculator.

First, we make a guess. Let's guess 4. Using a calculator we find 4^2, which is 16. Since $16 < 27$, we know that the square root is at *least* 4.

Next we try 5. We find 5^2, which is 25. This is also less than 27, so the square root is at least 5.

Next we try 6: $6^2 = 36$, which is greater than 27, so we know that $\sqrt{27}$ is between 5 and 6. Thus, $\sqrt{27}$ is 5 plus some fraction: $\sqrt{27} = 5.\ldots$.

Now we find the tenths digit by trying 5.1, 5.2, 5.3, and so on.

$(5.1)^2 = 26.01$ This is less than 27.
$(5.2)^2 = 27.04$ This is greater than 27.
The tenths digit is 1, so $\sqrt{27} = 5.1\ \ldots$.

Now we find the hundredths digit by trying 5.11, 5.12, 5.13, and so on.

$(5.11)^2 = 26.1121$
$(5.12)^2 = 26.2144$
$\ \vdots \qquad\quad \vdots$ All of these are less than 27, so the hundredths
$(5.18)^2 = 26.8324$ digit is 9.
The hundredths digit is 9, so $\sqrt{27} = 5.19\ldots$.

Now we find the thousandths digit by trying 5.191, 5.192, and so on.

$(5.196)^2 = 26.998416$ This is less than 27.
$(5.197)^2 = 27.008809$ This is greater than 27.
The thousandths digit is 6, so $\sqrt{27} \approx 5.196\ldots$.

Now we find the ten-thousandths digit by trying 5.1961, 5.1962, and so on.

$(5.1961)^2 = 26.99945521$ This is less than 27.
$(5.1962)^2 = 27.00049444$ This is greater than 27.
$\sqrt{27} \approx 5.1961\ldots$

How far do we continue? It depends on the desired accuracy. We decide to stop here and round back to three decimal places. Thus we get

$\sqrt{27} \approx 5.196$

as a final answer. Check this in Table 1.

Tables can be made this way, but there are other methods.

EXERCISES

Approximate these square roots to three decimal places using a calculator.

1. $\sqrt{17}$ 2. $\sqrt{80}$

3. $\sqrt{110}$ 4. $\sqrt{69}$

5. $\sqrt{200}$ 6. $\sqrt{10.5}$

7. $\sqrt{890}$ 8. $\sqrt{265.78}$

9. $\sqrt{2}$ 10. $\sqrt{0.2344}$

11. $\sqrt{\pi}$ 12. $\sqrt{2 + \sqrt{3}}$

NAME   CLASS   ANSWERS

EXERCISE SET 9.7

⚫ Divide and simplify.

1. $\dfrac{\sqrt{18}}{\sqrt{2}}$ 2. $\dfrac{\sqrt{20}}{\sqrt{5}}$ 3. $\dfrac{\sqrt{60}}{\sqrt{15}}$ 4. $\dfrac{\sqrt{108}}{\sqrt{3}}$

5. $\dfrac{\sqrt{75}}{\sqrt{15}}$ 6. $\dfrac{\sqrt{18}}{\sqrt{3}}$ 7. $\dfrac{\sqrt{12}}{\sqrt{75}}$ 8. $\dfrac{\sqrt{18}}{\sqrt{32}}$

9. $\dfrac{\sqrt{8x}}{\sqrt{2x}}$ 10. $\dfrac{\sqrt{18b}}{\sqrt{2b}}$ 11. $\dfrac{\sqrt{63y^3}}{\sqrt{7y}}$ 12. $\dfrac{\sqrt{48x^3}}{\sqrt{3x}}$

13. $\dfrac{\sqrt{15x^5}}{\sqrt{3x}}$ 14. $\dfrac{\sqrt{30a^5}}{\sqrt{5a}}$ 15. $\dfrac{\sqrt{3x}}{\sqrt{\dfrac{3x}{4}}}$ 16. $\dfrac{\sqrt{5x}}{\sqrt{\dfrac{5x}{9}}}$

1. _____

2. _____

3. _____

4. _____

5. _____

6. _____

7. _____

8. _____

9. _____

10. _____

11. _____

12. _____

13. _____

14. _____

15. _____

16. _____

Copyright © 1979, Philippines copyright 1979, by Addison-Wesley Publishing Company, Inc. All rights reserved.

ANSWERS

●● Rationalize the denominator.

17. _____

18. _____

19. _____

20. _____

17. $\dfrac{\sqrt{2}}{\sqrt{5}}$

18. $\dfrac{\sqrt{3}}{\sqrt{2}}$

19. $\dfrac{2}{\sqrt{2}}$

20. $\dfrac{3}{\sqrt{3}}$

21. _____

22. _____

23. _____

24. _____

21. $\dfrac{\sqrt{48}}{\sqrt{32}}$

22. $\dfrac{\sqrt{56}}{\sqrt{40}}$

23. $\dfrac{\sqrt{450}}{\sqrt{18}}$

24. $\dfrac{\sqrt{224}}{\sqrt{14}}$

25. _____

26. _____

27. _____

25. $\dfrac{\sqrt{3}}{\sqrt{x}}$

26. $\dfrac{\sqrt{2}}{\sqrt{y}}$

27. $\dfrac{4y}{\sqrt{3}}$

28. $\dfrac{8x}{\sqrt{5}}$

28. _____

29. _____

30. _____

29. $\dfrac{\sqrt{a^3}}{\sqrt{8}}$

30. $\dfrac{\sqrt{x^3}}{\sqrt{27}}$

31. $\dfrac{\sqrt{16a^4b^6}}{\sqrt{128a^6b^6}}$

32. $\dfrac{\sqrt{45mn^2}}{\sqrt{32m}}$

31. _____

32. _____

9.8 RADICALS AND RIGHT TRIANGLES

▮• RIGHT TRIANGLES

In a right triangle, the longest side is called the *hypotenuse*. The other two sides are called *legs*. We usually use the letters a and b for the lengths of the legs and c for the length of the hypotenuse. They are related as follows.

> *The Pythagorean Property of Right Triangles:* **In any right triangle if a and b are the lengths of the legs and c is the length of the hypotenuse, then**
>
> $$a^2 + b^2 = c^2.$$

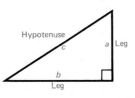

If we know the lengths of any two sides, we can find the length of the third side.

Example 1   Find the length of the hypotenuse of this right triangle.

$$4^2 + 5^2 = c^2$$
$$16 + 25 = c^2$$
$$41 = c^2$$
$$c = \sqrt{41}$$
$$c \approx 6.403 \quad \text{Table 1}$$

Example 2   Find the length of the leg of this right triangle.

$$10^2 + b^2 = 12^2$$
$$100 + b^2 = 144$$
$$b^2 = 144 - 100$$
$$b^2 = 44$$
$$b = \sqrt{44}$$
$$b = 6.633 \quad \text{Table 1}$$

DO EXERCISES 1 AND 2.

OBJECTIVES

After finishing Section 9.8, you should be able to:

▮• Given the lengths of any two sides of a right triangle, find the length of the third side.

▮•▮ Solve applied problems involving right triangles.

1. Find the length of the hypotenuse in this right triangle.

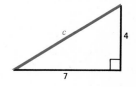

2. Find the length of the leg of this right triangle.

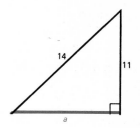

3. Find the length of the leg of this right triangle.

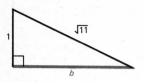

4. Find the length of the leg of this right triangle.

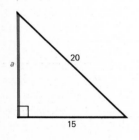

5. How long is a guy wire reaching from the top of a 15-ft pole to a point on the ground 10 ft from the pole?

Example 3   Find the length of the leg of this right triangle.

$$1^2 + b^2 = (\sqrt{7})^2$$
$$1 + b^2 = 7$$
$$b^2 = 7 - 1$$
$$b^2 = 6$$
$$b = \sqrt{6}$$
$$b \approx 2.449$$

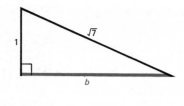

Example 4   Find the length of the leg of this right triangle.

$$a^2 + 10^2 = 15^2$$
$$a^2 + 100 = 225$$
$$a^2 = 225 - 100$$
$$a^2 = 125$$
$$a = \sqrt{125}$$
$$a = \sqrt{25 \cdot 5}$$
$$a = \sqrt{25} \cdot \sqrt{5}$$
$$a = 5\sqrt{5}$$
$$a \approx 5 \times 2.236$$
$$a \approx 11.18$$

DO EXERCISES 3 AND 4.

■■ **APPLIED PROBLEMS**

Example 5   A 12-foot ladder is leaning against a building. The bottom of the ladder is 7 ft from the building. How high is the top of the ladder?

a) We first make a drawing. In it we see a right triangle.

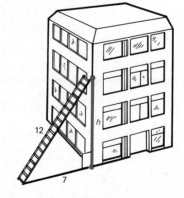

b) Now $7^2 + h^2 = 12^2$. We solve this equation.

$$49 + h^2 = 144$$
$$h^2 = 144 - 49$$
$$h^2 = 95$$
$$h = \sqrt{95}$$
$$h \approx 9.747 \text{ ft}$$

DO EXERCISE 5.

NAME   CLASS

EXERCISE SET 9.8

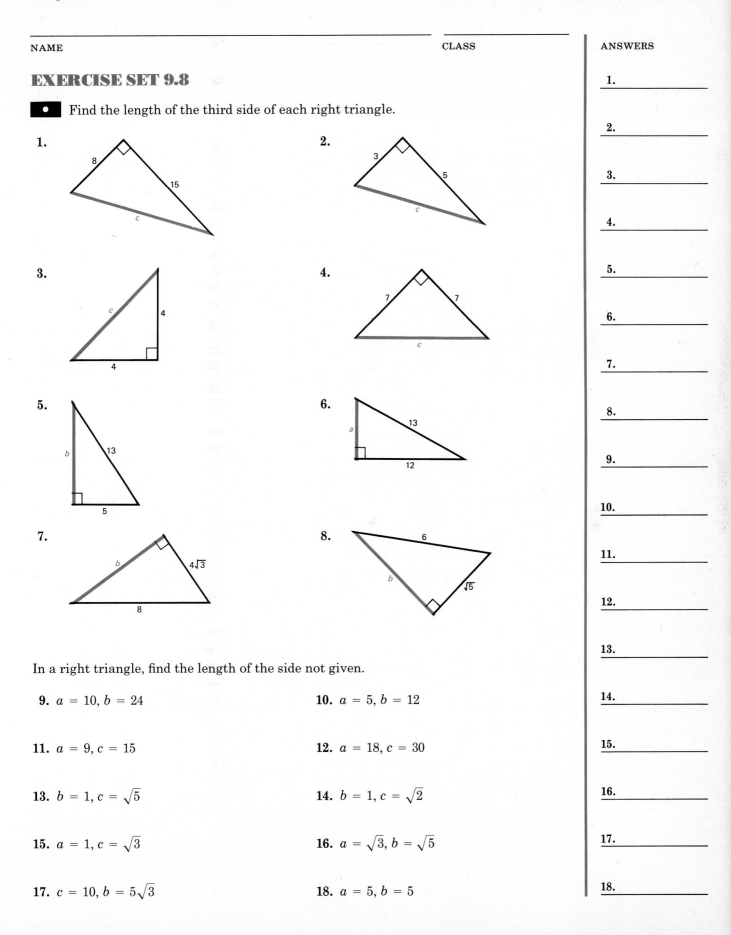

● Find the length of the third side of each right triangle.

1.

2.

3.

4.

5.

6.

7.

8.

In a right triangle, find the length of the side not given.

9. $a = 10, b = 24$

10. $a = 5, b = 12$

11. $a = 9, c = 15$

12. $a = 18, c = 30$

13. $b = 1, c = \sqrt{5}$

14. $b = 1, c = \sqrt{2}$

15. $a = 1, c = \sqrt{3}$

16. $a = \sqrt{3}, b = \sqrt{5}$

17. $c = 10, b = 5\sqrt{3}$

18. $a = 5, b = 5$

1. _____

2. _____

3. _____

4. _____

5. _____

6. _____

7. _____

8. _____

9. _____

10. _____

11. _____

12. _____

13. _____

14. _____

15. _____

16. _____

17. _____

18. _____

Copyright © 1979, Philippines copyright 1979, by Addison-Wesley Publishing Company, Inc. All rights reserved.

⬤ ⬤ Solve. Don't forget to make drawings.

19. A 10-meter ladder is leaning against a building. The bottom of the ladder is 5 m from the building. How high is the top of the ladder?

19. _____

20. Find the length of a diagonal of a square whose sides are 3 cm long.

21. How long is a guy wire reaching from the top of a 12-ft pole to a point 8 ft from the pole?

20. _____

22. How long must a wire be to reach from the top of a 13-m telephone pole to a point on the ground 9 m from the foot of the pole?

23. A slow-pitch softball diamond is actually a square 60 ft on a side. How far is it from home to second base?

21. _____

60 ft

22. _____

24. A baseball diamond is actually a square 90 ft on a side. How far is it from first to third base?

90 ft

23. _____

24. _____

An *equilateral* triangle is shown at the right.

25. _____

25. Find an expression for its height h in terms of a.

26. _____

26. Find an expression for its area A in terms of a.

9.9   EQUATIONS WITH RADICALS

◼◦ SOLVING EQUATIONS WITH RADICALS

To solve equations with radicals, we first convert them to equations without radicals. We do this by squaring both sides of the equation. The following new principle is used.

The Principle of Squaring: **If an equation** $a = b$ **is true, then the equation** $a^2 = b^2$ **is true.**

Example 1   Solve: $\sqrt{2x} - 4 = 7$.

$$\sqrt{2x} = 11 \qquad \text{Adding 4, to get the radical alone on one side}$$

$$(\sqrt{2x})^2 = 11^2 \qquad \text{Squaring both sides}$$

$$2x = 121$$

$$x = \frac{121}{2}$$

Check:

$$\frac{2x - 4 = 7}{\begin{array}{c|c} \sqrt{2 \cdot \dfrac{121}{2} - 4} & 7 \\ \sqrt{121} - 4 & \\ 11 - 4 & \\ 7 & \end{array}}$$

It is important to check! When we square both sides of an equation, the new equation may have solutions that the first one does not. For example, the equation

$$x = 1$$

has just one solution, the number 1. When we square both sides we get

$$x^2 = 1,$$

which has two solutions, 1 and -1.

DO EXERCISE 1.

Example 2   Solve: $\sqrt{x + 1} = \sqrt{2x - 5}$.

$$(\sqrt{x + 1})^2 = (\sqrt{2x - 5})^2 \qquad \text{Squaring both sides}$$

$$x + 1 = 2x - 5$$

$$x = 6$$

6 checks, so it is the solution.

DO EXERCISES 2 AND 3.

OBJECTIVES

After finishing Section 9.9, you should be able to:

◼◦ Solve equations with radicals.

◼◼ Solve applied problems involving radical equations.

Solve.

1. $\sqrt{3x} - 5 = 3$

Solve.

2. $\sqrt{3x + 1} = \sqrt{2x + 3}$

3. $\sqrt{x - 2} - 5 = 3$

(*Hint:* Get $\sqrt{x - 2}$ alone on one side.)

4. How far can you see to the horizon through an airplane window at a height of 8000 m?

How far can you see from a given height? There is a formula for this. At a height of h meters you can see V kilometers to the horizon. These numbers are related as follows:

$$V = 3.5\sqrt{h} \tag{1}$$

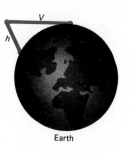

Earth

Example 3 How far to the horizon can you see through an airplane window at a height, or altitude, of 9000 meters?

We substitute 9000 for h in equation (1) and simplify.

$$V = 3.5\sqrt{9000}$$
$$V = 3.5\sqrt{900 \cdot 10}$$
$$V = 3.5 \times 30 \times \sqrt{10}$$
$$V \approx 3.5 \times 30 \times 3.162$$
$$V \approx 332.010 \text{ km}$$

DO EXERCISES 4 AND 5.

5. How far can a sailor see to the horizon from the top of a 20-m mast?

Example 4 A person can see 50.4 kilometers to the horizon from the top of a cliff. How high is the cliff?

We substitute 50.4 for V in equation (1) and solve.

$$50.4 = 3.5\sqrt{h}$$
$$\frac{50.4}{3.5} = \sqrt{h}$$
$$14.4 = \sqrt{h}$$
$$(14.4)^2 = (\sqrt{h})^2$$
$$(14.4)^2 = h$$
$$207.36 \text{ m} = h$$

The cliff is 207.36 meters high.

DO EXERCISE 6.

6. A sailor can see 91 km to the horizon from the top of a mast. How high is the mast?

NAME

CLASS

EXERCISE SET 9.9

⬛● Solve.

1. $\sqrt{x} = 5$

2. $\sqrt{x} = 7$

3. $\sqrt{x} = 6.2$

4. $\sqrt{x} = 4.3$

5. $\sqrt{x + 3} = 20$

6. $\sqrt{x + 4} = 11$

7. $\sqrt{2x + 4} = 25$

8. $\sqrt{2x + 1} = 13$

9. $3 + \sqrt{x - 1} = 5$

10. $4 + \sqrt{y - 3} = 11$

11. $6 - 2\sqrt{3n} = 0$

12. $8 - 4\sqrt{5n} = 0$

13. $\sqrt{5x - 7} = \sqrt{x + 10}$

14. $\sqrt{4x - 5} = \sqrt{x + 9}$

15. $\sqrt{x} = -7$

16. $\sqrt{x} = -5$

17. $\sqrt{2y + 6} = \sqrt{2y - 5}$

18. $2\sqrt{3x - 2} = \sqrt{2x - 3}$

1. _____

2. _____

3. _____

4. _____

5. _____

6. _____

7. _____

8. _____

9. _____

10. _____

11. _____

12. _____

13. _____

14. _____

15. _____

16. _____

17. _____

18. _____

Copyright © 1979. Philippines copyright 1979 by Addison-Wesley Publishing Company, Inc. All rights reserved.

ANSWERS

■ ■ Solve.

Use $V = 3.5\sqrt{h}$ for Exercises 19–22.

19. How far can you see to the horizon through an airplane window at a height of 9800 m?

19. _____

20. How far can a sailor see to the horizon from the top of a 24-m mast?

20. _____

21. A person can see 371 km to the horizon from an airplane window. How high is the airplane?

21. _____

22. A sailor can see 99.4 km to the horizon from the top of a mast. How high is the mast?

22. _____

The formula $r = 2\sqrt{5L}$ can be used to approximate the speed r, in mph, of a car that has left a skid mark of length L, in feet.

23. How far will a car skid at 50 mph? at 70 mph?

23. _____

24. How far will a car skid at 60 mph? at 100 mph?

24. _____

25. _____

The formula $T = 2\pi\sqrt{L/32}$ can be used to find the period T, in seconds, of a pendulum of length L, in feet.

25. ▦ What is the length of a pendulum that has a period of 1.6 sec? Use 3.14 for π.

26. ▦ What is the length of a pendulum that has a period of 3 sec? Use 3.14 for π.

26. _____

NAME	SCORE	ANSWERS

TEST OR REVIEW—CHAPTER 9

If you miss an item, review the indicated section and objective.

[9.1, ●]   Find the following.

1. $\sqrt{36}$ **2.** $-\sqrt{81}$

[9.1, ●●]   Identify the radicand.

3. $3x\sqrt{\dfrac{x}{2+x}}$

[9.1, ■■]   Simplify.

4. $\sqrt{m^2}$ **5.** $\sqrt{49t^2}$ **6.** $\sqrt{(x-4)^2}$

[9.2, ●]   Identify the rational numbers and the irrational numbers.

7. $\sqrt{3}$ **8.** $\sqrt{36}$ **9.** $-\sqrt{12}$ **10.** $-\sqrt{4}$

[9.3, ●]   Multiply.

11. $\sqrt{3}\sqrt{7}$ **12.** $\sqrt{x-3}\sqrt{x+3}$ **13.** $\sqrt{-2}\sqrt{-3}$

[9.3, ●●]   Factor. Simplify where possible.

14. $\sqrt{48}$ **15.** $\sqrt{x^2-49}$

[9.3, ●●●]   Approximate to three decimal places. Use Table 1.

16. $\sqrt{108}$

[9.4, ●●●]   Multiply and simplify.

17. $\sqrt{5}\sqrt{8}$ **18.** $\sqrt{5x}\sqrt{10xy^2}$

$\sqrt{5 \cdot 5 \cdot 2 \cdot x^2 \cdot y^2}$

ANSWERS

1. 6

2. −9

3. $\dfrac{x}{2+x}$

4. m

5. 7T

6. X−4

7. irrational

8. rational

9. irrational

10. rational

11. $\sqrt{21}$

12. $\sqrt{x^2-9}$

13. ~~X−11~~

14. $4\sqrt{3}$

15. can't do

16.

17. $2\sqrt{10}$

18. $5XY\sqrt{2}$

Copyright © 1979, Philippines copyright 1979, by Addison-Wesley Publishing Company, Inc. All rights reserved.

ANSWERS

19. $\dfrac{5}{8}$

20. $\dfrac{2\sqrt{5}}{3\sqrt{5}}$

21. $\dfrac{7}{T}$

22. $\dfrac{1}{2}\sqrt{2}$

23.

24.

25.

26.

27. $16\sqrt{3}$

28. $\sqrt{5}$

29. $\dfrac{1}{2}\sqrt{2}$

30.

31.

32.

33.

34.

35.

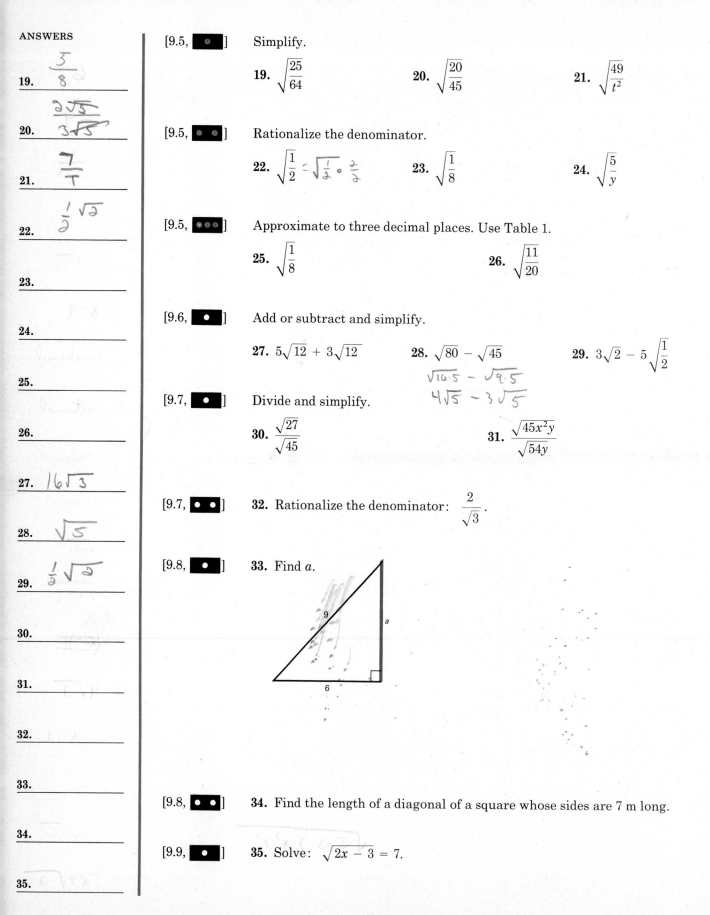

[9.5, ▨]   Simplify.

19. $\sqrt{\dfrac{25}{64}}$

20. $\sqrt{\dfrac{20}{45}}$

21. $\sqrt{\dfrac{49}{t^2}}$

[9.5, ◉◉]   Rationalize the denominator.

22. $\sqrt{\dfrac{1}{2}}$ $=\sqrt{\dfrac{1}{2}}\circ\dfrac{2}{2}$

23. $\sqrt{\dfrac{1}{8}}$

24. $\sqrt{\dfrac{5}{y}}$

[9.5, ◉◉◉]   Approximate to three decimal places. Use Table 1.

25. $\sqrt{\dfrac{1}{8}}$

26. $\sqrt{\dfrac{11}{20}}$

[9.6, ▨]   Add or subtract and simplify.

27. $5\sqrt{12}+3\sqrt{12}$

28. $\sqrt{80}-\sqrt{45}$

$\sqrt{16.5}-\sqrt{9.5}$

$4\sqrt{5}-3\sqrt{5}$

29. $3\sqrt{2}-5\sqrt{\dfrac{1}{2}}$

[9.7, ▨]   Divide and simplify.

30. $\dfrac{\sqrt{27}}{\sqrt{45}}$

31. $\dfrac{\sqrt{45x^2y}}{\sqrt{54y}}$

[9.7, ◉◉]   **32.** Rationalize the denominator: $\dfrac{2}{\sqrt{3}}$.

[9.8, ▨]   **33.** Find a.

[9.8, ◉◉]   **34.** Find the length of a diagonal of a square whose sides are 7 m long.

[9.9, ▨]   **35.** Solve: $\sqrt{2x-3}=7$.

QUADRATIC
EQUATIONS

READINESS CHECK: SKILLS FOR CHAPTER 10

1. Multiply: $(3x + 1)^2$.

2. What must be added to $x^2 - 8x$ to make it the square of a binomial?

3. Factor by completing the square:

$x^2 - 8x + 7$.

4. Find $\sqrt{9}$.

Simplify.

5. $\sqrt{20}$

6. $\sqrt{\dfrac{2890}{2560}}$

7. Approximate. Round to the nearest hundredth: $\sqrt{17}$.

8. Rationalize the denominator:

$\sqrt{\dfrac{7}{3}}$.

Solve.

9. $x^2 - 6x = 0$

10. $x^2 - 5x + 6 = 0$

11. Graph: $5x + 3y = 15$.

10.1 INTRODUCTION TO QUADRATIC EQUATIONS

■ WRITING QUADRATIC EQUATIONS IN STANDARD FORM

The following are quadratic equations. Each is of second degree.

$$4x^2 + 7x - 5 = 0, \quad 3x^2 - \frac{1}{2}x = 9, \quad 5y^2 = -6y$$

> An equation of the type $ax^2 + bx + c = 0$, **where** a, b, **and** c **are real number constants and** $a > 0$, **is called the** *standard form of a quadratic equation.*

Examples   Write standard form and determine a, b, and c.

1. $4x^2 + 7x - 5 = 0$ The equation is already in standard form.

$a = 4;\ b = 7;\ c = -5$

2. $3x^2 - 0.5x = 9$

$3x^2 - 0.5x - 9 = 0$ Adding -9. This is standard form.

$a = 3;\ b = 0.5;\ c = -9$

3. $-4y^2 = 5y$

$-4y^2 - 5y = 0$ Adding $-5y$

Not positive!

$4y^2 + 5y = 0$ Multiplying by -1. This is standard form.

$a = 4;\ b = 5;\ c = 0$

DO EXERCISES 1–3.

OBJECTIVES

After finishing Section 10.1, you should be able to:

■ Write a quadratic equation in standard form $ax^2 + bx + c = 0$, $a > 0$, **and determine the coefficients** a, b, **and** c.

■■ Solve quadratic equations of the type $ax^2 = k$, $a \neq 0$.

Write standard form and determine a, b, and c.

1. $x^2 = 7x$

2. $3 - x^2 = 9x$

3. $3x + 5x^2 = x^2 - 4 + x$

● ● **SOLVING EQUATIONS OF THE TYPE** $ax^2 = k$

When b is 0, we solve for x^2 and take the principal square root.

Example 4 Solve: $5x^2 = 15$.

$$x^2 = 3 \qquad \text{Solving for } x^2; \text{ multiplying by } \tfrac{1}{5}$$

$$|x| = \sqrt{3}$$

Since $|x|$ is either x or $-x$,

$$x = \sqrt{3} \quad \text{or} \quad -x = \sqrt{3}$$
$$x = \sqrt{3} \quad \text{or} \quad x = -\sqrt{3}.$$

Check: For $\sqrt{3}$:   For: $-\sqrt{3}$:

$$
\begin{array}{c|c}
5x^2 = 15 & \\
\hline
5(\sqrt{3})^2 & 15 \\
5 \cdot 3 & \\
15 &
\end{array}
\qquad
\begin{array}{c|c}
5x^2 = 15 & \\
\hline
5(-\sqrt{3})^2 & 15 \\
5 \cdot 3 & \\
15 &
\end{array}
$$

The solutions are $\sqrt{3}$ and $-\sqrt{3}$.

DO EXERCISE 4.

Example 5 Solve: $\dfrac{1}{3}x^2 = 0$.

$$x^2 = 0 \qquad \text{Multiplying by 3}$$
$$|x| = \sqrt{0} \qquad \text{Taking principal square root}$$
$$|x| = 0$$

The only number with absolute value 0 is 0. It checks, so it is the solution.

DO EXERCISE 5.

Example 6 Solve: $-3x^2 + 7 = 0$.

$$-3x^2 = -7 \qquad \text{Adding } -7$$

$$x^2 = \frac{-7}{-3} \qquad \text{Multiplying by } -\frac{1}{3}$$

$$x^2 = \frac{7}{3}$$

$$|x| = \sqrt{\frac{7}{3}} \qquad \text{Taking square root}$$

$$x = \sqrt{\frac{7}{3}} \quad \text{or} \quad x = -\sqrt{\frac{7}{3}}$$

$$x = \sqrt{\frac{7}{3} \cdot \frac{3}{3}} \quad \text{or} \quad x = -\sqrt{\frac{7}{3} \cdot \frac{3}{3}} \qquad \text{Rationalizing the denominators}$$

$$x = \frac{\sqrt{21}}{3} \quad \text{or} \quad x = -\frac{\sqrt{21}}{3}$$

Solve.

4. $4x^2 = 20$

Solve.

5. $2x^2 = 0$

ANSWERS ON PAGE A-30

Solve.

6. $2x^2 - 3 = 0$

7. $4x^2 - 9 = 0$

Check: For $\dfrac{\sqrt{21}}{3}$:   For: $-\dfrac{\sqrt{21}}{3}$:

$$
\begin{array}{c|c}
-3x^2 + 7 = 0 & \\
\hline
-3\left(\dfrac{\sqrt{21}}{3}\right)^2 + 7 & 0 \\
-3 \cdot \dfrac{21}{9} + 7 & \\
-7 + 7 & \\
& 0
\end{array}
\qquad
\begin{array}{c|c}
-3x^2 + 7 = 0 & \\
\hline
-3\left(-\dfrac{\sqrt{21}}{3}\right)^2 + 7 & 0 \\
-3 \cdot \dfrac{21}{9} + 7 & \\
-7 + 7 & \\
& 0
\end{array}
$$

The solutions are $\dfrac{\sqrt{21}}{3}$ and $-\dfrac{\sqrt{21}}{3}$.

DO EXERCISES 6 AND 7.

SOMETHING EXTRA

CALCULATOR CORNER

The length d of a diagonal of a rectangular solid is given by

$$d = \sqrt{a^2 + b^2 + c^2},$$

where a, b, and c are the lengths of the sides.

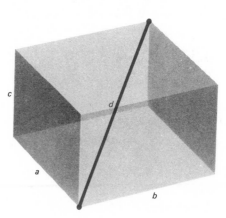

EXERCISES

1. Find the length of a diagonal of a rectangular solid whose sides have lengths $a = 2$, $b = 4$, and $c = 5$.

2. Will a baton of length 57 cm fit inside a briefcase that is 12.5 cm by 45 cm by 31.5 cm?

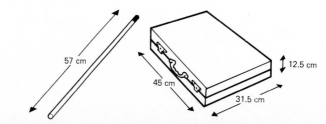

NAME   CLASS

EXERCISE SET 10.1

● Write standard form and determine a, b, and c.

1. $x^2 = 3x + 2$ **2.** $2x^2 = 3$

3. $7x^2 = 4x - 3$ **4.** $5 = -2x^2 + 3x$

5. $2x - 1 = 3x^2 + 7$ **6.** $4x^2 - 3x + 2 = 3x^2 + 7x - 5$

●● Solve.

7. $x^2 = 4$ **8.** $x^2 = 1$ **9.** $x^2 = 49$ **10.** $x^2 = 16$

11. $x^2 = 7$ **12.** $x^2 = 11$ **13.** $3x^2 = 30$ **14.** $5x^2 = 35$

Copyright © 1979, Philippines copyright 1979, by Addison-Wesley Publishing Company, Inc. All rights reserved.

ANSWERS

15. $3x^2 = 24$ **16.** $5x^2 = 60$ **17.** $4x^2 - 25 = 0$ **18.** $9x^2 - 4 = 0$

15. _____

16. _____

17. _____

18. _____

19. $3x^2 - 49 = 0$ **20.** $5y^2 - 16 = 0$ **21.** $4y^2 - 3 = 9$ **22.** $5x^2 - 100 = 0$

19. _____

20. _____

21. _____

22. _____

23. *Falling object.* The distance S traveled by a body falling freely from rest, in feet, in t seconds is given by

$$S = 16t^2.$$

This formula is an approximation since it does not account for air resistance. How far will an object fall in 1 sec, 2 sec, 5 sec, and 20 sec?

$S = 16t^2$

23. _____

24. ▦ Solve: $4.82x^2 = 12{,}000$.

24. _____

10.2 SOLVING BY FACTORING

⬛• EQUATIONS OF THE TYPE $ax^2 + bx = 0$

When c is 0 (and $b \neq 0$), we can factor and use the principle of zero products.

Example 1 Solve: $7x^2 + 2x = 0$.

$$x(7x + 2) = 0 \qquad \text{Factoring}$$

$x = 0$ or $7x + 2 = 0$ Principle of zero products

$x = 0$ or $7x = -2$

$x = 0$ or $x = -\frac{2}{7}$

Check: For 0:   For $-\frac{2}{7}$:

$$\begin{array}{c|c} 7x^2 + 2x = 0 \\ \hline 7 \cdot 0^2 + 2 \cdot 0 & 0 \\ 0 \end{array} \qquad \begin{array}{c|c} 7x^2 + 2x = 0 \\ \hline 7\left(-\dfrac{2}{7}\right)^2 + 2\left(-\dfrac{2}{7}\right) & 0 \\ 7\left(\dfrac{4}{49}\right) - \dfrac{4}{7} \\ \dfrac{4}{7} - \dfrac{4}{7} \\ 0 \end{array}$$

The solutions are 0 and $-\frac{2}{7}$.

When we use the principle of zero products, we need not check except to detect errors in solving.

Example 2 Solve: $20x^2 - 15x = 0$.

$$5x(4x - 3) = 0$$

$5x = 0$ or $4x - 3 = 0$

$x = 0$ or $4x = 3$

$x = 0$ or $x = \frac{3}{4}$

The solutions are 0 and $\frac{3}{4}$.

A quadratic equation of this type will always have 0 as one solution and a nonzero number as the other solution.

DO EXERCISES 1 AND 2.

⬛⬛• EQUATIONS OF THE TYPE $ax^2 + bx + c = 0$

When neither b nor c is 0, we can sometimes solve by factoring.

Example 3 Solve: $5x^2 - 8x + 3 = 0$.

$$(5x - 3)(x - 1) = 0 \qquad \text{Factoring}$$

$5x - 3 = 0$ or $x - 1 = 0$

$5x = 3$ or $x = 1$

$x = \frac{3}{5}$ or $x = 1$

The solutions are $\frac{3}{5}$ and 1.

OBJECTIVES

After finishing Section 10.2, you should be able to:

⬛• Solve equations of the type $ax^2 + bx = 0$, $a \neq 0$, $b \neq 0$, by factoring.

⬛⬛• Solve quadratic equations of the type $ax^2 + bx + c = 0$, $a \neq 0$, $b \neq 0$, $c \neq 0$, by factoring.

⬛⬛⬛ Solve applied problems involving quadratic equations.

Solve.

1. $3x^2 + 5x = 0$

2. $10x^2 - 6x = 0$

Solve.

3. $3x^2 + x - 2 = 0$

4. $(x - 1)(x + 1) = 5(x - 1)$

5. Use $d = \dfrac{n^2 - 3n}{2}$.

a) A heptagon has 7 sides. How many diagonals does it have?

b) A polygon has 44 diagonals. How many sides does it have?

Example 4   Solve: $(y - 3)(y - 2) = 6(y - 3)$.

We write standard form and then try to factor.

$$y^2 - 5y + 6 = 6y - 18$$
$$y^2 - 11y + 24 = 0 \qquad \text{Standard form}$$
$$(y - 8)(y - 3) = 0$$
$$y - 8 = 0 \quad \text{or} \quad y - 3 = 0$$
$$y = 8 \quad \text{or} \qquad y = 3$$

The solutions are 8 and 3.

DO EXERCISES 3 AND 4.

●●● SOLVING PROBLEMS

Example 5   The number of diagonals, d, of a polygon of n sides is given by

$$d = \frac{n^2 - 3n}{2}.$$

(a) An octagon has 8 sides. How many diagonals does it have? **(b)** A polygon has 27 diagonals. How many sides does it have?

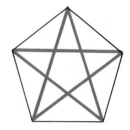

a) We substitute 8 for n:

$$d = \frac{n^2 - 3n}{2}$$

$$d = \frac{8^2 - 3 \cdot 8}{2} \qquad \text{Substituting 8 for } n$$

$$d = \frac{64 - 24}{2} = 20.$$

An octagon has 20 diagonals.

b) We substitute 27 for d and solve for n:

$$\frac{n^2 - 3n}{2} = d$$

$$\frac{n^2 - 3n}{2} = 27 \qquad \text{Substituting 27 for } d$$

$$n^2 - 3n = 54 \qquad \text{Multiplying by 2 to clear of fractions}$$
$$n^2 - 3n - 54 = 0$$
$$(n - 9)(n + 6) = 0$$
$$n - 9 = 0 \quad \text{or} \quad n + 6 = 0$$
$$n = 9 \quad \text{or} \qquad n = -6.$$

Since the number of sides cannot be negative, -6 cannot be a solution. But 9 checks, so the polygon has 9 sides (it is a nonagon).

DO EXERCISE 5.

NAME _____   CLASS _____   ANSWERS

EXERCISE SET 10.2

▪️ Solve.

1. $x^2 + 7x = 0$ **2.** $x^2 - 5x = 0$ **3.** $3x^2 + 2x = 0$ **4.** $5x^2 + 2x = 0$

5. $4x^2 + 4x = 0$ **6.** $10x^2 - 30x = 0$ **7.** $55x^2 - 11x = 0$ **8.** $14x^2 - 3x = 0$

▪️▪️ Solve.

9. $x^2 - 16x + 48 = 0$ **10.** $x^2 + 7x + 6 = 0$

11. $x^2 + 4x - 21 = 0$ **12.** $x^2 - 9x + 14 = 0$

13. $x^2 + 10x + 25 = 0$ **14.** $x^2 - 2x + 1 = 0$

15. $2x^2 - 13x + 15 = 0$ **16.** $3a^2 - 10a - 8 = 0$

1. _____

2. _____

3. _____

4. _____

5. _____

6. _____

7. _____

8. _____

9. _____

10. _____

11. _____

12. _____

13. _____

14. _____

15. _____

16. _____

Copyright © 1979, Philippines copyright 1979, by Addison-Wesley Publishing Company, Inc. All rights reserved.

ANSWERS

17. $3x^2 - 7x = 20$

18. $2x^2 + 12x = -10$

17. _____

19. $t(t - 5) = 14$

20. $3y^2 + 8y = 12y - 15$

18. _____

21. $t(9 + t) = 4(2t + 5)$

22. $(2x - 3)(x + 1) = 4(2x - 3)$

19. _____

20. _____

●●● Solve.

Use $d = \dfrac{n^2 - 3n}{2}$ for Exercises 23–26.

21. _____

23. A hexagon has 6 sides. How many diagonals does it have?

22. _____

24. A decagon has 10 sides. How many diagonals does it have?

23. _____

24. _____

25. A polygon has 14 diagonals. How many sides does it have?

25. _____

26. A polygon has 9 diagonals. How many sides does it have?

26. _____

27. _____

27. Solve for x.

$ax^2 + bx = 0$

28. 🖩 Solve.

$0.0025x^2 + 70,400x = 0$

28. _____

10.3   SQUARES OF BINOMIALS

● ● SOLVING EQUATIONS OF THE TYPE $(x + k)^2 = d$

In equations of the type $(x + k)^2 = d$, we have the square of a binomial equal to a constant.

Example 1   Solve: $(x - 5)^2 = 9$.

$$|x - 5| = \sqrt{9} \qquad \text{Taking principal square root}$$
$$|x - 5| = 3$$

$$x - 5 = 3 \quad \text{or} \quad x - 5 = -3$$
$$x = 8 \quad \text{or} \qquad x = 2$$

The solutions are 8 and 2.

Example 2   Solve: $(x + 2)^2 = 7$.

$$|x + 2| = \sqrt{7}$$

$$x + 2 = \sqrt{7} \qquad \text{or} \quad x + 2 = -\sqrt{7}$$
$$x = -2 + \sqrt{7} \quad \text{or} \qquad x = -2 - \sqrt{7}$$

The solutions are $-2 + \sqrt{7}$ and $-2 - \sqrt{7}$, or simply $-2 \pm \sqrt{7}$ (read "-2 plus or minus $\sqrt{7}$").

DO EXERCISES 1–3.

● ● MAKING $x^2 + bx$ THE SQUARE OF A BINOMIAL

We can make $x^2 + 10x$ the square of a binomial by adding the proper number to it.*

$$x^2 + 10x$$
$$\frac{10}{2} = 5 \qquad \text{Taking half the } x\text{-coefficient}$$

$$5^2 = 25 \qquad \text{Squaring}$$
$$x^2 + 10x + 25 \qquad \text{Adding}$$

The trinomial $x^2 + 10x + 25$ is the square of $x + 5$.

Examples   Make each the square of a binomial by adding the proper number to it.

3. $x^2 - 12x$

$$\left(\frac{-12}{2}\right)^2 = (-6)^2 = 36 \qquad \begin{array}{l}\text{Taking half the } x\text{-coefficient and}\\ \text{squaring}\end{array}$$

$$x^2 - 12x + 36$$

The trinomial $x^2 - 12x + 36$ is the square of $x - 6$.

---

*See p. 217.

OBJECTIVES

After finishing Section 10.3, you should be able to:

● ● **Solve equations of the type $(x + k)^2 = d$.**

● ● **Make $x^2 + bx$ the square of a binomial.**

● ● ● **Use quadratic equations to solve certain interest problems.**

Solve.

1. $(x - 3)^2 = 16$

2. $(x + 3)^2 = 10$

3. $(x - 1)^2 = 5$

ANSWERS ON PAGE A–31

Make each the square of a binomial.

4. $x^2 - 8x$

5. $x^2 + 10x$

6. $x^2 + 7x$

7. $x^2 - 3x$

Solve.

8. $1000 invested at 8% compounded annually for 2 years will grow to what amount?

4. $x^2 - 5x$

$$\left(\frac{-5}{2}\right)^2 = \frac{25}{4}$$

$$x^2 - 5x + \frac{25}{4}$$

The trinomial $x^2 - 5x + \frac{25}{4}$ is the square of $x - \frac{5}{2}$.

DO EXERCISES 4–7.

▪▪▪ APPLICATIONS: INTEREST PROBLEMS

If you put money in a savings account, the bank will pay you interest. At the end of a year, the bank will start paying you interest on both the original amount and the interest. This is called *compounding interest annually.*

> An amount of money P is invested at interest rate r. In t years it will grow to the amount A given by
> $$A = P(1 + r)^t.$$

Example 5 $1000 invested at 7% for 2 years compounded annually will grow to what amount?

$A = P(1 + r)^t$
$A = 1000(1 + 0.07)^2$ Substituting into the formula
$A = 1000(1.07)^2$
$A = 1000(1.1449)$
$A = 1144.90$ Computing

The amount is $1144.90.

DO EXERCISE 8.

Example 6 $2560 is invested at interest rate r compounded annually. In 2 years it grows to $2890. What is the interest rate?

We substitute 2560 for P, 2890 for A, and 2 for t in the formula, and solve for r.

$$A = P(1 + r)^t$$
$$2890 = 2560(1 + r)^2$$
$$\frac{2890}{2560} = (1 + r)^2$$
$$\frac{289}{256} = (1 + r)^2$$
$$\sqrt{\frac{289}{256}} = |1 + r|$$ Taking square root
$$\frac{17}{16} = |1 + r|$$

$$\frac{17}{16} = 1 + r \quad \text{or} \quad -\frac{17}{16} = 1 + r$$

$$-\frac{16}{16} + \frac{17}{16} = r \quad \text{or} \quad -\frac{16}{16} - \frac{17}{16} = r$$

$$\frac{1}{16} = r \quad \text{or} \quad -\frac{33}{16} = r$$

Since the interest rate cannot be negative,

$$0.0625 = r \qquad \left(\frac{1}{16} = 0.0625\right)$$
or $6.25\% = r.$

The interest rate must be 6.25% for \$2560 to grow to \$2890 in 2 years.

DO EXERCISE 9.

Example 7   For \$2000 to double itself in 2 years, what would the interest rate have to be?

We substitute 2000 for P, 4000 for A, and 2 for t in the formula.

$$A = P(1 + r)^t$$

$$4000 = 2000(1 + r)^2$$

$$\frac{4000}{2000} = (1 + r)^2$$

$$2 = (1 + r)^2$$

$$\sqrt{2} = |1 + r| \qquad \text{Taking square root}$$

$$\sqrt{2} = 1 + r \quad \text{or} \qquad -\sqrt{2} = 1 + r$$

$$-1 + \sqrt{2} = r \qquad \text{or} \quad -1 - \sqrt{2} = r$$

Since the interest rate cannot be negative,

$$-1 + (1.414) \approx r \qquad \text{Using Table 1}$$

$$0.414 \approx r$$

$$41.4\% \approx r$$

The interest rate would have to be 41.4% for the \$2000 to double itself. Such a rate would be hard to get.

DO EXERCISE 10.

9. Suppose \$2560 is invested at interest rate r compounded annually, and grows to \$3240 in 2 years. What is the interest rate?

10. Suppose \$1000 is invested at interest rate r compounded annually, and grows to \$3000 (it triples) in 2 years. What is the interest rate?

SOMETHING EXTRA

CALCULATOR CORNER: COMPOUND INTEREST

We have considered the formula

$$A = P(1 + r)^t$$

for interest compounded annually. If interest is compounded more than once a year, say quarterly, we can find a formula like the one above as follows:

$$A = P(1 + r)^t$$

The number of times interest is compounded goes from t to $4t$.

Each time interest is compounded the rate used is $r/4$.

$$\rightarrow A = P\left(1 + \frac{r}{4}\right)^{4t}$$

In general:

> **If principal P is invested at interest rate r, compounded n times a year, in t years it will grow to the amount A given by**
>
> $$A = P\left(1 + \frac{r}{n}\right)^{nt}.$$

Example   Suppose $1000 is invested at 8%, compounded quarterly. How much is in the account at the end of 2 years?

Substituting 1000 for P, 0.08 for r, 4 for n, and 2 for t, we get

$$A = 1000\left(1 + \frac{0.08}{4}\right)^{4 \cdot 2} = 1000(1.02)^8$$

$$\approx 1000(1.17166) \approx \$1171.66.$$

EXERCISES

Use your calculator. A power key $\boxed{a^b}$ will be needed for many of the exercises. Use 365 days for one year.

1. Suppose $1000 is invested at 8%, compounded semiannually ($n = 2$). How much is in the account at the end of 2 years?

2. Suppose $1000 is invested at 7%, compounded quarterly. How much is in the account at the end of 3 years?

3. Suppose $1000 is invested at 8%. How much is in the account at the end of 1 year, if interest is compounded (a) annually? (b) semiannually? (c) quarterly? (d) daily? (e) hourly?

4. Suppose $1 is invested at the interest rate of 100%, even though it would be hard to get such a rate. How much is in the account at the end of 1 year, if interest is compounded (a) annually? (b) semiannually? (c) quarterly? (d) daily? (e) hourly?

NAME   CLASS   ANSWERS

EXERCISE SET 10.3

• Solve.

1. $(x + 2)^2 = 25$ **2.** $(x - 2)^2 = 49$ **3.** $(x + 1)^2 = 6$

4. $(x + 3)^2 = 21$ **5.** $(x - 3)^2 = 6$ **6.** $(x + 13)^2 = 8$

• • Make each the square of a binomial by adding the proper number to it.

7. $x^2 - 2x$ **8.** $x^2 - 4x$ **9.** $x^2 + 18x$ **10.** $x^2 + 22x$

11. $x^2 - x$ **12.** $x^2 + x$ **13.** $x^2 + 5x$ **14.** $x^2 - 9x$

• • • Solve. Use $A = P(1 + r)^t$ for Exercises 15–25. What is the interest rate?

15. $1000 grows to $1210 in 2 years.   **16.** $1000 grows to $1440 in 2 years.

ANSWERS
1.
2.
3.
4.
5.
6.
7.
8.
9.
10.
11.
12.
13.
14.
15.
16.

Copyright © 1979, Philippines copyright 1979, by Addison-Wesley Publishing Company, Inc. All rights reserved.

17. _____

18. _____

19. _____

20. _____

21. _____

22. _____

23. _____

24. _____

25. _____

17. $2560 grows to $3610 in 2 years.

18. $4000 grows to $4410 in 2 years.

19. $6250 grows to $7290 in 2 years.

20. $6250 grows to $6760 in 2 years.

21. $2500 grows to $3600 in 2 years.

22. $1600 grows to $2500 in 2 years.

23. [▦] $1000 is invested at interest rate r. In 2 years it grows to $1188.10. What is the interest rate?

24. [▦] $4000 is invested at interest rate r. In 2 years it grows to $4928.40. What is the interest rate?

25. [▦] In two years you want to have $3000. How much do you need to invest now if you can get an interest rate of 8.75% compounded annually?

10.4 COMPLETING THE SQUARE

●   If we can write a quadratic equation in the form $(x + k)^2 = d$, we can solve it by taking the principal square root.

Example 1   Solve: $x^2 + 6x + 8 = 0$.
$$x^2 + 6x \qquad = -8 \qquad \text{Adding } -8$$

We take half of 6 and square it, to get 9. Then we add 9 on *both* sides of the equation. This makes the left side the square of a binomial.

$$x^2 + 6x + \boxed{9} = -8 + \boxed{9}$$
$$(x + 3)^2 = 1$$
$$|x + 3| = \sqrt{1} \qquad \text{Taking square root}$$
$$|x + 3| = 1$$
$$x + 3 = 1 \quad \text{or} \quad x + 3 = -1$$
$$x = -2 \quad \text{or} \qquad x = -4$$

The solutions are -2 and -4.

This method of solving is called *completing the square*.

Example 2   Solve $x^2 - 4x - 7 = 0$ by completing the square.
$$x^2 - 4x \qquad = 7 \qquad \text{Adding } 7$$
$$x^2 - 4x + \boxed{4} = 7 + \boxed{4} \qquad \text{Adding } 4: \left(\frac{-4}{2}\right)^2 = (-2)^2 = 4$$
$$(x - 2)^2 = 11$$
$$|x - 2| = \sqrt{11}$$
$$x - 2 = \sqrt{11} \qquad \text{or} \quad x - 2 = -\sqrt{11}$$
$$x = 2 + \sqrt{11} \quad \text{or} \qquad x = 2 - \sqrt{11}$$

The solutions are $2 \pm \sqrt{11}$.

DO EXERCISES 1–3.

Example 3   Solve by completing the square.
$$x^2 + 3x - 10 = 0$$
$$x^2 + 3x \qquad = 10$$
$$x^2 + 3x + \boxed{\frac{9}{4}} = 10 + \boxed{\frac{9}{4}} \qquad \text{Adding } \frac{9}{4}: \left(\frac{3}{2}\right)^2 = \frac{9}{4}$$
$$\left(x + \frac{3}{2}\right)^2 = \frac{40}{4} + \frac{9}{4}$$
$$\left(x + \frac{3}{2}\right)^2 = \frac{49}{4}$$
$$\left|x + \frac{3}{2}\right| = \sqrt{\frac{49}{4}} = \frac{7}{2}$$

OBJECTIVE

After finishing Section 10.4, you should be able to:

● Solve quadratic equations by completing the square.

Solve by completing the square.

1. $x^2 + 8x + 12 = 0$

2. $x^2 - 10x + 22 = 0$

3. $x^2 + 6x - 1 = 0$

Solve by completing the square.

4. $x^2 - 3x - 10 = 0$

5. $x^2 + 5x - 14 = 0$

Solve by completing the square.

6. $2x^2 + 3x - 3 = 0$

7. $3x^2 - 2x - 3 = 0$

$$x + \frac{3}{2} = \frac{7}{2} \quad \text{or} \quad x + \frac{3}{2} = -\frac{7}{2}$$

$$x = \frac{4}{2} \quad \text{or} \quad x = -\frac{10}{2}$$

$$x = 2 \quad \text{or} \quad x = -5$$

The solutions are 2 and -5.

DO EXERCISES 4 AND 5.

When the coefficient of x^2 is not 1, we can make it 1.

Example 4   Solve $2x^2 - 3x - 1 = 0$ by completing the square.

$$x^2 - \frac{3}{2}x - \frac{1}{2} = 0 \qquad \text{Multiplying by } \frac{1}{2} \text{ to make the } x^2\text{-coefficient 1}$$

$$x^2 - \frac{3}{2}x = \frac{1}{2}$$

$$x^2 - \frac{3}{2}x + \boxed{\frac{9}{16}} = \frac{1}{2} + \boxed{\frac{9}{16}} \qquad \text{Adding } \frac{9}{16}: \left[\frac{1}{2}\left(-\frac{3}{2}\right)\right]^2 = \left[-\frac{3}{4}\right]^2 = \frac{9}{16}$$

$$\left(x - \frac{3}{4}\right)^2 = \frac{8}{16} + \frac{9}{16}$$

$$\left(x - \frac{3}{4}\right)^2 = \frac{17}{16}$$

$$\left|x - \frac{3}{4}\right| = \sqrt{\frac{17}{16}}$$

$$\left|x - \frac{3}{4}\right| = \frac{\sqrt{17}}{4}$$

$$x - \frac{3}{4} = \frac{\sqrt{17}}{4} \qquad \text{or} \qquad x - \frac{3}{4} = -\frac{\sqrt{17}}{4}$$

$$x = \frac{3}{4} + \frac{\sqrt{17}}{4} \quad \text{or} \qquad x = \frac{3}{4} - \frac{\sqrt{17}}{4}$$

The solutions are $\dfrac{3 \pm \sqrt{17}}{4}$.

DO EXERCISES 6 AND 7.

EXERCISE SET 10.4

  ■●   Solve by completing the square. Show your work.

1. $x^2 - 6x - 16 = 0$ **2.** $x^2 + 8x + 15 = 0$ **3.** $x^2 + 22x + 21 = 0$

1. _____

2. _____

3. _____

4. $x^2 + 14x - 15 = 0$ **5.** $x^2 - 2x - 5 = 0$ **6.** $x^2 - 4x - 11 = 0$

4. _____

5. _____

6. _____

7. $x^2 - 18x + 74 = 0$ **8.** $x^2 - 22x + 102 = 0$ **9.** $x^2 + 7x - 18 = 0$

7. _____

8. _____

9. _____

ANSWERS

10. $x^2 + 5x - 6 = 0$ **11.** $x^2 + x - 6 = 0$ **12.** $x^2 + 10x - 4 = 0$

10. _____

11. _____

13. $x^2 - 7x - 2 = 0$ **14.** $x^2 + 3x - 28 = 0$ **15.** $x^2 + \dfrac{3}{2}x - \dfrac{1}{2} = 0$

12. _____

13. _____

14. _____

16. $2x^2 + 3x - 17 = 0$ **17.** $3x^2 + 4x - 1 = 0$ **18.** $2x^2 - 9x - 5 = 0$

15. _____

16. _____

17. _____

19. $4x^2 + 12x - 7 = 0$ **20.** $9x^2 - 6x - 9 = 0$

18. _____

19. _____

20. _____

10.5 THE QUADRATIC FORMULA

◼◦ SOLVING BY USING THE QUADRATIC FORMULA

Each time you solve by completing the square, you do about the same thing. In situations like this in mathematics, when we do about the same kind of computation many times, we look for a formula so we can speed up our work. Consider any quadratic equation in standard form:

$$ax^2 + bx + c = 0, \quad a > 0.$$

Let's solve by completing the square.

$$x^2 + \frac{b}{a}x + \frac{c}{a} = 0 \qquad \text{Multiplying by } \frac{1}{a}$$

$$x^2 + \frac{b}{a}x \quad\;\; = -\frac{c}{a} \qquad \text{Adding } -\frac{c}{a}$$

Half of $\frac{b}{a}$ is $\frac{b}{2a}$. The square is $\frac{b^2}{4a^2}$. We add $\boxed{\frac{b^2}{4a^2}}$ on both sides.

$$x^2 + \frac{b}{a}x + \boxed{\frac{b^2}{4a^2}} = -\frac{c}{a} + \boxed{\frac{b^2}{4a^2}}$$

$$\left(x + \frac{b}{2a}\right)^2 = -\frac{4ac}{4a^2} + \frac{b^2}{4a^2}$$

$$\left(x + \frac{b}{2a}\right)^2 = \frac{b^2 - 4ac}{4a^2}$$

$$\left|x + \frac{b}{2a}\right| = \sqrt{\frac{b^2 - 4ac}{4a^2}} \qquad \text{Taking square root}$$

$$x + \frac{b}{2a} = \sqrt{\frac{b^2 - 4ac}{4a^2}} \quad \text{or} \quad x + \frac{b}{2a} = -\sqrt{\frac{b^2 - 4ac}{4a^2}}$$

Since $a > 0$, $|a| = a$. Then

$$x + \frac{b}{2a} = \frac{\sqrt{b^2 - 4ac}}{2a} \quad \text{or} \quad x + \frac{b}{2a} = -\frac{\sqrt{b^2 - 4ac}}{2a}.$$

Thus,

$$x + \frac{b}{2a} = \pm\frac{\sqrt{b^2 - 4ac}}{2a},$$

so

$$x = -\frac{b}{2a} + \frac{\sqrt{b^2 - 4ac}}{2a} \quad \text{or} \quad x = -\frac{b}{2a} - \frac{\sqrt{b^2 - 4ac}}{2a}.$$

The solutions are given by:

> *The Quadratic Formula:* $x = \dfrac{-b \pm \sqrt{b^2 - 4ac}}{2a}.$

The solutions of a quadratic equation can always be found using the quadratic formula. They cannot always be found by factoring.

OBJECTIVES

After finishing Section 10.5, you should be able to:

◼◦ Solve quadratic equations using the quadratic formula.

◼◼ Find approximate solutions using a square root table.

Solve using the quadratic formula.

1. $2x^2 = 4 - 7x$

Example 1   Solve $5x^2 - 8x = -3$ using the quadratic formula.

First find standard form and determine a, b, and c.

$$5x^2 - 8x + 3 = 0$$
$$a = 5, b = -8, c = 3$$

Then use the quadratic formula.

$$x = \frac{-b \pm \sqrt{b^2 - 4ac}}{2a}$$

$$x = \frac{-(-8) \pm \sqrt{(-8)^2 - 4 \cdot 5 \cdot 3}}{2 \cdot 5}$$

$$x = \frac{8 \pm \sqrt{64 - 60}}{10}$$

$$x = \frac{8 \pm \sqrt{4}}{10}$$

$$x = \frac{8 \pm 2}{10}$$

$$x = \frac{8 + 2}{10} \quad \text{or} \quad x = \frac{8 - 2}{10}$$

$$x = \frac{10}{10} \quad \text{or} \quad x = \frac{6}{10}$$

$$x = 1 \quad \text{or} \quad x = \frac{3}{5}$$

The solutions are 1 and $\frac{3}{5}$.

DO EXERCISE 1.

> When $b^2 - 4ac \geqslant 0$, the equation has solutions. When $b^2 - 4ac < 0$, the equation has no real-number solutions. The expression $b^2 - 4ac$ is called the *discriminant*.

When using the quadratic formula, it is wise to compute the discriminant first. If it is negative, there are no solutions.

Example 2   Solve $3x^2 = 7 - 2x$ using the quadratic formula.

Find standard form and determine a, b, and c.

$$3x^2 + 2x - 7 = 0$$
$$a = 3, b = 2, c = -7$$

We compute the discriminant.

$$b^2 - 4ac = 2^2 - 4 \cdot 3 \cdot (-7)$$
$$= 4 + 84$$
$$= 88$$

This is positive, so there are solutions. They are given by

$$x = \frac{-2 \pm \sqrt{88}}{6} \qquad \text{Substituting into the quadratic formula}$$

$$x = \frac{-2 \pm \sqrt{4 \cdot 22}}{6}$$

$$x = \frac{-2 \pm 2\sqrt{22}}{6}$$

$$x = \frac{2(-1 \pm \sqrt{22})}{2 \cdot 3} \qquad \text{Factoring out 2 in the numerator and denominator}$$

$$x = \frac{-1 \pm \sqrt{22}}{3}.$$

The solutions are $\dfrac{-1 + \sqrt{22}}{3}$ and $\dfrac{-1 - \sqrt{22}}{3}$.

DO EXERCISE 2.

●● APPROXIMATING SOLUTIONS

A square root table can be used to approximate solutions.

Example 3   Approximate the solutions to the equation in Example 2.

From Table 1,

$$\sqrt{22} \approx 4.690$$

$$\frac{-1 + \sqrt{22}}{3} \approx \frac{-1 + 4.690}{3}$$

$$\approx \frac{3.69}{3}$$

$$\approx 1.2 \text{ to the nearest tenth;}$$

$$\frac{-1 - \sqrt{22}}{3} \approx \frac{-1 - 4.690}{3}$$

$$\approx \frac{-5.69}{3}$$

$$\approx -1.9 \text{ to the nearest tenth.}$$

DO EXERCISE 3.

Solve using the quadratic formula.

2. $5x^2 - 8x = 3$

3. Approximate the solutions to the equation in Exercise 2 above. Round to the nearest tenth.

SOMETHING EXTRA

CALCULATOR CORNER: ANNUAL PERCENTAGE RATE (APR)

Consider a car loan.

Situation: Car loan of $1000 at 7% for 1 year

Question: Couldn't the borrower put the $1000 in a savings account at 7.5% and make money?

Car loans are examples of what lending institutions call *add-on interest.* The nominal, or stated, interest rate is 7%. This is *not* the true rate, the *annual percentage rate*, APR. Lenders use the simple interest formula, $I = Prt$, and figure the loan will earn interest of 1000 × 0.07 × 1, or $70. They "add on" the $70, so you have to pay back $1070. For simplicity, suppose you pay back the loan in 4 payments. Each payment is 1070 ÷ 4, or $267.50. Your loan decreases as follows:

$1070, $802.50, $535, $267.50.

What's the catch? The lending institution *does not* allow you the full use of the $1070 for the year. The average principal you have is

$$\frac{1070 + 802.50 + 535 + 267.50}{4} = \$668.75.$$

To find the APR, we substitute 668.75 for P, 70 for I, 1 for t, and APR for r in the simple interest formula $I = Prt$. Then we solve for APR.

$$I = Prt$$
$$70 = 668.75 \cdot \text{APR} \cdot 1$$
$$\frac{70}{668.75} = \text{APR}$$
$$10.5\% = \text{APR}$$

APR = (Total interest) ÷ (Average principal)

For 12 payments in the above situation, the APR would have been 12.1%. In either case the true interest rate, or APR, is almost double the stated rate. You would not save money by putting the money in the bank. The Truth In Lending Law *requires* lenders to inform you of the APR.

EXERCISES

Find the APR. Assume these are car loans at the given add-on interest rate for 1 year and 12 payments.

1. Loan = $1000
   Add-on rate = 8%

2. Loan = $4000
   Add-on rate = 9%

NAME   CLASS   ANSWERS

EXERCISE SET 10.5

■● Solve using the quadratic formula.

1. $x^2 - 4x = 21$ **2.** $x^2 + 7x = 18$ **3.** $x^2 = 6x - 9$

1. _____

2. _____

3. _____

4. $x^2 = 8x - 16$ **5.** $3y^2 - 2y - 8 = 0$ **6.** $4y^2 + 12y = 7$

4. _____

5. _____

7. $x^2 - 9 = 0$ **8.** $x^2 - 4 = 0$ **9.** $y^2 - 10y + 26 = 4$

6. _____

7. _____

8. _____

10. $x^2 + 4x + 4 = 7$ **11.** $x^2 - 2x = 2$ **12.** $x^2 + 6x = 1$

9. _____

10. _____

11. _____

12. _____

Copyright © 1979, Philippines copyright 1979, by Addison-Wesley Publishing Company, Inc. All rights reserved.

ANSWERS

13. $4y^2 + 3y + 2 = 0$ **14.** $2t^2 + 6t + 5 = 0$ **15.** $3p^2 + 2p = 3$

13. _____

14. _____

15. _____

16. $3z^2 - 2z = 2$ **17.** $(y + 4)(y + 3) = 15$ **18.** $x^2 + (x + 2)^2 = 7$

16. _____

17. _____

● ● ●   Use Table 1 to approximate the solutions to the nearest tenth.

18. _____

19. $x^2 - 4x - 7 = 0$ **20.** $x^2 = 5$ **21.** $y^2 - 6y - 1 = 0$

19. _____

20. _____

21. _____

22. $4x^2 + 4x = 1$ **23.** $3x^2 + 4x - 2 = 0$ **24.** $2y^2 + 2y - 3 = 0$

22. _____

23. _____

24. _____

10.6 FRACTIONAL AND RADICAL EQUATIONS

● FRACTIONAL EQUATIONS

We can solve some fractional equations by first deriving a quadratic equation.

Example 1   Solve: $\dfrac{3}{x - 1} + \dfrac{5}{x + 1} = 2$.

The LCM is $(x - 1)(x + 1)$. We multiply by this.

$$(x - 1)(x + 1) \cdot \left(\frac{3}{x - 1} + \frac{5}{x + 1} \right) = 2 \cdot (x - 1)(x + 1)$$

We use the distributive law on the left.

$$(x - 1)(x + 1) \cdot \frac{3}{x - 1} + (x - 1)(x + 1) \cdot \frac{5}{x + 1} = 2(x - 1)(x + 1)$$

$$3(x + 1) + 5(x - 1) = 2(x - 1)(x + 1)$$

$$3x + 3 + 5x - 5 = 2(x^2 - 1)$$

$$8x - 2 = 2x^2 - 2$$

$$-2x^2 + 8x = 0$$

$$2x^2 - 8x = 0 \qquad \text{Multiplying by } -1$$

$$2x(x - 4) = 0 \qquad \text{Factoring}$$

$$2x = 0 \quad \text{or} \quad x - 4 = 0$$

$$x = 0 \quad \text{or} \qquad x = 4$$

Check: For 0:   For 4:

$$\dfrac{3}{x - 1} + \dfrac{5}{x + 1} = 2 \qquad \dfrac{3}{x - 1} + \dfrac{5}{x + 1} = 2$$

$$\begin{array}{c|c} \dfrac{3}{0 - 1} + \dfrac{5}{0 + 1} & 2 \\[2mm] \dfrac{3}{-1} + \dfrac{5}{1} & \\[2mm] -3 + 5 & \\[2mm] 2 & \end{array} \qquad \begin{array}{c|c} \dfrac{3}{4 - 1} + \dfrac{5}{4 + 1} & 2 \\[2mm] \dfrac{3}{3} + \dfrac{5}{5} & \\[2mm] 1 + 1 & \\[2mm] 2 & \end{array}$$

Both numbers check. The solutions are 0 and 4.

DO EXERCISE 1.

After finishing Section 10.6, you should be able to:

■ Solve certain fractional equations by first deriving a quadratic equation.

■■ Solve certain radical equations by first using the principle of squaring to derive a quadratic equation.

Solve.

1. $\dfrac{20}{x + 5} - \dfrac{1}{x - 4} = 1$

Solve.

2. $\sqrt{x + 2} = 4 - x$

●● **RADICAL EQUATIONS**

We can solve some radical equations by first using the principle of squaring to find a quadratic equation. When we do this we must be sure to check.

Example 2   Solve: $x - 5 = \sqrt{x + 7}$.

$$(x - 5)^2 = (\sqrt{x + 7})^2 \qquad \text{Principle of squaring}$$

$$x^2 - 10x + 25 = x + 7$$

$$x^2 - 11x + 18 = 0$$

$$(x - 9)(x - 2) = 0$$

$$x - 9 = 0 \quad \text{or} \quad x - 2 = 0$$

$$x = 9 \quad \text{or} \qquad x = 2$$

Check: For 9:   For 2:

$x - 5 = \sqrt{x + 7}$	$x - 5 = \sqrt{x + 7}$
$9 - 5 \mid \sqrt{9 + 7}$	$2 - 5 \mid \sqrt{2 + 7}$
$4 \mid 4$	$-3 \mid 3$

The number 9 checks, but 2 does not. Thus, the solution is 9.

DO EXERCISE 2.

Solve.

3. $\sqrt{30 - 3x} + 4 = x$

Example 3   Solve: $\sqrt{27 - 3x} + 3 = x$.

$$\sqrt{27 - 3x} = x - 3 \qquad \text{Adding } -3 \text{ to get the radical alone on one side}$$

$$(\sqrt{27 - 3x})^2 = (x - 3)^2 \qquad \text{Principle of squaring}$$

$$27 - 3x = x^2 - 6x + 9$$

$$0 = x^2 - 3x - 18 \qquad \text{We can have 0 on the left.}$$

$$0 = (x - 6)(x + 3) \qquad \text{Factoring}$$

$$x - 6 = 0 \quad \text{or} \quad x + 3 = 0$$

$$x = 6 \quad \text{or} \qquad x = -3$$

Check: For 6:   For -3:

$\sqrt{27 - 3x} + 3 = x$	$\sqrt{27 - 3x} + 3 = x$
$\sqrt{27 - 3 \cdot 6} + 3 \mid 6$	$\sqrt{27 - 3 \cdot (-3)} + 3 \mid -3$
$\sqrt{9} + 3$	$\sqrt{27 + 9} + 3$
$3 + 3$	$\sqrt{36} + 3$
6	$6 + 3$
	9

There is only one solution, 6.

DO EXERCISE 3.

NAME   CLASS

EXERCISE SET 10.6

● Solve.

1. $\dfrac{8}{x+2} + \dfrac{8}{x-2} = 3$

2. $\dfrac{24}{x-2} + \dfrac{24}{x+2} = 5$

3. $\dfrac{1}{x} + \dfrac{1}{x+6} = \dfrac{1}{4}$

4. $\dfrac{1}{x} + \dfrac{1}{x+9} = \dfrac{1}{20}$

5. $1 + \dfrac{12}{x^2 - 4} = \dfrac{3}{x-2}$

6. $\dfrac{5}{t-3} - \dfrac{30}{t^2 - 9} = 1$

7. $\dfrac{r}{r-1} + \dfrac{2}{r^2 - 1} = \dfrac{8}{r+1}$

8. $\dfrac{x+2}{x^2 - 2} = \dfrac{2}{3-x}$

9. $\dfrac{4-x}{x-4} + \dfrac{x+3}{x-3} = 0$

10. $\dfrac{y+2}{y} = \dfrac{1}{y+2}$

11. $\dfrac{x^2}{x-4} - \dfrac{7}{x-4} = 0$

12. $\dfrac{x^2}{x+3} - \dfrac{5}{x+3} = 0$

13. $x + 2 = \dfrac{3}{x+2}$

14. $x - 3 = \dfrac{5}{x-3}$

ANSWERS
1.
2.
3.
4.
5.
6.
7.
8.
9.
10.
11.
12.
13.
14.

Copyright © 1979. Philippines copyright 1979 by Addison-Wesley Publishing Company, Inc. All rights reserved.

ANSWERS

15. $\dfrac{1}{x} + \dfrac{1}{x+6} = \dfrac{1}{5}$

16. $\dfrac{1}{x} + \dfrac{1}{x+1} = \dfrac{1}{3}$

15. _____

16. _____

17. _____

●● Solve.

17. $x - 7 = \sqrt{x-5}$

18. $\sqrt{x+7} = x - 5$

19. $\sqrt{x+18} = x - 2$

18. _____

19. _____

20. _____

20. $x - 9 = \sqrt{x-3}$

21. $2\sqrt{x-1} = x - 1$

22. $x + 4 = 4\sqrt{x+1}$

21. _____

22. _____

23. $\sqrt{5x+21} = x + 3$

24. $\sqrt{27-3x} = x - 3$

25. $x = 1 + 6\sqrt{x-9}$

23. _____

24. _____

25. _____

26. $\sqrt{2x-1} + 2 = x$

27. $\sqrt{x^2+6} - x + 3 = 0$

28. $\sqrt{x^2+5} - x + 2 = 0$

26. _____

27. _____

29. $\sqrt{(p+6)(p+1)} - 2 = p + 1$

30. $\sqrt{(4x+5)(x+4)} = 2x + 5$

28. _____

29. _____

30. _____

10.7 FORMULAS

■ To solve a formula for a given letter, we try to get the letter alone on one side.

Example 1 Solve for h: $V = 3.5\sqrt{h}$ (the distance to the horizon).

$V^2 = (3.5)^2(\sqrt{h})^2$ Squaring both sides

$V^2 = 12.25h$

$\dfrac{V^2}{12.25} = h$ Multiplying by $\dfrac{1}{12.25}$ to get h alone

DO EXERCISE 1.

Example 2 Solve for g: $T = 2\pi\sqrt{\dfrac{L}{g}}$ (the period of a pendulum).

$T^2 = (2\pi)^2\left(\sqrt{\dfrac{L}{g}}\right)^2$ Squaring both sides

$T^2 = 4\pi^2\dfrac{L}{g}$

$T^2 = \dfrac{4\pi^2 L}{g}$

$gT^2 = 4\pi^2 L$ Multiplying by g, to clear of fractions

$g = \dfrac{4\pi^2 L}{T^2}$ Multiplying by $\dfrac{1}{T^2}$, to get g alone

DO EXERCISES 2 AND 3.

In most formulas the letters represent nonnegative numbers, so you don't need to use absolute values when taking square roots.

Example 3 (*Torricelli's theorem*) In hydrodynamics the speed v of a liquid leaving a tank from an orifice is related to the height h of the top of the water above the orifice by the formula

$h = \dfrac{v^2}{2g}.$

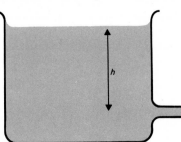

Solve for v.

$2gh = v^2$ Multiplying by $2g$, to clear of fractions

$\sqrt{2gh} = v$ Taking square root

DO EXERCISE 4.

OBJECTIVE

After finishing Section 10.7, you should be able to:

■ Solve a formula for a given letter.

1. Solve for L: $r = 2\sqrt{5L}$.

   (A formula for the speed of a skidding car)

2. Solve for L: $T = 2\pi\sqrt{\dfrac{L}{g}}$.

3. Solve for m: $c = \sqrt{\dfrac{E}{m}}$.

4. Solve for r: $A = \pi r^2$.

   (The area of a circle)

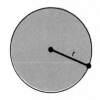

5. Solve for d: $C = P(d - 1)^2$.

6. Solve for n: $N = n^2 - n$.

7. Solve for t: $h = vt + 8t^2$.

Example 4 Solve for r:

$$A = P(1 + r)^2 \quad \text{(a compound interest formula)}.$$

$$\frac{A}{P} = (1 + r)^2 \qquad \text{Multiplying by } \frac{1}{P}$$

$$\sqrt{\frac{A}{P}} = 1 + r \qquad \text{Taking square root}$$

$$-1 + \sqrt{\frac{A}{P}} = r \qquad \text{Adding } -1, \text{ to get } r \text{ alone}$$

DO EXERCISE 5.

Sometimes you need to use the quadratic formula to solve for a given letter.

Example 5 Solve for n:

$$d = \frac{n^2 - 3n}{2} \quad \text{(the number of diagonals of a polygon)}.$$

$$n^2 - 3n = 2d \qquad \text{Multiplying by 2, to clear of fractions}$$

$$n^2 - 3n - 2d = 0 \qquad \text{Finding standard form}$$

$$a = 1, \quad b = -3, \quad c = -2d \qquad \text{All letters are considered constants except } n.$$

$$n = \frac{-b \pm \sqrt{b^2 - 4ac}}{2a}$$

$$n = \frac{-(-3) \pm \sqrt{(-3)^2 - 4 \cdot 1 \cdot (-2d)}}{2 \cdot 1} \qquad \text{Substituting into the quadratic formula}$$

$$n = \frac{3 \pm \sqrt{9 + 8d}}{2}$$

DO EXERCISE 6.

Example 6 Solve for t: $S = gt + 16t^2$.

$$16t^2 + gt - S = 0 \qquad \text{Finding standard form}$$

$$a = 16, \quad b = g, \quad c = -S$$

$$t = \frac{-b \pm \sqrt{b^2 - 4ac}}{2a}$$

$$t = \frac{-g \pm \sqrt{g^2 - 4 \cdot 16 \cdot (-S)}}{2 \cdot 16} \qquad \text{Substituting into the quadratic formula}$$

$$t = \frac{-g \pm \sqrt{g^2 + 64S}}{32}$$

DO EXERCISE 7.

NAME   CLASS

EXERCISE SET 10.7

█●█ Solve for the indicated letter.

1.

1. $N = 2.5\sqrt{A}$; A

2. $T = 2\pi\sqrt{\dfrac{L}{32}}$; L

3. $Q = \sqrt{\dfrac{aT}{c}}$; T

2.

3.

4.

4. $v = \sqrt{\dfrac{2gE}{m}}$; E

5. $E = mc^2$; c

6. $S = 4\pi r^2$; r

5.

6.

7. $Q = ad^2 - cd$; d

8. $P = kA^2 + mA$; A

9. $c^2 = a^2 + b^2$; a

7.

8.

9.

10. $c = \sqrt{a^2 + b^2}$; b

11. $S = \dfrac{1}{2}gt^2$; t

12. $V = \pi r^2 h$; r

10.

11.

12.

Copyright © 1979, Philippines copyright 1979, by Addison-Wesley Publishing Company, Inc. All rights reserved.

ANSWERS

13. $A = \pi r^2 + 2\pi rh; \quad r$

14. $A = 2\pi r^2 + 2\pi rh; \quad r$

13. _____

14. _____

15. $A = \dfrac{\pi r^2 S}{360}; \quad r$

16. $H = \dfrac{D^2 N}{2.5}; \quad D$

15. _____

16. _____

17. a) _____

b) _____

17. The circumference C of a circle is given by $C = 2\pi r$.

a) Solve $C = 2\pi r$ for r.

b) Express the area $A = \pi r^2$ in terms of the circumference C.

18. a) _____

18. The distance S, in feet, traveled by a body falling from rest in t seconds is given by $S = 16t^2$.

a) Solve for t.

b) 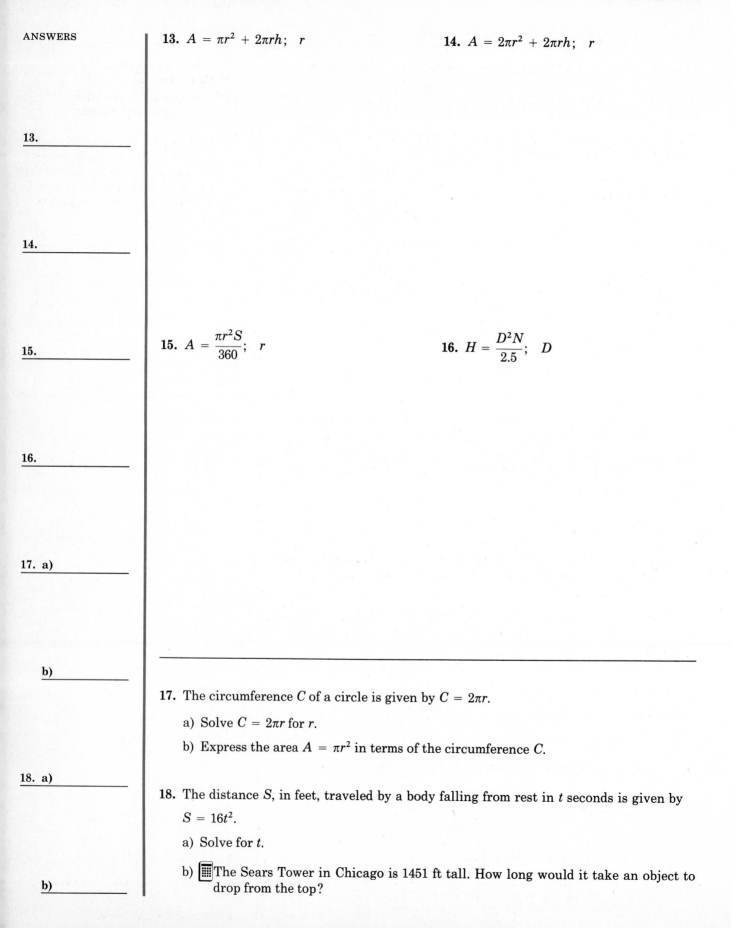The Sears Tower in Chicago is 1451 ft tall. How long would it take an object to drop from the top?

b) _____

10.8   APPLIED PROBLEMS

• We now use quadratic equations to solve more applied problems.

Example 1   A picture frame measures 20 cm by 14 cm. 160 square centimeters of picture shows. Find the width of the frame.

We first make a drawing. Let x = the width of the frame. To translate, we recall that area is length × width. Thus,

$$A = lw = 160$$
$$(20 - 2x)(14 - 2x) = 160.$$

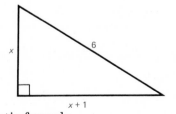

We solve.

$$280 - 68x + 4x^2 = 160$$
$$4x^2 - 68x + 120 = 0$$
$$x^2 - 17x + 30 = 0 \qquad \text{Multiplying by } \tfrac{1}{4}$$
$$(x - 15)(x - 2) = 0 \qquad \text{Factoring}$$
$$x - 15 = 0 \quad \text{or} \quad x - 2 = 0 \qquad \text{Principle of zero products}$$
$$x = 15 \quad \text{or} \qquad x = 2$$

We check in the original problem. 15 is not a solution because when $x = 15$, $20 - 2x = -10$, and the length of the picture cannot be negative. When $x = 2$, $20 - 2x = 16$. This is the length. When $x = 2$, $14 - 2x = 10$. This is the width. The area is 16 × 10, or 160. This checks, so the width of the frame is 2 cm.

DO EXERCISE 1.

Example 2   The hypotenuse of a right triangle is 6 m long. One leg is 1 m longer than the other. Find the lengths of the legs. Round to the nearest tenth.

We first make a drawing. Let x = the length of one leg. Then $x + 1$ is the length of the other leg. To translate we use the Pythagorean property.

$$x^2 + (x + 1)^2 = 6^2$$
$$x^2 + x^2 + 2x + 1 = 36$$
$$2x^2 + 2x - 35 = 0$$

Since we cannot factor, we use the quadratic formula.

$$a = 2, \quad b = 2, \quad c = -35$$
$$x = \frac{-b \pm \sqrt{b^2 - 4ac}}{2a} = \frac{-2 \pm \sqrt{2^2 - 4 \cdot 2(-35)}}{2 \cdot 2}$$
$$= \frac{-2 \pm \sqrt{4 + 280}}{4} = \frac{-2 \pm \sqrt{284}}{4}$$
$$= \frac{-2 \pm \sqrt{4 \cdot 71}}{4} = \frac{-2 \pm 2 \cdot \sqrt{71}}{2 \cdot 2} = \frac{-1 \pm \sqrt{71}}{2}$$

OBJECTIVE

After finishing Section 10.8, you should be able to:

• **Use quadratic equations to solve applied problems.**

1. A rectangular garden is 80 m by 60 m. Part of the garden is torn up to install a strip of lawn around the garden. The new area of the garden is 800 m². How wide is the strip of lawn?

2. The hypotenuse of a right triangle is 4 cm long. One leg is 1 cm longer than the other. Find the lengths of the legs. Round to the nearest tenth.

The square root table gives an approximation: $\sqrt{71} \approx 8.426$.

$$\frac{-1 + \sqrt{71}}{2} \approx 3.7 \qquad \frac{-1 - \sqrt{71}}{2} \approx -4.7$$

Since the length of a leg cannot be negative, -4.7 does not check; 3.7 does check: $3.7^2 + 4.7^2 = 13.69 + 22.09 = 35.78$ and $\sqrt{35.78} \approx 5.98 \approx 6$. Thus, one leg is about 3.7 m long and the other is about 4.7 m long.

DO EXERCISE 2.

Example 3   The current in a stream moves at a speed of 2 km/h. A boat travels 24 km upstream and 24 km downstream in a total time of 5 hours. What is the speed of the boat in still water?

First make a drawing. The distances are the same. Let r = speed of the boat in still water. Then, when traveling upstream the speed of the boat is $r - 2$. When traveling downstream, the speed of the boat is $r + 2$. We summarize in a chart.

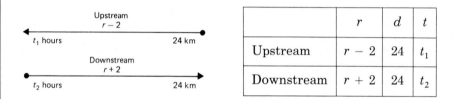

	r	d	t
Upstream	$r - 2$	24	t_1
Downstream	$r + 2$	24	t_2

3. The speed of a boat in still water is 12 km/h. The boat travels 45 km upstream and 45 km downstream in a total time of 8 hours. What is the speed of the stream? (*Hint:* Let s = the speed of the stream. Then $12 - s$ is the speed upstream and $12 + s$ is the speed downstream.)

Since $d = rt$, we know that $t = d/r$. Thus, $t_1 = 24/(r - 2)$ and $t_2 = 24/(r + 2)$. Since the total time is 5 hours, $t_1 + t_2 = 5$, and we have

$$\frac{24}{r - 2} + \frac{24}{r + 2} = 5.$$

The translation is complete. Now we solve. The LCM $= (r - 2)(r + 2)$.

$$(r - 2)(r + 2) \cdot \frac{24}{r - 2} + \frac{24}{r + 2} = (r - 2)(r + 2) \cdot 5 \qquad \text{Multiplying by the LCM}$$

$$(r - 2)(r + 2) \cdot \frac{24}{r - 2} + (r - 2)(r + 2) \cdot \frac{24}{r + 2} = (r^2 - 4)5$$

$$24(r + 2) + 24(r - 2) = 5r^2 - 20$$

$$24r + 48 + 24r - 48 = 5r^2 - 20$$

$$-5r^2 + 48r + 20 = 0$$

$$5r^2 - 48r - 20 = 0 \qquad \text{Multiplying by } -1$$

$$(5r + 2)(r - 10) = 0 \qquad \text{Factoring}$$

$$5r + 2 = 0 \quad \text{or} \quad r - 10 = 0 \qquad \text{Principle of zero products}$$

$$5r = -2 \quad \text{or} \qquad r = 10$$

$$r = -\frac{2}{5} \quad \text{or} \qquad r = 10$$

Since speed cannot be negative, $-\frac{2}{5}$ cannot be a solution. But 10 checks, so the speed of the boat in still water is 10 km/h

DO EXERCISE 3.

NAME   CLASS   ANSWERS

EXERCISE SET 10.8

■ ● ■ Solve.

1. A picture frame is 20 cm by 12 cm. There are 84 cm^2 of picture showing. Find the width of the frame.

1. _____

2. A picture frame is 18 cm by 14 cm. There are 192 cm^2 of picture showing. Find the width of the frame.

2. _____

3. The hypotenuse of a right triangle is 25 ft long. One leg is 17 ft longer than the other. Find the lengths of the legs.

3. _____

4. The hypotenuse of a right triangle is 26 yd long. One leg is 14 yd longer than the other. Find the lengths of the legs.

4. _____

5. The length of a rectangle is 2 cm greater than the width. The area is 80 cm^2. Find the length and width.

5. _____

6. The length of a rectangle is 3 m greater than the width. The area is 70 m^2. Find the length and width.

6. _____

Copyright © 1979, Philippines copyright 1979, by Addison-Wesley Publishing Company, Inc. All rights reserved.

ANSWERS

7. The width of a rectangle is 4 cm less than the length. The area is 320 cm². Find the length and width.

7. _____

8. The width of a rectangle is 3 cm less than the length. The area is 340 cm². Find the length and width.

8. _____

9. The length of a rectangle is twice the width. The area is 50 m². Find the length and width.

9. _____

10. The length of a rectangle is twice the width. The area is 32 cm². Find the length and width.

10. _____

Give approximate answers for Exercises 11–16. Round to the nearest tenth.

11. The hypotenuse of a right triangle is 8 m long. One leg is 2 m longer than the other. Find the lengths of the legs.

11. _____

12. The hypotenuse of a right triangle is 5 cm long. One leg is 2 cm longer than the other. Find the lengths of the legs.

12. _____

13. The length of a rectangle is 2 in. greater than the width. The area is 20 in². Find the length and width.

14. The length of a rectangle is 3 ft greater than the width. The area is 15 ft². Find the length and width.

15. The length of a rectangle is twice the width. The area is 10 m². Find the length and width.

16. The length of a rectangle is twice the width. The area is 20 cm². Find the length and width.

17. The current in a stream moves at a speed of 3 km/h. A boat travels 40 km upstream and 40 km downstream in a total time of 14 hours. What is the speed of the boat in still water?

18. The current in a stream moves at a speed of 3 km/h. A boat travels 45 km upstream and 45 km downstream in a total time of 8 hours. What is the speed of the boat in still water?

13. _____

14. _____

15. _____

16. _____

17. _____

18. _____

Copyright © 1979, Philippines copyright 1979, by Addison-Wesley Publishing Company, Inc. All rights reserved.

ANSWERS

19. The current in a stream moves at a speed of 4 mph. A boat travels 4 mi upstream and 12 mi downstream in a total time of 2 hours. What is the speed of the boat in still water?

19. _____

20. The current in a stream moves at a speed of 4 mph. A boat travels 5 mi upstream and 13 mi downstream in a total time of 2 hours. What is the speed of the boat in still water?

20. _____

21. The speed of a boat in still water is 10 km/h. The boat travels 12 km upstream and 28 km downstream in a total time of 4 hours. What is the speed of the stream?

21. _____

22. The speed of a boat in still water is 8 km/h. A boat travels 60 km upstream and 60 km downstream in a total time of 16 hours. What is the speed of the stream?

22. _____

23. An airplane flies 738 mi against the wind and 1062 mi with the wind in a total time of 9 hours. The speed of the airplane in still air is 200 mph. What is the speed of the wind?

23. _____

24. An airplane flies 520 km against the wind and 680 km with the wind in a total time of 4 hours. The speed of the airplane in still air is 300 km/h. What is the speed of the wind?

24. _____

Find the break-even point for each pair of total revenue and total cost equations. (See Chapter 6, pp. 281–282.)

25. $R = 2x^2 + 6x + 5,$
    $C = x^2 + 10x + 1$

26. $R = 3x^2 + 4x + 12,$
    $C = 2x^2 + 10x + 3$

Find the equilibrium point for each pair of demand and supply equations. (See Chapter 6, pp. 282–284.)

27. $D = (p - 4)^2,$
    $S = p^2 + 2p + 6$

28. $D = (p - 3)^2,$
    $S = p^2 + 2p + 1$

25. _____

26. _____

27. _____

28. _____

Copyright © 1979, Philippines copyright 1979, by Addison-Wesley Publishing Company, Inc. All rights reserved.

29. Find r in this figure. Round to the nearest hundredth.

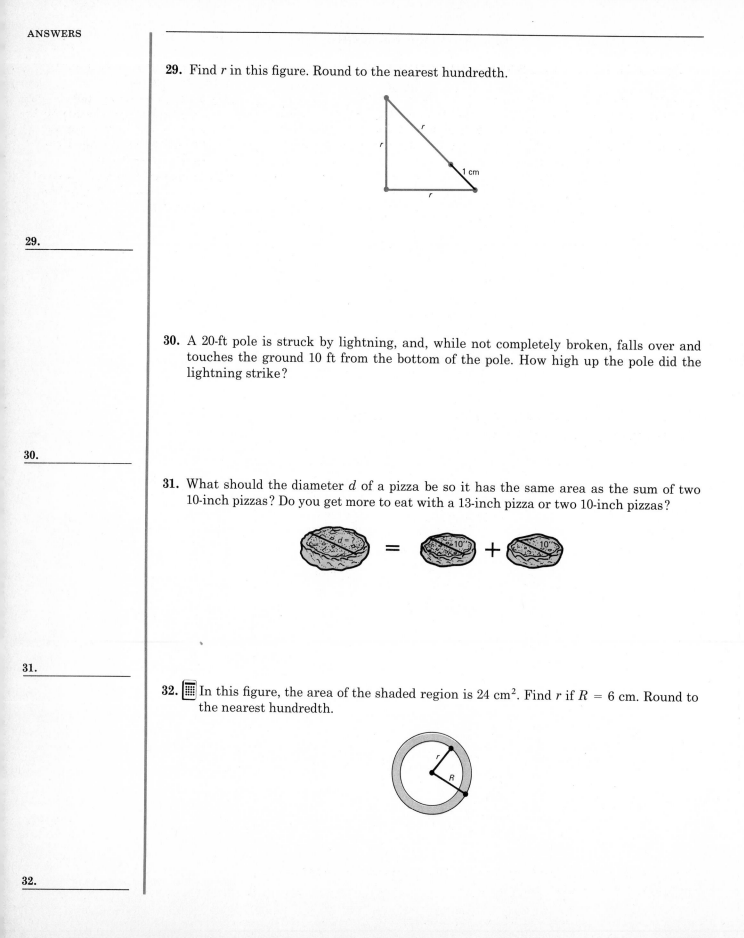

29. _____

30. A 20-ft pole is struck by lightning, and, while not completely broken, falls over and touches the ground 10 ft from the bottom of the pole. How high up the pole did the lightning strike?

30. _____

31. What should the diameter d of a pizza be so it has the same area as the sum of two 10-inch pizzas? Do you get more to eat with a 13-inch pizza or two 10-inch pizzas?

31. _____

32. In this figure, the area of the shaded region is 24 cm². Find r if $R = 6$ cm. Round to the nearest hundredth.

32. _____

10.9 GRAPHS OF QUADRATIC EQUATIONS

◖•◗ GRAPHING QUADRATIC EQUATIONS, $y = ax^2 + bx + c$

Graphs of quadratic equations, $y = ax^2 + bx + c$ (where $a \neq 0$), are always cup-shaped. They all have a *line of symmetry* like the dashed line shown in the figures below. If you fold on this line, the two halves will match exactly. The arrows show that the curve goes on forever.

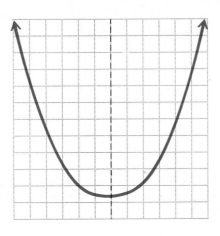

These curves are called *parabolas*. Some parabolas are thin and others are flat, but they all have the same general shape.

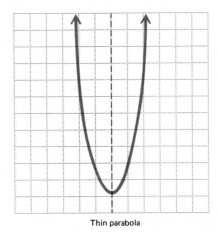

Thin parabola

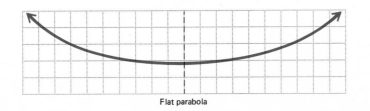

Flat parabola

To graph a quadratic equation, choose some numbers for x and compute the corresponding values of y.

OBJECTIVES

After finishing Section 10.9, you should be able to:

◖•◗ Without graphing, tell whether the graph of an equation of the type $y = ax^2 + bx + c$ opens upward or downward. Then graph the equation.

◖••◗ Approximate the solutions of $0 = ax^2 + bx + c$ by graphing.

Example 1   Graph $y = x^2 + 2x - 3$.

x	y
1	0
0	-3
-1	-4
-2	-3
-3	0
-4	5
2	5

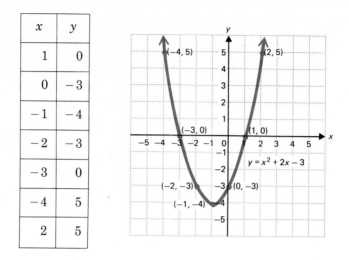

Example 2   Graph $y = -2x^2 + 3$.

x	y
0	3
1	1
-1	1
2	-5
-2	-5

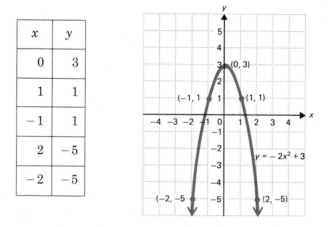

The graph in Example 1 opens upward and the coefficient of x^2 is 1, which is positive. The graph in Example 2 opens downward and the coefficient of x^2 is -2, which is negative.

> Graphs of quadratic equations $y = ax^2 + bx + c$ are all parabolas. They are *smooth* cup-shaped symmetric curves, with no sharp points or kinks in them.
>
> The graph of $y = ax^2 + bx + c$ opens upward if $a > 0$. It opens downward if $a < 0$.
>
> In drawing parabolas, be sure to plot enough points to see the general shape of each graph.

If your graphs look like any of the following, they are incorrect.

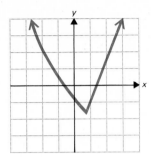

a) Sharp point is wrong.

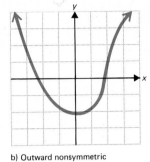

b) Outward nonsymmetric curve is wrong.

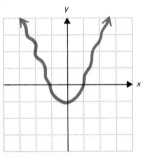

c) Kinks are wrong.

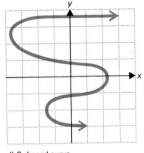

d) S-shaped curve is wrong.

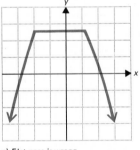

e) Flat nose is wrong.

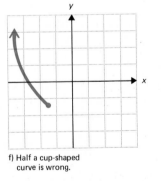

f) Half a cup-shaped curve is wrong.

DO EXERCISES 1 AND 2.

 APPROXIMATING SOLUTIONS OF $0 = ax^2 + bx + c$

Graphing can be used to approximate the solutions of

$0 = ax^2 + bx + c$.

We graph the equation $y = ax^2 + bx + c$. If the graph crosses the x-axis, the points of crossing will give us solutions.

1. a) Without graphing tell whether the graph of

   $y = x^2 + 6x + 9$

   opens upward or downward.

   b) Graph the equation.

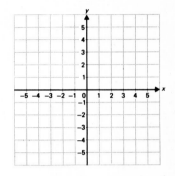

2. a) Without graphing tell whether the graph of

   $y = -3x^2 + 6x$

   opens upward or downward.

   b) Graph the equation.

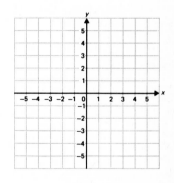

ANSWERS ON PAGE A–32

Approximate the solutions by graphing.

3. $x^2 - 4x + 4 = 0$

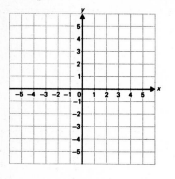

4. $-2x^2 - 4x + 1 = 0$

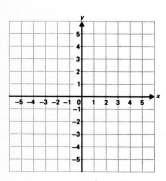

Example 3 Approximate the solutions of

$$-2x^2 + 3 = 0$$

by graphing.

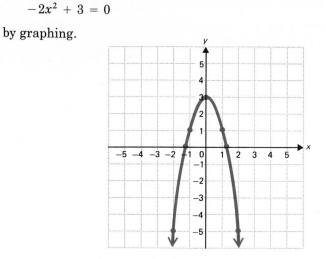

The graph was found in Example 2. The graph crosses the x-axis at about $(-1.2,0)$ and $(1.2,0)$. So the solutions are about -1.2 and 1.2.

DO EXERCISES 3 AND 4.

SOMETHING EXTRA

CALCULATOR CORNER: FINDING THE MEAN AND STANDARD DEVIATION

Statistics is used in many fields. Chances are good that you will take a statistics course later in your college program. The notions of *mean* X and *standard deviation s* are used in statistics. We show here how to find them using a calculator.

Example   Consider the data 15, 16, 22, 43, 35.

$\overline{X}$ = *mean* = the arithmetic average

$$= \frac{15 + 16 + 22 + 43 + 35}{5} = 26.2$$

To find the standard deviation we first need the average of the squares, $\overline{S}$, given by

$$\overline{S} = \frac{15^2 + 16^2 + 22^2 + 43^2 + 35^2}{5} = 807.8.$$

Then the standard deviation is the square root of $\overline{S}$ minus the square of the mean:

$$s = \sqrt{\overline{S} - (\overline{X})^2} = \sqrt{807.8 - (26.2)^2} = \sqrt{121.36} \approx 11.02.$$

EXERCISES

Find the mean and standard deviation of each set of data.

1. 16, 17, 23, 44, 36   **2.** 1, 4, 9, 9, 16, 16, 25, 25, 25, 25, 49, 49, 49, 64, 81

NAME

CLASS

EXERCISE SET 10.9

■ ● Without graphing, tell whether the graph of the equation opens upward or downward. Then graph the equation.

1. $y = x^2$

2. $y = 2x^2$

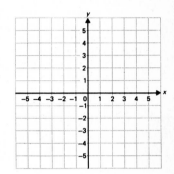

3. $y = -1 \cdot x^2$

4. $y = x^2 - 1$

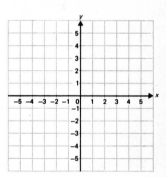

5. $y = -x^2 + 2x$

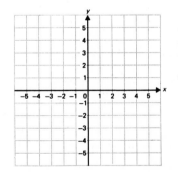

6. $y = x^2 + x - 6$

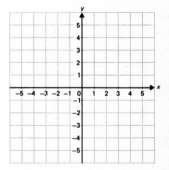

7. $y = 8 - x - x^2$

8. $y = x^2 + 2x + 1$

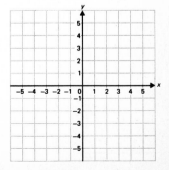

Copyright © 1979, Philippines copyright 1979, by Addison-Wesley Publishing Company, Inc. All rights reserved.

ANSWERS

15. _____

16. _____

17. _____

18. _____

19. _____

20. _____

21. _____

22. _____

9. $y = x^2 - 2x + 1$

10. $y = -\dfrac{1}{2}x^2$

11. $y = -x^2 + 2x + 3$

12. $y = -x^2 - 2x + 3$

13. $y = -2x^2 - 4x + 1$

14. $y = 2x^2 + 4x - 1$

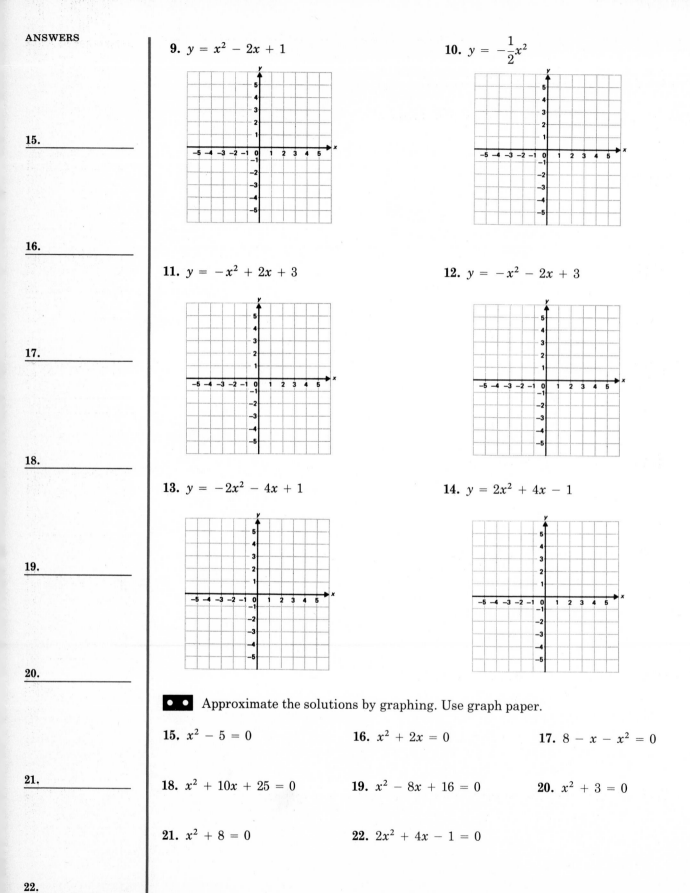

●● Approximate the solutions by graphing. Use graph paper.

15. $x^2 - 5 = 0$

16. $x^2 + 2x = 0$

17. $8 - x - x^2 = 0$

18. $x^2 + 10x + 25 = 0$

19. $x^2 - 8x + 16 = 0$

20. $x^2 + 3 = 0$

21. $x^2 + 8 = 0$

22. $2x^2 + 4x - 1 = 0$

NAME _____   SCORE _____

TEST OR REVIEW—CHAPTER 10

If you miss an item, review the indicated section and objective.

Solve.

[10.1, ● ●]   **1.** $8x^2 = 24$

1. _____

[10.2, ●]   **2.** $5x^2 - 7x = 0$

2. _____

[10.2, ● ●]   **3.** $3y^2 + 5y = 2$

3. _____

[10.3, ●]   **4.** $(x + 8)^2 = 13$

4. _____

[10.4, ●]   Solve by completing the square. Show your work.

     5. $x^2 - 2x - 10 = 0$

5. _____

[10.5, ●]   Solve using the quadratic formula.

     6. $x^2 + 6x - 9 = 0$

6. _____

[10.5, ● ●]   Use Table 1 to approximate solutions to the nearest tenth.

     7. $x^2 + 6x - 9 = 0$

7. _____

Copyright © 1979. Philippines copyright 1979. by Addison-Wesley Publishing Company, Inc. All rights reserved

ANSWERS

Solve.

[10.6, ⬛]   **8.** $\dfrac{1}{4-r} - \dfrac{1}{4+r} = \dfrac{1}{3}$

8. _____

[10.6, ⬛]   **9.** $\sqrt{x+5} = x - 1$

9. _____

[10.7, ⬛]   **10.** Solve for T: $V = 5\sqrt{\dfrac{aT}{L}}$.

10. _____

[10.3, ⬛]   **11.** \$1000 is invested at interest rate r, compounded annually. In 2 years it grows to \$1690. What is the interest rate?

11. _____

[10.8, ⬛]   **12.** The hypotenuse of a right triangle is 5 m long. One leg is 3 m longer than the other. Find the lengths of the legs. Round to the nearest tenth.

12. _____

[10.8, ⬛]   **13.** The current in a stream moves at a speed of 2 km/h. A boat travels 56 km upstream and 64 km downstream in a total time of 4 hours. What is the speed of the boat in still water?

13. _____

[10.9, ⬛]   **14.** a) Without graphing, tell whether the graph of $y = x^2 - 4x - 2$ opens upward or downward.
     b) Graph the equation.

14. a) _____

b) See graph.

INEQUALITIES
AND SETS

READINESS CHECK: SKILLS FOR CHAPTER 11

1. Graph $\dfrac{5}{3}$ on a number line.

Use the proper symbol $<$, $>$, or $=$.

2. $\dfrac{5}{2}$ $\dfrac{8}{2}$ **3.** $\dfrac{4}{9}$ $\dfrac{5}{11}$

Find the absolute value of each integer.

4. 9   **5.** 0   **6.** -8

Solve and check.

7. $x + 8 = -2$ **8.** $y - 5 = 1$

9. $-\dfrac{3}{4}x = \dfrac{1}{8}$ **10.** $-2y - 3 = 11$

11. Graph: $2x - 3y = 6$.

OBJECTIVE

After finishing Section 11.1, you should be able to:

◐ **Solve inequalities using the addition principle.**

11.1 USING THE ADDITION PRINCIPLE

◉ THE ADDITION PRINCIPLE

Consider the true inequality

$3 < 7$.

Add $\boxed{2}$ on both sides and we get another true inequality:

$3 + \boxed{2} < 7 + \boxed{2}$ or $5 < 9$.

Similarly, if we add $\boxed{-3}$ to both numbers we get another true inequality:

$3 + (\boxed{-3}) < 7 + (\boxed{-3})$ or $0 < 4$.

> *The Addition Principle for Inequalities:* **If any number is added on both sides of a true inequality, we get another true inequality.**

The addition principle holds whether the inequality contains $<$ or $>$. The symbol $\leq$ means *less than or equal to*. The symbol $\geq$ means *greater than or equal to*. The addition principle also holds for inequalities containing $\leq$ or $\geq$.

Let's see how we use the addition principle to solve inequalities.

Example 1 Solve $x + 2 > 8$.

We use the addition principle, adding $\boxed{-2}$:

$x + 2 + \boxed{-2} > 8 + \boxed{-2}$

$x > 6.$

Any number greater than 6 makes the last sentence true, hence is a solution of that sentence. Any such number is also a solution of the original sentence. Thus, we have it solved.

We cannot check all the solutions of an inequality* by substitution, as we can check solutions of equations. There are too many of them. However, we don't really need to check. Let us see why. Consider the first and last inequalities

$$x + 2 > 8 \quad \text{and} \quad x > 6.$$

Any number that makes the first one true must make the last one true. We know this by the addition principle. Now the question is, will any number that makes the last one true also be a solution of the first one? Let us use the addition principle again, adding 2 :

$$x > 6$$
$$x + \boxed{2} > 6 + \boxed{2}$$
$$x + 2 > 8.$$

Now we know that any number that makes $x > 6$ true also makes $x + 2 > 8$ true. Therefore, the sentences $x > 6$ and $x + 2 > 8$ have the same solutions. Any time we use the addition principle a similar thing happens. Thus, whenever we use the principle with inequalities the first and last sentences will have the same solutions.

Example 2 Solve: $3x + 1 < 2x - 3$.

$$3x + 1 \ \boxed{- 1} \ < 2x - 3 \ \boxed{- 1} \qquad \text{Adding } -1$$
$$3x < 2x - 4 \qquad \text{Simplifying}$$
$$3x \ \boxed{- 2x} \ < 2x - 4 \ \boxed{- 2x} \qquad \text{Adding } -2x$$
$$x < -4 \qquad \text{Simplifying}$$

Any number less than -4 is a solution. The following are some of the solutions:

$$-5, \quad -6, \quad -4.1, \quad -2045, \quad -18\pi, \quad -\sqrt{30}.$$

To describe all the solutions we use the set notation

$$\{x | x < -4\},$$

which is read:

The set of all x such that x is less than -4.

DO EXERCISES 1–3.

---

* A partial check could be done by substituting a number greater than 6, say 7, into the original inequality:

$$\begin{array}{c|c} x + 2 > 8 \\ \hline 7 + 2 & 8 \\ 9 & \end{array}$$

Since $9 > 8$ is true, 7 is a solution.

Solve. Write set notation for your answer.

1. $x + 3 > 5$

2. $x - 5 < 8$

3. $5x + 1 < 4x - 2$

Solve.

4. $x + \dfrac{2}{3} \le \dfrac{4}{5}$

Example 3   Solve: $x + \dfrac{1}{3} \ge \dfrac{5}{4}$.

$$x + \dfrac{1}{3} \boxed{-\dfrac{1}{3}} \ge \dfrac{5}{4} \boxed{-\dfrac{1}{3}} \qquad \text{Adding } -\dfrac{1}{3}$$

$$x \ge \dfrac{5}{4} \cdot \boxed{\dfrac{3}{3}} - \dfrac{1}{3} \cdot \boxed{\dfrac{4}{4}} \qquad \begin{array}{l}\text{Finding a common}\\ \text{denominator}\end{array}$$

$$x \ge \dfrac{15}{12} - \dfrac{4}{12}$$

$$x \ge \dfrac{11}{12}$$

Any number greater than or equal to $\frac{11}{12}$ is a solution. We say that the *solution set* is

$$\left\{ x \,\middle|\, x \ge \dfrac{11}{12} \right\}.$$

Here is a handy way to help read and write set notation:

$$\left\{ x \,\middle|\, x \ge \dfrac{11}{12} \right\}$$

The set of ←
all x ←
such that ←

x is greater than or equal to $\dfrac{11}{12}$ ←

DO EXERCISES 4 AND 5.

5. $5y + 2 \le -1 + 4y$

SOMETHING EXTRA

CALCULATOR CORNER: ESTIMATING $\sqrt{2}$

The following is an estimate for $\sqrt{2}$. Find it on your calculator. Give decimal notation to six decimal places.

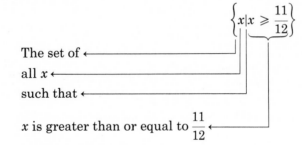

Start here: $\dfrac{1}{2} = 0.5$.

Then add 2: 2.5. Then take the reciprocal: $\dfrac{1}{2.5} = 0.4$. Then add 2, and so on.

NAME   CLASS

EXERCISE SET 11.1

● Solve using the addition principle. Write set notation for answers.

1. $x + 7 > 2$ **2.** $x + 6 > 3$ **3.** $y + 5 > 8$ **4.** $y + 7 > 9$

5. $x + 8 \leqslant -10$ **6.** $a + 12 < 6$ **7.** $x - 7 \leqslant 9$ **8.** $x - 6 > 2$

9. $y - 7 > -12$ **10.** $2x + 3 > x + 5$ **11.** $2x + 4 > x + 7$

12. $3x + 9 \leqslant 2x + 6$ **13.** $3x - 6 \geqslant 2x + 7$ **14.** $3x - 9 \geqslant 2x + 11$

15. $5x - 6 < 4x - 2$ **16.** $6x - 8 < 5x - 9$ **17.** $-7 + c > 7$

1. _____

2. _____

3. _____

4. _____

5. _____

6. _____

7. _____

8. _____

9. _____

10. _____

11. _____

12. _____

13. _____

14. _____

15. _____

16. _____

17. _____

Copyright © 1979, Philippines copyright 1979, by Addison-Wesley Publishing Company, Inc. All rights reserved.

ANSWERS

18. _____

19. _____

20. _____

21. _____

22. _____

23. _____

24. _____

25. _____

26. _____

27. _____

28. _____

29. _____

30. _____

31. _____

18. $-9 + b > 9$

19. $y + \dfrac{1}{4} \leqslant \dfrac{1}{2}$

20. $y + \dfrac{1}{3} \leqslant \dfrac{5}{6}$

21. $x - \dfrac{1}{3} > \dfrac{1}{4}$

22. $x - \dfrac{1}{8} > \dfrac{1}{2}$

23. $-14x + 21 > 21 - 15x$

24. $-10x + 15 > 18 - 11x$

25. $3(r + 2) < 2r + 4$

26. $4(r + 5) \geqslant 3r + 7$

27. $0.8x + 5 \geqslant 6 - 0.2x$

28. $0.7x + 6 \leqslant 7 - 0.3x$

---

29. A student is taking an introductory algebra course in which four tests are to be given. To get a B, a total of 320 points is needed. The student got scores of 81, 76, and 73 on the first three tests. Determine (in terms of an inequality) those scores that will allow the student to get a B.

30. If $2x - 5 \geqslant 8$, is $2x - 3 \geqslant 9$?

31. ▦ Solve: $17x + 9{,}479{,}756 \leqslant 16x - 8{,}579{,}243.$

11.2   USING THE MULTIPLICATION PRINCIPLE

● THE MULTIPLICATION PRINCIPLE

Consider the true inequality

$3 < 7.$

Multiply both numbers by 2 and we get another true inequality:

$3 \cdot 2 < 7 \cdot 2$

or

$6 < 14.$

Multiply both numbers by -3 and we get the false inequality

$3 \cdot (-3) < 7 \cdot (-3)$

or

$-9 < -21.$ False

However, if we reverse the inequality symbol we get a true inequality:

$-9 > -21.$ True

> *The Multiplication Principle for Inequalities:* **If we multiply on both sides of a true inequality by a positive number, we get another true inequality. If we multiply by a negative number and the inequality symbol is reversed, we get another true inequality.**

The multiplication principle also holds for inequalities containing $\geqslant$ or $\leqslant$.

Example 1   Solve: $4x < 28.$

$$\frac{1}{4} \cdot 4x < \frac{1}{4} \cdot 28 \qquad \text{Multiplying by } \frac{1}{4}$$

⎣——— The symbol stays the same.

$\qquad x < 7$ Simplifying

The solution set is $\{x | x < 7\}$.

DO EXERCISES 1 AND 2.

Example 2   Solve: $-2y < 18.$

$$-\frac{1}{2}(-2y) > -\frac{1}{2} \cdot 18 \qquad \text{Multiplying by } -\frac{1}{2}$$

⎣——— The symbol has to be reversed!

$\qquad y > -9$ Simplifying

The solution set is $\{y | y > -9\}$.

DO EXERCISES 3 AND 4.

OBJECTIVES

After finishing Section 11.2, you should be able to:

● Solve inequalities using the multiplication principle.

●● Solve inequalities using the addition and multiplication principles together.

Solve.

1. $8x < 64$

2. $5y \geqslant 160$

Solve.

3. $-4x \leqslant 24$

4. $-5y > 13$

Solve.

5. $7 - 4x < 8$

USING THE PRINCIPLES TOGETHER

We use the addition and multiplication principles together in solving inequalities in much the same way as in solving equations. We usually use the addition principle first.

Example 3   Solve: $6 - 5y > 7$.

$$\boxed{-6} + 6 - 5y > \boxed{-6} + 7 \qquad \text{Adding } -6$$

$$-5y > 1 \qquad \text{Simplifying}$$

$$\boxed{-\frac{1}{5}} \cdot (-5y) < \boxed{-\frac{1}{5}} \cdot 1 \qquad \text{Multiplying by } -\frac{1}{5}$$

$$\underset{\text{be reversed!}}{\underline{\hspace{1cm}} \text{The symbol has to}}$$

$$y < -\frac{1}{5}$$

The solution set is $\{y | y < -\frac{1}{5}\}$.

DO EXERCISE 5.

Solve.

6. $13x + 5 \leqslant 12x + 4$

Example 4   Solve: $5x + 9 \leqslant 4x + 3$.

$$5x + 9 \boxed{-9} \leqslant 4x + 3 \boxed{-9} \qquad \text{Adding } -9$$

$$5x \leqslant 4x - 6 \qquad \text{Simplifying}$$

$$5x \boxed{-4x} \leqslant 4x - 6 \boxed{-4x} \qquad \text{Adding } -4x$$

$$x \leqslant -6 \qquad \text{Simplifying}$$

The solution set is $\{x | x \leqslant -6\}$.

DO EXERCISE 6.

Solve.

7. $24 - 7y \geqslant 11y - 14$

Example 5   Solve: $17 - 5y < 8y - 5$.

$$17 - 5y \boxed{-17} < 8y - 5 \boxed{-17} \qquad \text{Adding } -17$$

$$-5y < 8y - 22 \qquad \text{Simplifying}$$

$$\boxed{-8y} - 5y < \boxed{-8y} + 8y - 22 \qquad \text{Adding } -8y$$

$$-13y < -22 \qquad \text{Simplifying}$$

$$\boxed{-\frac{1}{13}} \cdot (-13y) > \boxed{-\frac{1}{13}} \cdot (-22) \qquad \text{Multiplying by } -\frac{1}{13}$$

$$y > \frac{22}{13}$$

The solution set is $\{y | y > \frac{22}{13}\}$.

DO EXERCISE 7.

NAME

CLASS

ANSWERS

EXERCISE SET 11.2

▪ Solve using the multiplication principle.

1. $5x < 35$

2. $8x \geqslant 32$

3. $9y \leqslant 81$

4. $10x > 240$

5. $7x < 13$

6. $8y < 17$

7. $12x > -36$

8. $16x < -64$

9. $5y \geqslant -2$

10. $7x > -4$

11. $-2x \leqslant 12$

12. $-3y \leqslant 15$

13. $-4y \geqslant -16$

14. $-7x < -21$

15. $-3x < -17$

16. $-5y > -23$

17. $-2y > \dfrac{1}{7}$

18. $-4x \leqslant \dfrac{1}{9}$

19. $-\dfrac{6}{5} \leqslant -4x$

20. $-\dfrac{7}{8} > -56t$

1. _____

2. _____

3. _____

4. _____

5. _____

6. _____

7. _____

8. _____

9. _____

10. _____

11. _____

12. _____

13. _____

14. _____

15. _____

16. _____

17. _____

18. _____

19. _____

20. _____

Copyright © 1979, Philippines copyright 1979, by Addison-Wesley Publishing Company, Inc. All rights reserved.

ANSWERS

⚫⚫ Solve using the addition and multiplication principles.

21. $2x + 5 < 3$ **22.** $3x - 4 > 5$

21. _____

22. _____

23. $-3x + 7 \geqslant -2$ **24.** $-4x - 3 \leqslant -5$

23. _____

24. _____

25. $6t - 8 \leqslant 4t + 1$ **26.** $5b + 7 \geqslant b - 1$

25. _____

26. _____

27. $3 - 6c > 15$ **28.** $5 - 3y < -4$

27. _____

28. _____

29. $15x - 21 \geqslant 8x + 7$ **30.** $21 - 15x < -8x - 7$

29. _____

30. _____

31. Badger Rent-a-Car rents compact cars for $13.95 plus 10¢ per mile. Thirsty Rent-a-Car rents compact cars for $12.95 plus 12¢ per mile. For what mileages would the cost of renting a compact be cheaper at Badger?

31. _____

32. Badger Rent-a-Car rents an intermediate-size car at a daily rate of $14.95 plus $0.10 per mile. A businessperson is not to exceed a daily car rental budget of $76. Determine (in terms of an inequality) those mileages that will allow the businessperson to stay within the budget.

32. _____

11.3   GRAPHS OF INEQUALITIES

⬤ INEQUALITIES IN ONE VARIABLE

We graph inequalities in one variable on a number line.

Example 1   Graph $x < 2$.

The solutions of $x < 2$ are those numbers less than 2. They are shown on the graph by shading all points to the left of 2.

Example 2   Graph $y \geqslant -3$.

The solutions of $y \geqslant -3$ are shown by shading the point for -3 and all points to the right of -3.

DO EXERCISES 1 AND 2.

Example 3   Graph $3x + 2 < 5x - 1$.

First solve: $2 < 2x - 1$ Adding $-3x$

$3 < 2x$ Adding 1

$\dfrac{3}{2} < x$ Multiplying by $\dfrac{1}{2}$

DO EXERCISES 3 AND 4.

⬤⬤ GRAPHING INEQUALITIES WITH ABSOLUTE VALUE

Example 4   Graph $|x| < 3$.

The absolute value of a number is its distance from 0 on a number line. For the absolute value of a number to be less than 3 it must be between 3 and -3. Therefore, we shade the points between these two numbers.

DO EXERCISES 5 AND 6.

OBJECTIVES

After finishing Section 11.3, you should be able to:

⬤ Graph inequalities in one variable on a number line.

⬤⬤ Graph inequalities that contain absolute values on a number line.

⬤⬤⬤ Graph linear inequalities in two variables on a plane.

Graph on a number line.

1. $x < 4$

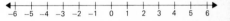

2. $y \geqslant -5$

Graph on a number line.

3. $x + 2 > 1$

4. $4x + 6 < 7x - 3$

Graph on a number line.

5. $|x| < 5$

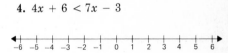

6. $|x| \leqslant 4$

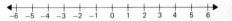

Graph on a number line.

7. $|x| \geqslant 3$

8. $|y| > 5$

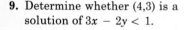

9. Determine whether $(4,3)$ is a solution of $3x - 2y < 1$.

Example 5   Graph $|x| \geqslant 2$.

For the absolute value of a number to be greater than or equal to 2 its distance from 0 must be 2 or more. Thus, the number must be 2 or greater, or it must be less than or equal to -2. Therefore, we shade the point for 2 and all points to its right. We also shade the point for -2 and all points to its left.

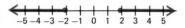

DO EXERCISES 7 AND 8.

●●● INEQUALITIES IN TWO VARIABLES

The solutions of inequalities in two variables are ordered pairs.

Example 6   Determine whether $(-3,2)$ is a solution of $5x - 4y < 13$.

We use alphabetical order of variables. We replace x by -3 and y by 2.

$$
\begin{array}{c|c}
5x - 4y < 13 & \\
\hline
5(-3) - 4 \cdot 2 & 13 \\
-15 - 8 & \\
-23 &
\end{array}
$$

Since $-23 < 13$ is true, $(-3,2)$ is a solution.

DO EXERCISE 9.

Example 7   Graph $y > x$.

We first graph the line $y = x$ for comparison. Every solution of $y = x$ is an ordered pair such as $(3,3)$. The first and second coordinates are the same. The graph of $y = x$ is shown to the left below.

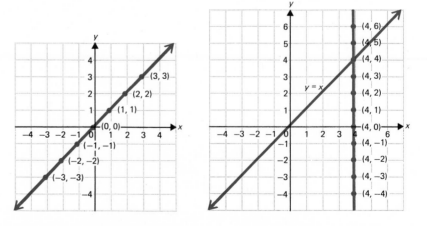

Now look at the graph to the right above. We consider a vertical line and ordered pairs on it. For all points above $y = x$, the second co-ordinate is greater than the first, $y > x$. For all points below the line, $y < x$. The same thing happens for any vertical line. Then for all points above $y = x$, the ordered pairs are solutions. We shade the half-plane above $y = x$. This is the graph of $y > x$. Points on $y = x$ are not in the graph, so we draw it dashed.

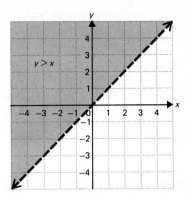

DO EXERCISE 10.

Example 8 Graph $y \leqslant x - 1$.

First sketch the line $y = x - 1$. Points on the line $y = x - 1$ are also in the graph of $y \leqslant x - 1$, so we draw the line solid. For points above the line, $y > x - 1$. These points are not in the graph. For points below the line, $y < x - 1$. These are in the graph, so we shade the lower half-plane.

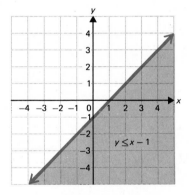

DO EXERCISE 11.

Example 9 Graph $6x - 2y < 10$.

We first solve for y.

$$-2y < -6x + 10 \qquad \text{Adding } -6x$$

$$y > -\frac{1}{2}(-6x + 10) \qquad \text{Multiplying by } -\frac{1}{2}$$

               Here the symbol has to be reversed.

$$y > 3x - 5$$

10. Graph $y < x$.

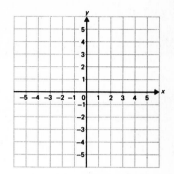

11. Graph $y \geqslant x + 2$.

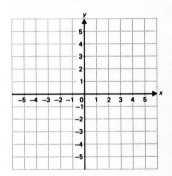

12. Graph $2x + 4y < 8$.

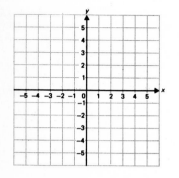

13. Graph $3x - 5y < 15$.

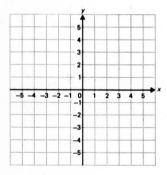

14. Graph $2x + 3y \geqslant 12$.

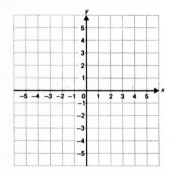

Now we graph $y = 3x - 5$ using a dashed line. We shade the half-plane above the line.

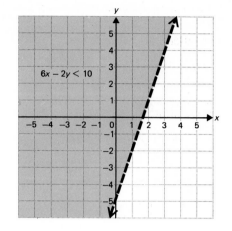

DO EXERCISE 12.

A *linear inequality* is one that we can get from a linear equation by changing the equals symbol to an inequality symbol. Every linear equation has a graph that is a straight line. The graph of a linear inequality is a half-plane, sometimes including the line along the edge. In the following example we give a different method of graphing. We graph the line using intercepts. Then we determine which side to shade by substituting a point from either half-plane.

Example 10   Graph $2x + 3y < 6$.

a) First graph the line $2x + 3y = 6$. The intercepts are $(3,0)$ and $(0,2)$. We use a dashed line for the graph since we have $<$.

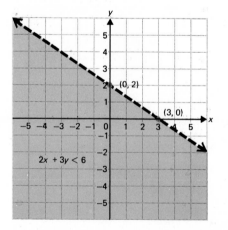

b) Pick a point that does not belong to the line. Substitute to determine whether this point is a solution. The origin $(0,0)$ is usually an easy one to use: $2 \cdot 0 + 3 \cdot 0 < 6$ is true, so the origin is a solution. This means we shade the lower half-plane. Had the substitution given us a false inequality we would have shaded the other half-plane.

If the line goes through the origin, then we must test some other point not on the line. The point $(1,1)$ is often a good one to try.

DO EXERCISES 13 AND 14.

EXERCISE SET 11.3

■ • ■ Graph on a number line.

1. $x < 5$

$$\xleftarrow{\quad} \underset{-6}{+} \underset{-5}{+} \underset{-4}{+} \underset{-3}{+} \underset{-2}{+} \underset{-1}{+} \underset{0}{+} \underset{1}{+} \underset{2}{+} \underset{3}{+} \underset{4}{+} \underset{5}{+} \underset{6}{+} \xrightarrow{\quad}$$

2. $x < 3$

$$\xleftarrow{\quad} \underset{-6}{+} \underset{-5}{+} \underset{-4}{+} \underset{-3}{+} \underset{-2}{+} \underset{-1}{+} \underset{0}{+} \underset{1}{+} \underset{2}{+} \underset{3}{+} \underset{4}{+} \underset{5}{+} \underset{6}{+} \xrightarrow{\quad}$$

3. $y \geqslant -4$

$$\xleftarrow{\quad} \underset{-6}{+} \underset{-5}{+} \underset{-4}{+} \underset{-3}{+} \underset{-2}{+} \underset{-1}{+} \underset{0}{+} \underset{1}{+} \underset{2}{+} \underset{3}{+} \underset{4}{+} \underset{5}{+} \underset{6}{+} \xrightarrow{\quad}$$

4. $x \geqslant -7$

$$\xleftarrow{\quad} \underset{-6}{+} \underset{-5}{+} \underset{-4}{+} \underset{-3}{+} \underset{-2}{+} \underset{-1}{+} \underset{0}{+} \underset{1}{+} \underset{2}{+} \underset{3}{+} \underset{4}{+} \underset{5}{+} \underset{6}{+} \xrightarrow{\quad}$$

5. $t - 3 \leqslant -7$

$$\xleftarrow{\quad} \underset{-6}{+} \underset{-5}{+} \underset{-4}{+} \underset{-3}{+} \underset{-2}{+} \underset{-1}{+} \underset{0}{+} \underset{1}{+} \underset{2}{+} \underset{3}{+} \underset{4}{+} \underset{5}{+} \underset{6}{+} \xrightarrow{\quad}$$

6. $x - 4 \leqslant -8$

$$\xleftarrow{\quad} \underset{-6}{+} \underset{-5}{+} \underset{-4}{+} \underset{-3}{+} \underset{-2}{+} \underset{-1}{+} \underset{0}{+} \underset{1}{+} \underset{2}{+} \underset{3}{+} \underset{4}{+} \underset{5}{+} \underset{6}{+} \xrightarrow{\quad}$$

7. $2x + 6 < 14$

$$\xleftarrow{\quad} \underset{-6}{+} \underset{-5}{+} \underset{-4}{+} \underset{-3}{+} \underset{-2}{+} \underset{-1}{+} \underset{0}{+} \underset{1}{+} \underset{2}{+} \underset{3}{+} \underset{4}{+} \underset{5}{+} \underset{6}{+} \xrightarrow{\quad}$$

8. $4x - 8 \geqslant 12$

$$\xleftarrow{\quad} \underset{-6}{+} \underset{-5}{+} \underset{-4}{+} \underset{-3}{+} \underset{-2}{+} \underset{-1}{+} \underset{0}{+} \underset{1}{+} \underset{2}{+} \underset{3}{+} \underset{4}{+} \underset{5}{+} \underset{6}{+} \xrightarrow{\quad}$$

9. $4y + 9 > 11y - 12$

$$\xleftarrow{\quad} \underset{-6}{+} \underset{-5}{+} \underset{-4}{+} \underset{-3}{+} \underset{-2}{+} \underset{-1}{+} \underset{0}{+} \underset{1}{+} \underset{2}{+} \underset{3}{+} \underset{4}{+} \underset{5}{+} \underset{6}{+} \xrightarrow{\quad}$$

10. $6x + 11 \leqslant 14x + 7$

$$\xleftarrow{\quad} \underset{-6}{+} \underset{-5}{+} \underset{-4}{+} \underset{-3}{+} \underset{-2}{+} \underset{-1}{+} \underset{0}{+} \underset{1}{+} \underset{2}{+} \underset{3}{+} \underset{4}{+} \underset{5}{+} \underset{6}{+} \xrightarrow{\quad}$$

■ ■ Graph on a number line.

11. $|x| < 2$

$$\xleftarrow{\quad} \underset{-6}{+} \underset{-5}{+} \underset{-4}{+} \underset{-3}{+} \underset{-2}{+} \underset{-1}{+} \underset{0}{+} \underset{1}{+} \underset{2}{+} \underset{3}{+} \underset{4}{+} \underset{5}{+} \underset{6}{+} \xrightarrow{\quad}$$

12. $|t| \leqslant 1$

$$\xleftarrow{\quad} \underset{-6}{+} \underset{-5}{+} \underset{-4}{+} \underset{-3}{+} \underset{-2}{+} \underset{-1}{+} \underset{0}{+} \underset{1}{+} \underset{2}{+} \underset{3}{+} \underset{4}{+} \underset{5}{+} \underset{6}{+} \xrightarrow{\quad}$$

13. $|a| \geqslant 4$

$$\xleftarrow{\quad} \underset{-6}{+} \underset{-5}{+} \underset{-4}{+} \underset{-3}{+} \underset{-2}{+} \underset{-1}{+} \underset{0}{+} \underset{1}{+} \underset{2}{+} \underset{3}{+} \underset{4}{+} \underset{5}{+} \underset{6}{+} \xrightarrow{\quad}$$

14. $|y| > 5$

$$\xleftarrow{\quad} \underset{-6}{+} \underset{-5}{+} \underset{-4}{+} \underset{-3}{+} \underset{-2}{+} \underset{-1}{+} \underset{0}{+} \underset{1}{+} \underset{2}{+} \underset{3}{+} \underset{4}{+} \underset{5}{+} \underset{6}{+} \xrightarrow{\quad}$$

Copyright © 1979, Philippines copyright 1979, by Addison-Wesley Publishing Company, Inc. All rights reserved.

Graph on a plane.

15. $y > x - 2$

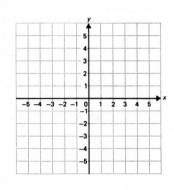

16. $y \leqslant x - 3$

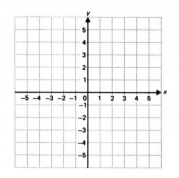

17. $6x - 2y \leqslant 12$

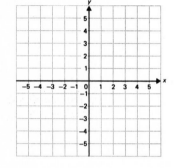

18. $2x + 3y < 12$

19. $3x - 5y \geqslant 15$

20. $5x + 2y > 10$

21. $y - 2x < 4$

22. $2x - y \leqslant 4$

11.4 SETS

◼◻ NAMING SETS

We have discussed notation like the following:

$\{x | x < 6\}$,

where we were considering the real numbers. If we were just considering whole numbers, the above set could also be named, or symbolized,

$\{0, 1, 2, 3, 4, 5\}$.

This way of naming a set is known as the *roster* method. In words, it is

The set of whole numbers 0 through 5.

DO EXERCISES 1 AND 2.

◼◼ MEMBERSHIP

The symbol $\in$ means *is a member of* or *belongs to*. Thus

$x \in A$

means

x is a member of A

or

x belongs to A.

Examples Determine whether true or false.

1. $1 \in \{1, 2, 3\}$ True

2. $1 \in \{2, 3\}$ False

3. $4 \in \{x | x$ is an even whole number$\}$ True

4. $5 \in \{x | x$ is an even whole number$\}$ False

Set membership can be illustrated with a diagram as in the figure below.

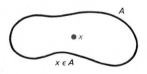

DO EXERCISES 3–6.

◼◼◼ INTERSECTIONS

The *intersection* of two sets A and B is the set of members common to both sets and is indicated by the symbol

$A \cap B$.

Thus,

$\{0, 1, 3, 5, 25\} \cap \{2, 3, 4, 5, 6, 7, 9\}$

represents the set

$\{3, 5\}$.

OBJECTIVES

After finishing Section 11.4, you should be able to:

◼◻ Name sets using the roster method.

◼◼ Determine whether a given object is a member of a set.

◼◼◼ Find the intersection of two sets.

◼◼ Find the union of two sets.

Name each set using the roster method.

1. The set of whole numbers 2 through 10.

2. The set of even whole numbers between 30 and 40.

Determine whether true or false.

3. $4 \in \{a, b, 3, 4, 5\}$

4. $c \in \{a, b, 3, 4, 5\}$

5. $\dfrac{2}{3} \in \{x | x$ is a rational number$\}$

6. Heads $\in$ The set of outcomes of flipping a penny

ANSWERS ON PAGE A–36

Find.

7. $\{a, 1, 0, t, 6, 9, 14\} \cap \{2, 1, s, a, 9\}$

8. $\{1, 2\} \cap \{2, 3\}$

Find.

9. $\{1, 2\} \cap \{3, 4\}$

10. $A \cap B$, where

$A = \{x | x$ is a rational number$\}$

and

$B = \{x | x$ is an irrational number$\}$

Find.

11. $\{0, 1, 2, 3, 4\} \cup \{2, 3, 4, 5, 6, 7\}$

12. $A \cup B$, where

$A = \{x | x$ is an even integer$\}$

and

$B = \{x | x$ is an odd integer$\}$

Set intersection is illustrated by the diagram below.

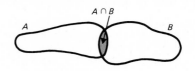

The solution of the system of equations

$$x + 2y = 7,$$
$$x - y = 4$$

is the ordered pair (5, 1). It is the intersection of the graphs of the two lines.

DO EXERCISES 7 AND 8.

The set without members is known as the *empty set*, and is often named $\varnothing$. Each of the following is a description of the empty set:

The set of all six-eyed algebra teachers;

$\{2, 3\} \cap \{5, 6, 7\}$;

$\{x | x$ is an even whole number$\} \cap \{x | x$ is an odd whole number$\}$.

The system of equations

$$x + 2y = 7,$$
$$x + 2y = 9$$

has no solution. The lines are parallel. Their intersection is empty.

DO EXERCISES 9 AND 10.

8·8 UNIONS

Two sets A and B may be combined to form a new set. It contains the members of A as well as those of B. The new set is called the *union* of A and B, and is represented by the symbol

$A \cup B$.

Thus,

$\{0, 5, 7, 13, 27\} \cup \{0, 2, 3, 4, 5\}$

represents the set

$\{0, 2, 3, 4, 5, 7, 13, 27\}$.

Set union is illustrated by the diagram below.

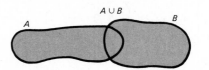

The solution set of the equation $(x - 3)(x + 2) = 0$ is $\{3, -2\}$. This set is the union of the solution sets of the equations $x - 3 = 0$ and $x + 2 = 0$, which are $\{3\}$ and $\{-2\}$.

DO EXERCISES 11 AND 12.

| NAME | CLASS | ANSWERS |

EXERCISE SET 11.4

■ ●   Name each set using the roster method.

1. _____

1. The set of whole numbers 3 through 8

2. _____

2. The set of whole numbers 101 through 107

3. _____

3. The set of odd numbers between 40 and 50

4. _____

4. The set of multiples of 5 between 10 and 40

5. _____

5. $\{x|x$ is a square root of 3$\}$ **6.** $\{x|x$ is the cube of 0.2$\}$

6. _____

7. _____

■ ■ ●   Determine whether true or false.

8. _____

7. $2 \in \{x|x$ is an odd number$\}$ **8.** $7 \in \{x|x$ is an odd number$\}$

9. _____

9. Elton John $\in$ The set of all rock stars   **10.** Apple $\in$ The set of all fruit

10. _____

11. $-3 \in \{-4, -3, 0, 1\}$ **12.** $0 \in \{-4, -3, 0, 1\}$

11. _____

12. _____

Copyright © 1979, Philippines copyright 1979, by Addison-Wesley Publishing Company, Inc. All rights reserved.

ANSWERS

●●● Find each intersection.

13. $\{a, b, c, d, e\} \cap \{c, d, e, f, g\}$

14. $\{a, e, i, o, u\} \cap \{q, u, i, c, k\}$

13. _____

14. _____

15. $\{1, 2, 5, 10\} \cap \{0, 1, 7, 10\}$

16. $\{0, 1, 7, 10\} \cap \{0, 1, 2, 5\}$

15. _____

16. _____

17. $\{1, 2, 5, 10\} \cap \{3, 4, 7, 8\}$

18. $\{a, e, i, o, u\} \cap \{m, n, f, g, h\}$

17. _____

18. _____

●●●● Find each union.

19. $\{a, e, i, o, u\} \cup \{q, u, i, c, k\}$

20. $\{a, b, c, d, e\} \cup \{c, d, e, f, g\}$

19. _____

20. _____

21. $\{0, 1, 7, 10\} \cup \{0, 1, 2, 5\}$

22. $\{1, 2, 5, 10\} \cup \{0, 1, 7, 10\}$

21. _____

22. _____

23. $\{a, e, i, o, u\} \cup \{m, n, f, g, h\}$

24. $\{1, 2, 5, 10\} \cup \{a, b\}$

23. _____

24. _____

NAME   SCORE

TEST OR REVIEW—CHAPTER 11

If you miss an item, review the indicated section and objective.

1. _____

[11.1, 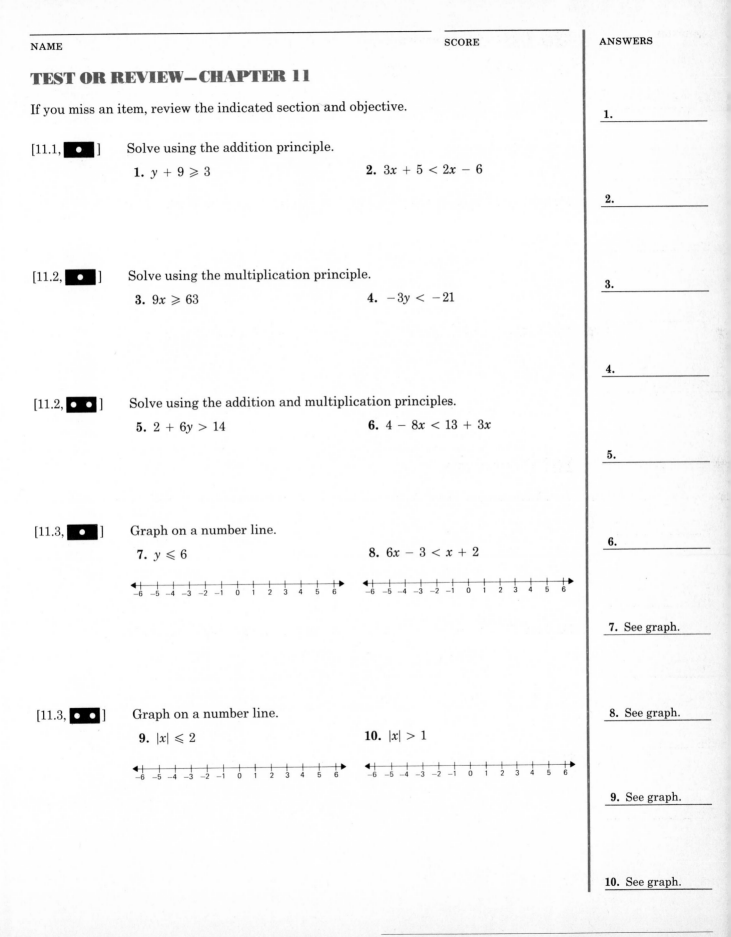]   Solve using the addition principle.

     1. $y + 9 \geqslant 3$ **2.** $3x + 5 < 2x - 6$

2. _____

[11.2, ●]   Solve using the multiplication principle.

     3. $9x \geqslant 63$ **4.** $-3y < -21$

3. _____

4. _____

[11.2, ●●]   Solve using the addition and multiplication principles.

     5. $2 + 6y > 14$ **6.** $4 - 8x < 13 + 3x$

5. _____

[11.3, ●]   Graph on a number line.

     7. $y \leqslant 6$ **8.** $6x - 3 < x + 2$

6. _____

7. See graph.

[11.3, ●●]   Graph on a number line.

     9. $|x| \leqslant 2$ **10.** $|x| > 1$

8. See graph.

9. See graph.

10. See graph.

Copyright © 1979, Philippines copyright 1979, by Addison-Wesley Publishing Company, Inc. All rights reserved.

[11.3, ●●●]   Graph these inequalities on a plane.

11. $y < x + 3$

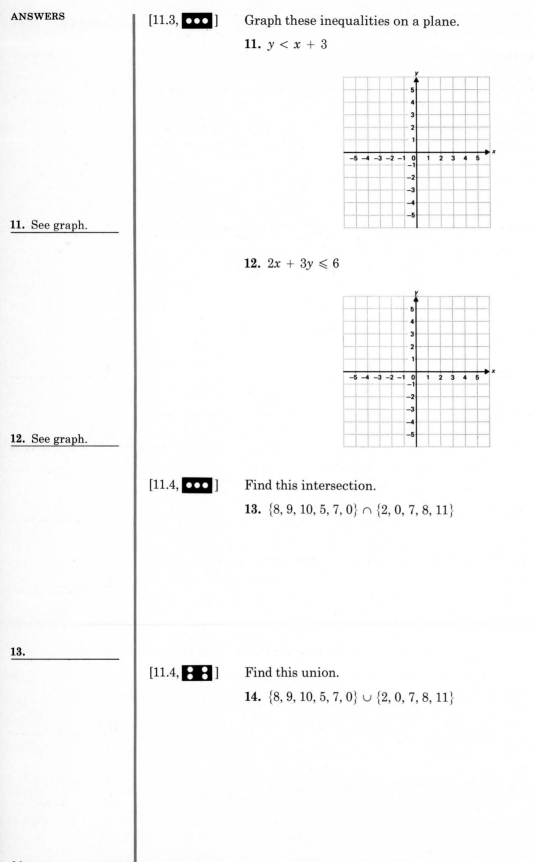

11. See graph.

12. $2x + 3y \leqslant 6$

12. See graph.

[11.4, ●●●]   Find this intersection.

13. $\{8, 9, 10, 5, 7, 0\} \cap \{2, 0, 7, 8, 11\}$

13. _____

[11.4, ⬛]   Find this union.

14. $\{8, 9, 10, 5, 7, 0\} \cup \{2, 0, 7, 8, 11\}$

14. _____

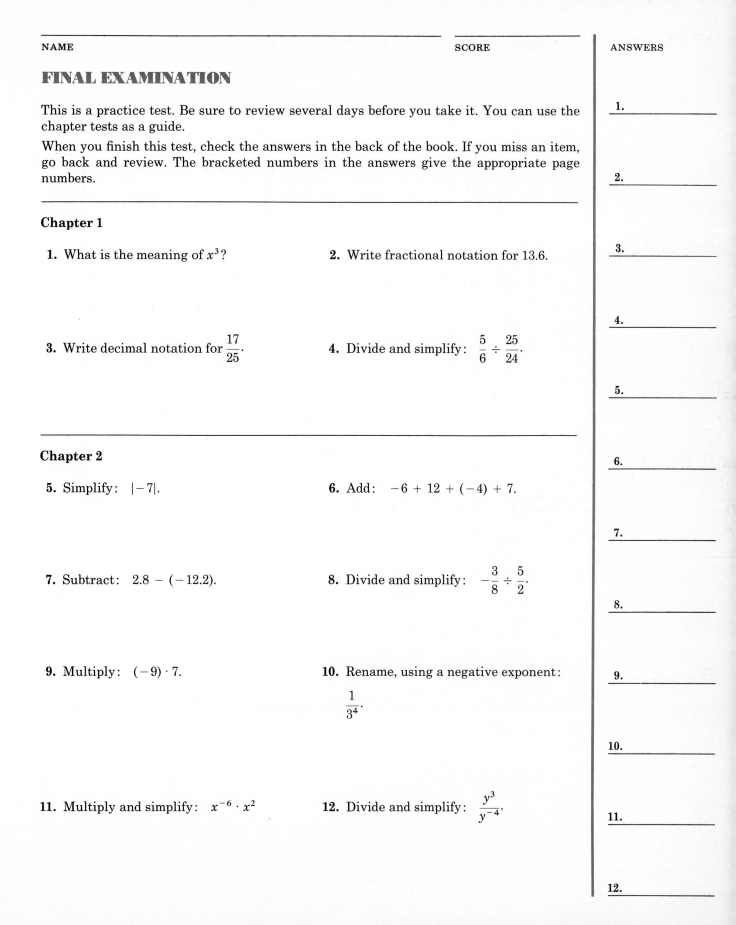

| NAME | SCORE | ANSWERS |

FINAL EXAMINATION

This is a practice test. Be sure to review several days before you take it. You can use the chapter tests as a guide.

When you finish this test, check the answers in the back of the book. If you miss an item, go back and review. The bracketed numbers in the answers give the appropriate page numbers.

Chapter 1

1. What is the meaning of x^3?

2. Write fractional notation for 13.6.

3. Write decimal notation for $\dfrac{17}{25}$.

4. Divide and simplify: $\dfrac{5}{6} \div \dfrac{25}{24}$.

Chapter 2

5. Simplify: $|-7|$.

6. Add: $-6 + 12 + (-4) + 7$.

7. Subtract: $2.8 - (-12.2)$.

8. Divide and simplify: $-\dfrac{3}{8} \div \dfrac{5}{2}$.

9. Multiply: $(-9) \cdot 7$.

10. Rename, using a negative exponent:

$\dfrac{1}{3^4}$.

11. Multiply and simplify: $x^{-6} \cdot x^2$

12. Divide and simplify: $\dfrac{y^3}{y^{-4}}$.

ANSWERS

1. _____

2. _____

3. _____

4. _____

5. _____

6. _____

7. _____

8. _____

9. _____

10. _____

11. _____

12. _____

Copyright © 1979, Philippines copyright 1979, by Addison-Wesley Publishing Company, Inc. All rights reserved.

ANSWERS

13. _____

14. _____

15. _____

16. _____

17. _____

18. _____

19. _____

20. _____

21. _____

22. _____

23. _____

24. _____

Chapter 3

13. Solve: $3x = -24$.

14. Solve: $3x + 7 = 2x - 5$.

15. Solve: $3(y - 1) - 2(y + 2) = 0$.

16. Solve for t: $A = \dfrac{4b}{t}$.

17. The sum of two consecutive integers is 49. What are the integers?

Chapter 4

18. Collect like terms and arrange in descending order.

$2x - 3 + 5x^3 - 2x^3 + 7x^3 + x$

19. Add: $(4x^3 + 3x^2 - 5) + (3x^3 - 5x^2 + 4x - 12)$

20. Subtract: $(6x^2 - 4x + 1) - (-2x^2 + 7)$.

21. Multiply: $-2x^2(4x^2 - 3x + 1)$.

22. Multiply: $(2x - 3)(3x^2 - 4x + 2)$.

23. Multiply: $(t - 5)(t + 5)$.

24. Multiply: $(3m - 2)^2$.

ANSWERS

Chapter 5

Factor completely.

25. $8x^2 - 4x$ **26.** $25x^2 - 4$ **27.** $6x^2 - 5x - 6$ **28.** $x^2 - 8x + 16$

29. $x(x - 8) - 5(x - 8)$ **30.** Solve: $x^2 - 8x + 15 = 0$.

31. The square of a number plus the number itself is 20. Find the number.

32. Factor by completing the square. Show your work.

$x^2 - 8x + 12$

Chapter 6

33. Graph: $y = \dfrac{1}{3}x - 2$.

34. Graph: $2x + 3y = -6$.

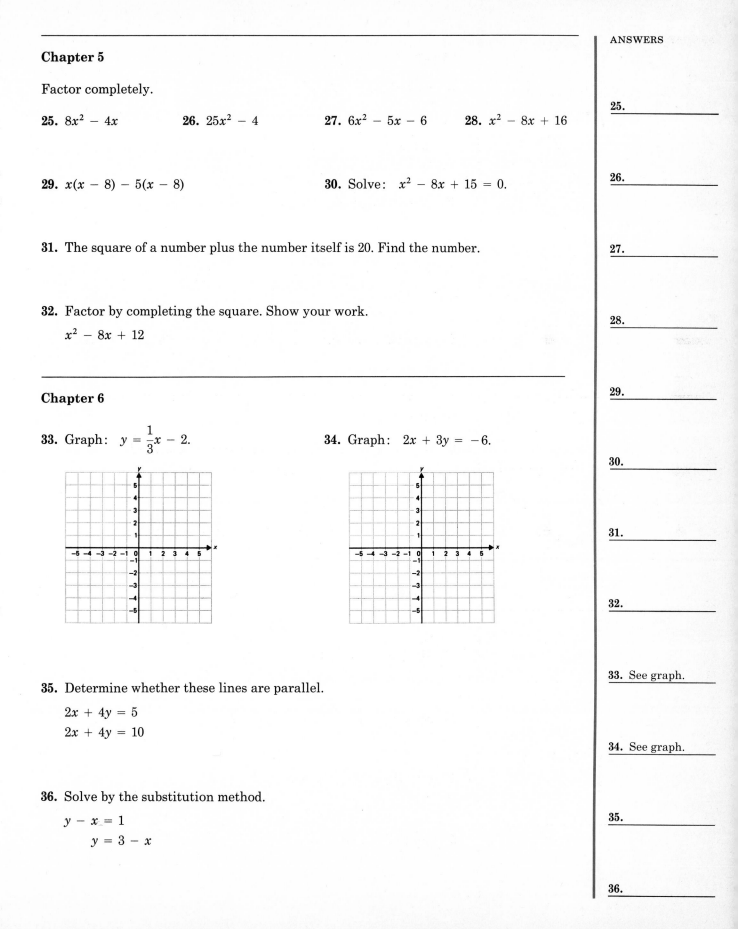

35. Determine whether these lines are parallel.

$2x + 4y = 5$
$2x + 4y = 10$

36. Solve by the substitution method.

$y - x = 1$
$\quad\ y = 3 - x$

25. _____

26. _____

27. _____

28. _____

29. _____

30. _____

31. _____

32. _____

33. See graph.

34. See graph.

35. _____

36. _____

Copyright © 1979, Philippines copyright 1979, by Addison-Wesley Publishing Company, Inc. All rights reserved.

ANSWERS

Solve by the addition method.

37. $x + y = 17$
    $x - y = 7$

38. $4x - 3y = 3$
    $3x - 2y = 4$

37. _____

39. The sum of two numbers is 24. Three times the first number minus the second number is 20. Find the numbers.

38. _____

_____

Chapter 7

39. _____

40. Collect like terms.

   $5x^2yz^3 - 3xy^2 + 2x^2yz^3 + 5x^2y^2 - 2xy^2$

40. _____

41. Arrange in descending powers of x.

   $x^2 + 3ax^3 + 5a^4x^4 + 2a$

41. _____

42. Add.

   $(x^3y - 5x^2y^2 + xy^3 + 2) + (x^3y - 2x^2y^2 + 4)$

42. _____

43. Subtract.

   $(15x^2y^3 + 10xy^2 + 5) - (5xy^2 - x^2y^2 - 2)$

43. _____

Multiply.

44. _____

44. $(x^2 - 2y)(x^2 + 2y)$

45. $(3x + 4y^2)^2$

45. _____

Factor.

46. $25a^2b^2 - 1$

47. $(x - y)^2 - 25$

46. _____

48. $9x^2 + 30xy + 25y^2$

49. $15x^2 + 14xy - 8y^2$

47. _____

48. _____

50. $2ac - 6ab - 3db + dc$

51. $16x^4 - y^8$

49. _____

50. _____

52. $3a^4 + 6a^2 - 72$

53. Solve for m: $\dfrac{1}{t} = \dfrac{1}{m} - \dfrac{1}{n}$.

51. _____

52. _____

Chapter 8

54. Multiply and simplify.

$$\frac{4}{2x - 6} \cdot \frac{x - 3}{x + 3}$$

55. Divide and simplify.

$$\frac{3a^4}{a^2 - 1} \div \frac{2a^3}{a^2 - 2a + 1}$$

53. _____

54. _____

56. Add: $\dfrac{3}{3x - 1} + \dfrac{4}{5x}$.

57. Subtract: $\dfrac{2}{x^2 - 16} - \dfrac{x - 3}{x^2 - 9x + 20}$

55. _____

56. _____

58. Solve: $\dfrac{8}{x} - \dfrac{3}{x + 5} = \dfrac{5}{3}$.

57. _____

58. _____

Copyright © 1979, Philippines copyright 1979, by Addison-Wesley Publishing Company, Inc. All rights reserved.

ANSWERS

59. _____

60. _____

61. _____

62. _____

63. _____

64. _____

65. _____

66. _____

67. _____

68. _____

69. _____

70. _____

Chapter 9

Simplify.

59. $\sqrt{50}$

60. $\dfrac{\sqrt{72}}{\sqrt{45}}$

61. Add and simplify: $4\sqrt{12} + 2\sqrt{12}$.

62. Solve: $3 - x = \sqrt{x^2 - 3}$.

Chapter 10

63. Solve: $x^2 - x - 6 = 0$.

64. Solve: $x^2 + 3x = 5$.

65. Graph: $y = x^2 + 2x + 1$.

66. The length of a rectangle is 7 m more than the width. The length of a diagonal is 13 m. Find the length.

Chapter 11

67. Solve: $5 - 9x \leqslant 19 + 5x$.

68. Graph: $4x - 3y > 12$.

69. Find this intersection.

$\{a, b, c, d, e\} \cap \{g, h, c, d, f\}$

70. Find this union.

$\{a, b, c, d, e\} \cup \{g, h, c, d, f\}$

CHAPTER 1

MARGIN EXERCISES, SECTION 1.1, pp. 2–5

Answers may vary in Exercises 1 through 7.

1. $\dfrac{4}{6}, \dfrac{10}{15}, \dfrac{6}{9}$ **2.** $\dfrac{5}{10}, \dfrac{2}{4}, \dfrac{3}{6}$ **3.** $\dfrac{6}{10}, \dfrac{9}{15}, \dfrac{12}{20}$ **4.** $\dfrac{2}{2}, \dfrac{5}{5}, \dfrac{14}{14}$ **5.** $\dfrac{12}{3}, \dfrac{4}{1}, \dfrac{20}{5}$ **6.** $\dfrac{8}{10}, \dfrac{12}{15}, \dfrac{24}{30}$ **7.** $\dfrac{16}{14}, \dfrac{40}{35}, \dfrac{24}{21}$ **8.** $\dfrac{2}{3}$ **9.** $\dfrac{19}{9}$ **10.** $\dfrac{8}{7}$

11. $\dfrac{1}{2}$ **12.** 4   **13.** $\dfrac{5}{2}$ **14.** $\dfrac{35}{16}$ **15.** $\dfrac{23}{15}$ **16.** $\dfrac{7}{12}$ **17.** $\dfrac{2}{15}$ **18.** $\dfrac{7}{36}$

EXERCISE SET 1.1, pp. 7–8

1. $\dfrac{8}{6}$, etc.   **3.** $\dfrac{12}{22}$, etc.   **5.** $\dfrac{4}{22}$, etc.   **7.** $\dfrac{25}{5}, \dfrac{20}{4}$, etc.   **9.** $\dfrac{4}{3}$ **11.** $\dfrac{1}{2}$ **13.** 2   **15.** $\dfrac{10}{3}$ **17.** $\dfrac{1}{8}$ **19.** $\dfrac{51}{8}$ **21.** 1   **23.** $\dfrac{7}{6}$

25. $\dfrac{5}{6}$ **27.** $\dfrac{1}{2}$ **29.** $\dfrac{5}{18}$ **31.** $\dfrac{31}{60}$

MARGIN EXERCISES, SECTION 1.2, pp. 9–12

1. $2000 + 300 + 70 + 4$; $2 \times 1000 + 3 \times 100 + 7 \times 10 + 4$; $2 \times 10 \times 10 \times 10 + 3 \times 10 \times 10 + 7 \times 10 + 4$

2. $30{,}000 + 4000 + 800 + 9$; $3 \times 10{,}000 + 4 \times 1000 + 8 \times 100 + 9$; $3 \times 10 \times 10 \times 10 \times 10 + 4 \times 10 \times 10 \times 10 + 8 \times 10 \times 10 + 9$

3. 10^4 **4.** $2 \times 10^4 + 3 \times 10^2 + 4 \times 10 + 7$ **5.** $7 \times 10^5 + 5 \times 10^4 + 2 \times 10^3 + 3 \times 10^2 + 9 \times 10 + 8$ **6.** $5 \cdot 5 \cdot 5 \cdot 5$

7. $x \cdot x \cdot x \cdot x \cdot x$ **8.** n^5 **9.** yyy **10.** 140   **11.** 8   **12.** $3 \times 10^3 + 5 \times 10^2 + 6 \times 10^1 + 2 \times 10^0$ **13.** 4   **14.** 1

15. $2 \times 10 + 3 + 6 \times \dfrac{1}{10} + 7 \times \dfrac{1}{100} + 8 \times \dfrac{1}{1000}$ **16.** $2 \times \dfrac{1}{10} + 7 \times \dfrac{1}{100}$ **17.** $4 + 6 \times \dfrac{1}{1000} + 7 \times \dfrac{1}{10{,}000}$ **18.** $\dfrac{162}{100}$

19. $\dfrac{35{,}431}{1000}$ **20.** 0.875   **21.** 0.8   **22.** $0.81\overline{81}$ **23.** 2.55

SOMETHING EXTRA—CALCULATOR CORNER: NUMBER PATTERNS, p. 12

1. 9; 1089; 110,889; 11,108,889; 1,111,088,889   **2.** 54; 6534; 665,334; 66,653,334; 6,666,533,334   **3.** 111; 1221; 12,321; 123,321; 1,233,321   **4.** 111; 222; 333; 444; 555

EXERCISE SET 1.2, pp. 13–14

1. $5000 + 600 + 70 + 7$; $5 \times 1000 + 6 \times 100 + 7 \times 10 + 7$; $5 \times 10 \times 10 \times 10 + 6 \times 10 \times 10 + 7 \times 10 + 7$

3. $900{,}000 + 8000 + 500 + 60 + 3$; $9 \times 100{,}000 + 8 \times 1000 + 5 \times 100 + 6 \times 10 + 3$; $9 \times 10 \times 10 \times 10 \times 10 \times 10 + 8 \times 10 \times 10 \times 10 + 5 \times 10 \times 10 + 6 \times 10 + 3$

5. $5 \cdot 5$ **7.** $m \cdot m \cdot m$ **9.** $4 \cdot 4 \cdot 4 \cdot 4 \cdot 4$ **11.** $x \cdot x \cdot x \cdot x \cdot x \cdot x$ **13.** n^3 **15.** r^2 **17.** xxx **19.** $yyyy$

21. $10 \cdot 10$, or 100   **23.** $10 \cdot 10 \cdot 10 \cdot 10 \cdot 10$, or 100,000   **25.** 0.0036   **27.** 120   **29.** 1   **31.** p

33. $3 \times 10^2 + 7 \times 10^1 + 8 \times 10^0$ **35.** $8 \times 10^4 + 5 \times 10^2 + 6 \times 10^1 + 4 \times 10^0$ **37.** $\dfrac{291}{10}$ **39.** $\dfrac{467}{100}$ **41.** $3 + 6 \times \dfrac{1}{10}$

43. $1 \times 10 + 8 + 7 \times \dfrac{1}{10} + 8 \times \dfrac{1}{100} + 9 \times \dfrac{1}{1000}$ **45.** 0.25   **47.** 0.6   **49.** $0.2\overline{2}$ **51.** 0.125   **53.** $0.45\overline{45}$

55. Answers may vary; for a nine-digit readout, 6^{12} is too large.

MARGIN EXERCISES, SECTION 1.3, pp. 15–18

1. 22   **2.** 30   **3.** 26   **4.** 30   **5.** 27   **6.** 27   **7.** 27   **8.** 1820   **9.** 1820   **10.** 23   **11.** 25   **12.** Comm. law, mult.
13. Assoc. law, add.   **14.** Assoc. law, add.   **15.** Assoc. law, mult.   **16.** (a) 28; (b) 28   **17.** (a) 77; (b) 77   **18.** 23
19. 15   **20.** (a) 240; (b) 240   **21.** $4(x + y)$ **22.** $5(a + b)$ **23.** $7(p + q + r)$ **24.** $5(x + y)$; 35   **25.** $7(x + y)$; 49

EXERCISE SET 1.3, pp. 19–20

1. 22   **3.** 89   **5.** 41   **7.** 59   **9.** Comm. law, add.   **11.** Assoc. law, add.   **13.** 52   **15.** $9(x + y)$ **17.** $\frac{1}{2}(a + b)$
19. $1.5(x + z)$ **21.** $4(x + y + z)$ **23.** $\frac{4}{7}(a + b + c + d)$ **25.** $9(x + y)$; 135   **27.** $10(x + y)$; 150   **29.** $5(a + b)$; 45
31. $20(a + b)$; 180   **33.** 7   **35.** 10, 110, 1110, 11,110; 11,111,110

MARGIN EXERCISES, SECTION 1.4, pp. 21–22

1. $5(x + 2)$ **2.** $3(4 + x)$ **3.** $3(2x + 4 + 3y)$ **4.** $5(x + 2y + 5)$ **5.** $Q(1 + ab)$ **6.** $5y + 15$ **7.** $4x + 8y + 20$
8. $8m + 24n + 32p$ **9.** $8y$ **10.** $5x$ **11.** $1.03x$ **12.** $18p + 9q$ **13.** $11x + 8y$ **14.** $16y$ **15.** $7s + 13w$ **16.** $9x + 10y$
17. $6a + 1.07b$

EXERCISE SET 1.4, pp. 23–24

1. $2(x + 2)$ **3.** $6(x + 4)$ **5.** $3(3x + y)$ **7.** $7(2x + 3y)$ **9.** $5(1 + 2x + 3y)$ **11.** $8(a + 2b + 8)$ **13.** $3(x + 6y + 5z)$
15. $3x + 3$ **17.** $4 + 4y$ **19.** $36t + 27z$ **21.** $7x + 28 + 42y$ **23.** $15x + 45 + 35y$ **25.** $5x$ **27.** $11a$ **29.** $8x + 9z$
31. $43a + 150c$ **33.** $1.09x + 1.2t$ **35.** $24u + 5t$ **37.** $50 + 6t + 8y$ **39.** b **41.** $\frac{13}{4}y$

MARGIN EXERCISES, SECTION 1.5, pp. 25–28

1. $\frac{4}{7}$ **2.** 1   **3.** $\frac{2}{3}$ **4.** 1   **5.** $\frac{11}{4}$ **6.** $\frac{7}{15}$ **7.** $\frac{1}{5}$ **8.** 3   **9.** $\frac{21}{20}$ **10.** $\frac{5}{6}$ **11.** $\frac{45}{28}$ **12.** $\frac{8}{21}$ **13.** $\frac{12}{35}$ **14.** $\frac{14}{45}$

15.   **16.**

17. True   **18.** True   **19.** False   **20.** $<$ **21.** $=$ **22.** $>$ **23.** $>$ **24.** $<$ **25.** $>$ **26.** $>$

SOMETHING EXTRA—CALCULATOR CORNER, p. 28

1. 2.6; 0.003077; 2.600065   **2.** 61.353033; 0.077760; 61.35264

EXERCISE SET 1.5, pp. 29–30

1. $\frac{4}{3}$ **3.** 8   **5.** 1   **7.** $\frac{35}{18}$ **9.** $\frac{10}{3}$ **11.** $\frac{1}{2}$ **13.** $\frac{5}{36}$ **15.** 500   **17.** $\frac{3}{40}$ **19.**

21.   **23.** $=$ **25.** $>$ **27.** $>$ **29.** $=$ **31.** $<$ **33.** 3.2   **35.** $<$

MARGIN EXERCISES, SECTION 1.6, pp. 31–34

Answers may vary in Exercises 1 through 3.

1. $2 + 3 = 5$; $8 - 5 = 3$; $4 + 5.8 = 9.8$ **2.** $2 + 3 = 6$; $8 - 5 = 4$; $4 + 5.8 = 8.8$ **3.** $x + 2 = 5$; $y - 1 = 8$; $10 - y = 3$

4. 4, 8, 10; any three numbers other than 7   **5.** 7   **6.** 6   **7.** 4   **8.** 5   **9.** 8   **10.** 273   **11.** 4.87   **12.** $\frac{9}{8}$ **13.** 7

14. $\frac{9}{2}$, or 4.5   **15.** 5.6   **16.** Yes   **17.** No   **18.** Yes   **19.** No   **20.** No   **21.** No

EXERCISE SET 1.6, pp. 35–36

1. 2   **3.** 9   **5.** 5   **7.** 20   **9.** 7   **11.** 5   **13** 19   **15.** 5.66   **17.** 2818   **19.** $\frac{5}{12}$ **21.** $\frac{1}{6}$ **23.** 4   **25.** $\frac{5}{4}$, or 1.25
27. 0.24   **29.** 3.1   **31.** 8.5   **33.** $\frac{140}{3}$ **35.** 470,188

MARGIN EXERCISES, SECTION 1.7, p. 37

1. 36   **2.** 66   **3.** $0.320 = (1\frac{1}{2})h$ **4.** $224 - x = \frac{3}{4} \cdot 224$

EXERCISE SET 1.7, pp. 39–42

1. 72   **3.** 17   **5.** 9   **7.** $78, 114 = 4A$ **9.** $35 = \frac{2}{5}t$ **11.** $78.3 = 13.5 + m$ **13.** $640 = 1.6s$ **15.** $1175 = 1.8c$

17. 19528.5 km^2 **19.** $87\frac{1}{2}$ words per min   **21.** 64.8°C   **23.** 400 kW-h   **25.** \$652.78   **27.** Salary now

---

MARGIN EXERCISES, SECTION 1.8, pp. 43–46

1. 0.462   **2.** 1   **3.** $\frac{67}{100}$ **4.** $\frac{456}{1000}$ **5.** $\frac{1}{400}$ **6.** 677%   **7.** 99.44%   **8.** 25%   **9.** 37.5%   **10.** $66.6\bar{6}\%$, or $66\frac{2}{3}\%$

11. 11.04   **12.** 10   **13.** 32%   **14.** 25%   **15.** 225   **16.** 50   **17.** $x = 19\% \cdot 586{,}400$; 111,416 sq mi

18. $x + 7\%x = 8988$; \$8400

EXERCISE SET 1.8, pp. 47–48

1. 0.76   **3.** 0.547   **5.** $\frac{20}{100}$ **7.** $\frac{786}{1000}$ **9.** 454%   **11.** 99.8%   **13.** 12.5%   **15.** 68%   **17.** 546   **19.** 85   **21.** 125%

23. $76 = x\% \cdot 88$; approx. 86.36%   **25.** $208 = 26\% \cdot x$; \$800   **27.** $x = 5\% \cdot 428.86$; \$21.44; \$450.30

29. $x + 9\%x = 8502$; \$7800   **31.** $(8\% - 7.4\%) \cdot 9600 = x$; \$57.60

EXERCISE SET 1.8A, pp. 49–52

1. 0.38   **3.** 0.721   **5.** 0.654   **7.** 0.0325   **9.** 0.0824   **11.** 0.0061   **13.** 0.0043   **15.** 0.00012   **17.** 0.00045   **19.** 0.000035

21. 1.25   **23.** 2.4   **25.** $\frac{30}{100}$ **27.** $\frac{70}{100}$ **29.** $\frac{135}{1000}$ **31.** $\frac{734}{1000}$ **33.** $\frac{32}{1000}$ **35.** $\frac{84}{1000}$ **37.** $\frac{120}{100}$ **39.** $\frac{250}{100}$ **41.** $\frac{35}{10{,}000}$

43. $\frac{48}{10{,}000}$ **45.** $\frac{42}{100{,}000}$ **47.** $\frac{83}{100{,}000}$ **49.** 62%   **51.** 85%   **53.** 62.3%   **55.** 81.2%   **57.** 720%   **59.** 350%   **61.** 200%

63. 400%   **65.** 7.2%   **67.** 1.3%   **69.** 0.13%   **71.** 0.73%   **73.** 17%   **75.** 119%   **77.** 70%   **79.** 80%   **81.** 35%

83. 28%   **85.** 50%   **87.** 25%   **89.** 60%   **91.** 34%   **93.** $33\frac{1}{3}\%$ **95.** 37.5%   **97.** 95   **99.** 31.62   **101.** 25%   **103.** 175

MARGIN EXERCISES, SECTION 1.9, pp. 53–58

1. 21 m^2; 20 m   **2.** 25 m^2, 20 m   **3.** 60 cm^2 **4.** 38 m^2 **5.** 35 cm^2 **6.** 64^0 **7.** 200 m^3 **8.** 18 m   **9.** 56.52 cm

10. 254.34 ft^2

SOMETHING EXTRA—CALCULATOR CORNER: ESTIMATING π, p. 58

1. 3.1416   **2.** 3.3436734

EXERCISE SET 1.9, pp. 59–60

1. 1100 cm^2; 138 cm   **3.** 225 ft^2; 60 ft   **5.** 120 cm^2 **7.** 200 in.^2 **9.** 280 m^2 **11.** 74°   **13.** 36°   **15.** 384 m^3

17. 4.8 m; 15.072 m   **19.** $\frac{3}{2}$ in.; $\frac{33}{7}$ in.   **21.** 18.0864 m^2 **23.** $31{,}400 \text{ ft}^2$ **25.** 12.3 cm   **27.** 7.2 yd

TEST OR REVIEW—CHAPTER 1, pp. 61–62

1. $\dfrac{2}{3}$　**2.** $\dfrac{23}{15}$　**3.** $\dfrac{9}{32}$　**4.** $y \cdot y \cdot y \cdot y$　**5.** $\dfrac{569}{100}$　**6.** 0.375　**7.** 62　**8.** Assoc. law, mult.　**9.** $5(x + y)$; 50

10. $6(3x + y)$　**11.** $4(9x + 4 + y)$　**12.** $40a + 24b + 16$　**13.** $40y + 10a$　**14.** $10a + 1.23b$　**15.** $\dfrac{4}{3}$　**16.** $<$　**17.** 8.4

18. 6　**19.** $\dfrac{3}{5}x = 18$; 30　**20.** 0.458　**21.** 56%　**22.** $6\dfrac{2}{3}\%$　**23.** $64\,\text{ft}^2$　**24.** $26\,\text{m}^2$　**25.** $314\,\text{yd}^2$

CHAPTER 2

READINESS CHECK, p. 64

1. [1.1,●●●] $\dfrac{3}{5}$　**2.** [1.2, ●●] $w \cdot w \cdot w \cdot w$　**3.** [1.3, ●●●] Distributive　**4.** [1.4, ●●] $3(x + 3 + 4y)$

5. [1.4, ●●] $21z + 7y + 14$　**6.** [1.5, ●●] $\dfrac{28}{3}$

MARGIN EXERCISES, SECTION 2.1, pp. 64–67

1. $>$　**2.** $>$　**3.** $<$　**4.** $>$　**5.** $<$　**6.** $<$　**7.** 8　**8.** 10　**9.** 29　**10.** 6　**11.** 0　**12.** -3　**13.** -8　**14.** 4　**15.** 0　**16.** $4 + (-5) = -1$　**17.** $-2 + (-4) = -6$　**18.** $-3 + 8 = 5$　**19.** -11　**20.** -12　**21.** -34　**22.** -22　**23.** 2　**24.** -4　**25.** -2　**26.** 3　**27.** 0　**28.** 0　**29.** 0　**30.** 0　**31.** -12

EXERCISE SET 2.1, pp. 69–70

1. $>$　**3.** $<$　**5.** $<$　**7.** $<$　**9.** $>$　**11.** $<$　**13.** 3　**15.** 10　**17.** 0　**19.** 24　**21.** 53　**23.** 8　**25.** 6　**27.** 6　**29.** 3　**31.** -9　**33.** -5　**35.** 7　**37.** -3　**39.** 0　**41.** -5　**43.** -21　**45.** 2　**47.** -26　**49.** -22　**51.** 32　**53.** 0　**55.** 45　**57.** -198　**59.** 52　**61.** $30{,}937$

MARGIN EXERCISES, SECTION 2.2, pp. 71–73

1. 0　**2.** 0　**3.** 0　**4.** 0　**5.** 6　**6.** -8　**7.** 7　**8.** 0　**9.** 4　**10.** 20　**11.** -9　**12.** -12　**13.** 0　**14.** 38　**15.** 4　**16.** -6　**17.** -8　**18.** 1　**19.** -4　**20.** -8　**21.** -8　**22.** (a) -2; (b) -2　**23.** (a) -11; (b) -11　**24.** (a) 4; (b) 4　**25.** (a) -2; (b) -2　**26.** -6　**27.** 7　**28.** -11　**29.** 1　**30.** -4

SOMETHING EXTRA—CALCULATOR CORNER: NUMBER PATTERNS, p. 74

1. 1; 121; $12{,}321$; $1{,}234{,}321$; $123{,}454{,}321$　**2.** 24; 2904; $295{,}704$; $29{,}623{,}704$; $2{,}962{,}903{,}704$　**3.** 48; 408; 4008; $40{,}008$; $400{,}008$
4. 81; 9801; $998{,}001$; $99{,}980{,}001$; $9{,}999{,}800{,}001$

EXERCISE SET 2.2, pp. 75–76

1. 7　**3.** -1　**5.** 12　**7.** -70　**9.** 0　**11.** 1　**13.** -7　**15.** 14　**17.** 0　**19.** -4　**21.** -7　**23.** -6　**25.** 0　**27.** -7　**29.** -1　**31.** 18　**33.** -5　**35.** -3　**37.** -21　**39.** 5　**41.** -8　**43.** 12　**45.** -23　**47.** -68　**49.** -73　**51.** 116　**53.** 0　**55.** $-309{,}882$

MARGIN EXERCISES, SECTION 2.3, pp. 77–78

1. $20, 10, 0, -10, -20, -30$　**2.** -18　**3.** -100　**4.** -80　**5.** $-10, 0, 10, 20, 30$　**6.** 12　**7.** 32　**8.** 35　**9.** -2　**10.** 5　**11.** -3　**12.** 8　**13.** -6

EXERCISE SET 2.3, pp. 79–80

1. -16 **3.** -42 **5.** -24 **7.** -72 **9.** 16 **11.** 42 **13.** 24 **15.** 72 **17.** -120 **19.** 1000 **21.** 90 **23.** 200

25. -6 **27.** -13 **29.** -2 **31.** 4 **33.** -8 **35.** 2 **37.** -12 **39.** -8 **41.** $-9, -98, -987, -9876; -98,765$

MARGIN EXERCISES, SECTION 2.4, pp. 81–82

1. -8.7 **2.** $\frac{8}{9}$ **3.** 7.74 **4.** $\frac{10}{3}$ **5.** 8.32 **6.** $-\frac{5}{4}$ **7.** 12 **8.** $\frac{5}{6}$ **9.** -17.2 **10.** 4.1 **11.** $\frac{8}{3}$ **12.** 3.5 **13.** -6.2

14. $-\frac{2}{9}$ **15.** $-\frac{19}{20}$ **16.** 510.8 ml **17.** -9 **18.** $-\frac{1}{5}$ **19.** -7.5 **20.** $-\frac{1}{24}$

EXERCISE SET 2.4, pp. 83–84

1. 4.7 **3.** $-\frac{7}{2}$ **5.** 7 **7.** 26.9 **9.** $\frac{1}{3}$ **11.** $-\frac{7}{6}$ **13.** 9.3 **15.** -90.3 **17.** -12.4 **19.** $\frac{9}{10}$ **21.** 34.8 **23.** -567

25. 19.2 **27.** $\frac{2}{3}$ **29.** 89.3 **31.** $\frac{14}{3}$ **33.** -1.8 **35.** -8.1 **37.** $-\frac{1}{5}$ **39.** $-\frac{8}{7}$ **41.** $-\frac{3}{8}$ **43.** $-\frac{29}{35}$ **45.** $-\frac{11}{15}$

47. -6.3 **49.** -4.3 **51.** -4 **53.** $-\frac{1}{4}$ **55.** $\frac{1}{12}$ **57.** $-\frac{17}{12}$ **59.** $\frac{1}{8}$ **61.** 19.9 **63.** -9 **65.** -0.01 **67.** 67.85

MARGIN EXERCISES, SECTION 2.5, pp. 85–86

1. -30 **2.** $-\frac{10}{27}$ **3.** $-\frac{7}{10}$ **4.** -30.033 **5.** 64 **6.** $\frac{20}{63}$ **7.** $\frac{2}{3}$ **8.** 13.455 **9.** -30 **10.** -30.75 **11.** $-\frac{5}{3}$ **12.** 120

13. $\frac{3}{3}, \frac{-8}{-8}, \frac{\frac{2}{3}}{\frac{2}{3}}$, answers may vary **14.** $-\frac{4}{5}$ **15.** $-\frac{1}{3}$ **16.** -5 **17.** $-\frac{20}{21}$ **18.** $-\frac{12}{5}$ **19.** $\frac{16}{7}$

EXERCISE SET 2.5, pp. 87–88

1. -72 **3.** -12.4 **5.** 24 **7.** 21.7 **9.** $-\frac{2}{5}$ **11.** $\frac{1}{12}$ **13.** -17.01 **15.** $-\frac{5}{12}$ **17.** -90 **19.** 420 **21.** $\frac{2}{7}$ **23.** -60

25. 150 **27.** $-\frac{1}{5}$ **29.** 4 **31.** $-\frac{5}{7}$ **33.** $-\frac{11}{4}$ **35.** $-\frac{9}{8}$ **37.** $\frac{5}{3}$ **39.** $\frac{9}{14}$ **41.** $\frac{9}{64}$ **43.** -1 **45.** 1 **47.** -8

MARGIN EXERCISES, SECTION 2.6, pp. 89–90

1. (a) 8; (b) 8 **2.** (a) -4; (b) -4 **3.** (a) -25; (b) -25 **4.** (a) -20; (b) -20 **5.** $5x, -4y, 3$ **6.** $-4y, -2x, 3z$

7. $4(x - 2)$ **8.** $3(x - 2y + 3)$ **9.** $b(x + y - z)$ **10.** $3x - 15$ **11.** $5x - 5y + 20$ **12.** $-2x + 6$ **13.** $bx - 2by + 4bz$

14. $3x$ **15.** $6x$ **16.** $0.59x$ **17.** $3x + 3y$ **18.** $-4x + 2y - 3z$

EXERCISE SET 2.6, pp. 91–92

1. 12 **3.** -8 **5.** $8x, -1.4y$ **7.** $-5x, 3y, -14z$ **9.** $8(x - 3)$ **11.** $4(8 - x)$ **13.** $2(4x + 5y - 11)$ **15.** $a(x - 7)$

17. $a(x + y - z)$ **19.** $7x - 14$ **21.** $-7y + 14$ **23.** $-21 + 3t$ **25.** $-4x - 12y$ **27.** $-14x - 28y + 21$ **29.** $8x$

31. $16y$ **33.** $-11x$ **35.** $0.17x$ **37.** $4x + 2y$ **39.** $7x + y$ **41.** $0.8x + 0.5y$ **43.** $\frac{3}{5}x + \frac{3}{5}y$ **45.** $1.08P$

MARGIN EXERCISES, SECTION 2.7, pp. 93–94

1. $-x - 2$ **2.** $-5x + 2y + 8$ **3.** $2x - 9$ **4.** $3y + 2$ **5.** 6 **6.** 4 **7.** $5x - y - 8$

EXERCISE SET 2.7, pp. 95–96

1. $-2x - 7$ **3.** $-5x + 8$ **5.** $-4a + 3b - 7c$ **7.** $-6x - 8y - 5$ **9.** $-3x - 5y + 6$ **11.** $8x + 6y + 43$ **13.** $5x - 3$

15. $7a - 9$ **17.** $5x - 6$ **19.** $-5x - 2y$ **21.** $y - 5z$ **23.** -16 **25.** 56 **27.** 19 **29.** 22 **31.** $12x - 2$

33. $16x - 4$ **35.** $3x + 30$ **37.** $9x - 18$ **39.** $-4x - 64$

MARGIN EXERCISES, SECTION 2.8, pp. 97–100

1. $\frac{1}{4^3}$, or $\frac{1}{4 \cdot 4 \cdot 4}$, or $\frac{1}{64}$ **2.** $\frac{1}{5^2}$, or $\frac{1}{5 \cdot 5}$, or $\frac{1}{25}$ **3.** $\frac{1}{2^4}$, or $\frac{1}{2 \cdot 2 \cdot 2 \cdot 2}$, or $\frac{1}{16}$ **4.** 3^{-2} **5.** 5^{-4} **6.** 7^{-3} **7.** $\frac{1}{5^3}$ **8.** $\frac{1}{7^5}$

9. $\frac{1}{10^4}$ **10.** 3^8 **11.** 5^2 **12.** 6^{-7} **13.** 5^{-5} **14.** y^{-2} **15.** x^{-8} **16.** 4^3 **17.** 7^{-5} **18.** a^7 **19.** b **20.** x^3

21. y^{-7} **22.** 3^{20} **23.** x^{-12} **24.** y^{15} **25.** x^{-32} **26.** $16x^{20}y^{-12}$ **27.** $25x^{10}y^{-12}z^{-6}$ **28.** $27y^{-6}x^{-15}z^{24}$ **29.** $\$2382.03$

EXERCISE SET 2.8, pp. 101–102

1. $\frac{1}{3^2}$, or $\frac{1}{3 \cdot 3}$, or $\frac{1}{9}$ **3.** $\frac{1}{10^4}$, or $\frac{1}{10 \cdot 10 \cdot 10 \cdot 10}$, or $\frac{1}{10,000}$ **5.** 4^{-3} **7.** x^{-3} **9.** a^{-4} **11.** p^{-n} **13.** $\frac{1}{7^3}$ **15.** $\frac{1}{a^3}$ **17.** $\frac{1}{y^4}$

19. $\frac{1}{z^n}$ **21.** 2^7 **23.** 3^3 **25.** 4^{-2} **27.** x^7 **29.** x^{-13} **31.** 1 **33.** 7^3 **35.** 8^{-4} **37.** x^9 **39.** z^{-4} **41.** x^3 **43.** 1

45. 2^6 **47.** 5^{-6} **49.** x^{12} **51.** $x^{-12}y^{-15}$ **53.** $x^{24}y^8$ **55.** $9x^6y^{-16}z^{-6}$ **57.** $\$2289.80$ **59.** $\$15,281.01$

TEST OR REVIEW—CHAPTER 2, pp. 103–104

1. 9 **2.** $\frac{3}{4}$ **3.** $\frac{5}{6}$ **4.** -17 **5.** 5.8 **6.** $-\frac{5}{21}$ **7.** -7 **8.** 19.9 **9.** $-\frac{5}{4}$ **10.** 40 **11.** $-\frac{5}{8}$ **12.** -360 **13.** -2

14. $\frac{27}{49}$ **15.** $7(x - 4)$ **16.** $-15x + 6$ **17.** $-2a + 4b$ **18.** $13y - 8$ **19.** $5x - 6y + 30$ **20.** 5^{-5} **21.** $\frac{1}{6^4}$ **22.** y^{-4}

23. a^8 **24.** x^{-12} **25.** $\$1157.63$

CHAPTER 3

READINESS CHECK, p. 106

1. [2.1, ■■■] -11 **2.** [2.2, ■■■] 16 **3.** [2.5, ■□] $-\frac{5}{12}$ **4.** [2.5, ▨] $\frac{1}{3}$ **5.** [2.6, ▨] $3x - 15$ **6.** [2.6, ▨] $-7x$

7. [2.7, ■□] $3w + 5$

MARGIN EXERCISES, SECTION 3.1, p. 106

1. -5 **2.** 13.2 **3.** -2 **4.** 10.8

EXERCISE SET 3.1, pp. 107–108

1. 4 **3.** -20 **5.** $\frac{1}{3}$ **7.** $\frac{41}{24}$ **9.** 4 **11.** $\frac{1}{2}$ **13.** -5.1 **15.** 10.9 **17.** $-\frac{7}{4}$ **19.** -4 **21.** 16 **23.** -5 **25.** $1\frac{5}{6}$

27. $6\frac{2}{3}$ **29.** $-7\frac{6}{7}$ **31.** 342.246

MARGIN EXERCISES, SECTION 3.2, p. 109

1. 5 **2.** -12 **3.** $-\frac{7}{4}$ **4.** -3

SOMETHING EXTRA—CALCULATOR CORNER, p. 110

AEDG of length 893

EXERCISE SET 3.2, pp. 111–112

1. 6   **3.** -4 **5.** -6 **7.** 63   **9.** $\frac{3}{5}$ **11.** -20 **13.** 36   **15.** $\frac{3}{2}$ **17.** 10   **19.** -2 **21.** 8   **23.** 13.38   **25.** $\frac{9}{2}$

27. 15.9

MARGIN EXERCISES, SECTION 3.3, pp. 113–114

1. 5   **2.** 4   **3.** 4   **4.** $-\frac{3}{2}$ **5.** -3 **6.** 800   **7.** 1   **8.** 2   **9.** 2   **10.** $\frac{17}{2}$

EXERCISE SET 3.3, pp. 115–118

1. 5   **3.** 10   **5.** -8 **7.** 6   **9.** 5   **11.** -20 **13.** 6   **15.** 8400   **17.** 7   **19.** 7   **21.** 3   **23.** 5   **25.** 2   **27.** 10

29. 4   **31.** 8   **33.** $-\frac{3}{5}$ **35.** -4 **37.** $\frac{10}{7}$ **39.** 4.4233464

MARGIN EXERCISES, SECTION 3.4, p. 119

1. 2   **2.** 3   **3.** -2 **4.** $-\frac{1}{2}$

SOMETHING EXTRA—HANDLING DIMENSION SYMBOLS (PART I), p. 120

1. $5\,\frac{\text{mi}}{\text{hr}}$ **2.** $34\,\frac{\text{km}}{\text{hr}}$ **3.** $2.2\,\frac{\text{m}}{\text{sec}}$ **4.** $19\,\frac{\text{ft}}{\text{min}}$ **5.** 52 ft   **6.** 102 sec   **7.** 31 m   **8.** 32 hr   **9.** $\frac{23}{20}$ lb   **10.** 12 km

11. 22 g   **12.** $105\,\frac{\text{m}}{\text{sec}}$

EXERCISE SET 3.4, pp. 121–124

1. 6   **3.** 2   **5.** 6   **7.** 8   **9.** 4   **11.** 1   **13.** 17   **15.** -8 **17.** $-\frac{5}{3}$ **19.** -3 **21.** 2   **23.** 5   **25.** 434.08657

MARGIN EXERCISES, SECTION 3.5, p. 125

1. 3, -4 **2.** 7, 3   **3.** $-\frac{1}{4}, \frac{2}{3}$ **4.** 0, $\frac{17}{3}$

SOMETHING EXTRA—CALCULATOR CORNER: NUMBER PATTERNS, p. 126

1. 1, 4, 9, 16, 25, 36; 49, 64   **2.** $\frac{8 \cdot 9}{2}$ or 36, $\frac{9 \cdot 10}{2}$ or 45; $\frac{n(n + 1)}{2}$

EXERCISE SET 3.5, pp. 127–128

1. $-8, -6$ **3.** 3, -5 **5.** 12, 11   **7.** 0, 13   **9.** 0, -21 **11.** $-\frac{5}{2}, -4$ **13.** $\frac{1}{3}, -2$ **15.** 0, $\frac{2}{3}$ **17.** 0, 18   **19.** $\frac{1}{9}, \frac{1}{10}$

21. 2, 6   **23.** 0, 2   **25.** $\frac{1}{3}, 20$ **27.** $-10, \frac{10}{3}$

29. In (a), 4 checks so it is a solution; 11 is not a solution since it does not check. In (b) neither number checks, so neither is a solution. In both (a) and (b) the principle of zero products does not apply because the right side is 8 instead of 0.

MARGIN EXERCISES, SECTION 3.6, pp. 129–132

1. 3 ft, 5 ft   **2.** 5   **3.** $5700   **4.** 18, 20   **5.** 206.2 miles   **6.** Length is 20 m; width is 10 m   **7.** 30°, 60°, 90°

EXERCISE SET 3.6, pp. 133–138

1. 19   **3.** -10 **5.** 40   **7.** 20 m; 40 m; 120 m   **9.** 37; 39   **11.** 56; 58   **13.** 35; 36; 37   **15.** 61; 63; 65

17. Length is 90 m; width is 65 m   **19.** Length is 49 m; width is 27 m   **21.** 22.5°   **23.** $4400   **25.** $16   **27.** 450.5 mi

29. 28°; 68°; 84°   **31.** Approx. 3.92 billion   **33.** 1985   **35.** $6400

MARGIN EXERCISES, SECTION 3.7, pp. 139–140

1. $I = \dfrac{E}{R}$ **2.** $\dfrac{C}{\pi} = D$ **3.** $4A - a - b - d = c$ **4.** $I = \dfrac{9R}{A}$

EXERCISE SET 3.7, pp. 141–142

1. $b = \dfrac{A}{h}$ **3.** $r = \dfrac{d}{t}$ **5.** $P = \dfrac{I}{rt}$ **7.** $a = \dfrac{F}{m}$ **9.** $w = \dfrac{1}{2}(P - 2l)$ **11.** $r^2 = \dfrac{A}{\pi}$ **13.** $b = \dfrac{2A}{h}$ **15.** $m = \dfrac{E}{c^2}$

17. $3A - a - c = b$ **19.** $t = \dfrac{3k}{v}$ **21.** $b = \dfrac{2A - ah}{h}$ **23.** $D^2 = \dfrac{2.5H}{N}$ **25.** $S = \dfrac{360A}{\pi r^2}$ **27.** $t = \dfrac{R - 3.85}{-0.0075}$

TEST OR REVIEW—CHAPTER 3, pp. 143–144

1. 6.1   **2.** 1   **3.** 12   **4.** -7 **5.** 4   **6.** -5 **7.** 6   **8.** -6 **9.** $-9, 6$ **10.** $0, \dfrac{7}{4}$ **11.** 9   **12.** 57, 59

13. Width is 11 cm; length is 17 cm   **14.** 102°; 34°; 44°   **15.** $s = \dfrac{P}{4}$ **16.** $h = \dfrac{3V}{B}$

CHAPTER 4

READINESS CHECK, p. 146

1. [1.4, ▮▮] $3s + 3t + 24$ **2.** [2.6, ▮▮] $-7x - 28$ **3.** [2.6, ▮▮] $5x$ **4.** [2.1, ▮▮▮] -10 **5.** [2.2, ▮▮] 6

6. [2.2, ▮▮▮] 12   **7.** [2.3, ▮▮] 30   **8.** [2.6, ▮▮] $6x, -5y, 9z$ **9.** [2.7, ▮▮] $-4x + 7y - 2$ **10.** [2.7, ▮▮] $-4y - 2$

11. [2.8, ▮▮] x^{10}

MARGIN EXERCISES, SECTION 4.1, pp. 146–148

1. $x^2 + 2x - 8$; $x^3 + x^2 - 2x - 5$, $x - 7$, answers may vary   **2.** -19 **3.** -104 **4.** -13 **5.** 8   **6.** 15¢ per mile

7. $-9x^3 + (-4x^5)$ **8.** $-2x^3 + 3x^7 + (-7x)$ **9.** $3x^2, 6x, \dfrac{1}{2}$ **10.** $-4y^5, 7y^2, -3y, -2$ **11.** $4x^3$ and $-x^3$

12. $4t^4$ and $-7t^4$; $-9t^3$ and $10t^3$ **13.** $8x^2$ **14.** $2x^3 + 2$ **15.** $-\dfrac{1}{4}x^5 + 2x^2$ **16.** $-4x^3$ **17.** $5x^3$ **18.** $25 - 3x^5$

19. $6x$ **20.** $4x^3 + 4$ **21.** $-\dfrac{1}{4}x^3 + 4x^2 + 7$ **22.** $3x^2 + x^3 - 1$

EXERCISE SET 4.1, pp. 149–150

1. -18 **3.** 19   **5.** -12 **7.** Approx. 449   **9.** 2   **11.** 4   **13.** 11   **15.** $2, -3x, x^2$ **17.** $6x^2$ and $-3x^2$

19. $2x^4$ and $-3x^4$, $5x$ and $-7x$ **21.** $-3x$ **23.** $-8x$ **25.** $11x^3 + 4$ **27.** $x^3 - x$ **29.** $4b^5$ **31.** $\dfrac{3}{4}x^5 - 2x - 5$ **33.** x^4

35. $\dfrac{15}{16}x^3 - \dfrac{7}{6}x^2$

MARGIN EXERCISES, SECTION 4.2, pp. 151–152

1. $6x^7 + 3x^5 - 2x^4 + 4x^3 + 5x^2 + x$ **2.** $7x^5 - 5x^4 + 2x^3 + 4x^2 - 3$ **3.** $14t^7 - 10t^5 + 7t^2 - 14$ **4.** $-2x^2 - 3x + 2$

5. $10x^4 - 8x - \dfrac{1}{2}$ **6.** $4, 2, 1, 0; 4$ **7.** $5, 6, 1, -1, 4$ **8.** x **9.** x^3, x^2, x, x^0 **10.** x^2, x **11.** x^3 **12.** Monomial

13. None of these   **14.** Binomial   **15.** Trinomial

EXERCISE SET 4.2, pp. 153–154

1. $x^5 + 6x^3 + 2x^2 + x + 1$ **3.** $15x^9 + 7x^8 + 5x^3 - x^2 + x$ **5.** $-5y^8 + y^7 + 9y^6 + 8y^3 - 7y^2$ **7.** $x^6 + x^4$

9. $13x^3 - 9x + 8$ **11.** $-5x^2 + 9x$ **13.** $12x^4 - 2x + \dfrac{1}{4}$ **15.** $3, 2, 1, 0; 3$ **17.** $2, 1, 6, 4; 6$ **19.** $-3, 6$ **21.** $6, 7, -8, -2$

23. x^2, x **25.** x^3, x^2, x^0 **27.** None missing   **29.** Trinomial   **31.** None of these   **33.** Binomial   **35.** Monomial

37. $x^2 - 16$

MARGIN EXERCISES, SECTION 4.3, pp. 155–156

1. $x^2 + 7x + 3$ **2.** $-4x^5 + 7x^4 + 3x^3 + 2x^2 + 4$ **3.** $24x^4 + 5x^3 + x^2 + 1$ **4.** $2x^3 + \dfrac{10}{3}$ **5.** $2x^2 - 3x - 1$

6. $8x^3 - 2x^2 - 8x + \dfrac{5}{2}$ **7.** $\dfrac{7}{2}x^2$

SOMETHING EXTRA—APPLICATIONS, p. 156

1. \$1225.04   **2.** \$1259.71   **3.** \$1242.30   **4.** \$1250.98

EXERCISE SET 4.3, pp. 157–158

1. $-x + 5$ **3.** $x^2 - 5x - 1$ **5.** $3x^5 + 13x^2 + 6x - 3$ **7.** $-4x^4 + 6x^3 + 6x^2 + 2x + 4$ **9.** $12x^2 + 6$

11. $5x^4 - 2x^3 - 7x^2 - 5x$ **13.** $9x^8 + 8x^7 - 3x^4 + 2x^2 - 2x + 5$ **15.** $-\dfrac{1}{2}x^4 + \dfrac{2}{3}x^3 + x^2$

17. $0.01x^5 + x^4 - 0.2x^3 + 0.2x + 0.06$ **19.** $5x^2 + 4x$ **21.** $48.544x^6 - 0.795x^5 + 890x$

MARGIN EXERCISES, SECTION 4.4, pp. 159–160

1. $-12x^4 + 3x^2 - 4x$ **2.** $4x^4 - 3x^2 + 4x$ **3.** $13x^6 - 2x^4 + 3x^2 - x + \dfrac{5}{13}$ **4.** $7y^3 - 2y^2 + y - 3$ **5.** $-4x^3 + 6x - 3$

6. $-5x^4 - 3x^2 - 7x + 5$ **7.** $-14x^{10} + \dfrac{1}{2}x^5 - 5x^3 + x^2 - 3x$ **8.** $3x^2 + 3$ **9.** $2x^3 + 2x + 8$ **10.** $x^2 - 6x - 2$

11. $-8x^4 - 5x^3 + 8x^2 - 1$ **12.** $-2x^2 - 2x + 2$ **13.** $x^3 - x^2 - \dfrac{4}{3}x - 0.9$

EXERCISE SET 4.4, pp. 161–162

1. $5x$ **3.** $x^2 - 10x + 2$ **5.** $-12x^4 + 3x^3 - 3$ **7.** $-3x + 7$ **9.** $-4x^2 + 3x - 2$ **11.** $4x^4 + 6x^2 - \dfrac{3}{4}x + 8$

13. $2x^2 + 14$ **15.** $-2x^5 - 6x^4 + x + 2$ **17.** $9x^2 + 9x - 8$ **19.** $\dfrac{3}{4}x^3 - \dfrac{1}{2}x$ **21.** $0.06x^3 - 0.05x^2 + 0.01x + 1$

MARGIN EXERCISES, SECTION 4.5, p. 163

1. $-8x^4 + 4x^3 + 12x^2 + 5x - 8$ **2.** $-x^3 + x^2 + 3x + 3$ **3.** $2x^3 + 5x^2 - 2x - 5$ **4.** $-x^5 - 2x^3 + 3x^2 - 2x + 2$

EXERCISE SET 4.5, pp. 165–168

1. $-3x^4 + 3x^2 + 4x$ **3.** $3x^5 - 3x^4 - 3x^3 + x^2 + 3x$ **5.** $5x^3 - 9x^2 + 4x - 7$ **7.** $\dfrac{1}{4}x^4 - \dfrac{1}{4}x^3 + \dfrac{3}{2}x^2 + 6\dfrac{3}{4}x + \dfrac{1}{4}$

9. $-x^4 + 3x^3 + 2x + 1$ **11.** $x^4 + 4x^2 + 12x - 1$ **13.** $x^5 - 6x^4 + 4x^3 - x^2 + 1$ **15.** $7x^4 - 2x^3 + 7x^2 + 4x + 9$

17. $3x^5 + x^4 + 10x^3 + x^2 + 3x - 6$ **19.** $1.05x^4 + 0.36x^3 + 14.22x^2 + x + 0.97$ **21.** $13x^4 - 3x^3 - x^2 + 15x - 4$

23. $4x^4 + 3x^3 - 7x^2 - 3x + 6$ **25.** $-x^5 - 6x^4 + 6x^3 - 2x^2 - \frac{1}{2}x - \frac{1}{4}$ **27.** $5x^4 + 12x^3 - 9x^2 - 8x - 8$

29. $-4x^5 + 9x^4 + 6x^2 + 16x + 6$ **31.** $569.607x^3 - 15.168x$

MARGIN EXERCISES, SECTION 4.6, pp. 169–171

1. $-15x$ **2.** $-x^2$ **3.** x^2 **4.** $-x^5$ **5.** $12x^7$ **6.** $-8x^{11}$ **7.** $7y^5$ **8.** 0 **9.** $8x^2 + 16x$ **10.** $-15x^3 + 6x^2$

11. $x^2 + 13x + 40$ **12.** $x^2 + x - 20$ **13.** $5x^2 - 17x - 12$ **14.** $6x^2 - 19x + 15$ **15.** $x^4 + 3x^3 + x^2 + 15x - 20$

16. $6x^5 - 20x^3 + 15x^2 + 14x - 35$ **17.** $3x^3 + 13x^2 - 6x + 20$ **18.** $20x^4 - 16x^3 + 32x^2 - 32x - 16$

19. $6x^4 - x^3 - 18x^2 - x + 10$

SOMETHING EXTRA—CALCULATOR CORNER, p. 172

1. $1, 9, 36, 100; 225, 441$ **2.** 2^n: $2, 4, 8, 16, 32, 64, 128$; 2^{-n}: $0.5, 0.25, 0.125, 0.0625, 0.0313, 0.0156, 0.0078$

EXERCISE SET 4.6, pp. 173–174

1. $-12x$ **3.** $42x^2$ **5.** $30x$ **7.** $-2x^3$ **9.** x^6 **11.** $6x^6$ **13.** 0 **15.** $-0.02x^{10}$ **17.** $8x^2 - 12x$ **19.** $-6x^4 - 6x^3$

21. $4x^6 - 24x^5$ **23.** $-x^2 - 4x + 12$ **25.** $2x^2 - 15x + 25$ **27.** $9x^2 - 25$ **29.** $2x^2 + \frac{5}{2}x - \frac{3}{4}$ **31.** $x^3 + 7x^2 + 7x + 1$

33. $-10x^3 - 19x^2 - x + 3$ **35.** $3x^4 - 6x^3 - 7x^2 + 18x - 6$ **37.** $6t^4 + t^3 - 16t^2 - 7t + 4$ **39.** $x^4 - 1$

MARGIN EXERCISES, SECTION 4.7, p. 175

1. $8x^3 - 12x^2 + 16x$ **2.** $10y^6 + 8y^5 - 10y^4$ **3.** $x^2 + 7x + 12$ **4.** $x^2 - 2x - 15$ **5.** $2x^2 + 9x + 4$

6. $2x^3 - 4x^2 - 3x + 6$ **7.** $12x^5 + 6x^2 + 10x^3 + 5$ **8.** $y^6 - 49$ **9.** $-2x^7 + x^5 + x^3$

EXERCISE SET 4.7, pp. 177–178

1. $4x^2 + 4x$ **3.** $-3x^2 + 3x$ **5.** $x^5 + x^2$ **7.** $6x^3 - 18x^2 + 3x$ **9.** $x^3 + 3x + x^2 + 3$ **11.** $x^4 + x^3 + 2x + 2$

13. $x^2 - x - 6$ **15.** $9x^2 + 15x + 6$ **17.** $5x^2 + 4x - 12$ **19.** $9x^2 - 1$ **21.** $4x^2 - 6x + 2$ **23.** $x^2 - \frac{1}{16}$

25. $x^2 - 0.01$ **27.** $2x^3 + 2x^2 + 6x + 6$ **29.** $-2x^2 - 11x + 6$ **31.** $x^2 + 14x + 49$ **33.** $1 - x - 6x^2$

35. $x^5 - x^2 + 3x^3 - 3$ **37.** $x^3 - x^2 - 2x + 2$ **39.** $3x^6 - 6x^2 - 2x^4 + 4$ **41.** $6x^7 + 18x^5 + 4x^2 + 12$

43. $8x^6 + 65x^3 + 8$ **45.** $4x^3 - 12x^2 + 3x - 9$ **47.** $4x^6 + 4x^5 + x^4 + x^3$

MARGIN EXERCISES, SECTION 4.8, pp. 179–181

1. $x^2 - 25$ **2.** $4x^2 - 9$ **3.** $x^2 - 4$ **4.** $x^2 - 49$ **5.** $9t^2 - 25$ **6.** $4x^6 - 1$ **7.** $x^2 + 16x + 64$ **8.** $x^2 - 10x + 25$

9. $x^2 + 4x + 4$ **10.** $a^2 - 8a + 16$ **11.** $4x^2 + 20x + 25$ **12.** $16x^4 - 24x^3 + 9x^2$ **13.** $y^2 + 18y + 81$

14. $9x^4 - 30x^2 + 25$ **15.** $x^2 + 11x + 30$ **16.** $t^2 - 16$ **17.** $-8x^5 + 20x^4 + 40x^2$ **18.** $81x^4 + 18x^2 + 1$

19. $4a^2 + 6a - 40$ **20.** $4x^2 - 2x + \frac{1}{4}$

SOMETHING EXTRA—POLYNOMIALS AND MULTIPLICATION, p. 182

1. $(100 - 2)^2 = 9604$ **2.** $(100 - 3)^2 = 9409$ **3.** $(100 - 3)(100 + 3) = 9991$ **4.** $(100 - 4)(100 + 4) = 9984$

5. $(100 + 1)^2 = 10{,}201$ **6.** $(200 + 3)^2 = 41{,}209$ **7.** $(10{,}000 - 1)^2 = 998{,}001$ **8.** $(200 - 2)(200 + 2) = 39{,}996$

EXERCISE SET 4.8, pp. 183–184

1. $x^2 - 16$ **3.** $4x^2 - 1$ **5.** $25m^2 - 4$ **7.** $4x^4 - 9$ **9.** $9x^8 - 16$ **11.** $x^{12} - x^4$ **13.** $x^8 - 9x^2$ **15.** $x^{24} - 9$

17. $4x^{16} - 9$ **19.** $x^2 + 4x + 4$ **21.** $9x^4 + 6x^2 + 1$ **23.** $x^2 - x + \dfrac{1}{4}$ **25.** $9 + 6x + x^2$ **27.** $x^4 + 2x^2 + 1$

29. $4 - 12x^4 + 9x^8$ **31.** $25 + 60x^2 + 36x^4$ **33.** $9 - 12x^3 + 4x^6$ **35.** $4x^3 + 24x^2 - 12x$ **37.** $4x^4 - 2x^2 + \dfrac{1}{4}$

39. $-1 + 9p^2$ **41.** $15t^5 - 3t^4 + 3t^3$ **43.** $36x^8 + 48x^4 + 16$ **45.** $12x^3 + 8x^2 + 15x + 10$ **47.** $64 - 96x^4 + 36x^8$

49. $4567.0564x^2 + 435.891x + 10.400625$

TEST OR REVIEW—CHAPTER 4, pp. 185–186

1. $-5x^2 + 2x + 1$ **2.** $\dfrac{1}{5}a^3 - \dfrac{58}{5}a$ **3.** $2x^5 + 3x^4 - 6x^3 - 8x^2 + 2x - 11$ **4.** $3x^5 - 9x^4 + 3x^3 + 2$ **5.** $8x^2 - 4x - 6$

6. $x^5 - 3x^3 - 2x^2 + 8$ **7.** $6a^3 + 5a^2 - 8a - 3$ **8.** $-18y^4 + 30y^3 + 12y^2$ **9.** $x^2 - 81$ **10.** $x^2 - 18x + 81$

11. $2x^2 - 11x - 21$ **12.** $49x^2 - 14x + 1$ **13.** $6p^5 + 9p^3 - 10p^2 - 15$ **14.** $4x^6 - 49$ **15.** $9x^2 + 12x + 4$

16. $15x^7 - 40x^6 + 50x^5 + 10x^4$

CHAPTER 5

5

READINESS CHECK, p. 188

1. [1.4, ▪] $4(y + 7 + 3z)$ **2.** [2.6, ▪▪▪] $8(x - 4)$ **3.** [3.3, ▪▪] 2 **4.** [3.5, ▪] $-\dfrac{3}{4}, 7$ **5.** [4.6, ▪▪] $-12x^{13}$

6. [4.7, ▪] $16x^3 - 48x^2 + 8x$ **7.** [4.7, ▪▪] $x^2 + 2x - 24$ **8.** [4.7, ▪▪] $28w^2 + 17w - 6$ **9.** [4.8, ▪▪] $t^2 - 18t + 81$

10. [4.8, ▪▪] $25x^2 + 30x + 9$ **11.** [4.8, ▪] $p^2 - 16$

MARGIN EXERCISES, SECTION 5.1, pp. 188–190

1. (a) $12x^2$; (b) $3x \cdot 4x,\ 2x \cdot 6x$, answers may vary   **2.** (a) $16x^3$; (b) $(2x)(8x^2),\ (4x)(4x^2)$, answers may vary

3. $8x \cdot x^3$; $4x^2 \cdot 2x^2$; $2x^3 \cdot 4x$, answers may vary   **4.** $7x \cdot 3x,\ (-7x)(-3x),\ (21x)(x)$, answers may vary

5. $6x^4 \cdot x,\ (-2x^3)(-3x^2),\ (3x^3)(2x^2)$, answers may vary   **6.** (a) $3x + 6$; (b) $3(x + 2)$

7. (a) $2x^3 + 10x^2 + 8x$; (b) $2x(x^2 + 5x + 4)$ **8.** $x(x + 3)$ **9.** $x^2(3x^4 - 5x + 2)$ **10.** $3x^2(3x^2 - 5x + 1)$

11. $\dfrac{1}{4}(3x^3 + 5x^2 + 7x + 1)$ **12.** $7x^3(5x^4 - 7x^3 + 2x^2 - 9)$ **13.** $(x + 2)(x + 5)$ **14.** $(x + 3)(x - 4)$

15. $(2x - 3)(x + 4)$ **16.** $(4x - 3)(4x + 5)$

EXERCISE SET 5.1, pp. 191–192

1. $6x^2 \cdot x,\ 3x^2 \cdot 2x,\ (-3x^2)(-2x)$, answers may vary   **3.** $(-9x^4) \cdot x,\ (-3x^2)(3x^3),\ (-3x)(3x^4)$, answers may vary

5. $(8x^2)(3x^2),\ (-8x^2)(-3x^2),\ (4x^3)(6x)$, answers may vary   **7.** $x(x - 4)$ **9.** $x^2(x + 6)$ **11.** $8x^2(x^2 - 3)$

13. $17x(x^4 + 2x^2 + 3)$ **15.** $5(2x^3 + 5x^2 + 3x - 4)$ **17.** $\dfrac{x^3}{3}(5x^3 + 4x^2 + x + 1)$ **19.** $(y + 1)(y + 4)$ **21.** $(x - 1)(x - 4)$

23. $(2x + 3)(3x + 2)$ **25.** $(x - 4)(3x - 4)$ **27.** $(5x + 3)(7x - 8)$ **29.** $(2x - 3)(2x + 3)$ **31.** $(2x^2 + 5)(x^2 + 3)$

MARGIN EXERCISES, SECTION 5.2, pp. 193–197

1. $(x + 3)(x + 4)$ **2.** $(x - 7)(x - 5)$ **3.** $(x - 2)(x + 1)$ **4.** $(2x + 1)(3x + 2)$ **5.** $3(2x + 3)(x + 1)$

6. $2(x + 3)(x - 1)$ **7.** $2(2x + 3)(x - 1)$ **8.** $(2x - 1)(3x - 1)$ **9.** $(2x + 5)(x - 3)$ **10.** $(4x + 1)(3x - 5)$

11. $(3x - 4)(x - 5)$ **12.** $2(5x - 4)(2x - 3)$

EXERCISE SET 5.2, pp. 199–200

1. $(x + 3)(x + 5)$ **3.** $(x - 5)(x + 3)$ **5.** $(x + 3)(x + 4)$ **7.** $(x + 5)(x - 3)$ **9.** $(y + 8)(y + 1)$ **11.** $(x^2 + 2)(x^2 + 3)$

13. $(x - 4)(x + 7)$ **15.** $(x + 5)(x + 3)$ **17.** $(a - 11)(a - 1)$ **19.** $\left(x - \dfrac{1}{5}\right)\left(x - \dfrac{1}{5}\right)$ **21.** $(y - 0.4)(y + 0.2)$

23. $(3x + 1)(x + 1)$ **25.** $4(3x - 2)(x + 3)$ **27.** $(2x + 1)(x - 1)$ **29.** $(3x - 2)(3x + 8)$ **31.** $5(3x + 1)(x - 2)$

33. $(3x + 4)(4x + 5)$ **35.** $(7x - 1)(2x + 3)$ **37.** $(3x^2 + 2)(3x^2 + 4)$ **39.** $(3x - 7)(3x - 7)$ **41.** $2x(3x + 5)(x - 1)$

43. Not factorable

MARGIN EXERCISES, SECTION 5.3, pp. 201–204

1. a, b, d, f   **2.** $(x + 1)^2$ **3.** $(x - 1)^2$ **4.** $(x + 2)^2$ **5.** $(5x - 7)^2$ **6.** $(4x - 7)^2$ **7.** a, e, f   **8.** $(x - 3)(x + 3)$

9. $(x - 8)(x + 8)$ **10.** $5(1 - 2x^3)(1 + 2x^3)$ **11.** $(3x - 1)(3x + 1)(9x^2 + 1)$ **12.** $x^4(7 - 5x^3)(7 + 5x^3)$

SOMETHING EXTRA—CALCULATOR CORNER: A NUMBER PATTERN, p. 204

1. 1, 1, 1, 1, 1, 1   **2.** 1, 1, 1   **3.** $x^2 - (x + 1)(x - 1) = 1;\ x^2 - [x^2 - 1] = x^2 - x^2 + 1 = 1$

EXERCISE SET 5.3, pp. 205–206

1. Yes   **3.** No   **5.** $(x + 3)^2$ **7.** $2(x - 1)^2$ **9.** $x(x - 9)^2$ **11.** $(y - 6)^2$ **13.** $(8 - y)^2$ **15.** $3(2x + 3)^2$ **17.** $(x^3 - 8)^2$

19. $(7 - 3x)^2$ **21.** $5(y^2 + 1)^2$ **23.** No   **25.** Yes   **27.** $(x - 2)(x + 2)$ **29.** $2(2x - 7)(2x + 7)$ **31.** $(2x + 5)(2x - 5)$

33. $x(4 - 9x)(4 + 9x)$ **35.** $x^2(4 - 5x)(4 + 5x)$ **37.** $(7x^2 + 9)(7x^2 - 9)$ **39.** $a^2(a^5 + 2)(a^5 - 2)$

41. $(11a^4 + 10)(11a^4 - 10)$ **43.** $4(x^2 + 4)(x + 2)(x - 2)$ **45.** $(1 + y^4)(1 + y^2)(1 + y)(1 - y)$

47. $(x^4 + 9)(x^2 + 3)(x^2 - 3)$ **49.** $(4 + t^2)(2 + t)(2 - t)$ **51.** $\left(6a - \dfrac{5}{4}\right)^2$

53. $76^2 - 24^2 = (76 - 24)(76 + 24) = 52 \cdot 100 = 5200$

MARGIN EXERCISES, SECTION 5.4, pp. 207–210

1. $-2, 3$ **2.** $7, -4$ **3.** 3   **4.** $0, 4$ **5.** $4, -4$ **6.** $5, -5$ **7.** $7, 8$ **8.** $-4, 5$ **9.** Length is 5 cm; width is 3 cm

10. (a) 342; (b) 9

SOMETHING EXTRA—CALCULATOR CORNER: NESTED MULTIPLICATION, p. 210

1. $x(x(x(x(x + 3) - 1) + 1) - 1) + 9;\ 6432.0122;\ 170{,}616.6106;\ -5{,}584{,}092$

2. $x(x(x(5x - 17) + 2) - 1) + 11;\ 1476.2115;\ 43{,}101.696;\ 1{,}607{,}136$

3. $x(x(x(2x - 3) + 5) - 2) + 18;\ 1279.3152;\ 22{,}235.0912;\ 598{,}892$

4. $x(x(x(x(-2x + 4) + 8) - 4) - 3) + 24;\ -4120.9615;\ -208{,}075.1491;\ 13{,}892{,}691$

EXERCISE SET 5.4, pp. 211–216

1. $-1, -5$ **3.** $0, 5$ **5.** -3 **7.** $\dfrac{5}{3}, -1$ **9.** $4, -\dfrac{5}{3}$ **11.** $\dfrac{2}{3}, -\dfrac{1}{4}$ **13.** $0, \dfrac{3}{5}$ **15.** $-\dfrac{3}{4}, 1$ **17.** 13 and 14, or -13 and -14

19. 13 and 15; -13 and -15 **21.** 5   **23.** $6m$ **25.** 506   **27.** 12   **29.** 780   **31.** 20

MARGIN EXERCISES, SECTION 5.5, pp. 217–218

1. $(x + 5)(x - 3)$ **2.** $(x - 9)(x + 7)$ **3.** $(x + 7)(x + 1)$ **4.** $(x - 2)(x + 18)$ **5.** 36   **6.** 9   **7.** 49   **8.** 121

9. $(x + 3)(x - 1)$ **10.** $(x + 4)(x + 2)$ **11.** $(x - 9)(x + 1)$

EXERCISE SET 5.5, pp. 219–220

1. $(x + 13)(x - 1)$ **3.** $(x + 3)(x + 11)$ **5.** $(x - 8)(x - 10)$ **7.** $(x^2 + 8)(x^2 + 12)$ **9.** $3(x^3 + 9)(x^3 - 2)$

11. $(a^2 - 2)(a + 2)(a - 2)$ **13.** 4   **15.** 49   **17.** 100   **19.** $(x + 4)(x + 2)$ **21.** $(x - 6)(x - 2)$ **23.** $2(x + 8)(x - 16)$

25. $(a - 25)(a + 1)$ **27.** $10(b - 12)(b + 7)$ **29.** $(x - 18)(x + 4)$ **31.** $0.1(2x + 0.1)$

TEST OR REVIEW—CHAPTER 5, pp. 221–222

1. $4x^3(x + 5)$ **2.** $(x - 5)(x + 3)$ **3.** $(x - 5)(x - 3)$ **4.** $x(x - 6)(x + 5)$ **5.** $3(x + 8)(x - 2)$ **6.** $(10x + 1)(x + 10)$

7. $(x - 3)^2$ **8.** $3(2x + 5)^2$ **9.** $(3a - 2)(3a + 2)$ **10.** $2(x - 5)(x + 5)$ **11.** $(4t^2 + 1)(2t + 1)(2t - 1)$ **12.** $-7, 5$

13. 16 and 18, or -16 and -18 **14.** $(x - 9)(x + 1)$

CHAPTER 6

READINESS CHECK, p. 224

1. [3.1, 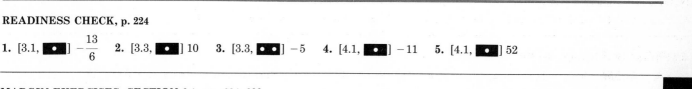] $-\dfrac{13}{6}$ **2.** [3.3,] 10   **3.** [3.3,] -5 **4.** [4.1,] -11 **5.** [4.1,] 52

MARGIN EXERCISES, SECTION 6.1, pp. 224–226

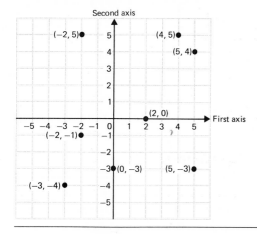

9. Both are negative numbers   **10.** First positive; second negative

11. I   **12.** III   **13.** IV   **14.** II   **15.** $B(-3,5)$; $C(-4,-3)$; $D(2,-4)$; $E(1,5)$; $F(-2,0)$; $G(0,3)$ **16.** No   **17.** Yes

SOMETHING EXTRA—AN APPLICATION: COORDINATES, p. 226

1. Latitude 32.5° north, longitude 64.5° west   **2.** Latitude 27° north, longitude 81° west

EXERCISE SET 6.1, pp. 227–228

1.

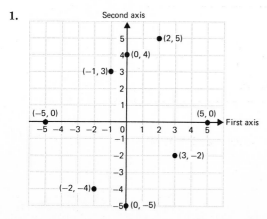

3. II   **5.** IV   **7.** III   **9.** I   **11.** Negative; negative

13. $A(3,3)$; $B(0,-4)$; $C(-5,0)$; $D(-1,-1)$; $E(2,0)$ **15.** Yes

17. No **19.** No **21.**

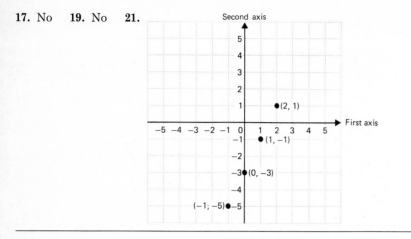

MARGIN EXERCISES, SECTION 6.2, pp. 229–233

1.

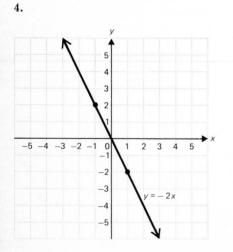

2.

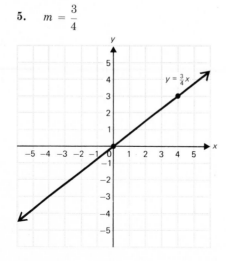

3.

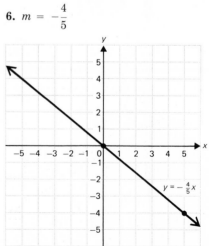

4.

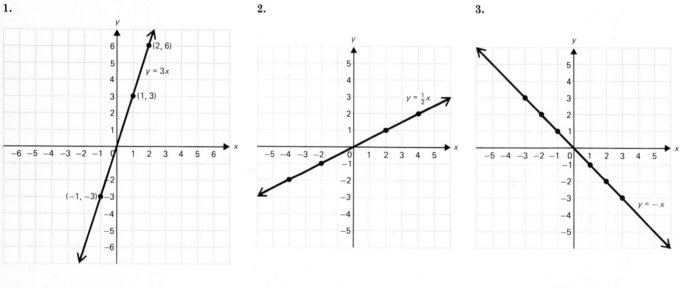

5. $m = \dfrac{3}{4}$

6. $m = -\dfrac{4}{5}$

7. $y = x + 3$ looks like $y = x$ moved *up* 3 units.

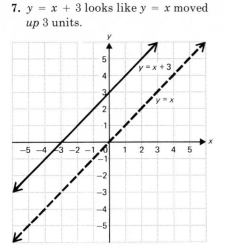

8. $y = x - 1$ looks like $y = x$ moved *down* 1 unit.

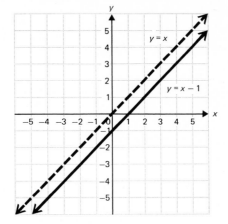

9. $y = 2x + 3$ looks like $y = 2x$ moved *up* 3 units.

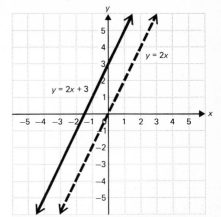

10. $m = \frac{3}{5}$; y-intercept is $(0,2)$

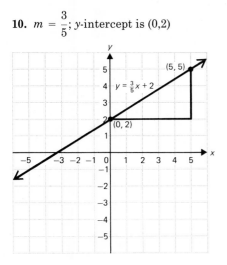

11. $m = \frac{3}{5}$; y-intercept is $(0, -2)$

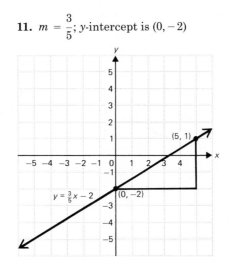

12. $m = -\frac{3}{5}$; y-intercept is $(0, -1)$

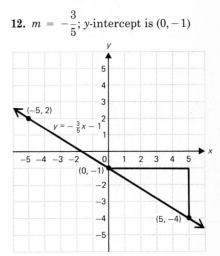

13. $m = -\frac{3}{5}$; y-intercept is $(0,4)$

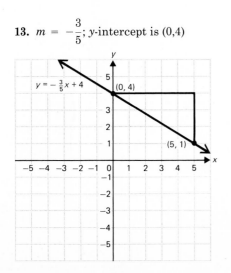

SOMETHING EXTRA—AN APPLICATION, p. 234

1. 57.848%   **2.** 62.6%   **3.** 66.56%   **4.** 70.52%

EXERCISE SET 6.2, pp. 235–236

1.

3.

5. $m = \dfrac{1}{3}$

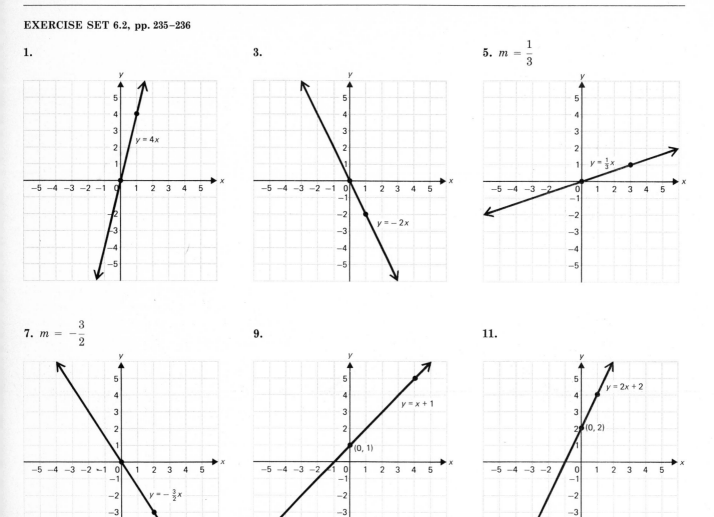

7. $m = -\dfrac{3}{2}$

9.

11.

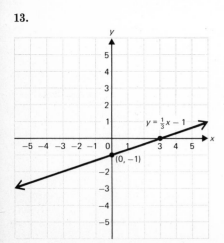

13.

15.

17.

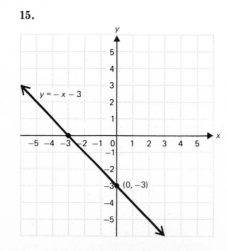

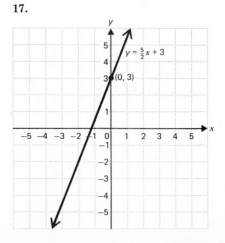

19.

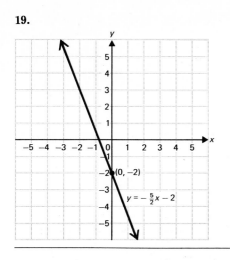

MARGIN EXERCISES, SECTION 6.3, pp. 237–240

1. $y = \frac{2}{3}x - 2; m = \frac{2}{3},$
  y-intercept is $(0, -2)$

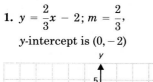

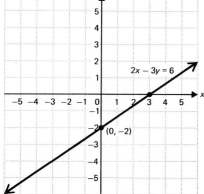

2. x-intercept is $(3,0)$;
  y-intercept is $(0,2)$

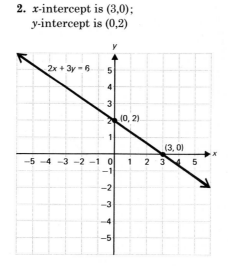

3. x-intercept is $(-3,0)$;
  y-intercept is $(0,4)$

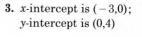

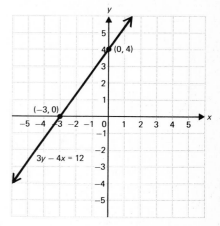

4.

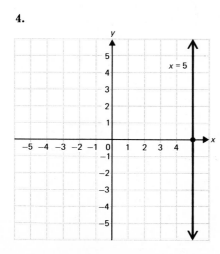

5.

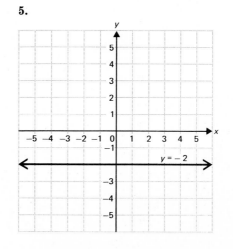

6.

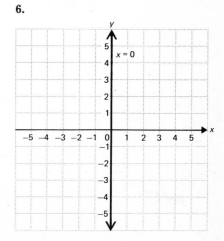

7.

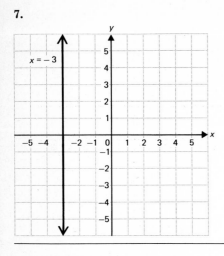

8. a, b, f   **9.** Yes   **10.** No

EXERCISE SET 6.3, pp. 241–242

1. $y = \frac{5}{3}x - 5$; $m = \frac{5}{3}$,
y-intercept is $(0, -5)$

3. $y = -2x + 4$; $m = -2$,
y-intercept is $(0,4)$

5. x-intercept is $(1,0)$;
y-intercept is $(0, -1)$

7. x-intercept is $\left(\frac{1}{2}, 0\right)$;
y-intercept is $(0, -1)$

9. x-intercept is $(3,0)$;
y-intercept is $(0, -4)$

11. x-intercept is $\left(\frac{6}{7}, 0\right)$;
y-intercept is $(0,3)$

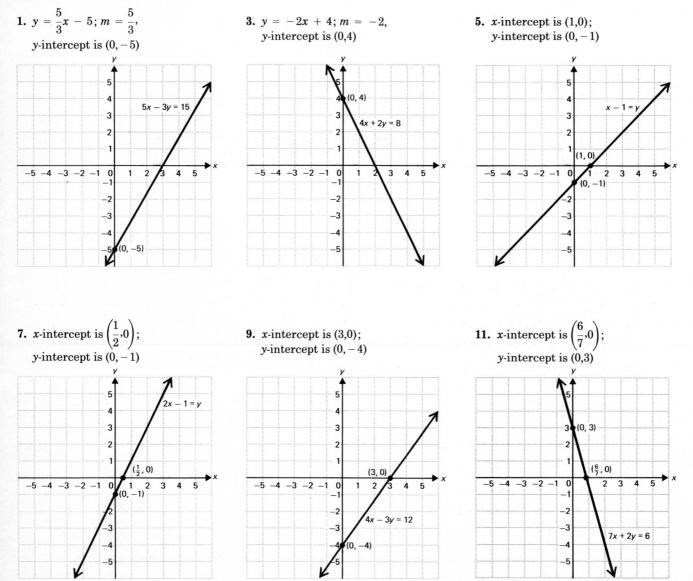

13. x-intercept is $(-1,0)$;
y-intercept is $(0,-4)$

15.

17.

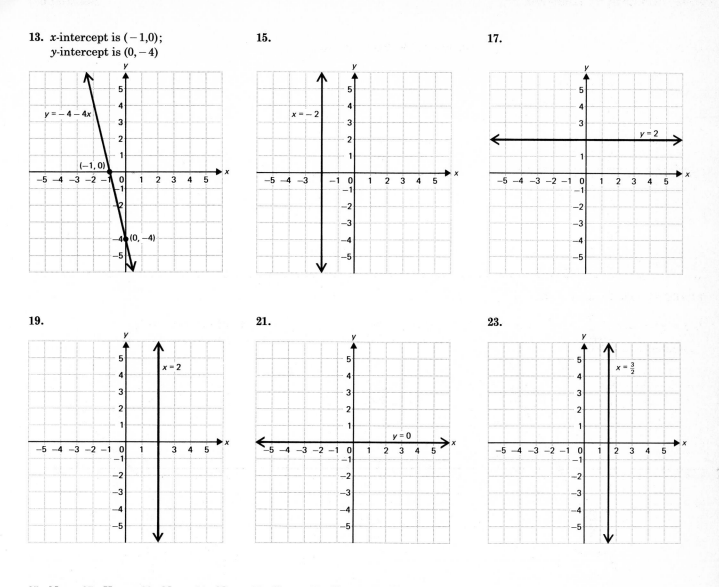

25. No   **27.** Yes   **29.** No   **31.** No   **33.** Yes   **35.** Yes   **37.** No

---

MARGIN EXERCISES, SECTION 6.4, pp. 243–244

1. $x + y = 115$, $x - y = 21$, where x is one number and y is the other

2. $21.95 + 0.23m = c$, $24.95 + 0.19m = c$, where $m = $ mileage and $c = $ cost

3. $2l + 2w = 76$, $l = w + 17$, where $l = $ length and $w = $ width

---

EXERCISE SET 6.4, pp. 245–246

1. $x + y = 58$, $x - y = 16$, where $x = $ one number and $y = $ the other number

3. $2l + 2w = 400$, $w = l - 40$, where $l = $ length and $w = $ width

5. $13.95 + 0.10m = c$, $14.95 + 0.17m = c$, where $m = $ mileage and $c = $ cost

7. $x - y = 16$, $3x = 7y$, where $x = $ the larger number and $y = $ the smaller

9. $x + y = 180$, $y = 3x + 8$, where x and y are the angles   **11.** $x + y = 90$, $x - y = 34$, where x and y are the angles

13. $x + y = 820$, $y = x + 140$, where $x = $ number of hectares of Riesling and $y = $ hectares of Chardonnay

15. $C = 6400 + 3\%S$, $C = S$, where $C = $ costs and $S = $ sales

MARGIN EXERCISES, SECTION 6.5, pp. 247–248

1. Yes **2.** No **3.** $(2, -3)$ **4.** No solution; lines are parallel.

EXERCISE SET 6.5, pp. 249–250

1. Yes **3.** No **5.** Yes **7.** $(-12, 11)$ **9.** $(4, 3)$ **11.** $(-6, -2)$

MARGIN EXERCISES, SECTION 6.6, pp. 251–253

1. $(3, 2)$ **2.** $(3, -1)$ **3.** $\left(\dfrac{24}{5}, -\dfrac{8}{5}\right)$ **4.** 68 and 47

SOMETHING EXTRA—DEPRECIATION: THE STRAIGHT-LINE METHOD (PART I), p. 254

1. $3250, $2762.50, $2275, $1787.50, $1300

EXERCISE SET 6.6, pp. 255–256

1. $(1, 3)$ **3.** $(1, 2)$ **5.** $(4, 3)$ **7.** $(-2, 1)$ **9.** $(-1, -3)$ **11.** $\left(\dfrac{17}{3}, \dfrac{16}{3}\right)$ **13.** $\left(\dfrac{25}{8}, -\dfrac{11}{4}\right)$ **15.** $(-3, 0)$ **17.** $(6, 3)$

19. 37 and 21 **21.** $(4.3821792, 4,3281211)$

MARGIN EXERCISES, SECTION 6.7, pp. 257–260

1. $(3, 2)$ **2.** $(1, -1)$ **3.** $(1, 4)$ **4.** $(1, 1)$ **5.** $(1, -1)$ **6.** $\left(\dfrac{17}{13}, -\dfrac{7}{13}\right)$ **7.** No solution

8. Length is 27.5 cm; width is 10.5 cm

EXERCISE SET 6.7, pp. 261–262

1. $(9, 1)$ **3.** $(5, 3)$ **5.** $\left(3, -\dfrac{1}{2}\right)$ **7.** $\left(-1, 1\dfrac{1}{5}\right)$ **9.** $(-3, -5)$ **11.** No solution **13.** $(2, -2)$ **15.** $(1, -1)$ **17.** $(-2, 3)$

19. $(8, 6)$ **21.** Length is 120 m; width is 80 m **23.** $(-1.8088508, 3.3817573)$

MARGIN EXERCISES, SECTION 6.8, pp. 263–264

1. 75 miles **2.** A is 47, B is 21

EXERCISE SET 6.8, pp. 265–268

1. 60 miles **3.** Sammy is 44; his daughter is 22 **5.** 28 and 12 **7.** $43°$ and $137°$ **9.** $62°$ and $28°$

11. 480 hectares Chardonnay; 340 hectares Riesling **13.** $6597.94 **15.** 23 pheasants; 12 rabbits

---

MARGIN EXERCISES, SECTION 6.9, pp. 269–270

1. 324 mi **2.** 3 hr **3.** 168 km **4.** 275 km/h

EXERCISE SET 6.9, pp. 271–274

1. 2 hr **3.** 4.5 hr **5.** $7\dfrac{1}{2}$ hr after the first train leaves **7.** 14 km/h **9.** 384 km **11.** 330 km/h **13.** 15 mi

15. 317.02702 km/h

MARGIN EXERCISES, SECTION 6.10, pp. 275–276

1. 7 quarters, 13 dimes **2.** 125 adults, 41 children **3.** $22\dfrac{1}{2}\,\ell$ of 50%, $7\dfrac{1}{2}\,\ell$ of 70% **4.** 30 lb of A, 20 lb of B

EXERCISE SET 6.10, pp. 277–280

1. 70 dimes, 33 quarters   **3.** 22 quarters, 29 half dollars   **5.** 300 nickels, 100 dimes   **7.** 203 adults, 226 children

9. 130 adults, 70 students   **11.** 40 g of A; 60 g of B **13.** 43.75 ℓ   **15.** 80 ℓ of 30%, 120 ℓ of 50%

17. 10 kg of chocolate, 5 kg of other   **19.** 100 kg of Brazilian, 200 kg of Turkish

MARGIN EXERCISES, SECTION 6.11, pp. 281–284

1. (a) $P = 17x - 100,000$; (b) 5883 units   **2.** ($125,700)

EXERCISE SET 6.11, pp. 285–286

1. (a) $P = 20x - 600,000$; (b) 30,000 units   **3.** (a) $P = 50x - 120,000$; (b) 2400 units

5. (a) $P = 80x - 10,000$; (b) 125 units   **7.** ($10, 1400)   **9.** ($22, 474)   **11.** ($50, 6250)

13. (a) $C = 40x + 22,500$; (b) $R = 85x$; (c) $P = 45x - 22,500$; (d) $112,500 profit; (e) 500 units

TEST OR REVIEW—CHAPTER 6, pp. 287–288

1. $m = -\dfrac{2}{5}$; y-intercept is (0,3)   **2.**   **3.**

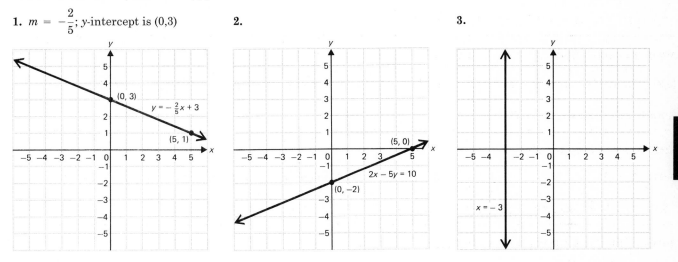

4. Yes   **5.** (0,5)   **6.** (1,4)   **7.** $(-2,4)$ **8.** 5 and -3 **9.** 135 km/h   **10.** 297 orchestra seats; 211 balcony seats

11. 40 ℓ of each   **12.** 7000 units

CHAPTER 7

READINESS CHECK, p. 290

1. [4.1, ▪] -15 **2.** [4.1, ▪▪] $3x^4 + 3x^3$ **3.** [4.3, ▪] $x^2 + 3x - 5$ **4.** [4.4, ▪▪▪] $7x^2 - 7x + 9$

5. [4.6, ▪▪] $3x^3 - 11x^2 - 22x + 10$ **6.** [4.7, ▪▪] $x^2 - x - 6$ **7.** [4.8, ▪▪] $9t^2 - 1$ **8.** [4.8, ▪▪] $p^2 + 8p + 16$

9. [5.1, ▪▪▪] $(x + 2)(x - 7)$ **10.** [5.2, ▪▪] $(3x + 1)(x - 2)$ **11.** [5.3, ▪▪▪] $(3t - 1)^2$ **12.** [5.3, ▪▪] $x^3(x + 4)(x - 4)$

13. [3.4, ▪] 3   **14.** [3.7, ▪] $r = \dfrac{d}{t}$

MARGIN EXERCISES, SECTION 7.1, pp. 290–292

1. -7940 **2.** -176 **3.** 32   **4.** $-3, 3, -2, 1, 2$ **5.** 3, 7, 1, 1, 0; 7   **6.** $2x^2y + 3xy$ **7.** $5pq + 4$

8. $5xy^4 - 3xy^3 - 7xy^2 + 3xy$ **9.** $-2 + 5xy^2z + 2x^2yz + 5x^3yz^2$

EXERCISE SET 7.1, pp. 293–294

1. -1 **3.** -7 **5.** \$12,597.12   **7.** 44.4624 in.^{2} **9.** Coefficients: 1, -2, 3, -5; degrees: 4, 2, 2, 0; 4

11. Coefficients: 17, -3, -7; degrees: 5, 5, 0; 5   **13.** $-a - 2b$ **15.** $3x^2y - 2xy^2 + x^2$ **17.** $8u^2v - 5uv^2$

19. $-8au + 10av$ **21.** a^3 **23.** $-y^3 - xy^2 + 5x^2y$ **25.** $-y^3 - xy^2 + x^3$ **27.** $4n^2 - 3mn^3 + 5m^2n - m^4n$

29. $3y^2 + 2xy + x^2$ **31.** $-11uv^2 + 5uv + 7u^2v$

MARGIN EXERCISES, SECTION 7.2, pp. 295–296

1. $14x^3y + 7x^2y - 3xy - 2y$ **2.** $-5p^2q^4 + 2p^2q^2 + 3q + 6pq^2 + 3p^2q + 5$ **3.** $-8s^4t + 6s^3t^2 + 2s^2t^3 - s^2t^2$

4. $-9p^4q + 10p^3q^2 - 4p^2q^3 - 9q^4$ **5.** $x^5y^5 + 2x^4y^2 + 3x^3y^3 + 6x^2$ **6.** $p^5q - 4p^3q^3 + 3pq^3 + 6q^4$

EXERCISE SET 7.2, pp. 297–298

1. $x^2 - 4xy + 3y^2$ **3.** $3r + 7$ **5.** $-x^2 - 8xy - y^2$ **7.** $2ab$ **9.** $-2a + 10b - 5c + 8d$ **11.** $6z^2 + 7zu - 3u^2$

13. $a^4b^2 - 7a^2b + 10$ **15.** $a^4 + a^3 - a^2y - ay + a + y - 1$ **17.** $a^6 - b^2c^2$ **19.** $y^6x + y^4x + y^4 + 2y^2 + 1$

21. $r^3 + r^2y + rs^2 + r^2s + rsy + s^3$ **23.** $5x^2y^2 + xy - 4$ **25.** $5y^2z^4 + 4yz^2 - 3$ **27.** $a^4 - b^4$ **29.** $\pi R^2 - \pi r^2$

MARGIN EXERCISES, SECTION 7.3, pp. 299–300

1. $3x^3y + 6x^2y^3 + 2x^3 + 4x^2y^2$ **2.** $2x^2 - 11xy + 15y^2$ **3.** $16x^2 + 40xy + 25y^2$ **4.** $9x^4 - 12x^3y^2 + 4x^2y^4$

5. $4x^2y^4 - 9x^2$ **6.** $16y^2 - 9x^2y^4$ **7.** $9y^2 + 24y + 16 - 9x^2$ **8.** $4a^2 - 25b^2 - 10bc - c^2$

EXERCISE SET 7.3, pp. 301–302

1. $12x^2y^2 + 2xy - 2$ **3.** $12 - c^2d^2 - c^4d^4$ **5.** $m^3 + m^2n - mn^2 - n^3$ **7.** $x^9y^9 - x^6y^6 + x^5y^5 - x^2y^2$

9. $x^2 + 2xh + h^2$ **11.** $r^6t^4 - 8r^3t^2 + 16$ **13.** $p^8 + 2m^2n^2p^4 + m^4n^4$ **15.** $4a^6 - 2a^3b^3 + \dfrac{1}{4}b^6$

17. $3a^3 - 12a^2b + 12ab^2$ **19.** $4a^2 - b^2$ **21.** $c^4 - d^2$ **23.** $a^2b^2 - c^2d^4$ **25.** $x^2 + 2xy + y^2 - 9$

27. $x^2 - y^2 - 2yz - z^2$ **29.** $a^2 - b^2 - 2bc - c^2$ **31.** $-2ab + 2b^2$ **33.** $4ab$ **35.** $4y - 4$ **37.** $a^2 - 2ab + b^2 - b + a$

39. $A^3 + 3A^2B + 3AB^2 + B^3$

MARGIN EXERCISES, SECTION 7.4, pp. 303–304

1. $x^2y(x^2y + 2x + 3)$ **2.** $2p^4q^2(5p^2 - 2pq + q^2)$ **3.** $(a - b)(2x + 5 + y^2)$ **4.** $(a + b)(x^2 + y)$

5. $(3a + 4x^2)(3a - 4x^2)$ **6.** $(5xy^2 + 2a)(5xy^2 - 2a)$ **7.** $5(1 - xy^3)(1 + xy^3)$ **8.** $y^2(4 + 9x^2)(2 + 3x)(2 - 3x)$

9. $3(a + b - 5)(a + b + 5)$

EXERCISE SET 7.4, pp. 305–306

1. $12n^2(1 + 2n)$ **3.** $9xy(xy - 4)$ **5.** $2\pi r(h + r)$ **7.** $(a + b)(2x + 1)$ **9.** $(x - 1 - y)(x + 1)$ **11.** $(n + p)(n + 2)$

13. $(2x + z)(x - 2)$ **15.** $(x - y)(x + y)$ **17.** $(ab - 3)(ab + 3)$ **19.** $(3x^2y - b)(3x^2y + b)$ **21.** $3(x + 4y)(x - 4y)$

23. $(8z - 5cd)(8z + 5cd)$ **25.** $7(p^2 + q^2)(p - q)(p + q)$ **27.** $(9a^2 + b^2)(3a - b)(3a + b)$ **29.** $2m^2(3m + 1)^2$

31. $(x - a + y - b)(x - a - y + b)$ **33.** $(b - 4 + a^3)(b - 4 - a^3)$ **35.** $(4a + 3b)(2a - 3b)$ **37.** $(s - t)(s + t - 4)$

39. $h(2x + h)$

MARGIN EXERCISES, SECTION 7.5, p. 307

1. $(x^2 + y^2)^2$ **2.** $-(2x - 3y)^2$ **3.** $(xy + 4)(xy + 1)$ **4.** $2(x^2y^3 + 5)(x^2y^3 - 2)$

SOMETHING EXTRA—THE STRAIGHT-LINE METHOD (PART II), p. 308

1. (a)

Year	Rate of depreciation	Annual depreciation	Book value (V)	Total depreciation
0			\$8700	
1	$\frac{1}{5}$ or 20%	\$1420	7280	\$1420
2	20%	1420	5860	2840
3	20%	1420	4440	4260
4	20%	1420	3020	5680
5	20%	1420	1600	7100

(b) $V = 8700 - 1420n$

2. (a)

Year	Rate of depreciation	Annual depreciation	Book value (V)	Total depreciation
0			\$9400	
1	$\frac{1}{10}$ or 10%	\$720	8680	\$ 720
2	10%	720	7960	1440
3	10%	720	7240	2160
4	10%	720	6520	2880
5	10%	720	5800	3600
6	10%	720	5080	4320
7	10%	720	4360	5040
8	10%	720	3640	5760
9	10%	720	2920	6480
10	10%	720	2200	7200

(b) $V = 9400 - 720n$

EXERCISE SET 7.5, pp. 309–310

1. $(x - y)^2$ **3.** $(3c + d)^2$ **5.** $(7m^2 - 8n)^2$ **7.** $(y^2 + 5z^2)^2$ **9.** $\left(\frac{1}{2}a + \frac{1}{3}b\right)^2$ **11.** $(a + b)(a - 2b)$

13. $(m + 20n)(m - 18n)$ **15.** $(mn - 8)(mn + 4)$ **17.** $a^3(ab + 5)(ab - 2)$ **19.** $(r - s - 1)^2$ **21.** $(t - 2 + s)(t - 2 - s)$

23. $-(k - 18l)^2$ **25.** $(1 + n + x)(1 - n - x)$ **27.** $(3b + a)(b - 6a)$ **29.** $(a + b)(a - b)(c + 2)(c - 2)$

31. $-(p - 10t)^2$ **33.** $2a(a - b)$ **35.** $(s^2 + t^2)(k + 3)(k - 3)$

MARGIN EXERCISES, SECTION 7.6, pp. 311–312

1. $y = \frac{3 + b}{a}$ **2.** $x = \frac{6}{b}$ **3.** 1   **4.** $x = \frac{2 - ab}{a + b}$ **5.** $x = \frac{a^2 + b^2}{2a}$

EXERCISE SET 7.6, pp. 313–314

1. $x = \frac{2a - 12}{3}$ **3.** $y = \frac{3b - 1}{2}$ **5.** $y = \frac{9}{2c}$ **7.** $x = a - b$ **9.** $z = \frac{3c - 5b + 12}{6}$ **11.** $x = \frac{b^2 + b - 2}{b - 1}$, or $b + 2$

13. $x = \frac{5 - 3a}{2a - 1}$ **15.** $x = \frac{7b}{4}$ **17.** $z = 2$ **19.** $x = \frac{b^2 + b - 12}{b - 3}$, or $b + 4$ **21.** $x = a + b$ **23.** $y = \frac{1}{a}$ **25.** $x = 3a$

27. $x = d - 1$ **29.** $x = \frac{c}{rt}$ **31.** $x = \frac{c}{2a - b}$ **33.** $x = \frac{5m}{2}$ **35.** $x = \frac{d^2 - c^2}{2d - 2c}$, or $\frac{d + c}{2}$

MARGIN EXERCISES, SECTION 7.7, p. 315

1. $M = \frac{fd^2}{km}$ **2.** $a^2 = \frac{6V - \pi h^3}{3\pi h}$

SOMETHING EXTRA—APPLICATIONS, p. 316

1. $D = \dfrac{1}{2}L - \dfrac{1}{4}C_1 - \dfrac{1}{4}C_2$

EXERCISE SET 7.7, pp. 317–318

1. $r = \dfrac{S}{2\pi h}$ **3.** $b = \dfrac{2A}{h}$ **5.** $n = \dfrac{S + 360}{180}$ **7.** $b = \dfrac{3V - kB - 4kM}{k}$ **9.** $r = \dfrac{S - a}{S - l}$ **11.** $h = \dfrac{2A}{b_1 + b_2}$ **13.** $a = \dfrac{v^2 pL}{r}$

15. $b_1 = \dfrac{2A - hb_2}{h}$ **17.** $E = \dfrac{180A}{\pi r^2}$ **19.** $M = \dfrac{-V - hB - hc}{4h}$ **21.** $L = \dfrac{ay}{v^2 p}$ **23.** $p = \dfrac{ar}{v^2 L}$ **25.** $n = \dfrac{a}{c(1 + b)}$

27. $F = \dfrac{9C + 160}{5}$ **29.** $q = \dfrac{mf + t}{m}$

TEST OR REVIEW—CHAPTER 7, pp. 319–320

1. $4x^2 yz^3 + x^2 y^2$ **2.** $a^3 + 3a^2 b + 3ab^2 + b^3$ **3.** $2a^3 b - 10a^2 b^2 - ab^3 + 11$ **4.** $4x + 3y + 5z$

5. $2y^4 z^3 - y^3 z^2 - 4y^2 z^2 - 4yz + 3$ **6.** $a^3 b^3 + 2a^3 b - 3a^2 b^2 - 6a^2$ **7.** $a^4 + 4a^2 b + 4b^2$ **8.** $c^4 - d^2$

9. $pq(p^3 - p^2 - 1)$ **10.** $(2u + v)(w - 3x)$ **11.** $(4xy + 1)(4xy - 1)$ **12.** $(4c^2 + d^2)(2c - d)(2c + d)$

13. $(c + 3 - 4x)(c + 3 + 4x)$ **14.** $(3p + 7q)^2$ **15.** $(5x - y)(7x - 3y)$ **16.** $x = \dfrac{r^2 + 2s^2 - r - s}{r}$ **17.** $P = \dfrac{A}{1 + rt}$

CHAPTER 8

READINESS CHECK, p. 322

1. $[1.1, \blacksquare\blacksquare\blacksquare]$ $\dfrac{1}{16}$ **2.** $[1.1, \blacksquare\blacksquare\blacksquare]$ 1 **3.** $[1.5, \blacksquare\blacksquare\blacksquare]$ $\dfrac{5}{2}$ **4.** $[1.5, \blacksquare\blacksquare\blacksquare]$ $\dfrac{3}{2}$ **5.** $[1.1, \blacksquare\blacksquare\blacksquare]$ $\dfrac{9}{10}$ **6.** $[1.1, \blacksquare\blacksquare\blacksquare]$ $\dfrac{37}{45}$

7. $[4.4, \blacksquare\blacksquare\blacksquare]$ $9x + 12$ **8.** $[4.8, \blacksquare\blacksquare]$ $x^2 - 4$ **9.** $[5.2, \blacksquare\blacksquare]$ $(x + 2)(x + 1)$ **10.** $[5.3, \blacksquare\blacksquare]$ $(x - 3)^2$ **11.** $[3.3, \blacksquare\blacksquare]$ $\dfrac{12}{5}$

12. $[5.4, \blacksquare\blacksquare]$ $2, 3$

MARGIN EXERCISES, SECTION 8.1, pp. 322–326

1. $\dfrac{x^2 + 5x + 6}{5x + 20}$ **2.** $\dfrac{-12}{4x^2 - 1}$ **3.** $\dfrac{2x^2 + x}{3x^2 - 2x}$ **4.** $\dfrac{x^2 + 3x + 2}{x^2 - 4}$ **5.** $\dfrac{8 - x}{y - x}$ **6.** 5 **7.** $\dfrac{x}{3}$ **8.** $\dfrac{2x + 1}{3x + 2}$ **9.** $\dfrac{x + 1}{2x + 1}$

10. $x + 2$ **11.** $\dfrac{y + 2}{4}$ **12.** $\dfrac{a - 2}{a - 3}$ **13.** $\dfrac{x - 5}{2}$

EXERCISE SET 8.1, pp. 327–328

1. $\dfrac{x^2 - 4x + 4}{x^2 - 25}$ **3.** $\dfrac{c^2 - 9d^2}{c^2 - d^2}$ **5.** $\dfrac{6a^2 - 5a + 1}{6a^2 + a - 2}$ **7.** $\dfrac{3x + 2}{3x - 2}$ **9.** $\dfrac{a + b}{a - b}$ **11.** $\dfrac{t - 5}{t - 4}$ **13.** $\dfrac{x + 2}{2(x - 4)}$ **15.** $\dfrac{a - 3}{a - 4}$

17. $\dfrac{6}{x - 3}$ **19.** $\dfrac{a^2 + 1}{a + 1}$ **21.** $\dfrac{t}{t + 2}$ **23.** $\dfrac{12a}{a - 2}$ **25.** $\dfrac{1}{a}$ **27.** $\dfrac{21{,}375}{x^2 - 119{,}094.01}$

MARGIN EXERCISES, SECTION 8.2, pp. 329–330

1. $\dfrac{2}{7}$ **2.** $\dfrac{2x^3 - 1}{x^2 + 5}$ **3.** $\dfrac{1}{x - 5}$ **4.** $x^2 - 3$ **5.** $\dfrac{6}{35}$ **6.** $\dfrac{x^2 - 5x + 6}{x^2 + 10x + 25}$ **7.** $\dfrac{x - 3}{x + 2}$ **8.** $\dfrac{y + 1}{y - 1}$

EXERCISE SET 8.2, pp. 331–332

1. $\dfrac{x}{4}$ **3.** $\dfrac{1}{x^2 - y^2}$ **5.** $\dfrac{x^2 - 4x + 7}{x^2 + 2x - 5}$ **7.** $\dfrac{3}{10}$ **9.** $\dfrac{1}{4}$ **11.** $\dfrac{y^2}{x}$ **13.** $\dfrac{(a + 2)(a + 3)}{(a - 3)(a - 1)}$ **15.** $\dfrac{(x - 1)^2}{x}$ **17.** $\dfrac{1}{2}$ **19.** $\dfrac{3}{2}$

21. $\dfrac{(x + y)^2}{x^2 + y}$ **23.** $\dfrac{x + 3}{x - 5}$ **25.** $\dfrac{1}{(c - 5)^2}$ **27.** $\dfrac{t + 5}{t - 5}$ **29.** $\dfrac{2s}{r + 2s}$ **31.** $\dfrac{d - 1}{-5d}$, or $\dfrac{1 - d}{5d}$

MARGIN EXERCISES, SECTION 8.3, p. 333

1. $\dfrac{7}{9}$ **2.** $\dfrac{3 + x}{x - 2}$ **3.** $\dfrac{6x + 4}{x - 1}$ **4.** $\dfrac{x - 5}{4}$ **5.** $\dfrac{x - 1}{x - 3}$

SOMETHING EXTRA—AN APPLICATION: HANDLING DIMENSION SYMBOLS (PART II), p. 334

1. 4 yd   **2.** 96 oz   **3.** 27 km   **4.** 60 m   **5.** 3 g   **6.** 18 mi   **7.** 2.347 km   **8.** 550 mm   **9.** 0.7 kg^2/m^2

10. 720 lb-mi^2/ft-hr^2 **11.** 14 m-kg/sec^2

EXERCISE SET 8.3, pp. 335–336

1. 1   **3.** $\dfrac{6}{3 + x}$ **5.** $\dfrac{2x + 3}{x - 5}$ **7.** $\dfrac{1}{4}$ **9.** $-\dfrac{1}{t}$ **11.** $\dfrac{-x + 7}{x - 6}$ **13.** $y + 3$ **15.** $\dfrac{2b - 14}{b^2 - 16}$ **17.** $-\dfrac{1}{y + z}$ **19.** $\dfrac{5x + 2}{x - 5}$

21. -1 **23.** $\dfrac{-x^2 + 9x - 14}{(x - 3)(x + 3)}$

MARGIN EXERCISES, SECTION 8.4, p. 337

1. $\dfrac{4}{11}$ **2.** $\dfrac{x^2 + 2x + 1}{2x + 1}$ **3.** $\dfrac{3x - 1}{3}$ **4.** $\dfrac{4x - 3}{x - 2}$

SOMETHING EXTRA—AN APPLICATION: HANDLING DIMENSION SYMBOLS (PART III), p. 338

1. 6 ft   **2.** 1020 min   **3.** 172,800 sec   **4.** 0.1 hr   **5.** 600 g/cm   **6.** 30 mi/hr   **7.** 2,160,000 cm^2 **8.** 0.81 ton/yd^3

9. 150¢/hr   **10.** 60 person-days   **11.** Approx 5,865,696,000,000 mi/yr   **12.** 6,570,000 mi/yr

---

EXERCISE SET 8.4, pp. 339–340

1. $\dfrac{1}{2}$ **3.** 1   **5.** $\dfrac{4}{x - 1}$ **7.** $\dfrac{8}{3}$ **9.** $\dfrac{13}{a}$ **11.** $\dfrac{4x - 5}{4}$ **13.** $\dfrac{x - 2}{x - 7}$ **15.** $\dfrac{2x - 16}{x^2 - 16}$ **17.** $\dfrac{2x - 4}{x - 9}$ **19.** $\dfrac{-9}{2x - 3}$ **21.** $\dfrac{18x + 5}{x - 1}$

23. 0   **25.** $\dfrac{20}{2y - 1}$

MARGIN EXERCISES, SECTION 8.5, pp. 341–342

1. 144   **2.** 12   **3.** 10   **4.** 120   **5.** $\dfrac{35}{144}$ **6.** $\dfrac{1}{4}$ **7.** $\dfrac{11}{10}$ **8.** $\dfrac{9}{40}$ **9.** $60x^3y^2$ **10.** $(y + 1)^2(y + 4)$ **11.** $7(t^2 + 16)(t - 2)$

12. $3x(x + 1)^2(x - 1)$, or $3x(x + 1)^2(1 - x)$

EXERCISE SET 8.5, pp. 343–344

1. 108   **3.** 72   **5.** 126   **7.** 360   **9.** 420   **11.** 300; $\dfrac{59}{300}$ **13.** 120; $\dfrac{71}{120}$ **15.** 180; $\dfrac{23}{180}$ **17.** $8a^2b^2$ **19.** c^3d^2

21. $8(x - 1)$, or $8(1 - x)$ **23.** $(a + 1)(a - 1)^2$ **25.** $(3k + 2)(3k - 2)$ or $(3k + 2)(2 - 3k)$ **27.** $18x^3(x - 2)^2(x + 1)$

29. Every 60 years

MARGIN EXERCISES, SECTION 8.6, pp. 345–346

1. $\dfrac{x^2 + 6x - 8}{(x + 2)(x - 2)}$ 2. $\dfrac{4x^2 - x + 3}{x(x - 1)(x + 1)^2}$ 3. $\dfrac{8x + 88}{(x + 16)(x + 1)(x + 8)}$ 4. $\dfrac{2a^2 - 5a + 15}{(a + 2)(a - 9)}$

EXERCISE SET 8.6, pp. 347–348

1. $\dfrac{2x + 5}{x^2}$ 3. $\dfrac{x^2 + 4xy + y^2}{x^2 y^2}$ 5. $\dfrac{4x}{(x - 1)(x + 1)}$ 7. $\dfrac{x^2 + 6x}{(x + 4)(x - 4)}$ 9. $\dfrac{3x - 1}{(x - 1)^2}$ 11. $\dfrac{x^2 + 5x + 1}{(x + 1)^2(x + 4)}$

13. $\dfrac{2x^2 - 4x + 34}{(x - 5)(x + 3)}$ 15. $\dfrac{3a + 2}{(a + 1)(a - 1)}$ 17. $\dfrac{2x + 6y}{(x + y)(x - y)}$ 19. $\dfrac{3x^2 + 19x - 20}{(x + 3)(x - 2)^2}$

MARGIN EXERCISES, SECTION 8.7, pp. 349–350

1. $\dfrac{-x - 7}{15x}$ 2. $\dfrac{6x^2 - 2x - 2}{3x(x + 1)}$

EXERCISE SET 8.7, pp. 351–352

1. $\dfrac{-(x + 4)}{6}$ 3. $\dfrac{7z - 12}{12z}$ 5. $\dfrac{4x^2 - 13xt + 9t^2}{3x^2 t^2}$ 7. $\dfrac{2x - 40}{(x + 5)(x - 5)}$ 9. $\dfrac{3 - 5t}{2t(t - 1)}$ 11. $\dfrac{2s - st - s^2}{(t + s)(t - s)}$ 13. $\dfrac{2}{y(y - 1)}$

15. $\dfrac{z - 3}{2z - 1}$ 17. $\dfrac{1 - 3x}{(2x - 3)(x + 1)}$ 19. $\dfrac{1}{2c - 1}$

MARGIN EXERCISES, SECTION 8.8, pp. 353–354

1. $\dfrac{20}{21}$ 2. $\dfrac{2(6 + x)}{5}$ 3. $\dfrac{7x^2}{3(2 - x^2)}$ 4. $b - x$

EXERCISE SET 8.8, pp. 355–356

1. $\dfrac{25}{4}$ 3. $\dfrac{1}{3}$ 5. $\dfrac{1 + 3x}{1 - 5x}$ 7. -1 9. $\dfrac{5}{3y^2}$ 11. c 13. $-\dfrac{5}{ab}$ 15. $\dfrac{x + y}{x}$ 17. $\dfrac{x - 2}{x - 3}$

MARGIN EXERCISES, SECTION 8.9, pp. 357–358

1. $x^2 + 3x + 2$ 2. $2x^2 + x - \dfrac{2}{3}$ 3. $4x^2 - \dfrac{3}{2}x + \dfrac{1}{2}$ 4. $2x^2 - 3x + 5$ 5. $x - 2$ 6. $x + 4$

7. $x + 4, \text{R} -2$; or $x + 4 + \dfrac{-2}{x + 3}$ 8. $x^2 + x + 1$

EXERCISE SET 8.9, pp. 359–360

1. $1 - 2u - u^4$ 3. $5t^2 + 8t - 2$ 5. $-4x^4 + 4x^2 + 1$ 7. $1 - 2x^2 y + 3x^4 y^5$ 9. $x - 5 + \dfrac{-50}{x - 5}$; or $x - 5, \text{R} -50$

11. $x + 2$ 13. $x - 2 + \dfrac{-2}{x + 6}$; or $x - 2, \text{R} -2$ 15. $x^4 + x^3 + x^2 + x + 1$ 17. $t^2 + 1$ 19. $x^3 - 6$

MARGIN EXERCISES, SECTION 8.10, pp. 361–362

1. $\dfrac{33}{2}$ 2. 3 3. 1 4. 2 5. $-\dfrac{1}{8}$

EXERCISE SET 8.10, pp. 363–364

1. 3 3. 10 5. 5 7. 3 9. $\dfrac{17}{2}$ 11. No solution   13. $\dfrac{5}{3}$ 15. $\dfrac{2}{9}$ 17. $\dfrac{1}{2}$ 19. No solution

MARGIN EXERCISES, SECTION 8.11, pp. 365–367

1. -3 **2.** 40 km/h, 50 km/h   **3.** $\dfrac{24}{7}$, or $3\dfrac{3}{7}$ hr   **4.** $p = \dfrac{n}{2 - m}$ **5.** $f = \dfrac{pq}{p + q}$

SOMETHING EXTRA—A NUMBER PATTERN, p. 368

1. $\dfrac{7}{8}$ **2.** $\dfrac{100}{101}$ **3.** $\dfrac{n}{n + 1}$

EXERCISE SET 8.11, pp. 369–370

1. $\dfrac{20}{9}$, or $2\dfrac{2}{9}$ **3.** 20 and 15   **5.** 30 km/h, 70 km/h   **7.** 20 mph   **9.** Passenger: 80 km/h, freight: 66 km/h   **11.** $2\dfrac{2}{9}$ hr

13. $5\dfrac{1}{7}$ hr   **15.** $p = \dfrac{qf}{q - f}$ **17.** $A = P(1 + r)$ **19.** $R = \dfrac{r_1 r_2}{r_1 + r_2}$ **21.** $D = \dfrac{BC}{A}$ **23.** $h_2 = \dfrac{p(h_1 - q)}{q}$

MARGIN EXERCISES, SECTION 8.12, pp. 371–373

1. 58 km/ℓ   **2.** 0.280   **3.** 124 km/h   **4.** 2.4 fish/yd^2 **5.** 3.45   **6.** 42   **7.** 2074   **8.** \$0.20; \$0.008$\dfrac{1}{3}$ **9.** $10\dfrac{2}{3}$

SOMETHING EXTRA—CALCULATOR CORNER: PRICE EARNINGS RATIO, p. 374

1. 6.0   **2.** 11.9   **3.** 4.7   **4.** 9.4

EXERCISE SET 8.12, pp. 375–376

1. 9   **3.** 2.3 km/h   **5.** $581\dfrac{9}{11}$ **7.** 702 km   **9.** 1.92 g   **11.** 287   **13.** (a) 1.92 tons; (b) 14.4 kg   **15.** 2500   **17.** \$1500

19. 32 kg

TEST OR REVIEW—CHAPTER 8, pp. 377–378

1. $\dfrac{7x + 3}{x - 3}$ **2.** $\dfrac{a - 6}{5}$ **3.** $\dfrac{2x^2 - 2x}{x + 1}$ **4.** $\dfrac{3}{x - 2}$ **5.** -1 **6.** $\dfrac{2x + 3}{x - 2}$ **7.** $\dfrac{-x^2 + x + 26}{(x - 5)(x + 5)(x + 1)}$ **8.** 2

9. $2x^2 - 8x + 25 + \dfrac{-79}{x + 3}$; or $2x^2 - 8x + 25$, R -79 **10.** 1   **11.** 30 km/h, 20 km/h   **12.** $R = \dfrac{Er - er}{e}$ **13.** 100

CHAPTER 9

READINESS CHECK, p. 380

1. [1.2, ●●] $5 \cdot 5$ **2.** [2.3, ●] 25   **3.** [1.1, ●●●] $\dfrac{16}{9}$ **4.** [1.1, ●●] $\dfrac{9}{25}$ **5.** [1.1, ●●] 3   **6.** [1.2, ⦂⦂] 0.875

7. [1.2, ⦂⦂] $0.45\overline{45}$ **8.** [2.1, ●●] 8   **9.** [2.8, ●●] x^6 **10.** [5.1, ●●] $x^2(x - 1)$ **11.** [5.3, ●●] $(x + 1)^2$

12. [3.3, ●●] 6

MARGIN EXERCISES, SECTION 9.1, pp. 380–382

1. $6, -6$ **2.** $8, -8$ **3.** $15, -15$ **4.** 4   **5.** 7   **6.** 10   **7.** -4 **8.** -7 **9.** -13 **10.** $45 + x$ **11.** $\dfrac{x}{x + 2}$

12. Yes   **13.** No   **14.** Yes   **15.** No   **16.** $|xy|$ **17.** $|xy|$ **18.** $|x - 1|$ **19.** $|x + 4|$ **20.** $5|x|$

EXERCISE SET 9.1, pp. 383–384

1. $1, -1$ **3.** $4, -4$ **5.** $10, -10$ **7.** $13, -13$ **9.** 2 **11.** -3 **13.** -8 **15.** 15 **17.** 19 **19.** 18 **21.** $a - 4$

23. $t^2 + 1$ **25.** $\dfrac{3}{x + 2}$ **27.** Yes   **29.** No   **31.** Yes   **33.** No   **35.** $|x|$ **37.** $2|a|$ **39.** 5 **41.** $3|b|$ **43.** $|x - 7|$

45. $|x + 1|$ **47.** $|2x - 5|$

MARGIN EXERCISES, SECTION 9.2, pp. 385–386

1. An irrational number is a number that cannot be named by fractional notation $\dfrac{a}{b}$, where a and b are integers and $b \neq 0$.

2. Irrational   **3.** Rational   **4.** Irrational   **5.** Irrational   **6.** Rational   **7.** Rational   **8.** Rational   **9.** Rational

10. Irrational   **11.** $\dfrac{7}{128}$ **12.** -0.6781 **13.** 5.69895

14. The set of real numbers consists of the rational and the irrational numbers.   **15.** 2.646 **16.** 8.485

EXERCISE SET 9.2, pp. 387–388

1. Irrational   **3.** Irrational   **5.** Rational   **7.** Irrational   **9.** Rational   **11.** Rational   **13.** Rational   **15.** Irrational

17. Rational   **19.** Irrational   **21.** $\dfrac{1}{2}$ **23.** $\dfrac{41}{60}$ **25.** $6\dfrac{7}{24}$ **27.** -1.451 **29.** 0.3125 **31.** 2.236 **33.** 9 **35.** 9.644

37. 7.937 **39.** Rational; integers   **41.** $12.5, 15, 17.5, 20$

MARGIN EXERCISES, SECTION 9.3, pp. 389–390

1. (a) 8; (b) 8 **2.** $\sqrt{21}$ **3.** 5 **4.** $\sqrt{x^2 + x}$ **5.** $\sqrt{x^2 - 1}$ **6.** $4\sqrt{2}$ **7.** $\sqrt{x + 9}\sqrt{x - 9}$ **8.** $5|x|$ **9.** $6|m|$ **10.** $2\sqrt{19}$

11. $\sqrt{x + 1}\sqrt{x - 1}$ **12.** $8|t|$ **13.** $10|a|$ **14.** 16.585 **15.** 10.099

EXERCISE SET 9.3, pp. 391–392

1. $\sqrt{6}$ **3.** 3 **5.** $3\sqrt{2}$ **7.** $\sqrt{\dfrac{3}{10}}$ **9.** $\sqrt{2x}$ **11.** $\sqrt{x^2 - 3x}$ **13.** $\sqrt{x^2 + 3x + 2}$ **15.** $\sqrt{x^2 - y^2}$ **17.** Meaningless

19. $2\sqrt{3}$ **21.** $5\sqrt{3}$ **23.** $10\sqrt{2x}$ **25.** $4|a|$ **27.** $7|t|$ **29.** $|x|\sqrt{x - 2}$ **31.** $\sqrt{2x + 3}\sqrt{x - 4}$ **33.** $\sqrt{a - b}\sqrt{a + b}$

35. 11.180 **37.** 18.972 **39.** 17.320 **41.** 11.043 **43.** 20 mph, 37.42 mph, 42.43 mph

---

MARGIN EXERCISES, SECTION 9.4, pp. 393–394

1. $4\sqrt{2}$ **2.** $5|h|\sqrt{2}$ **3.** $\sqrt{3}|x - 1|$ **4.** $|x^4|$ **5.** $|(x + 2)^7|$ **6.** $|x^7| \cdot \sqrt{x}$ **7.** $3\sqrt{2}$ **8.** 10 **9.** $4|x^3 y^2|$

10. $5|xy^2| \cdot \sqrt{2xy}$

EXERCISE SET 9.4, pp. 395–396

1. $2\sqrt{6}$ **3.** $2\sqrt{10}$ **5.** $5\sqrt{7}$ **7.** $4\sqrt{3x}$ **9.** $2|x|\sqrt{7}$ **11.** $\sqrt{2}|2x + 1|$ **13.** $|t^3|$ **15.** $|x^2| \cdot \sqrt{x}$ **17.** $|(y - 2)^4|$

19. $6|m|\sqrt{m}$ **21.** $8|x^3 y|\sqrt{7y}$ **23.** $3\sqrt{6}$ **25.** $6\sqrt{7}$ **27.** 10 **29.** $5|b|\sqrt{3}$ **31.** $|a|\sqrt{bc}$ **33.** $6|xy^3| \cdot \sqrt{3xy}$

35. $10|ab^2| \cdot \sqrt{5ab}$

MARGIN EXERCISES, SECTION 9.5, pp. 397–398

1. $\dfrac{4}{3}$ **2.** $\dfrac{1}{5}$ **3.** $\dfrac{1}{3}$ **4.** $\dfrac{3}{4}$ **5.** $\dfrac{15}{16}$ **6.** $\dfrac{1}{5}\sqrt{15}$ **7.** $\dfrac{1}{4}\sqrt{10}$ **8.** 0.535 **9.** 0.791

EXERCISE SET 9.5, pp. 399–400

1. $\dfrac{3}{7}$ **3.** $\dfrac{1}{6}$ **5.** $-\dfrac{4}{9}$ **7.** $\dfrac{8}{17}$ **9.** $\dfrac{13}{14}$ **11.** $\dfrac{6}{|a|}$ **13.** $\dfrac{3|a|}{25}$ **15.** $\dfrac{1}{5}\sqrt{10}$ **17.** $\dfrac{1}{4}\sqrt{6}$ **19.** $\dfrac{1}{2}\sqrt{2}$ **21.** $\dfrac{1}{|x|}\sqrt{3x}$ **23.** 0.655

25. 0.577　**27.** 0.592　**29.** 1.549　**31.** 1.57 sec, 3.14 sec, 8.87992 sec, 11.0999 sec

---

MARGIN EXERCISES, SECTION 9.6, pp. 401–402

1. $12\sqrt{2}$　**2.** $5\sqrt{5}$　**3.** $-12\sqrt{10}$　**4.** $5\sqrt{6}$　**5.** $\sqrt{x+1}$　**6.** $\frac{3}{2}\sqrt{2}$　**7.** $\frac{2}{15}\sqrt{15}$

SOMETHING EXTRA—AN APPLICATION: WIND CHILL TEMPERATURE, p. 402

1. $0°$　**2.** $-10°$　**3.** $-22°$　**4.** $-64°$

EXERCISE SET 9.6, pp. 403–404

1. $7\sqrt{2}$　**3.** $-8\sqrt{a}$　**5.** $8\sqrt{3}$　**7.** $\sqrt{3}$　**9.** $13\sqrt{2}$　**11.** $-24\sqrt{2}$　**13.** $(2+9|x|)\sqrt{x}$　**15.** $3\sqrt{2x+2}$

17. $(3x|y| - x|x| + y|y|)\sqrt{xy}$　**19.** $\frac{2}{3}\sqrt{3}$　**21.** $\frac{13}{2}\sqrt{2}$　**23.** $\frac{1}{18}\sqrt{3}$

25. Any pairs of numbers a,b such that $a = 0$, $b \geqslant 0$; or $a \geqslant 0$, $b = 0$.

MARGIN EXERCISES, SECTION 9.7, p. 405

1. $\frac{1}{3}$　**2.** $\frac{\sqrt{3}}{3}$　**3.** $|x|\sqrt{6x}$　**4.** $\frac{\sqrt{35}}{7}$　**5.** $\frac{\sqrt{xy}}{|y|}$

SOMETHING EXTRA—CALCULATOR CORNER: FINDING SQUARE ROOTS ON A CALCULATOR, p. 406

1. 4.123　**2.** 8.944　**3.** 10.488　**4.** 8.307　**5.** 14.142　**6.** 3.240　**7.** 29.833　**8.** 16.303　**9.** 1.414　**10.** 0.484

11. 1.772　**12.** 1.932

EXERCISE SET 9.7, pp. 407–408

1. 3　**3.** 2　**5.** $\sqrt{5}$　**7.** $\frac{2}{5}$　**9.** 2　**11.** $3|y|$　**13.** $|x^2| \cdot \sqrt{5}$　**15.** 2　**17.** $\frac{\sqrt{10}}{5}$　**19.** $\sqrt{2}$　**21.** $\frac{\sqrt{6}}{2}$　**23.** 5　**25.** $\frac{\sqrt{3x}}{|x|}$

27. $\frac{4y\sqrt{3}}{3}$　**29.** $\frac{|a|\sqrt{2a}}{4}$　**31.** $\frac{\sqrt{2}}{4|a|}$

MARGIN EXERCISES, SECTION 9.8, pp. 409–410

1. $c = \sqrt{65} \approx 8.062$　**2.** $a = \sqrt{75} \approx 8.660$　**3.** $b = \sqrt{10} \approx 3.162$　**4.** $a = \sqrt{175} = 5\sqrt{7} \approx 13.23$　**5.** $5\sqrt{13} \approx 18.03$

EXERCISE SET 9.8, pp. 411–412

1. $c = 17$　**3.** $c = \sqrt{32} \approx 5.657$　**5.** $b = 12$　**7.** $b = 4$　**9.** $c = 26$　**11.** $b = 12$　**13.** $a = 2$　**15.** $b = \sqrt{2} \approx 1.414$

17. $a = 5$　**19.** $\sqrt{75} \approx 8.660$ m　**21.** $\sqrt{208} = 4\sqrt{13} \approx 14.424$ ft　**23.** $60\sqrt{2} \approx 84.84$ ft　**25.** $h = \frac{a}{2}\sqrt{3}$

MARGIN EXERCISES, SECTION 9.9, pp. 413–414

1. $\frac{64}{3}$　**2.** 2　**3.** 66　**4.** Approx. 313.040 km　**5.** Approx. 15.652 km　**6.** 676 m

---

EXERCISE SET 9.9, pp. 415–416

1. 25　**3.** 38.44　**5.** 397　**7.** $\frac{621}{2}$　**9.** 5　**11.** 3　**13.** $\frac{17}{4}$　**15.** No solution　**17.** No solution　**19.** 346.43 km, approx.

21. 11,236 m　**23.** 125 ft, 245 ft　**25.** 2.0772 sec, approx.

TEST OR REVIEW—CHAPTER 9, pp. 417–418

1. 6　**2.** -9　**3.** $\frac{x}{2+x}$　**4.** $|m|$　**5.** $7|t|$　**6.** $|x-4|$　**7.** Irrational　**8.** Rational　**9.** Irrational　**10.** Rational

11. $\sqrt{21}$ **12.** $\sqrt{x^2 - 9}$ **13.** Meaningless   **14.** $4\sqrt{3}$ **15.** $\sqrt{x-7}\sqrt{x+7}$ **16.** 10.392   **17.** $2\sqrt{10}$ **18.** $5|xy|\sqrt{2}$

19. $\dfrac{5}{8}$ **20.** $\dfrac{2}{3}$ **21.** $\dfrac{7}{|t|}$ **22.** $\dfrac{1}{2}\sqrt{2}$ **23.** $\dfrac{1}{4}\sqrt{2}$ **24.** $\dfrac{1}{|y|}\sqrt{5y}$ **25.** 0.354   **26.** 0.742   **27.** $16\sqrt{3}$ **28.** $\sqrt{5}$ **29.** $\dfrac{1}{2}\sqrt{2}$

30. $\dfrac{\sqrt{15}}{5}$ **31.** $\dfrac{|x|\sqrt{30}}{6}$ **32.** $\dfrac{2\sqrt{3}}{3}$ **33.** $a = \sqrt{45} \approx 6.708$ **34.** $\sqrt{98} \approx 9.899$ m   **35.** 26

CHAPTER 10

READINESS CHECK, p. 420

1. [4.8, 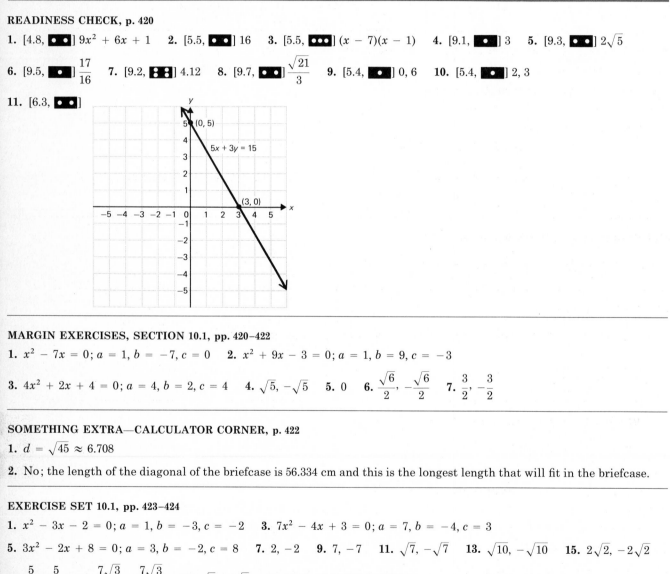] $9x^2 + 6x + 1$ **2.** [5.5, ●●] 16   **3.** [5.5, ●●●] $(x-7)(x-1)$ **4.** [9.1, ●] 3   **5.** [9.3, ●●] $2\sqrt{5}$

6. [9.5, ●●] $\dfrac{17}{16}$ **7.** [9.2, ●●] 4.12   **8.** [9.7, ●●] $\dfrac{\sqrt{21}}{3}$ **9.** [5.4, ●●] $0, 6$ **10.** [5.4, ●●] $2, 3$

11. [6.3, ●●]

MARGIN EXERCISES, SECTION 10.1, pp. 420–422

1. $x^2 - 7x = 0;\ a = 1, b = -7, c = 0$ **2.** $x^2 + 9x - 3 = 0;\ a = 1, b = 9, c = -3$

3. $4x^2 + 2x + 4 = 0;\ a = 4, b = 2, c = 4$ **4.** $\sqrt{5}, -\sqrt{5}$ **5.** 0   **6.** $\dfrac{\sqrt{6}}{2}, -\dfrac{\sqrt{6}}{2}$ **7.** $\dfrac{3}{2}, -\dfrac{3}{2}$

SOMETHING EXTRA—CALCULATOR CORNER, p. 422

1. $d = \sqrt{45} \approx 6.708$

2. No; the length of the diagonal of the briefcase is 56.334 cm and this is the longest length that will fit in the briefcase.

EXERCISE SET 10.1, pp. 423–424

1. $x^2 - 3x - 2 = 0;\ a = 1, b = -3, c = -2$ **3.** $7x^2 - 4x + 3 = 0;\ a = 7, b = -4, c = 3$

5. $3x^2 - 2x + 8 = 0;\ a = 3, b = -2, c = 8$ **7.** $2, -2$ **9.** $7, -7$ **11.** $\sqrt{7}, -\sqrt{7}$ **13.** $\sqrt{10}, -\sqrt{10}$ **15.** $2\sqrt{2}, -2\sqrt{2}$

17. $\dfrac{5}{2}, -\dfrac{5}{2}$ **19.** $\dfrac{7\sqrt{3}}{3}, -\dfrac{7\sqrt{3}}{3}$ **21.** $\sqrt{3}, -\sqrt{3}$ **23.** 16 ft, 64 ft, 400 ft, 6400 ft

MARGIN EXERCISES, SECTION 10.2, pp. 425–426

1. $0, -\dfrac{5}{3}$ **2.** $0, \dfrac{3}{5}$ **3.** $\dfrac{2}{3}, -1$ **4.** $4, 1$ **5.** (a) 14; (b) 11

EXERCISE SET 10.2, pp. 427–428

1. $0, -7$　**3.** $0, -\dfrac{2}{3}$　**5.** $0, -1$　**7.** $0, \dfrac{1}{5}$　**9.** $4, 12$　**11.** $3, -7$　**13.** -5　**15.** $\dfrac{3}{2}, 5$　**17.** $4, -\dfrac{5}{3}$　**19.** $-2, 7$　**21.** $4, -5$

23. 9　**25.** 7　**27.** $0, -\dfrac{b}{a}$

MARGIN EXERCISES, SECTION 10.3, pp. 429–431

1. $7, -1$　**2.** $-3 \pm \sqrt{10}$　**3.** $1 \pm \sqrt{5}$　**4.** $x^2 - 8x + 16 = (x - 4)^2$　**5.** $x^2 + 10x + 25 = (x + 5)^2$

6. $x^2 + 7x + \dfrac{49}{4} = \left(x + \dfrac{7}{2}\right)^2$　**7.** $x^2 - 3x + \dfrac{9}{4} = \left(x - \dfrac{3}{2}\right)^2$　**8.** \$1166.40　**9.** 12.5%　**10.** 73.2%

SOMETHING EXTRA—CALCULATOR CORNER: COMPOUND INTEREST, p. 432

1. \$1169.86　**2.** \$1231.44　**3.** (a) \$1080.00; (b) \$1081.60; (c) \$1082.43; (d) \$1083.2775; (e) \$1083.2866

4. (a) \$2; (b) \$2.25; (c) \$2.44; (d) \$2.7145; (e) \$2.7181

EXERCISE SET 10.3, pp. 433–434

1. $-7, 3$　**3.** $-1 \pm \sqrt{6}$　**5.** $3 \pm \sqrt{6}$　**7.** $x^2 - 2x + 1 = (x - 1)^2$　**9.** $x^2 + 18x + 81 = (x + 9)^2$

11. $x^2 - x + \dfrac{1}{4} = \left(x - \dfrac{1}{2}\right)^2$　**13.** $x^2 + 5x + \dfrac{25}{4} = \left(x + \dfrac{5}{2}\right)^2$　**15.** 10%　**17.** 18.75%　**19.** 8%　**21.** 20%　**23.** 9%

25. \$2536.66

MARGIN EXERCISES, SECTION 10.4, pp. 435–436

1. $-2, -6$　**2.** $5 \pm \sqrt{3}$　**3.** $-3 \pm \sqrt{10}$　**4.** $5, -2$　**5.** $-7, 2$　**6.** $\dfrac{-3 \pm \sqrt{33}}{4}$　**7.** $\dfrac{1 \pm \sqrt{10}}{3}$

EXERCISE SET 10.4, pp. 437–438

1. $-2, 8$　**3.** $-21, -1$　**5.** $1 \pm \sqrt{6}$　**7.** $9 \pm \sqrt{7}$　**9.** $-9, 2$　**11.** $-3, 2$　**13.** $\dfrac{7 \pm \sqrt{57}}{2}$　**15.** $\dfrac{-3 \pm \sqrt{17}}{4}$

17. $\dfrac{-2 \pm \sqrt{7}}{3}$　**19.** $-\dfrac{7}{2}, \dfrac{1}{2}$

MARGIN EXERCISES, SECTION 10.5, pp. 439–441

1. $\dfrac{1}{2}, -4$　**2.** $\dfrac{4 \pm \sqrt{31}}{5}$　**3.** $-0.3, 1.9$

SOMETHING EXTRA—CALCULATOR CORNER: APR, p. 442

1. 13.7%　**2.** 15.2%

EXERCISE SET 10.5, pp. 443–444

1. $-3, 7$　**3.** 3　**5.** $-\dfrac{4}{3}, 2$　**7.** $-3, 3$　**9.** $5 \pm \sqrt{3}$　**11.** $1 \pm \sqrt{3}$　**13.** No real-number solutions　**15.** $\dfrac{-1 \pm \sqrt{10}}{3}$

17. $\dfrac{-7 \pm \sqrt{61}}{2}$　**19.** $-1.3, 5.3$　**21.** $-0.2, 6.2$　**23.** $-1.7, 0.4$

MARGIN EXERCISES, SECTION 10.6, pp. 445–446

1. $13, 5$　**2.** 2　**3.** 7

EXERCISE SET 10.6, pp. 447–448

1. $6, -\dfrac{2}{3}$ 3. $6, -4$ 5. 1 7. $5, 2$ 9. No solution   11. $\sqrt{7}, -\sqrt{7}$ 13. $-2 \pm \sqrt{3}$ 15. $2 \pm \sqrt{34}$ 17. 9 19. 7

21. $1, 5$ 23. 3 25. $25, 13$ 27. No solution   29. 3

MARGIN EXERCISES, SECTION 10.7, pp. 449–450

1. $L = \dfrac{r^2}{20}$ 2. $L = \dfrac{T^2 g}{4\pi^2}$ 3. $m = \dfrac{E}{c^2}$ 4. $r = \sqrt{\dfrac{A}{\pi}}$ 5. $d = \sqrt{\dfrac{C}{P} + 1}$ 6. $n = \dfrac{1 \pm \sqrt{1 + 4N}}{2}$ 7. $t = \dfrac{-v \pm \sqrt{v^2 + 32h}}{16}$

EXERCISE SET 10.7, pp. 451–452

1. $A = \dfrac{N^2}{6.25}$ 3. $T = \dfrac{cQ^2}{a}$ 5. $c = \sqrt{\dfrac{E}{m}}$ 7. $d = \dfrac{c \pm \sqrt{c^2 + 4aQ}}{2a}$ 9. $a = \sqrt{c^2 - b^2}$ 11. $t = \sqrt{\dfrac{2S}{g}}$

13. $r = \dfrac{-\pi h \pm \sqrt{\pi^2 h^2 + \pi A}}{\pi}$ 15. $r = 6\sqrt{\dfrac{10A}{\pi S}}$ 17. (a) $r = \dfrac{C}{2\pi}$; (b) $A = \dfrac{C^2}{4\pi}$

MARGIN EXERCISES, SECTION 10.8, pp. 453–454

1. 20 m   2. 2.3 cm; 3.3 cm   3. 3 km/h

EXERCISE SET 10.8, pp. 455–460

1. 3 cm   3. 7 ft; 24 ft   5. Width 8 cm; length 10 cm   7. Length 20 cm; width 16 cm   9. Width 5 m; length 10 m

11. 4.6 m; 6.6 m   13. Width 3.6 in.; length 5.6 in.   15. Width 2.2 m; length 4.4 m   17. 7 km/h   19. 8 mph

21. 4 km/h   23. 36 mph   25. 2 units   27. ($\$1, 9$)   29. $1 + \sqrt{2} \approx 2.41$ 31. $d = 10\sqrt{2} = 14.14$; two 10-inch pizzas

---

MARGIN EXERCISES, SECTION 10.9, pp. 461–464

1. (a) Upward; (b)   2. (a) Downward; (b)

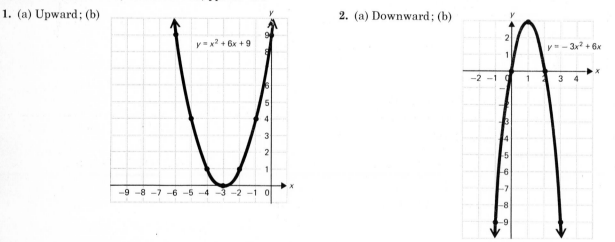

3. 2 4. $0.2, -0.2$

SOMETHING EXTRA—CALCULATOR CORNER: FINDING THE MEAN AND STANDARD DEVIATION, p. 464

1. $\bar{X} = 27.2$; $s = 11.02$ 2. $\bar{X} = 29.8$; $s = 22.70$

EXERCISE SET 10.9, pp. 465–466

1. $y = x^2$

3. $y = -1 \cdot x^2$

5. $y = -x^2 + 2x$

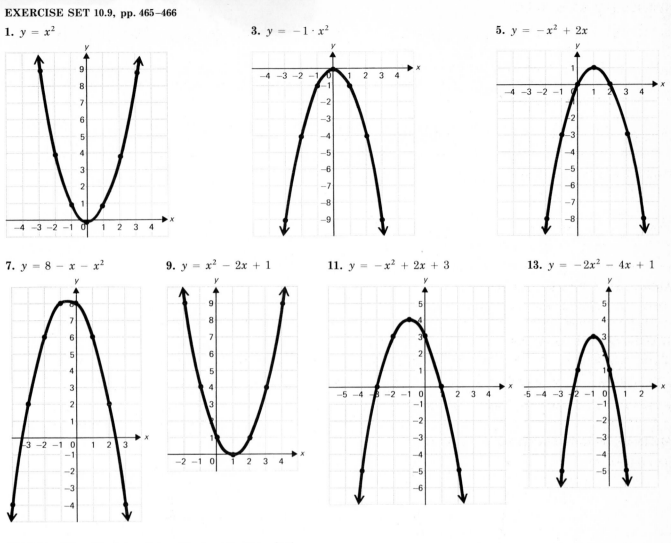

7. $y = 8 - x - x^2$ **9.** $y = x^2 - 2x + 1$ **11.** $y = -x^2 + 2x + 3$ **13.** $y = -2x^2 - 4x + 1$

15. $2.2, -2.2$ **17.** $2.4, -3.4$ **19.** 4 **21.** No solution

---

TEST OR REVIEW—CHAPTER 10, pp. 467–468

1. $\sqrt{3}, -\sqrt{3}$ **2.** $0, \dfrac{7}{5}$ **3.** $\dfrac{1}{3}, -2$ **4.** $-8 \pm \sqrt{13}$ **5.** $1 \pm \sqrt{11}$ **6.** $-3 \pm 3\sqrt{2}$ **7.** $1.2, -7.2$ **8.** $2, -8$ **9.** 4

10. $T = \dfrac{LV^2}{25a}$ **11.** 30% **12.** 1.7 m; 4.7 m   **13.** 30 km/h   **14.** (a) Upward;
(b) $y = x^2 - 4x - 2$

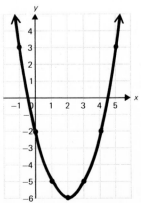

CHAPTER 11

READINESS CHECK, p. 470

1. [1.5, 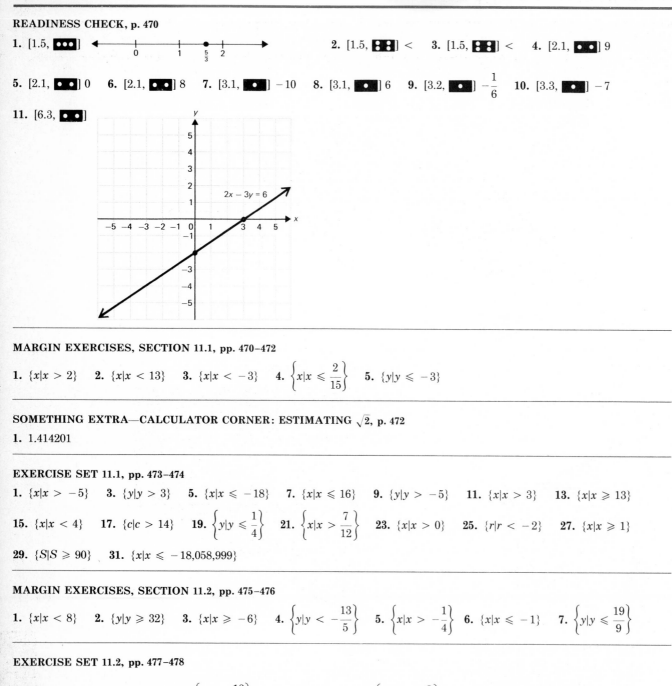]

2. [1.5, ▨] <   **3.** [1.5, ▨] <   **4.** [2.1, ▨] 9

5. [2.1, ▨] 0   **6.** [2.1, ▨] 8   **7.** [3.1, ▨] −10   **8.** [3.1, ▨] 6   **9.** [3.2, ▨] $-\frac{1}{6}$ **10.** [3.3, ▨] −7

11. [6.3, ▨]

$2x - 3y = 6$

MARGIN EXERCISES, SECTION 11.1, pp. 470–472

1. $\{x|x > 2\}$ **2.** $\{x|x < 13\}$ **3.** $\{x|x < -3\}$ **4.** $\left\{x\middle|x \leqslant \frac{2}{15}\right\}$ **5.** $\{y|y \leqslant -3\}$

SOMETHING EXTRA—CALCULATOR CORNER: ESTIMATING $\sqrt{2}$, p. 472

1. 1.414201

EXERCISE SET 11.1, pp. 473–474

1. $\{x|x > -5\}$ **3.** $\{y|y > 3\}$ **5.** $\{x|x \leqslant -18\}$ **7.** $\{x|x \leqslant 16\}$ **9.** $\{y|y > -5\}$ **11.** $\{x|x > 3\}$ **13.** $\{x|x \geqslant 13\}$

15. $\{x|x < 4\}$ **17.** $\{c|c > 14\}$ **19.** $\left\{y\middle|y \leqslant \frac{1}{4}\right\}$ **21.** $\left\{x\middle|x > \frac{7}{12}\right\}$ **23.** $\{x|x > 0\}$ **25.** $\{r|r < -2\}$ **27.** $\{x|x \geqslant 1\}$

29. $\{S|S \geqslant 90\}$ **31.** $\{x|x \leqslant -18,058,999\}$

MARGIN EXERCISES, SECTION 11.2, pp. 475–476

1. $\{x|x < 8\}$ **2.** $\{y|y \geqslant 32\}$ **3.** $\{x|x \geqslant -6\}$ **4.** $\left\{y\middle|y < -\frac{13}{5}\right\}$ **5.** $\left\{x\middle|x > -\frac{1}{4}\right\}$ **6.** $\{x|x \leqslant -1\}$ **7.** $\left\{y\middle|y \leqslant \frac{19}{9}\right\}$

EXERCISE SET 11.2, pp. 477–478

1. $\{x|x < 7\}$ **3.** $\{y|y \leqslant 9\}$ **5.** $\left\{x\middle|x < \frac{13}{7}\right\}$ **7.** $\{x|x > -3\}$ **9.** $\left\{y\middle|y \geqslant -\frac{2}{5}\right\}$ **11.** $\{x|x \geqslant -6\}$ **13.** $\{y|y \leqslant 4\}$

15. $\left\{x\middle|x > \frac{17}{3}\right\}$ **17.** $\left\{y\middle|y < -\frac{1}{14}\right\}$ **19.** $\left\{x\middle|\frac{3}{10} \geqslant x\right\}$ **21.** $\{x|x < -1\}$ **23.** $\{x|x \leqslant 3\}$ **25.** $\left\{t\middle|t \leqslant \frac{9}{2}\right\}$ **27.** $\{c|c < -2\}$

29. $\{x|x \geqslant 4\}$ **31.** $\{m|m > 50\}$

MARGIN EXERCISES, SECTION 11.3, pp. 479–482

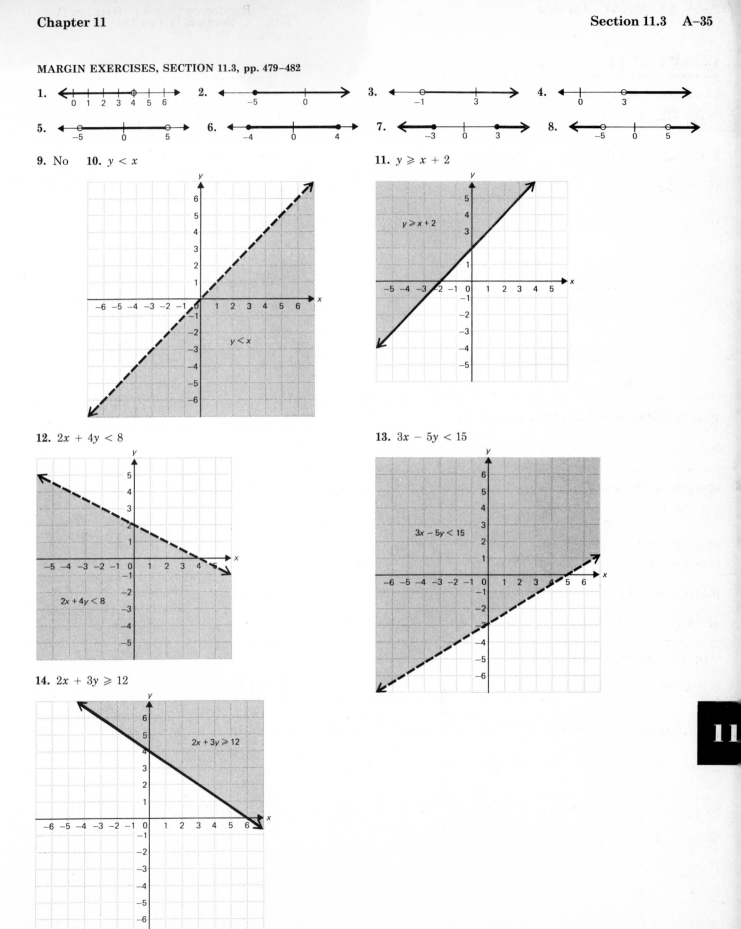

1. number line, open circle at 4, shaded left, marks 0 1 2 3 4 5 6

2. number line, closed circle at −5, shaded left, marks −5 0

3. number line, open circle at −1, shaded right, marks −1 3

4. number line, open circle at 3, shaded right, marks 0 3

5. number line, open circles at −5 and 5, shaded between, marks −5 0 5

6. number line, closed circles at −4 and 4, shaded between, marks −4 0 4

7. number line, closed circles at −3 and 3, shaded outside, marks −3 0 3

8. number line, open circles at −5 and 5, shaded outside, marks −5 0 5

9. No   **10.** $y < x$

11. $y \geqslant x + 2$

$y \geqslant x + 2$

$y < x$

12. $2x + 4y < 8$

$2x + 4y < 8$

13. $3x - 5y < 15$

$3x - 5y < 15$

14. $2x + 3y \geqslant 12$

$2x + 3y \geqslant 12$

11

EXERCISE SET 11.3, pp. 483-484

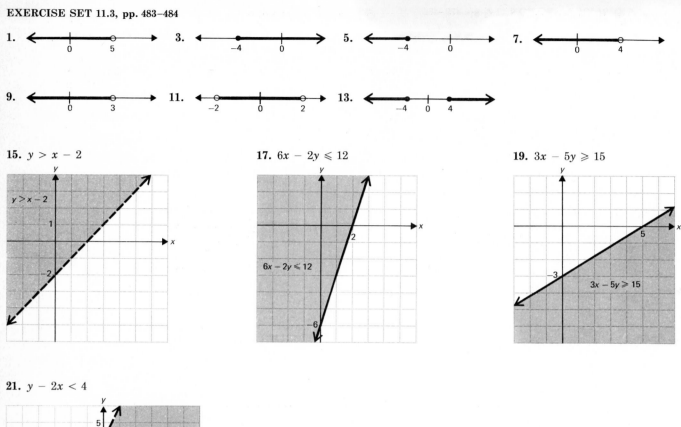

15. $y > x - 2$

17. $6x - 2y \leqslant 12$

19. $3x - 5y \geqslant 15$

21. $y - 2x < 4$

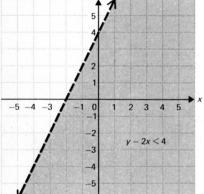

MARGIN EXERCISES, SECTION 11.4, pp. 485-486

1. $\{2,3,4,5,6,7,8,9,10\}$ **2.** $\{32,34,36,38\}$ **3.** True   **4.** False   **5.** True   **6.** True   **7.** $\{a,1,9\}$ **8.** $\{2\}$ **9.** $\varnothing$ **10.** $\varnothing$

11. $\{0,1,2,3,4,5,6,7\}$ **12.** The set of integers

EXERCISE SET 11.4, pp. 487-488

1. $\{3,4,5,6,7,8\}$ **3.** $\{41,43,45,47,49\}$ **5.** $\{\sqrt{3}, -\sqrt{3}\}$ **7.** False   **9.** True   **11.** True   **13.** $\{c,d,e\}$ **15.** $\{1,10\}$ **17.** $\varnothing$

19. $\{a,e,i,o,u,q,c,k\}$ **21.** $\{0,1,2,5,7,10\}$ **23.** $\{a,e,i,o,u,m,n,f,g,h\}$

TEST OR REVIEW—CHAPTER 11, pp. 489-490

1. $\{y|y \geqslant -6\}$ **2.** $\{x|x < -11\}$ **3.** $\{x|x \geqslant 7\}$ **4.** $\{y|y > 7\}$ **5.** $\{y|y > 2\}$ **6.** $\left\{x|x > -\dfrac{9}{11}\right\}$

7. 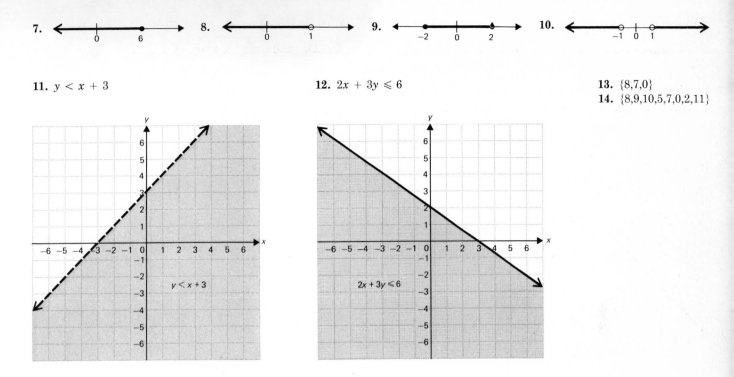 **8.**   **9.**   **10.**

11. $y < x + 3$ **12.** $2x + 3y \leqslant 6$ **13.** $\{8,7,0\}$
                                                                          14. $\{8,9,10,5,7,0,2,11\}$

FINAL EXAMINATION, pp. 491–496

1. [9] $x \cdot x \cdot x$ **2.** [11] $\dfrac{136}{10}$ **3.** [11] 0.68 **4.** [26] $\dfrac{4}{5}$ **5.** [65] 7   **6.** [66] 9   **7.** [82] 15   **8.** [86] $-\dfrac{3}{20}$ **9.** [77] -63

10. [97] 3^{-4} **11.** [98] x^{-4} **12.** [98] y^7 **13.** [109] -8 **14.** [114] -12 **15.** [119] 7   **16.** [140] $t = \dfrac{4b}{A}$

17. [130] 24 and 25   **18.** [151] $10x^3 + 3x - 3$ **19.** [155] $7x^3 - 2x^2 + 4x - 17$ **20.** [159] $8x^2 - 4x - 6$

21. [175] $-8x^4 + 6x^3 - 2x^2$ **22.** [170] $6x^3 - 17x^2 + 16x - 6$ **23.** [179] $t^2 - 25$ **24.** [180] $9m^2 - 12m + 4$

25. [189] $4x(2x - 1)$ **26.** [203] $(5x - 2)(5x + 2)$ **27.** [194] $(3x + 2)(2x - 3)$ **28.** [202] $(x - 4)^2$

29. [190] $(x - 5)(x - 8)$ **30.** [207] 3, 5   **31.** [208] 4, -5 **32.** [218] $(x - 2)(x - 6)$

33. [233]   **34.** [237]

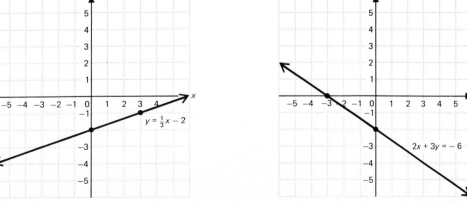

35. [240] Yes   **36.** [251] (1,2)   **37.** [257] (12,5)   **38.** [259] (6,7)   **39.** [253] 11 and 13   **40.** [292] $7x^2yz^3 + 5x^2y^2 - 5xy^2$

41. [292] $5a^4x^4 + 3ax^3 + x^2 + 2a$ **42.** [295] $2x^3y - 7x^2y^2 + xy^3 + 6$ **43.** [295] $15x^2y^3 + x^2y^2 + 5xy^2 + 7$

44. [300] $x^4 - 4y^2$ **45.** [299] $9x^2 + 24xy^2 + 16y^4$ **46.** [304] $(5ab - 1)(5ab + 1)$ **47.** [304] $(x - y - 5)(x - y + 5)$

48. [307] $(3x + 5y)^2$ **49.** [307] $(5x - 2y)(3x + 4y)$ **50.** [303] $(2a + d)(c - 3b)$ **51.** [304] $(2x + y^2)(2x - y^2)(4x^2 + y^4)$

52. [304] $3(a^2 + 6)(a + 2)(a - 2)$ **53.** [315] $m = \dfrac{tn}{t + n}$ **54.** [325] $\dfrac{2}{x + 3}$ **55.** [330] $\dfrac{3a(a - 1)}{2(a + 1)}$ **56.** [345] $\dfrac{27x - 4}{5x(3x - 1)}$

57. [349] $\dfrac{-x^2 + x + 2}{(x + 4)(x - 4)(x - 5)}$ **58.** [362] 4, -6 **59.** [390] $5\sqrt{2}$ **60.** [405] $\dfrac{2\sqrt{10}}{5}$ **61.** [401] $12\sqrt{3}$ **62.** [413] 2

63. [425] 3, -2 **64.** [440] $\dfrac{-3 \pm \sqrt{29}}{2}$

65. [461]

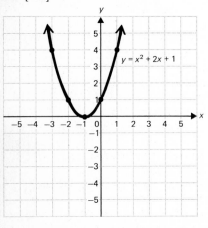

66. [453] 12 m
67. [476] $\{x | x \geqslant -1\}$

68. [480] $4x - 3y > 12$

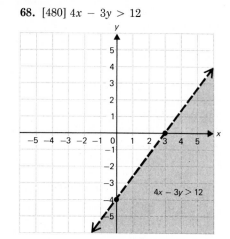

69. [485] $\{c,d\}$ **70.** [486] $\{a,b,c,d,e,f,g,h\}$